PACKET FORWARDING
TECHNOLOGIES

PACKET FORWARDING

TECHNOLOGIES

WEIDONG WU

CRC Press
Taylor & Francis Group
Boca Raton London New York

CRC Press is an imprint of the
Taylor & Francis Group, an **informa** business

AN AUERBACH BOOK

First published 2008 Auerbach Publications

Published 2019 by CRC Press
Taylor & Francis Group
6000 Broken Sound Parkway NW, Suite 300
Boca Raton, FL 33487-2742

First issued in paperback 2019

No claim to original U.S. Government works

ISBN 13: 978-0-367-45280-3 (pbk)
ISBN 13: 978-0-8493-8057-0 (hbk)

Visit the Taylor & Francis Web site at
http://www.taylorandfrancis.com

and the CRC Press Web site at
http://www.crcpress.com

Library of Congress Cataloging-in-Publication Data

Wu, Weidong.
   Packet forwarding technologies / by Weidong Wu.
      p. cm.
   Includes bibliographical references and index.
   ISBN-13: 978-0-8493-8057-0
   ISBN-10: 0-8493-8057-X
   1. Packet switching (Data transmission) 2. Routers (Computer networks) I. Title.

TK5105.W83 2008
621.39′81--dc22   2007026355

Contents

Preface

This book mainly targets high-speed packet networking. As Internet traffic grows exponentially, there is a great need to build multi-terabit Internet protocol (IP) routers. The forwarding engine in routers is the most important part of the high-speed router.

Packet forwarding technologies have been investigated and researched intensively for almost two decades, but there are very few appropriate textbooks describing it. Many engineers and students have to search for technical papers and read them in an *ad-hoc* manner. This book is the first that explains packet forwarding concepts and implementation technologies in broad scope and great depth.

This book addresses the data structure, algorithms, and architectures to implement high-speed routers. The basic concepts of packet forwarding are described and new technologies are discussed. The book will be a practical guide to aid understanding of IP routers.

We have done our best to accurately describe packet forwarding technologies. If any errors are found, please send an email to wuweidong@wust.edu.cn. We will correct them in future editions.

Audience

This book can be used as a reference book for industry people whose job is related to IP networks and router design. It is also intended to help engineers from network equipment and Internet service providers to understand the key concepts of high-speed packet forwarding. This book will also serve as a good text for senior and graduate students in electrical engineering, computer engineering, and computer science. Using it, students will understand the technology trend in IP networks so that they can better position themselves when they graduate and look for jobs in the high-speed networking field.

Organization of the Book

The book is organized as follows:

Chapter 1 introduces the basic concept and functionalities of the IP router. It also discusses the evolution of the IP router and the characteristics of its key components.

Chapter 2 explains the background of IP-address lookup by briefly describing the evolution of the Internet addressing architecture, the characteristics of the routing table, and the complexity of IP-address lookup. It discusses the design criteria and the performance requirements of high-speed routers.

Chapter 3 introduces basic schemes, such as linear search, cache replacement algorithm, binary trie, path-compressed trie, dynamic prefix trie, and others. We describe the problems of the algorithms proposed before 1996.

Chapter 4 discusses the multibit trie, in which the search operation requires simultaneous inspection of several bits. We describe the principles involved in constructing an efficient multibit trie and examine some schemes in detail.

Chapter 5 discusses the pipelined ASIC architecture that can produce significant savings in cost, complexity, and space for the high-end router.

Chapter 6 discusses the dynamic data structure of the bursty access pattern. We examine the designs of the data structure and show how to improve the throughput by turning it according to lookup biases.

Chapter 7 introduces the advance caching techniques that speed up packet forwarding. We discuss the impact of traffic locality, cache size, and the replacement algorithm on the miss ratio.

Chapter 8 discusses the improved hash schemes that can be used for Internet address lookups. We examine the binary search of hash tables, parallel hashing, multiple hashing, and the use of Bloom filter.

Chapter 9 discusses the forwarding engine based on TCAM. We examine route update algorithms and power efficient schemes.

Chapter 10 discusses the partitioning techniques based on the properties of the forwarding table.

Acknowledgments

This book could not have been published without the help of many people. We thank Pankaj Gupta, Srinivasan Vankatachary, Sartaj Sahni, Geoff Huston, Isaac Keslassy, Mikael Degermark, Will Eatherton, Haoyu Song, Marcel Waldvogel, Soraya Kasnavi, Vincent C. Gaudet, H. Jonathan Chao, Vittorio Bilo, Michele Flammini, Ernst W. Biersack, Willibald Doeringer, Gunnar Karlsson, Rama Sangireddy, Mikael Sundstrom, Anindya Basu, Girija Narlikar, Gene Cheung, Funda Ergun, Tzi-cker Chiueh, Mehrdad Nourani, Nian-Feng Tzeng, Hyesook Lim, Andrei Broder, Michael Mitzenmacher, Sarang Dharmapurika, Masayoshi Kobayashi, Samar Sharma, V.C. Ravikumar, Rabi Mahapatra, Kai Zheng, B. Lampson, Haibin Lu, Yiqiang Q. Zhao, and others.

We would like to thank Jianxun Chen and Xiaolong Zhang (Wuhan University of Science and Technology) for their support and encouragement. Weidong Wu wants to thank his wife and his child for their love, support, patience, and perseverance.

Acknowledgments

About the Author

Weidong Wu received his PhD in electronics and information engineering from Huazhong University of Science and Technology, China. In 2006, he joined Wuhan University of Science and Technology. His research involves algorithms to improve Internet router performance, network management, network security, and traffic engineering.

Chapter 1

Introduction

1.1 Introduction

The Internet comprises a mesh of routers interconnected by links, in which routers forward packets to their destinations, and physical links transport packets from one router to another. Because of the scalable and distributed nature of the Internet, there are more and more users connected to it and more and more intensive applications over it. The great success of the Internet thus leads to exponential increases in traffic volumes, stimulating an unprecedented demand for the capacity of the core network. The trend of such exponential growth is not expected to slow down, mainly because data-centric businesses and consumer networking applications continue to drive global demand for broadband access solutions. This means that packets have to be transmitted and forwarded at higher and higher rates. To keep pace with Internet traffic growth, researchers are continually exploring transmission and forwarding technologies.

Advances in fiber throughput and optical transmission technologies have enabled operators to deploy capacity in a dramatic fashion. For example, dense wavelength division multiplexing (DWDM) equipment can multiplex the signals of 300 channels of 11.6 Gbit/s to achieve a total capacity of more than 3.3 Tbit/s on a single fiber and transmit them over 7000 km [1]. In the future, DWDM networks will widely support 40 Gbit/s (OC-768) for each channel, and link capacities are keeping pace with the demand for bandwidth.

Historically, network traffic doubled every year [2], and the speed of optical transmissions (such as DWDM) every seven months [3]. However, the capacity of routers has doubled every 18 months [3], laging behind network traffic and the increasing speed of optical transmission. Therefore, the router becomes the bottleneck of the Internet.

In the rest of this chapter, we briefly describe the router including the basic concept, its functionalities, architecture, and key components.

1.2 Concept of Routers

The Internet can be described as a collection of networks interconnected by routers using a set of communications standards known as the Transmission Control Protocol/Internet Protocol (TCP/IP) suite. TCP/IP is a layered model with logical levels: the application layer, the transport layer, the network layer, and the data link layer. Each layer provides a set of services that can be used by the layer above [4]. The network layer provides the services needed for Internetworking, that is, the transfer of data from one network to another. Routers operate at the network layer, and are sometimes called IP routers.

Routers knit together the constituent networks of the global Internet, creating the illusion of a unified whole. In the Internet, a router generally connects with a set of input links through which a packet can come in and a set of output links through which a packet can be sent out. Each packet contains a destination IP address; the packet has to follow a path through the Internet to its destination. Once a router receives a packet at an input link, it must determine the appropriate output link by looking at the destination address of the packet. The packet is transferred router by router so that it eventually ends up at its destination. Therefore, the primary functionality of the router is to transfer packets from a set of input links to a set of output links. This is true for most of the packets, but there are also packets received at the router that require special treatment by the router itself.

1.3 Basic Functionalities of Routers

Generally, routers consist of the following basic components: several network interfaces to the attached networks, processing module(s), buffering module(s), and an internal interconnection unit (or switch fabric). Typically, packets are received at an inbound network interface, processed by the processing module and, possibly, stored in the buffering module. Then, they are forwarded through the internal interconnection unit to the outbound interface that transmits them to the next hop on their journey to the final destination. The aggregate packet rate of all attached network interfaces needs to be processed, buffered, and relayed. Therefore, the processing and memory modules may be replicated either fully or partially on the network interfaces to allow for concurrent operations.

A generic architecture of an IP router is given in Figure 1.1. Figure 1.1a shows the basic architecture of a typical router: the controller card [which holds the central processing unit (CPU)], the router backplane, and interface cards. The CPU in the router typically performs such functions as path computations, routing table maintenance, and reachability propagation. It runs whichever routing protocols are needed in the router. The interface cards consist of adapters that perform inbound and outbound packet forwarding (and may even cache routing table entries or have extensive packet processing capabilities). The router backplane is responsible for transferring packets between the cards. The basic functionalities in an IP router can be categorized as: route processing, packet forwarding, and router special services. The two key functionalities are route processing (i.e., path computation, routing table maintenance, and reachability propagation) and packet forwarding, shown in Figure 1.1b. We discuss the three functionalities in more detail subsequently.

1.3.1 Route Processing

Routing protocols are the means by which routers gain information about the network. Routing protocols map network topology and store their view of that topology in the routing table. Thus, route processing includes routing table construction and maintenance using routing protocols,

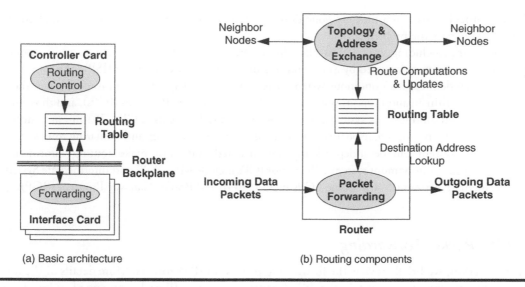

Figure 1.1   **Generic architecture of a router. (From Aweya, J.,** *Journal of Systems Architecture,* **46, 6, 2000. With permission.)**

such as the Routing Information Protocol (RIP) and Open Shortest Path First (OSPF) [5–7]. The routing table consists of routing entries that specify the destination and the next-hop router through which the packets should be forwarded to reach the destination. Route calculation consists of determining a route to the destination: network, subnet, network prefix, or host.

In static routing, the routing table entries are created by default when an interface is configured (for directly connected interfaces), added by, for example, the route command (normally from a system bootstrap file), or created by an Internet Control Message Protocol (ICMP) redirect (usually when the wrong default is used) [8]. Once configured, the network paths will not change. With static routing, a router may issue an alarm when it recognizes that a link has gone down, but will not automatically reconfigure the routing table to reroute the traffic around the disabled link. Static routing, used in LANs over limited distances, requires basically the network manager to configure the routing table. Thus, static routing is fine if the network is small, there is a single connection point to other networks, and there are no redundant routes (where a backup route can be used if a primary route fails). Dynamic routing is normally used if any of these three conditions do not hold true.

Dynamic routing, used in Internetworking across wide area networks, automatically reconfigures the routing table and recalculates the least expensive path. In this case, routers broadcast advertisement packets (signifying their presence) to all network nodes and communicate with other routers about their network connections, the cost of connections, and their load levels. Convergence, or reconfiguration of the routing tables, must occur quickly, before routers with incorrect information misroute data packets into dead ends. Some dynamic routers can also rebalance the traffic load.

The use of dynamic routing does not change the way an IP forwarding engine performs routing at the IP layer. What changes is the information placed in the routing table—instead of coming from the route commands in bootstrap files, the routes are added and deleted dynamically by a routing protocol, as routes change over time. The routing protocol adds a routing policy to the system, choosing which routes to place in the routing table. If the protocol finds multiple routes to a destination, the protocol chooses which route is the best, and which one to insert in the table.

If the protocol finds that a link has gone down, it can delete the affected routes or add alternate routes that bypass the problem.

A network (including several networks administered as a whole) can be defined as an autonomous system. A network owned by a corporation, an Internet Service Provider (ISP), or a university campus often defines an autonomous system. There are two principal routing protocol types: those that operate within an autonomous system, or the Interior Gateway Protocols (IGPs), and those that operate between autonomous systems, or Exterior Gateway Protocols (EGPs). Within an autonomous system, any protocol may be used for route discovery, propagating, and validating routes. Each autonomous system can be independently administered and must make routing information available to other autonomous systems. The major IGPs include RIP, OSPF, and Intermediate System to Intermediate System (IS–IS). Some EGPs include EGP and Border Gateway Protocol (BGP).

1.3.2 Packet Forwarding

In this section, we briefly review the forwarding process in IPv4 routers. More details of the forwarding requirements are given in Ref. [9]. A router receives an IP packet on one of its interfaces and then forwards the packet out of another of its interfaces (or possibly more than one, if the packet is a multicast packet), based on the contents of the IP header. As the packet is forwarded hop by hop, the packet's (original) network layer header (IP header) remains relatively unchanged, containing the complete set of instructions on how to forward the packet (IP tunneling may call for prepending the packet with other IP headers in the network). However, the data-link headers and physical-transmission schemes may change radically at each hop to match the changing media types.

Suppose that the router receives a packet from one of its attached network segments, the router verifies the contents of the IP header by checking the protocol version, header length, packet length, and header checksum fields. The protocol version must be equal to 4 for IPv4, for which the header length must be greater than or equal to the minimum IP header size (20 bytes). The length of the IP packet, expressed in bytes, must also be larger than the minimum header size. In addition, the router checks that the entire packet has been received by checking the IP packet length against the size of the received Ethernet packet, for example, in the case where the interface is attached to an Ethernet network. To verify that none of the fields of the header have been corrupted, the 16-bit ones-complement checksum of the entire IP header is calculated and verified to be equal to 0xffff. If any of these basic checks fail, the packet is deemed to be malformed and is discarded without sending an error indication back to the packet's originator.

Next, the router verifies that the time-to-live (TTL) field is greater than 1. The purpose of the TTL field is to make sure that packets do not circulate forever when there are routing loops. The host sets the packet's TTL field to be greater than or equal to the maximum number of router hops expected on the way to the destination. Each router decrements the TTL field by 1 when forwarding; when the TTL field is decremented to 0, the packet is discarded, and an ICMP TTL exceeded message is sent back to the host. On decrementing the TTL, the router must update the packet's header checksum. RFC1624 [10] contains implementation techniques for computing the IP checksum. Because a router often changes only the TTL field (decrementing it by 1), it can incrementally update the checksum when it forwards a received packet, instead of calculating the checksum over the entire IP header again.

The router then looks at the destination IP address. The address indicates a single destination host (unicast), a group of destination hosts (multicast), or all hosts on a given network segment (broadcast). Unicast packets are discarded if they were received as data-link broadcasts or as multicasts; otherwise, multiple routers may attempt to forward the packet, possibly contributing to a broadcast storm. In packet forwarding, the destination IP address is used as a key for the routing table lookup. The best-matching routing table entry is returned, indicating whether to forward the packet and, if so, the interface to forward the packet out of and the IP address of the next IP router (if any) in the packet's path. The next-hop IP address is used at the output interface to determine the link address of the packet, in case the link is shared by multiple parties [such as an Ethernet, Token Ring, or Fiber Distributed Data Interface (FDDI) network], and is consequently not needed if the output connects to a point-to-point link.

In addition to making forwarding decisions, the forwarding process is responsible for making packet classifications for quality of service (QoS) control and access filtering. Flows can be identified based on source IP address, destination IP address, TCP/UDP port numbers as well as IP type of service (TOS) field. Classification can even be based on higher layer packet attributes.

If the packet is too large to be sent out of the outgoing interface in one piece [i.e., the packet length is greater than the outgoing interface's Maximum Transmission Unit (MTU)], the router attempts to split the packet into smaller fragments. Fragmentation, however, can affect performance adversely [11]. The host may instead wish to prevent fragmentation by setting the Don't Fragment (DF) bit in the fragmentation field. In this case, the router does not fragment the packet, but instead drops it and sends an ICMP Destination Unreachable (subtype fragmentation needed and DF set) message back to the host. The host uses this message to calculate the minimum MTU along the packet's path [12], which in turn is used to size future packets.

The router then prepends the appropriate data-link header for the outgoing interface. The IP address of the next hop is converted to a data-link address, usually using the Address Resolution Protocol (ARP) [13] or a variant of ARP, such as Inverse ARP [14] for Frame Relay subnets. The router then sends the packet to the next hop, where the process is repeated.

An application can also modify the handling of its packets by extending the IP headers of its packets with one or more IP options. IP options are used infrequently for regular data packets, because most Internet routers are heavily optimized for forwarding packets having no options. Most IP options (such as the record-route and timestamp options) are used to aid in statistics collection, but do not affect a packet's path. However, the strict-source route and the loose-source route options can be used by an application to control the path its packets take. The strict-source route option is used to specify the exact path that the packet will take, router by router. The utility of a strict-source route is limited by the maximum size of the IP header (60 bytes), which limits to 9 the number of hops specified by the strict-source route option. The loose-source route is used to specify a set of intermediate routers (again, up to 9) through which the packet must go on the way to its destination. Loose-source routing is used mainly for diagnostic purposes, for instance, as an aid to debugging Internet routing problems.

1.3.3   *Router Special Services*

Besides dynamically finding the paths for packets to take toward their destinations, routers also implement other functions. Anything beyond core routing functions falls into this category, for example, authentication and access services, such as packet filtering for security/firewall purposes. Companies often put a router between their company network and the Internet and then configure the router to prevent unauthorized access to the company's resources from the Internet. This configuration may consist of certain patterns (e.g., source and destination address and TCP port) whose matching packets should not be forwarded or of more complex rules to deal with protocols that vary their port numbers over time, such as the File Transfer Protocol (FTP). Such routers are called firewalls. Similarly, ISPs often configure their routers to verify the source address in all packets received from the ISP's customers. This foils certain security attacks and makes other attacks easier to trace back to their source. Similarly, ISPs providing dial-in access to their routers typically use Remote Authentication Dial-In User Service (RADIUS) [15] to verify the identity of the person dialing in.

Often, other functions less directly related to packet forwarding also get incorporated into IP routers. Examples of these nonforwarding functions include network management components, such as Simple Network Management Protocol (SNMP) and Management Information Bases (MIBs). Routers also play an important role in TCP/IP congestion control algorithms. When an IP network is congested, routers cannot forward all the packets they receive. By simply discarding some of their received packets, routers provide feedback to TCP congestion control algorithms, such as the TCP slow-start algorithm [16,17]. Early Internet routers simply discarded excess packets instead of queuing them onto already full transmit queues; these routers are termed drop-tail gateways. However, this discard behavior was found to be unfair, favoring applications that send larger and more bursty data streams. Modern Internet routers employ more sophisticated, and fairer, drop algorithms, such as Random Early Detection (RED) [18].

Algorithms also have been developed that allow routers to organize their transmit queues so as to give resource guarantees to certain classes of traffic or to specific applications. These queuing or link scheduling algorithms include Weighted Fair Queuing (WFQ) [19] and Class Based Queuing (CBQ) [20]. A protocol called Resource Reservation Protocol (RSVP) [21] has been developed that allows hosts to dynamically signal to routers which applications should get special queuing treatment. However, RSVP has not yet been deployed, with some people arguing that queuing preference could more simply be indicated by using the TOS bits in the IP header [22,23].

Some vendors allow collection of traffic statistics on their routers: for example, how many packets and bytes are forwarded per receiving and transmitting interface on the router. These statistics are used for future capacity planning. They can also be used by ISPs to implement usage-based charging schemes for their customers.

Therefore, IP routers' functions can be classified into two types: datapath functions and control functions. Datapath functions are performed on every packet that passes through the router. These include forwarding decisions, switching through the backplane, and output link scheduling. These are most often implemented in special purpose hardware, called a forwarding engine.

Control functions include system configuration, management, and exchange of routing table information with neighboring routers. These are performed relatively infrequently. The route controller exchanges topology information with other routers and constructs a routing table based on a routing protocol (e.g., RIP, OSPF, and BGP). It can also create a forwarding table for the forwarding engine. Control functions are not processed for each arriving packet, because speed is not critical, they are implemented in software.

Therefore, the state of a router is maintained by the control function, the per-packet performance of a router is determined by its datapath functions. In this book, we will focus only on datapath functions (forwarding engine) and will not cover control functions, such as system configuration, management, routing mechanisms, and routing protocol. For further information on routing protocols see Refs. [24–27].

1.4 Evolution of Router Architecture

Routers are the core equipment in the Internet, and are found at every level in the Internet. Routers in access networks allow homes and small businesses to connect to an ISP. Routers in enterprise networks link tens of thousands of computers within a campus or enterprise. Routers in the backbone link together ISPs and enterprise networks with long distance trunks.

The rapid growth of the Internet has created different challenges for routers in backbone, enterprise, and access networks. The backbone needs routers capable of routing at high speeds on a few links. Enterprise routers should have a low cost per port, a large number of ports, be easy to configure, and support QoS. Finally, access routers should support many heterogeneous, high-speed ports, a variety of protocols at each port, and so on. All of these challenges drive the improvement of the routers in datapath functions and control functions.

The Internet has been in operation since the 1970s, and routers have gone through several design generations over the decades. The evolution of routers is often described in terms of three generations of architecture by Aweya [27] until 1999. Nick McKeown proposes the fourth generation and the future of router architecture [28,29].

1.4.1 First Generation—Bus-Based Router Architectures with Single Processor

The earliest routers (until the mid-to-late 1980s) were based on software implementations on a CPU. These routers consist of a general-purpose processor and multiple interface cards interconnected through a shared bus, as depicted in Figure 1.2.

Packets arriving at the interfaces are forwarded to the CPU, which determines the next-hop address and sends them back to the appropriate outgoing interface(s). Data are usually buffered in a centralized data memory, which leads to the disadvantage of having the data cross the bus twice, making it the major system bottleneck. Packet processing and node management software (including routing protocol operations, routing table maintenance, routing table lookups, and other control and management protocols such as ICMP and SNMP) are also implemented on

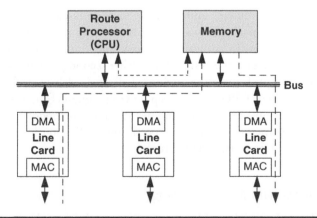

Figure 1.2 Traditional bus-based router architecture. (From Aweya, J., *Journal of Systems Architecture*, 46, 6, 2000. With permission.)

the central processor. Unfortunately, this simple architecture yields low performance for the following reasons:

■ The central processor has to process all packets flowing through the router (as well as those destined to it). This represents a serious processing bottleneck.
■ Some major packet processing tasks in a router involve memory-intensive operations (e.g., table lookups), which limits the effectiveness of processor power upgrades in boosting the router packet processing throughput. Routing table lookups and data movements are the major consumers of overall packet processing cycles. Packet processing time does not decrease linearly if faster processors are used because of the sometimes dominating effect of the memory access rate.
■ Moving data from one interface to the other (either through main memory or not) is a time consuming operation that often exceeds the packet header processing time. In many cases, the computer input/output (I/O) bus quickly becomes a severe limiting factor to overall router throughput.

Because routing table lookup is a time-consuming process of packet forwarding, some traditional software-based routers cache the IP destination-to-next-hop association in a separate database that is consulted as the front end to the routing table before the routing table lookup. The justification for route caching is that packet arrivals are temporally correlated, so that if a packet belonging to a new flow arrives, then more packets belonging to the same flow can be expected to arrive in the near future. Route caching of IP destination/next-hop address pairs will decrease the average processing time per packet if locality exists for packet addresses [30]. Still, the performance of the traditional bus-based router depends heavily on the throughput of the shared bus and on the forwarding speed of the central processor. This architecture cannot scale to meet the increasing throughput requirements of multigigabit network interface cards.

1.4.2 Second Generation—Bus-Based Router Architectures with Multiple Processors

For the second generation IP routers, improvement in the shared-bus router architecture was introduced by distributing the packet forwarding operations. In some architectures, distributing fast processors and route caches, in addition to receive and transmit buffers, over the network interface cards reduces the load on the system bus. Other second generation routers remedy this problem by employing multiple forwarding engines (dedicated solely to packet forwarding operation) in parallel because a single CPU cannot keep up with requests from high-speed input ports. An advantage of having multiple forwarding engines serving as one pool is the ease of balancing loads from the ports when they have different speeds and utilization levels. We review, in this section, these second generation router architectures.

1.4.2.1 Architectures with Route Caching

This architecture reduces the number of bus copies and speeds up packet forwarding by using a route cache of frequently seen addresses in the network interface, as shown in Figure 1.3. Packets are therefore transmitted only once over the shared bus. Thus, this architecture allows the network interface cards to process packets locally some of the time.

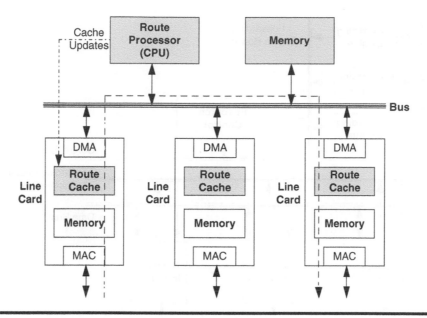

Figure 1.3 Reducing the number of bus copies using a route cache in the network interface. (From Aweya, J., *Journal of Systems Architecture,* **46, 6, 2000. With permission.)**

In this architecture, a router keeps a central master routing table and the satellite processors in the network interfaces each keep only a modest cache of recently used routes. If a route is not in a network interface processor's cache, it would request the relevant route from the central table. The route cache entries are traffic-driven in that the first packet to a new destination is routed by the main CPU (or route processor) via the central routing table information and as part of that forwarding operation, a route cache entry for that destination is then added in the network interface. This allows subsequent packet flows to the same destination network to be switched based on an efficient route cache match. These entries are periodically aged out to keep the route cache current and can be immediately invalidated if the network topology changes. At high speeds, the central routing table can easily become a bottleneck, because the cost of retrieving a route from the central table is many times more expensive than actually processing the packet local in the network interface.

A major limitation of this architecture is that it has a traffic-dependent throughput and also the shared bus is still a bottleneck. The performance of this architecture can be improved by enhancing each of the distributed network interface cards with larger memories and complete forwarding tables. The decreasing cost of high-bandwidth memories makes this possible. However, the shared bus and the general purpose CPU can neither scale to high-capacity links nor provide traffic pattern-independent throughput.

1.4.2.2 Architectures with Multiple Parallel Forwarding Engines

Another bus-based multiple processor router architecture is described in Ref. [31]. Multiple forwarding engines are connected in parallel to achieve high-packet processing rates as shown in Figure 1.4. The network interface modules transmit and receive data from links at the required rates. As a packet comes into a network interface, the IP header is stripped by a control circuitry, augmented with an identifying tag, and sent to a forwarding engine for validation and routing.

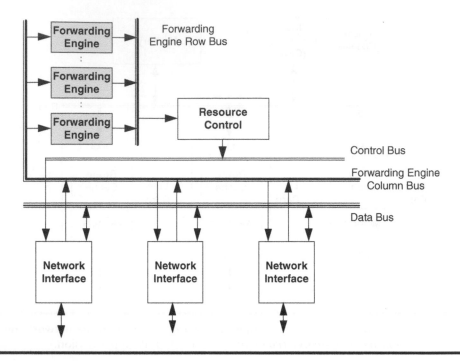

Figure 1.4   Bus-based router architecture with multiple parallel forwarding engines. (From Aweya, J., *Journal of Systems Architecture,* **46, 6, 2000. With permission.)**

While the forwarding engine is performing the routing function, the remainder of the packet is deposited in an input buffer (in the network interface) in parallel. The forwarding engine determines which outgoing link the packet should be transmitted on, and sends the updated header fields to the appropriate destination interface module along with the tag information. The packet is then moved from the buffer in the source interface module to a buffer in the destination interface module and eventually transmitted on the outgoing link.

The forwarding engines can each work on different headers in parallel. The circuitry in the interface modules peels the header off from each packet and assigns the headers to the forwarding engines in a round-robin fashion. Because in some (real time) applications packet order maintenance is an issue, the output control circuitry also goes round-robin, guaranteeing that packets will then be sent out in the same order as they were received. Better load-balancing may be achieved by having a more intelligent input interface, which assigns each header to the lightest loaded forwarding engine [31]. The output control circuitry would then have to select the next forwarding engine to obtain a processed header from by following the demultiplexing order followed at the input, so that order preservation of packets is ensured. The forwarding engine returns a new header (or multiple headers, if the packet is to be fragmented), along with routing information (i.e., the immediate destination of the packet). A route processor runs the routing protocols and creates a forwarding table that is used by the forwarding engines.

The choice of this architecture was premised on the observation that it is highly unlikely that all interfaces will be bottlenecked at the same time. Hence sharing of the forwarding engines can increase the port density of the router. The forwarding engines are only responsible for resolving next-hop addresses. Forwarding only IP headers to the forwarding engines eliminates an unnecessary packet payload transfer over the bus. Packet payloads are always directly transferred between the interface modules and they never go to either the forwarding engines or the route processor unless they are specifically destined to them.

1.4.3 Third Generation—Switch Fabric-Based Router Architecture

To alleviate the bottlenecks of the second generation of IP routers, the third generation of routers was designed with the shared bus replaced by a switch fabric. This provides sufficient bandwidth for transmitting packets between interface cards and allows throughput to be increased by several orders of magnitude. With the interconnection unit between interface cards not the bottleneck, the new bottleneck is packet processing.

The multigigabit router (MGR) is an example of this architecture [32]. The design has dedicated IP packet forwarding engines with route caches in them. The MGR consists of multiple linecards (each supporting one or more network interfaces) and forwarding engine cards, all connected to a high-speed (crossbar) switch as shown in Figure 1.5.

The design places forwarding engines on boards distinct from linecards. When a packet arrives at a linecard, its header is removed and passed through the switch to a forwarding engine. The remainder of the packet remains on the inbound linecard. The forwarding engine reads the header to determine how to forward the packet and then updates the header and sends the updated header and its forwarding instructions back to the inbound linecard. The inbound linecard integrates the new header with the rest of the packet and sends the entire packet to the outbound linecard for transmission. The MGR, like most routers, also has a control (and route) processor that provides basic management functions, such as generation of routing tables for the forwarding engines and link (up/down) management. Each forwarding engine has a set of the forwarding tables (which are a summary of the routing table data).

In the MGR, once headers reach the forwarding engine, they are placed in a request first-in-first-out (FIFO) queue for processing by the forwarding processor. The forwarding process can be roughly described by the following three stages [32].

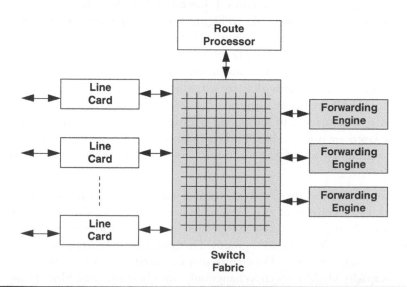

Figure 1.5 Switch-based router architecture with multiple forwarding engines. (From Aweya, J., *Journal of Systems Architecture*, 46, 6, 2000. With permission.)

The first stage includes the following that are done in parallel:

■ The forwarding engine does basic error checking to confirm that the header is indeed from an IPv4 datagram;
■ It confirms that the packet and header lengths are reasonable;
■ It confirms that the IPv4 header has no options.

In the second stage, the forwarding engine checks to see if the cached route matches the destination of the datagram (a cache hit). If not, the forwarding engine carries out an extended lookup of the forwarding table associated with it. In this case, the processor searches the routing table for the correct route, and generates a version of the route for the route cache. Because the forwarding table contains prefix routes and the route cache is a cache of routes for a particular destination, the processor has to convert the forwarding table entry into an appropriate destination-specific cache entry. Then, the forwarding engine checks the IP TTL field and computes the updated TTL and IP checksum, and determines if the packet is for the router itself.

In the third stage, the updated TTL and checksum are put in the IP header. The necessary routing information is extracted from the forwarding table entry and the updated IP header is written out along with link-layer information from the forwarding table.

1.4.4   Fourth Generation—Scaling Router Architecture Using Optics

Three generations of routers built around a single-stage crossbar and a centralized scheduler do not scale, and (in practice) do not provide the throughput guarantees that network operators need to make efficient use of their expensive long-haul links. Keslassy et al. propose a scaling router architecture using optics, shown in Figure 1.6 [33].

The router combines the massive information densities of optical communications, and the fast and flexible switching of the electronics. It has multiple racks that are connected by optical fibers, each rack has a group of linecards. In Figure 1.6, the architecture is arranged as G groups of L linecards. In the center, M statically configured $G \times G$ Micro-Electro-Mechanical Systems (MEMS) switches [34] interconnect the G groups. The MEMS switches are reconfigured only when a linecard is added or removed and provide the ability to create the needed paths to distribute the data to the linecards that are actually present. Each group of linecard spreads packets over the MEMS switches using an $L \times M$ electronic crossbar. Each output of the electronic crossbar is connected to a different MEMS switch over a dedicated fiber at a fixed wavelength (the lasers are not tunable). Packets from the MEMS switches are spread across the L linecards in a group by an $M \times L$ electronic crossbar. The architecture has the following advantages [33]:

1. Multirack routers spread the system power over multiple racks, reducing power density.
2. The switch fabric consists of three stages. It is the extension of the load-balanced router architecture [35] and has provably 100 percent throughput without a central scheduler.
3. All linecards are partitioned into G groups. The groups are connected together by M different $G \times G$ middle stage switches. The architecture can handle a very large number of linecards.
4. The high-capacity MEMS switches change only when linecards are added or moved. Only the lower-capacity local switches (crossbar) in each group need to be reconfigured frequently.

To design a 100 Tb/s router that implements the requirements of RFC 1812 [24], Keslassy et al. used the scalable router architecture. The router is assumed to occupy $G = 40$ multiple racks,

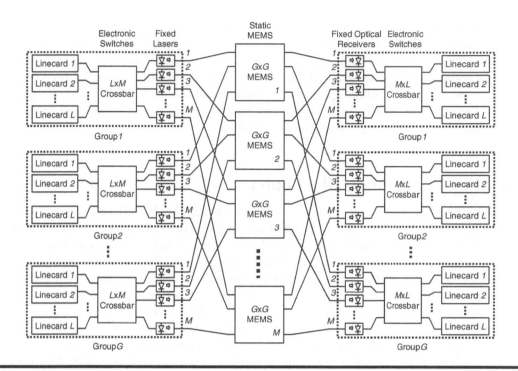

Figure 1.6 A hybrid optical-electrical router architecture. (From Isaac Keslassy, I. et al., *Proceedings of ACM SIGCOMM*, Karlsruhe, Germany, 2003, New York, ACM Press, 2003. With permission.)

as shown in Figure 1.7, with up to $L = 16$ linecards per rack. Each linecard operates at 160 Gb/s. Its input block performs address lookup, segments the variable length packet into one or more fixed length packets, and then forwards the packet to the local crossbar switch. Its output block receives packets from the local crossbar switch, collects them together, reassembles them into

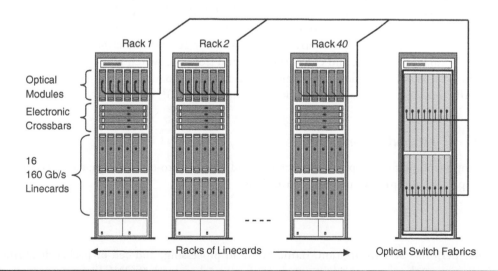

Figure 1.7 A 100Tb/s router example. (From Isaac Keslassy, I. et al., *Proceedings of ACM SIGCOMM*, Karlsruhe, Germany, 2003, New York, ACM Press, 2003. With permission.)

variable length packets and delivers them to the external line. Forty racks and 55 (= *L* + *G*) statically configured 40 × 40 MEMS switches are connected by optical fibers. In terms of optical technology, it is possible to multiplex and demultiplex 64 Wavelength-Division-Multiplexing channels onto a single optical fiber, and that each channel can operate at up to 10 Gb/s.

In future, as optical technology matures, it will be possible to replace the hybrid optical– electrical switch with an all-optical fabric. This has the potential to reduce power further by eliminating many electronic crossbars and serial links.

1.5 Key Components of a Router

From the first to the fourth generation, all routers must process headers of packets, switch packet-by-packet, and buffer packets during times of congestion. Therefore, the key components of a router are the forwarding engine to lookup IP address, the switch fabric to exchange packets between linecards, and the scheduler to manage the buffer.

As the router architectures change from centralized mode to distributed mode, more and more functions, such as buffer, IP-address lookup, and traffic management, are moved to linecards. Linecards become more complex and consume more power. To reduce power density, high-capacity routers use a multirack system with distributed, multistage switch fabrics. So, linecards and switch fabrics are the key components that implement the datapath functions. We will next discuss the line-card, network processor and switch fabric. Subsections 1.5.1 and 1.5.2 are from [36] (© 2002 IEEE).

1.5.1 Linecard

The linecards are the entry and exit points of data to and from a router. They provide the interface from the physical and higher layers to the switch fabric. The tasks provided by linecards are becoming more complex as new applications develop and protocols evolve.

Each linecard supports at least one full-duplex fiber connection on the network side, and at least one ingress and one egress connection to the switch fabric backplane. Generally speaking, for high-bandwidth applications, such as OC-48 (2.5 Gb/s) and above, the network connections support channelization for aggregation of lower-speed lines into a large pipe, and the switch fabric connections provide flow-control mechanisms for several thousand input and output queues to regulate the ingress and egress traffic to and from the switch fabric.

A linecard usually includes components such as a transponder, framer, network processor (NP), traffic manager (TM), and CPU, shown in Figure 1.8 [36].

1.5.1.1 Transponder/Transceiver

This component performs optical-to-electrical and electrical-to-optical signal conversions and serial-to-parallel and parallel-to-serial conversions.

1.5.1.2 Framer

A framer performs synchronization, frame overhead processing, and cell or packet delineation. For instance, on the transmit side, a synchronous optical network (SONET) framer generates a section, line, and path overhead. It performs framing pattern insertion (A1, A2) and scrambling.

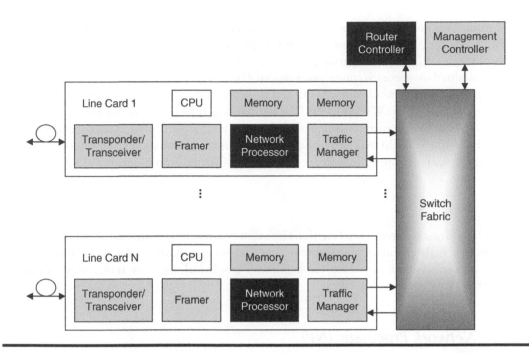

Figure 1.8   A typical router architecture. (From Chao, H., *Proceedings of the IEEE*, 90. With permission.)

It generates section, line, and path bit interleaved parity (B1/B2/B3) for far-end performance monitoring. On the receiver side, it processes the section, line, and path overhead. It performs frame delineation, descrambling, alarm detection, pointer interpretation, bit interleaved parity monitoring (B1/B2/B3), and error count accumulation for performance monitoring [37].

1.5.1.3   Network Processor (NP)

The NP mainly performs IP-address lookup, packet classification, and packet modification. It can perform at the line rate using external memory, such as static RAMs (SRAMs), DRAM, or content addressable memory (CAMs). The NPs are considered as fundamental a part of routers and other network equipment as a microprocessor is for personal computers. Various architectures for the NP are discussed in the next section.

1.5.1.4   Traffic Manager

To meet each connection and service class requirement, the traffic manager (TM) performs various control functions on packet streams, including traffic access control, buffer management, and packet scheduling. Traffic access control consists of a collection of specification techniques and mechanisms to: (*i*) specify the expected traffic characteristics and service requirements (e.g., peak rate, required delay bound, and loss tolerance) of a data stream; (*ii*) shape (i.e., delay) data streams (e.g., reducing their rates or burstiness); and (*iii*) police data streams and take corrective actions (e.g., discard, delay, or mark packets) when traffic deviates from its specification. The usage parameter control (UPC) in the asynchronous transfer mode (ATM) and the differentiated service (DiffServ) in the IP perform similar access control functions at the network edge. Buffer management performs packet discarding, according to loss requirements and priority levels, when the buffer exceeds a certain threshold. The proposed schemes include random early packet discard (RED), weighted RED, early packet discard (EPD), and partial packet discard (PPD). Packet scheduling ensures that packets are transmitted to meet each connection's allocated bandwidth/delay requirements. The proposed schemes include deficit round-robin, WFQ and its variants, such as shaped virtual clock [38] and worst-case fairness WFQ (WF²Q+) [39]. The last two algorithms achieve the worst-case fairness properties. Many QoS control techniques, algorithms, and implementation architectures can be found in Ref. [40]. The TM may also manage many queues to resolve contention among the inputs of a switch fabric, for example, hundreds or thousands of virtual output queues (VOQs).

1.5.1.5 CPU

The CPU performs control plane functions including connection setup/tear-down, forwarding table updates, register/buffer management, and exception handling. The CPU is usually not in-line with the fast-path on which the maximum-bandwidth network traffic moves between the interfaces and the switch fabric.

1.5.2 Network Processor (NP)

It is widely believed that the NP is the most effective solution to the challenges facing the communication industry regarding its ability to meet the time-to-market need with products at increasingly higher speed, while supporting the convergence and globalization trends of IP traffic. However, different router features and switch fabric specifications require a suitable NP with a high degree of flexibility to handle a wide variety of functions and algorithms. For instance, it is desirable for an NP to be universally applicable across a wide range of interfaces, protocols, and product types. This requires programmability at all levels of the protocol stack, from layer 2 through layer 7. However, this flexibility is a tradeoff with performance, such as speed and capacity.

Currently, a wide variety of the NPs on the market offer different functions and features. The way to select the proper NP depends on the applications, features, flexibility in protocols and algorithms, and scalability in the number of routes and flows. In general, NPs are classified by the achievable port speed, function list capability and programmability, hardware assisting functions, for example, hashing, tree structure, filtering, classifier for security, check sum or cyclic redundancy check (CRC) data, and operation speed (i.e., clock frequency of embedded processors).

With current router requirements, a single-processor system may not be able to meet router processing demands due to the growing gap between the link and processor speeds. With increasing port speeds, packets arrive faster than a single processor can process them. However, because packet streams have dependencies only among packets of the same flow and not across different flows, the processing of these packets can be easily distributed over several processors working in parallel. The current state of integrated circuit technology enables multiple processors to be built on a single silicon die. To support high performance, flexibility, and scalability, the NP architecture must effectively address efficient handling of I/O events (memory access and interrupts), scheduling process management, and provide a different set of instructions to each processor.

Several parallel processing schemes can be considered as prospective architectures for the NP. They are briefly discussed subsequently. With multiple instruction multiple data (MIMD) processing, multiple processors may perform different functions in parallel. The processors in this architecture can be of the reduced instruction set computing (RISC) type and are interconnected to a shared memory and I/O through a switch fabric. When packets of the same flow are processed in different processors, interprocessor communication is required. This causes memory dependencies and may limit the flexibility of partitioning the function across multiple processors.

Very long instruction word (VLIW) processing has a structure similar to MIMD processing, except that it uses multiple special-purpose coprocessors that can simultaneously perform different tasks. They are specifically designed for certain functions and thus can achieve high-data rates. Because these coprocessors are function-specific, adaptation of new functions and protocols is restricted.

According to the implementation style and the type of embedded processor, NPs can be classified into the following two broad groups:

■ *Configurable.* This kind of NP consists of multiple special-purpose coprocessors interconnected by a configurable network, and a manager handling the interconnect configuration, the memory access, and the set of instructions used by the coprocessors. Figure 1.9 shows an example of a configurable NP.

A coprocessor can perform a predefined set of functions (e.g., longest or exact prefix match instructions for table lookup or classification). The manager instructs the coprocessors what functions can be performed from the available set and selects a path along which packets flow among the coprocessors. When a packet arrives at the NP, the manager routes the packet to a classification and table lookup coprocessor. After the packet is processed by the coprocessor, it is passed to the next one (the packet analysis and modification unit) in the pipeline. After the packet has been modified, it is passed to the next coprocessor (switch fabric forwarding), where it may be segmented into cells and wait to be transmitted to the switch fabric (assuming no TM follows the NP). When the packet processing is completed, the manager schedules the time the packet exits the NP.

This NP is designed with a narrow set of function choices to optimize the chip area and speed. The advantage of this NP is that the embedded coprocessors can be designed for high performance.

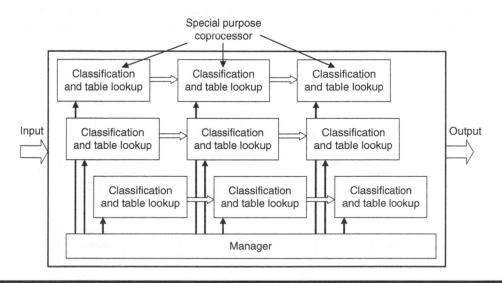

Figure 1.9 A configurable network processor. (From Chao, H., *Proceedings of the IEEE*, 90. With permission.)

The disadvantage is that this approach limits the NP adopting new applications and protocols and may make the NP obsolete in a short time. Configurable NPs are considered to be one of the VLIW processing architectures.

■ *Programmable.* This kind of NP has a main controller and multiple task units that are interconnected by a central switch fabric (e.g., a crossbar network). A task unit can be a cluster of (one or more) RISC processors or a special-purpose coprocessor. The controller handles the downloading of the instruction set to each RISC processor, the access of a RISC processor to special-purpose coprocessors and memory, and the configuration of the switch fabric.

Figure 1.10 depicts a simple general architecture for a programmable NP. When a packet arrives at the NP, the controller assigns an idle RISC processor to handle the processing of the packet. The RISC processor may perform the classification function by itself or forward the packet to the classification coprocessor. The latter approach allows a new function to be performed by the RISC processor and a specific function to be performed by a coprocessor. If coprocessor access is required, the RISC processor sends the request to the controller. It schedules the time when the request will be granted. After the packet is classified, the RISC processor may perform the packet modification or forward the packet to a modification coprocessor.

The processing of the packet continues until it is done. Then the task unit informs the controller, which schedules the departure time for the processed packet. This approach offers great flexibility because the executed functions and their processing order can be programmed. The disadvantage is that because of the flexibility, the design of the interconnection fabric, RISC processors, and coprocessors cannot be optimized for all functions. As a result, the processing of some functions takes more time and cannot meet the wire-speed requirement. This NP category is considered one of the MIMD processing architectures.

Because there may be up to 16 processors (either special-purpose coprocessors or general-purpose RISC processors) in the NP (there may be more in the future), how to effectively program the NP to support different applications at line rate is very challenging. Some companies specialize in creating machine codes based on the NP structure. The user just needs to build applications using a user interface based on state-machine definitions and never needs to look at the code.

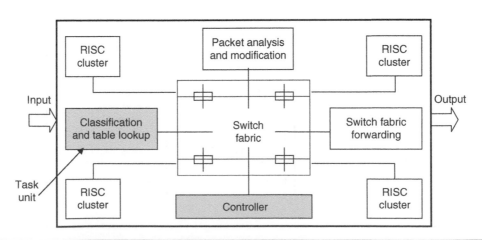

Figure 1.10 A network processor with multiple RISC clusters. (From Chao, H., *Proceedings of the IEEE*, 90. With permission.)

This also allows the applications created by the development environment to be completely portable from the old to the new generation NP as NP technology evolves. In general, the processing capacity of a programmable NP is a function of the following parameters: number of RISC processors, size of on-chip caches, and number of I/O channels. A potential research topic is the study of multithreading processing on multiple on-chip processors.

1.5.3 Switch Fabric

Switch fabric is a principal building block in a router. It connects each input with every output and allows a dynamic configuration of the connections. The manager that controls the dynamic connections is called the Scheduler. There are two main components in almost any router: the switch fabric and the scheduler. They are often implemented based on hardware and software, respectively. These two components are tightly related and improving one without the other fails to enhance the overall performance. The switch fabric determines the switching speed once the data is ready in the input of the switch and the scheduler delivers packets from the network input lines to the fabric and from the fabric to the network output lines. It is necessary for the scheduler to perform these deliveries taking into account various factors, such as fabric speed, sampling rate, buffer size, QoS, and so on.

There are many designs of switch fabric to build high-speed and large-capacity switches. Based on the multiplexing techniques, they can be classified into two groups: Time-Division Switching (TDS) and Space-Division Switching (SDS). And they can be further divided. Based on the buffer strategies, they can be classified into internally buffered switch, input-buffered switch, output-buffered switch, shared-buffer switch, VOQ switch, and so on. This section describes several popular switch architectures.

1.5.3.1 Shared Medium Switch

In a router, packets may be routed by means of a shared medium, for example, bus, ring, or dual bus. The simplest switch fabric is the bus. Bus-based routers implement a monolithic backplane comprising a single medium over which all intermodule traffic must flow. Data are transmitted across the bus using Time Division Multiplexing (TDM), in which each module is allocated a time slot in a continuously repeating transmission. However, a bus is limited in capacity and by the arbitration overhead for sharing this critical resource. The challenge is that, it is almost impossible to build a bus arbitration scheme fast enough to provide nonblocking performance at multigigabit speeds.

An example of a fabric using a TDM bus is shown in Figure 1.11. Incoming packets are sequentially broadcast on the bus (in a round-robin fashion). At each output, address filters examine the internal routing tag on each packet to determine if the packet is destined for that output. The address filters pass the appropriate packets through to the output buffers.

It is apparent that the bus must be capable of handling the total throughput. For discussion, we assume a router with N input ports and N output ports, with all port speeds equal to S (fixed size) packets per second. In this case, a packet time is defined as the time required to receive or transmit an entire packet at the port speed, that is, $1/S$ seconds. If the bus operates at a sufficiently high speed, at least NS packets/s, then there are no conflicts for bandwidth and all queuing occurs at the outputs. Naturally, if the bus speed is less than NS packets/s, some input queuing will probably be necessary.

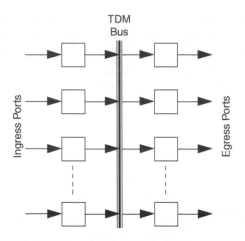

Figure 1.11   Shared medium switch fabric: a TDM bus. (From Aweya, J., *Journal of Systems Architecture*, 46, 6, 2000. With permission.)

In this architecture, the outputs are modular from each other, which has advantages in implementation and reliability. The address filters and output buffers are straightforward to implement. Also, the broadcast-and-select nature of this approach makes multicasting and broadcasting natural. For these reasons, the bus type switch fabric has found a lot of implementation in routers. However, the address filters and output buffers must operate at the speed of the shared medium, which could be up to N times faster than the port speed. There is a physical limit to the speed of the bus, the address filters, and the output buffers; these limit the scalability of this approach to large sizes and high speeds. Either size N or speed S can be large, but there is a physical limitation on the product NS. As with the shared memory approach (to be discussed next), this approach involves output queuing, which is capable of optimal throughput (compared to a simple FIFO input queuing). However, the output buffers are not shared, and hence this approach requires more total amount of buffers than the shared memory fabric for the same packet loss rate. Examples of shared-medium switches are IBM PARIS switch [41], ForeRunner ASX-100 switch [42].

1.5.3.2   Shared Memory Switch Fabric

Shared memory switch fabric is also based on TDS. A typical architecture of a shared memory fabric is shown in Figure 1.12.

Incoming packets are typically converted from serial to parallel form, which are then written sequentially into a (dual port) random access memory. Their packet headers with internal routing tags are typically delivered to a memory controller, which decides the order in which packets are read out of the memory. The outgoing packets are demultiplexed to the outputs, where they are converted from parallel to serial form. Functionally, this is an output queuing approach, where the output buffers all physically belong to a common buffer pool. The output buffered approach is attractive, because it can achieve a normalized throughput of one under a full load [43,44]. Sharing a common buffer pool has the advantage of minimizing the amount of buffer required to achieve a specified packet loss rate. The main idea is that a central buffer is most capable of taking advantage of statistical sharing. If the rate of traffic to one output port is high, it can draw upon more buffer space until the common buffer pool is (partially or) completely filled. For these reasons, it

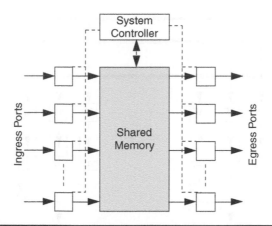

Figure 1.12   A shared memory switch fabric. (From Aweya, J., *Journal of Systems Architecture*, 46, 6, 2000. With permission.)

is a popular approach for router design (e.g., Cisco's catalyst 8510 architecture, Torrent IP9000 gigabit router).

Unfortunately, the approach has its disadvantages. As the packets must be written into and read out from the memory one at a time, the shared memory must operate at the total throughput rate. It must be capable of reading and writing a packet (assuming fixed size packets) in every $1/NS$ second, that is, N times faster than the port speed. As the access time of random access memories is physically limited, this speedup factor N limits the ability of this approach to scale up to large sizes and fast speeds. Moreover, the (centralized) memory controller must process (the routing tags of) packets at the same rate as the memory. This might be difficult if, for instance, the controller must handle multiple priority classes and complicated packet scheduling. Multicasting and broadcasting in this approach will also increase the complexity of the controller. Multicasting is not natural to the shared memory approach but can be implemented with additional control circuitry. A multicast packet may be duplicated in the memory or read multiple times from the memory. The first approach obviously requires more memory because multiple copies of the same packet are maintained in the memory. In the second approach, a packet is read multiple times from the same memory location [45–47]. The control circuitry must keep the packet in memory until it has been read to all the output ports in the multicast group.

A single point of failure is invariably introduced in the shared memory-based design because adding a redundant switch fabric to this design is very complex and expensive. As a result, shared memory switch fabrics are best suited for small capacity systems.

1.5.3.3   Distributed Output Buffered Switch Fabric

The distributed output buffered approach is shown in Figure 1.13. Independent paths exist between all N^2 possible pairs of inputs and outputs. In this design, arriving packets are broadcast on separate buses to all outputs. Address filters at each output determine if the packets are destined for that output. Appropriate packets are passed through the address filters to the output queues.

This approach offers many attractive features. Naturally there is no conflict among the N^2 independent paths between inputs and outputs, and hence all queuing occurs at the outputs.

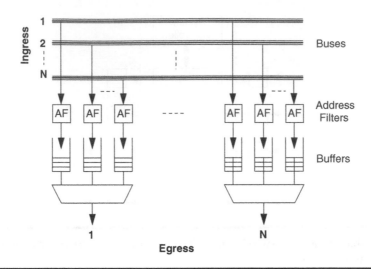

Figure 1.13 A distributed output buffered switch fabric. (From Aweya, J., *Journal of Systems Architecture*, 46, 6, 2000. With permission.)

As stated earlier, output queuing achieves the optimal normalized throughput compared to simple FIFO input queuing [43,44]. Like the shared medium approach, it is also broadcast-and-select in nature and, therefore, multicasting is natural. For multicasting, an address filter can recognize a set of multicast addresses as well as output port addresses. The address filters and output buffers are simple to implement. Unlike the shared medium approach, the address filters and buffers need to operate only at port speed. All of the hardware can operate at the same speed. There is no speedup factor to limit scalability in this approach. For these reasons, this approach has been adopted in some commercial router designs [45].

Unfortunately, the quadratic N^2 growth of buffers means that the size N must be limited for practical reasons. However, in principle, there is no severe limitation on S. The port speed S can be increased to the physical limits of the address filters and output buffers. Hence, this approach might realize a high total throughput (NS packets per second) by scaling up the port speed S. The knockout switch was an early prototype that suggested a tradeoff to reduce the amount of buffer at the cost of higher packet loss [46]. Instead of N buffers at each output, it was proposed to use only a fixed number of L buffers at each output (for a total of NL buffers, which is linear in N), based on the observation that the simultaneous arrival of more than L packets (cells) to any output was improbable. It was argued that $L = 8$ is sufficient under uniform random traffic conditions to achieve a cell loss rate of 10^{-6} for a large N.

1.5.3.4 Crossbar Switch

In TDS, the switch fabrics are shared by all input and output ports. In a time slot, only one input port can send packets to output ports. In fact, when the input port A send packets to the output port B, it may have an impact on a packet from the input port C to the output port D. In SDS, the switch fabric can allow packets from any input to any output port simultaneously if there is no conflict.

There are several switch fabrics based on SDS; one of the most popular switch fabrics is crossbar switch because of its (*i*) low cost, (*ii*) good scalability, and (*iii*) nonblocking properties. Crossbar switches have an architecture that, depending on the implementation, can scale to very high bandwidths. Considerations of cost and complexity are the primary constraints on the capacity of switches of this type, because the number of the crosspoints grows to N^2 for a router with N inputs and N outputs.

There are three possible locations for the buffers in a crossbar switch: (a) at the crosspoints in the switch fabric, (b) at the inputs of the switch, and (c) at inputs and outputs of the switch. Each one has its advantages and disadvantages [47].

Figure 1.14 is a simple example of the crossbar switch that can physically connect any of the N inputs to any of the N outputs. An input buffered crossbar switch has the crossbar fabric running at the link rate. In this architecture, buffering occurs at the inputs, and the speed of the memory does not need to exceed the speed of a single port. Given the current state of technology, this architecture is widely considered to be substantially more scalable than output buffered shared memory switches. However, the crossbar architecture presents many technical challenges that need to be overcome to provide bandwidth and delay guarantees. Examples of commercial routers that use crossbar switch fabrics are Cisco's 12000 Gigabit switch router, Ascend Communication's GRF400, and so on.

We start with the issue of providing bandwidth guarantees in the crossbar switch. For the case where there is a single FIFO queue at each input, it has long been known that a serious problem referred to as head-of-line (HOL) blocking [43] can substantially reduce achievable throughput. In particular, the well-known result of Ref. [43] is that for uniform random distribution of input traffic, the achievable throughput is only 58.6 percent. Moreover, Li [48] has shown that the maximum throughput of the switch decreases monotonically with increasing burst size. Considerable amount of work has been done in recent years to build input buffered switches, which match the performance of an output buffered switch. One way of reducing the effect of HOL blocking is to increase the speed of the I/O channel (i.e., the speedup of the switch fabric). Speedup is defined as the ratio of the switch fabric bandwidth to the bandwidth of the input links. There have been a number of studies, such as Refs. [49,50], which showed that an input buffered crossbar switch with a single FIFO at the input can achieve about 99 percent throughput under certain assumptions on the input traffic statistics for speedup in the range of 4 to 5. A more recent simulation study [51] suggested that speedup as low as 2 may be sufficient to obtain performance comparable to that of output buffered switches.

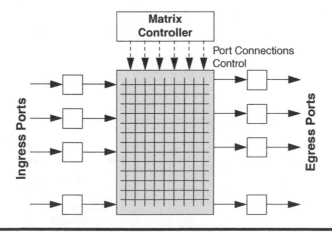

Figure 1.14 A crossbar switch. (From Aweya, J., *Journal of Systems Architecture,* **46, 6, 2000. With permission.)**

Another way of eliminating HOL blocking is by changing the queuing structure at the input. Instead of maintaining a single FIFO at the input, a separate queue for each output can be maintained at each input VOQs (Fig. 1.15).

At each input, a separate FIFO queue is maintained for each output, as shown in Figure 1.15. After a forwarding decision has been made, an arriving packet is placed in the queue corresponding to its output port. No packet is held up by a packet in front of it that is destined to a different output; HOL blocking is reduced. In theory, a scheduling algorithm can increase the throughput of a crossbar switch from 60 percent with FIFO queuing to a full 100 percent if VOQs are used [52]. Input queuing with VOQs is very popular, however, it requires a fast and intelligent scheduler at the crossbar, because at beginning of each time slot, the scheduler examines the contents of all the input queues, whose total number is N^2.

Because there could be contention at the inputs and outputs, an arbitration algorithm is required to schedule packets between various inputs and outputs (equivalent to the matching problem for bipartite graphs). It has been shown that an input buffered switch with VOQs can provide asymptotic 100 percent throughput using a maximum matching algorithm [52]. However, the complexity of the best known maximum match algorithm is too high for high-speed implementation. Moreover, under certain traffic conditions, maximum matching can lead to starvation. Over the years, a number of maximal matching algorithms have been proposed [53–55]; iSLIP is an example of one of the popular schemes.

iSLIP [56] has been proposed by N. McKeown. It provides high throughput, is starvation-free, and because of its simplicity is able to operate at high speed. This scheme requires an input arbiter per input port and an output arbiter per output port. An output arbiter selects a VOQ request coming from the inputs, and an input arbiter selects a grant coming from an output arbiter. Each arbiter has a pointer to indicate the priority list order for each arbiter. The iSLIP can have any number of iterations. In this scheme, the grant pointers update their positions only if their grants are accepted, that is, if a match is completed, in the first iteration. In this scheme, starvation is avoided because a recently matched pair gets the lowest priority. The steps for this scheme are as follows.

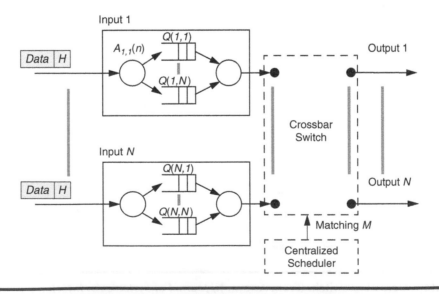

Figure 1.15 A model of an N-port input-queued switch with VOQ. (From McKeown, N., *Business Communications Review***, 1997. With permission.)**

1. Each unmatched input sends a request to every output for which it has a queued cell.
2. If an unmatched output receives multiple requests, it chooses the one that appears next in a fixed round-robin schedule, starting from the highest priority element. The output notifies each input whether or not its request was granted.
3. If an input receives multiple grants, it accepts the one that appears next in a fixed, round-robin schedule starting from the highest priority element. The pointer a_j is incremented (modulo N) to one location beyond the accepted output. The accept pointers are only updated in the first iteration.
4. The grant pointer g_j is incremented (module N) to one location beyond the granted input if and only if the grant is accepted in step 3 of the first iteration.

Because of the round-robin moving of the pointers, the algorithm provides a fair allocation of bandwidth among all flows. This scheme contains $2N$ arbiters, where each arbiter is implemented with low complexity. The throughput offered with this algorithm is 100 percent under traffic with uniform distribution for any number of iterations due to the desynchronization effect [52].

As stated above, increasing the speedup of the switch fabric can improve the performance of an input buffered switch. However, when the switch fabric has a higher bandwidth than the links, buffering is required at the outputs too. Thus a combination of input buffered and output buffered switch is required, that is, Combined Input and Output Buffered (CIOB). The goal of most designs then is to find the minimum speedup required to match the performance of an output buffered switch using a CIOB and VOQs. Chuang et al. [57] have shown that a CIOB switch with VOQs is always work-conserving if speedup is greater than $N/2$. Prabhakar et al. [58] showed that a speed of 4 is sufficient to emulate an output buffered switch (with an output FIFO) using a CIOB switch with VOQs.

Multicast in space division fabrics is simple to implement, but has some consequences. For example, a crossbar switch (with input buffering) is naturally capable of broadcasting one incoming packet to multiple outputs. However, this would aggravate HOL blocking at the input buffers. Approaches to alleviate the HOL blocking effect would increase the complexity of buffer control. Other inefficient approaches in crossbar switches require an input port to write out multiple copies of the packet to different output ports one at a time. This does not support the one-to-many transfers required for multicasting as in the shared bus and fully distributed output buffered architectures. The usual concern about making multiple copies is that it reduces effective switch throughput. Several approaches for handling multicasting in crossbar switches have been proposed [59]. Generally, multicasting increases the complexity of space division fabrics.

1.5.3.5 *Space-Time Division Switch*

The conventional switch fabric, that is, crossbar and shared memory switches, will not be able to practically satisfy the terabit data transfer demand. Nourani et al. propose a switch fabric based on mixing the space and time division multiplexing [60] called Space-Time Division Switch. These subsections are from [60] (© 2002 IEEE).

To achieve high performance over the system, Nourani et al. combine the best qualities of both a crossbar and a shared memory switch. When combined, the switch has the low latency of a crossbar switch, with the bandwidth and reconfigurability of a shared memory switch. The fundamental building block in the Space-Time Division Switch, called the port block, is shown in Figure 1.16. A port block essentially implements a $1 \times K$ or $K \times 1$ crossbar switch (e.g., $1 \times K$ DeMUX or $K \times 1$ MUX, respectively) and has the capability of storing data. This storage capability

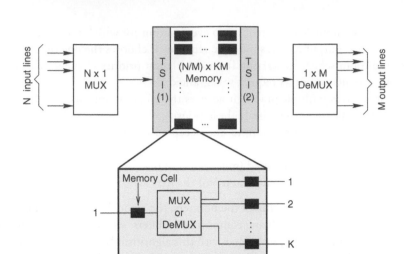

Figure 1.16   A mixed space-time switch architecture. (From Nourani, M., *Proceedings of IEEE Midwest Symposium on Circuits and Systems (MWSCAS)*, Fairborn, OH, 2001, New York, 2001. With permission.)

has been added to match the top level architecture, which is supposed to work as a shared memory. Additionally, having a small size crossbar with memory cells provides us with shorter latency and more flexibility in forwarding and scheduling packets without worrying that they maybe lost. The complete switching devices are formed using multiple port blocks with a single device. This allows for granularity in switching data as well as scalability.

Note carefully that we use two $K \times K$ time-slot interchangers that rearrange the data stream according to the desired switching pattern and store it in one row of memory [60]. The first one places the incoming packet into any outgoing time slot (i.e., any port block) on the fabric. The second one can locate the packet from any port block (i.e., any incoming time slot) and place it in any outgoing time slot, which will be distributed among the output lines. The two interchangers behave like a $K \times K$ nonblocking crossbar switch. For N input lines and M output lines, the number of rows cannot be less than $\lceil N/M \rceil$ and in general it depends on the queue/scheduling strategy. A combination of port blocks and time-slot interchangers may now be used to implement any kind of $N \times M$ switch of any architecture that one would like to build: Clos, Banyan, Butterfly, and so on. This is done by finding an optimal solution on how to map inputs to the available port blocks and eventually forwarding them to the outputs. One flexible technique is to map N inputs to NM port blocks and mapping their NMK outputs to the M outputs. The aggregate throughput now becomes MKf, which is an improvement by a factor of K compared to conventional strategies. A number of these switching devices, as shown in Figure 1.16, can be cascaded together to make N and M as large as the design calls for. The scalability is truly limited by the number of devices one can accommodate on a single board given the form factor requirements for the networking system. This can be scaled to larger switches by introducing fabrics that contain multiple switching devices. For example, if a 1024×1024 switch fabric with port blocks of size $K = 64$ is used in a router that runs at 100 MHz, it can run at 100 percent efficiency and achieve peak throughput on every cycle, which is $1024 \times 64 \times 100$ MHz = 6.5 Tbps.

The switch fabric is nonblocking, because it is still based on the shared memory fundamentals and can sustain inputs to it till all the port blocks in the switch fabric are exhausted. Because each device has upward of hundreds of port blocks, the fabric effectively implements a switch of a size that is hundreds of input ports by hundreds of output ports with each I/O port being 128 bits wide.

1.5.4 IP-Address Lookup: A Bottleneck

The explosive growth of link bandwidth requires fast IP-address lookup. For example, links running at 10 Gb/s can carry 31.25 million packets per second (Mp/s), assuming minimum sized 40-byte packets. After the deployment of DWDM, the link speed has doubled every seven months [61]. However, historically, the capacity of routers has doubled every 18 months [61]. Therefore, IP-address lookup becomes a bottleneck for the high-performance router.

To eliminate cumbersome functions, such as IP-address lookup from the core network, a Multi-Protocol Label Switching (MPLS) protocol was proposed. Once a flow has been identified at the network edge and mapped onto a label, there will be no more IP-address lookup along the path. IP-address lookup will be performed only at edge routers. However, before the MPLS is ubiquitously adopted in a core network, IP-address lookup remains a necessary and challenging issue. Hence, this book discusses the problem of IP-address lookup.

References

1. Vareille, G., Pitel, F., and Marcerou, J.F., 3-Tbit/s (300 × 11.6 Gbit/s) transmission over 7380 km using C+L band with 25 GHz channel spacing and NRZ format, *Proc. Optical Fiber Communication Conf. Exhibit*, 2001, OFC 2001, Volume 4, Issue, 17–22 March 2001 Page(s): PD22-1–PD22-3.
2. Odlyzko, A. M., Comments on the Larry Roberts and Caspian Networks study of Internet traffic growth, *The Cook Report on the Internet*, pp. 12–15, Dec. 2001.
3. McKeown, N., Internet routers, Stochastic Networks Seminar, February 22, 2002 http://tiny-tera. stanford.edu/~nickm/talks/Stochnets_Feb_2002.ppt.
4. Comer, D., *Internetworking with TCP/IP. Principle, Protocols and Architecture*. Englewood Cliffs, NJ: Prentice-Hall, 1989.
5. Perlman, R., *Interconnections: Bridges and Routers*, Addison-Wesley, Reading, MA, 1992.
6. Huitema, C., *Routing in the Internet*, Prentice Hall, Englewood Cliffs, NJ, 1996.
7. Moy, J., *OSPF: Anatomy of an Internet Routing Protocol*, Addison-Wesley, Reading, MA, 1998.
8. Stevens, W.R., *TCP/IP Illustrated*, vol. 1, The Protocols, Addison-Wesley, Reading, MA, 1994.
9. Baker, F., Requirements for IP Version 4 routers, *IETF RFC 1812*, June 1995.
10. Rijsinghani, A., Computation of the Internet checksum via incremental update, *IETF RFC1624*, May 1994.
11. Kent, C.A. and Mogul, J.C., Fragmentation considered harmful, *Computer Commun. Rev.* 1987; 17(5):390–401.
12. Mogul, J. and Deering, S., Path MTU discovery IETF *RFC 1191*, April 1990.
13. Plummer, D., Ethernet address resolution protocol: or converting network protocol addresses to 48-bit Ethernet addresses for transmission on ethernet hardware, *IETF RFC 826*, November 1982.
14. Bradley, T. and Brown, C., Inverse address resolution protocol, *IETF RFC 1293*, January 1992.
15. Rigney, C. et al., Remote Authentication Dial-In User Service (RADIUS), *IETF RFC 2138*, April 1997.
16. Jacobson, V., Congestion avoidance and control, *Comput. Commun. Rev.* 1988; 18(4):314–329.
17. Stevens, W., TCP slow start, congestion avoidance, fast retransmit and fast recovery algorithms, *IETF RFC 2001*, January 1997.

18. Floyd, S. and Jacobson, V., Random early detection gateways for congestion avoidance, *IEEE/ACM Transactions on Networking* 1993; 2:397–413.

19. Demers, A., Keshav, S. and Shenker, S., Analysis and simulation of a fair queueing algorithm, *Journal of Internetworking: Research and Experience* 1990; 1:3–26.

20. Floyd, S. and Jacobson, V., Link sharing and resource management models for packet networks, *IEEE/ACM Transactions on Networking* 1995; 3(4):365–386.

21. Zhang, L. et al., RSVP: A new resource reservation protocol, *IEEE Network* 1993; 7(5):8–18.

22. Nichols, K. and Blake, S., Differentiated services operational model and definitions, *IETF Work in Progress*.

23. Blake, S., An architecture for differentiated services, *IETF Work in Progress*.

24. Baker, F., Requirements for IP version 4 routers, *IETF RFC 1812*. June 1995.

25. Malkin, G., RIP Version 2, *IETF RFC 2453*. November 1998.

26. Rekhter, Y. and Li, T., A Border Gateway Protocol 4 (BGP-4), *IETF RFC 1771*, March 1995.

27. Aweya, J., On the design of IP routers Part 1: Router architectures, *Journal of Systems Architecture* 2000; 46(6):483–511.

28. McKeown, N., Optics inside routers, *Proc. of ECOC 2003*, Rimini, Italy, 2003; 2:168–171.

29. McKeown, N., Internet routers: past present and future. Lecture for British Computer Society, Ada Lovelace Award, June 2006, London, http://tiny-tera.stanford.edu/~nickm/talks/BCSv6.0.ppt.

30. Feldmeier, D.C., Improving gateway performance with a routing-table cache, *Proceedings of the IEEE Infocom'88*, New Orleans, LA, March 1988, pp. 298–307.

31. Asthana, S. et al., Towards a gigabit IP router, *Journal of High Speed Networks* 1992; 1(4):281–288.

32. Partridge, C. et al., A 50Gb/s IP router, *IEEE/ACM Transactions on Networking* 1998; 6(3):237–248.

33. Isaac Keslassy, I. et al., Scaling internet routers using optics, *Proceedings of ACM SIGCOMM*, Karlsruhe, Germany, Aug. 2003. New York, ACM Press, 2003, pp. 189–200.

34. Ryf, R. et al., 1296-port MEMS transparent optical crossconnect with 2.07 petabit/s switch capacity, *Proc. Optical Fiber Comm. Conf. and Exhibit (OFC) '01*, Vol. 4, pp. PD28–P1-3, 2001.

35. Valiant, L.G. and Brebner, G.J., Universal schemes for parallel communication, *Proceedings of the 13th ACM Symposium on Theory of Computation*, 1981, pp. 263–277.

36. Chao, H., Next generation routers, *Proceedings of the IEEE*, 90(9), September 2002, pp. 1518–1558.

37. Chow, M.-C., *Understanding SONET/SDH: Standards and Applications*. Holmdel, NJ: Andan Publishers, 1995.

38. Stiliadis, D. and Varma, A., A general methodology for design efficient traffic scheduling and shaping algorithms, *Proc. IEEE INFOCOM*, April 1997, pp. 326–335.

39. Bennett, J.C.R. and Zhang, H., Hierarchical packet fair queuing algorithms, *IEEE/ACM Transactions Networking* 1997; 5:675–689.

40. Chao, H.J. and Guo, X., *Quality of Service Control in High-Speed Networks*. New York: Wiley, 2001.

41. Cidon, I. et al., Real-time packet switching: a performance analysis, *IEEE Journal on Selected Areas in Communications* 1988; 6(9):1576–1586.

42. FORE systems, Inc., White paper: ATM switching architecture, November 1993.

43. Karol, M., Hluchyj, M., and Morgan, S., Input versus output queueing on a space-division packet switch, *IEEE Transactions on Communication* 1987; 35:1337–1356.

44. Hluchyj, M. and Karol, M., Queuing in high-performance packet switching, *IEEE Journal on Selected Areas in Communications* 1988; 6:1587–1597.

45. The integrated network services switch architecture and technology, whiter paper, Berkeley Networks, 1997.

46. Yeh, Y.S., Hluchyj, M., and Acampora, A.S., The knockout switch: a simple modular architecture for high-performance packet switching, *IEEE Journal on Selected Areas in Communications* 1987; 5(8):1274–1282.

47. Chao, H.J., Lam, C.H., and Oki, E., *Broadband Packet Switching Technologies: A Practical Guide to ATM Switches and IP Routers*. New York: Wiley, 2001.

48. Li, S.Q., Performance of a non-blocking space-division packet switch with correlated input traffic, *Proc. of the IEEE Globecom '89*, 1989, pp. 1754–1763.

49. Chang, C.Y., Paulraj, A.J., and Kailath, T., A broadband packet switch architecture with input and output queueing, *Proceedings of the Globecom '94*, 1994.

50. Iliadis, I. and Denzel, W., Performance of packet switches with input and output queueing, *Proc. ICC '90*, 1990, pp. 747–753.

51. Guerin, R. and Sivarajan, K.N., Delay and throughput performance of speed-up input-queueing packet switches, *IBM Research Report RC 20892*, June 1997.

52. McKeown, N., Mekkittikul, A., Anantharam, V., and Walrand, J., Achieving 100% throughput in input-queued switches, *IEEE Transactions on Communications* 1999; 47:1260–1267.

53. Anderson, T.E. et al., High speed switch scheduling for local area networks, *ACM Transactions Computer Systems* 1993; 11(4):319–352.

54. Stiliadis, D. and Verma, A., Providing bandwidth guarantees in an input-buffered crossbar switch, *Proc. IEEE Infocom '95* (1995) 960–968.

55. Lund, C., Phillips, S., and Reingold, N., Fair prioritized scheduling in an input-buffered switch, *Proceedings of the Broadband Communications*, 1996, pp. 358–369.

56. McKeown, N., The iSLIP scheduling algorithm for input-queued switches, *IEEE/ACM Transactions Networking* 1999; 7:188–200.

57. Chuang, S.T., Goel, A., Mckeown, N., Prabhakar, B., Matching output queueing with a combined input output queued switch, *IEEE Journal on Selected Areas in Communications* 1999; 17(16):1030–1039.

58. Prabhakar, B. and McKeown, N., On the speedup required for combined input and output queued switching, *Automatica* 1999; 35(12):1909–1920.

59. McKeown, N., A fast switched backplane for a gigabit switched router, *Business Communications Review*, 1997; 27(12).

60. Nourani, M., Kavipurapu, G., and Gadiraju, R., System requirements for super terabit routing, in: Robert Ewing, Carla Purdy, Hal Carter eds, *Proceedings of IEEE Midwest Symposium on Circuits and Systems (MWSCAS)*, Fairborn, OH, 2001, New York: IEEE Press 2001: 926–929.

61. McKeown, N., Growth in router capacity, IPAM Workshop, Lake Arrowhead, CA, October 2003, http://tiny-tera.stanford.edu/~nickm/talks/IPAM_Oct_2003.ppt.

Chapter 2

Concept of IP-Address Lookup and Routing Table

This chapter explains the background of the Internet Protocol (IP)-address lookup by briefly describing the evolution of the Internet addressing architecture, the characteristics of routing tables, and the complexity of IP-address lookup.

2.1 IP Address, Prefix, and Routing Table

IP is the suite of protocols that helps all these networks communicate with each other. IPv4 is the version of IP most widely used in the Internet today. The IP address is the unique identifier of a host in the Internet. IPv4 addresses are 32-bits long and when broken up into four groups of 8-bits, are commonly written in the dotted-decimal notation. For example, the address 10000010 01010110 00010000 01000010 corresponds in dotted-decimal notation to 130.86.16.66. It is sometimes useful to view an IPv4 address as 32-bit unsigned numbers on the number line, $[0 . . . (2^{32} - 1)]$, called the IP number line. For example, the IP address 130.86.16.66 represents the decimal number 2186678338 $(130 \times 2^{24} + 86 \times 2^{16} + 16 \times 2^{8} + 66)$.

One of the fundamental objectives of the IP is to interconnect networks, so it is a natural choice to route based on a network, rather than on a host. Thus, each IPv4 address is partitioned into two parts: netid and hostid, where netid identifies a network and hostid identifies a host on that network; this is called a two-level hierarchy. All hosts on the same network have the same netid but different hostids.

Because *netid* corresponds to the first bits of the IP address, it is called the address prefix (short for prefix). An address prefix corresponds the bit strings followed by a*. For example, the prefix 10000010 01010110 00010000* can represent all the 2^{8} addresses from 10000010 01010110 00010000 00000000 (130.86.16.0) to 10000010 01010110 00010000 11111111 (130.86.16.255), which are in the range [2186678272, 2186678527] on the IP number line. The prefix can be written in dotted-decimal notation as 130.86.16/24, where the number after the slash indicates the number of bits that the prefix contains, called the length of the prefix.

The IP suite of protocols has some routing protocols such as routing information protocol (RIP), open shortest path first (OSPF), or border gateway protocol (BGP). The routers perform one or more routing protocols to exchange protocol messages with neighboring routers. The protocol messages mainly contain a representation of the network topology state information and store the current information about the best-known paths to destination networks, called the reachability information. The reachability information is stored in a router as a table called the Routing Table. In the BGP routing table, an entry often has parameters: Network (Prefix), Next hop, AS path, Metric, LocPrf, Weight etc. The prefix is a unique identifier of a neighboring network: each prefix must have one or more next hops that can forward the packets to the neighboring network. If a prefix has only one next hop, then other parameters would not affect the BGP decision process. If a network has more then one next hop, then it is dependent on the other parameters to select one of next hops. The AS path is a list of autonomous systems (ASs) that packets must transit to reach the network (Prefix). The metric is usually multi-exit discriminator in cisco routers. Setting the local preference (LocPrf) also affects the BGP decision process. If multiple paths are available for the same prefix, then the AS path with the larger LocPrf is preferred. LocPrf is at the highest level of the BGP decision process (comes after the Cisco proprietary weight parameter); it is considered before the AS path length. A longer path with a higher LocPrf is preferred over a shorter path with a lower LocPrf [1].

The router typically maintains a version of the routing table in all line cards that is directly used for packet forwarding, called the forwarding table. The forwarding table is made from the routing table by the BGP decision process. In the forwarding table, each entry may only keep a prefix, a next hop, and some statistics. In mathematics, the forwarding table is a function F:

$$F : \{P_1, P_2, P_3, \ldots, P_n\} \rightarrow \{h_1, h_2, h_3, \ldots, h_m\}$$

where P_i is a prefix, n is the number of prefixes in a forwarding table, h is the next hop, and m is the number of next hops on a router.

2.2 Concept of IP-Address Lookup

As we have seen above, using address prefixes is a simple method to represent groups of contiguous IP addresses, that is to say, a prefix is a set of contiguous IP addresses.

Proposition 2.2.1: For an IP address $D = s_1 s_2 s_3 \ldots s_{32}$, ($s_i = 0, 1$) and a prefix $P = t_1 t_2 t_3 \ldots t_l^*$ ($t_i = 0, 1$). $D \in P$, if and only if $s_i = t_i$, $i = 1, 2, \ldots, l$.

Proof: If $D \in P$, then $P_1 = t_1 t_2 t_3 \ldots t_l \underbrace{000 \cdots 0}_{32-l} \leq D \leq P_2 = t_1 t_2 t_3 \ldots t_l \underbrace{111 \cdots 1}_{32-l}$. $\therefore s_k = t_k\, 1 \leq k \leq l$.

If $s_i = t_i$, $i = 1, 2, \ldots, l$, $\because 0 \leq s_i \leq 1$, $i = l+1, l+1, \ldots, 32$,

$\therefore P_1 = t_1 t_2 t_3 \ldots t_l \underbrace{000 \cdots 0}_{32-l} \leq D \leq P_2 = t_1 t_2 t_3 \ldots t_l \underbrace{111 \cdots 1}_{32-l}$, $D \in P$.

Definition 2.2.1:   For an IP address D and a prefix P, if $D \in P$, we say that D matches P.

For an IP packet with destination address D, the router will find a next hop to forward the packet to by looking up in the forwarding table the prefix P that matches D. That is to say, for a given IP address D, we must find the next hop $h_i = F(P_k)$, in which P_k is subject to $D \in P_k$. In this book, the main problem is to give the algorithm to find the appropriate prefix P_k, in which P_k is subject to $D \in P_k$, called IP-address Lookup.

Proposition 2.2.2:   For any two prefixes with different length P_1 and P_2, if there is an IP address D such that $D \in P_1$ and $D \in P_2$, then $D \in P_2 \subset P_1$ ($D \in P_1 \subset P_2$), or $P_1 \cap P_2 = \phi$.

Proof: Let there be two IP address prefixes: $P_1 = t_1 t_2 t_3 \ldots t_m$, $(t_i = 0$ or $1)$, $P_2 = s_1 s_2 s_3 \ldots s_n$, $(s_i = 0$ or $1)$ Suppose $m < n$.

$D = d_1 d_2 d_3 \ldots d_{32} \in P_1$, then $d_i = t_i$ $(1 \leq i \leq m)$.
$D = d_1 d_2 d_3 \ldots d_{32} \in P_2$, then $d_j = s_j$ $(1 \leq j \leq n)$.
$\therefore t_i = s_i$ $(1 \leq i \leq m)$
$\forall D_1 = s_1 s_2 s_3 \ldots s_m s_{m+1} \ldots s_n x_{n+1} x_{n+2} \ldots x_{32} \in P_2$
$D_1 = s_1 s_2 s_3 \ldots s_m s_{m+1} \ldots s_n x_{n+1} x_{n+2} \ldots x_{32} = t_1 t_2 t_3 \ldots t_m s_{m+1} \ldots s_n x_{n+1} x_{n+2} \ldots x_{32} \in P_1$
P_2 is a subset of P_1, $D \in P_2 \subset P_1$.
If $m > n$, then $D \in P_1 \subset P_2$.
If $P_1 \cap P_2 \neq \phi$, for example $D' \in P_1 \cap P_2$, then $D' \in P_1$, and $D' \in P_2$.
$\therefore$ If there is no IP address D such that $D \in P_1$ and $D \in P_2$, then $P_1 \cap P_2 = \phi$

Proposition 2.2.3:   For any two prefixes P_1 and P_2 with same length if there is an IP address D such that $D \in P_1$ and $D \in P_2$, then $P_1 = P_2$, or $P_1 \cap P_2 = \phi$.

2.3   Matching Techniques

Matching is most often associated with algorithms that search for a single data item in a large string of constant data. There are many existing matching techniques such as exact matching, substring matching, wildcard matching, pattern matching, range matching, and prefix matching [2].

In the exact matching problem, the result is identical with the input item. A large variety of techniques is known for exact matching in different data structures, such as linear search, binary search, hashing, etc. For example, hashing is a very prominent candidate among the exact matching group, because—on average—it can provide for $O(1)$ access in $O(1)$ memory per input item. Probably the most prominent representative is known as perfect hashing, providing for $O(1)$ worst-case access and memory. Unfortunately, finding a good hash function is difficult, and building and updating time of the data structure can take nonpolynomial time and lead to expensive hash functions.

Wildcard matching extends exact matching by providing for a "fall-back" or "match-all" entry, which is considered to be matched if no exact match is found. Its complexity is much more than that of exact matching.

Substring matching is a variation of exact matching, in which the database doses not consist of several independent entries, but of a single large sequence of symbols, a string, and the search item has to be compared with every possible substring. Knuth–Morris–Pratt [3] and Boyer–Moore [4] are the well-known algorithms used for substring matching problems.

Pattern matching is in wide use for matching wild-carded substrings in a large string. The best-known solutions in this field are variations on the Knuth–Morris–Pratt and Boyer–Moore algorithm [5].

Range matching is used to check whether a key belongs to a range. Many applications encounter range-matching problems. For example, in a distributed-storage networking scenario where many hosts might access shared data at very high speeds, the hosts protect their access by locking the accessed address range. Hosts must check whether the range is locked before beginning a memory access. There are two types of range searches: the point intersection problem that determines whether a set of ranges contains a query, and the range intersection problem that determines whether a query range intersects with any range in a set of ranges.

Prefix matching is a special case of range matching if a prefix does not overlap with any other prefixes. Prefixes have a limited range: the size of the range is a power of two and its start, and therefore also its end, is a multiple of its size. The problem to determine whether an IP-address matches with a prefix is the point intersection problem, which is one of exact matching. As soon as we introduce overlaps into the prefix database, we require a mechanism to differentiate between the overlapping prefixes. This implies that when allowing for overlaps, the search algorithms need to be more elaborate. If we assign a higher priority to the longer prefix, the longest prefix can be found. In this case, the prefix matching problem is called the Longest Prefix Matching (LPM) problem.

2.3.1 Design Criteria and Performance Requirement

There are a number of properties that are desirable in all IP lookup algorithms [6]:

High Speed. Increasing data rates of physical links require faster address lookups at routers. For example, links running at OC192c (approximately 10 Gbps) rates need the router to process 31.25 million packets per second (assuming minimum-sized 40 byte TCP or IP packets). We generally require algorithms to perform well in the worst case. If this were not the case, all packets (regardless of the flow they belong to) would need to be queued. It is hard to control the delay of a packet through the router. At the same time, a lookup algorithm that performs well in the average case may be acceptable, in fact desirable, because the average lookup performance can be much higher than the worst-case performance. For OC192c links (10.0 Gbps), the lookup algorithm needs to process packets at the rate of 3.53 million packets per second, assuming an average Internet packet size of approximately 354 bytes. Table 2.1 lists the lookup performance required in one router port.

- Average-case search time—Because IP routers are statistical devices anyway, that is, are equipped with buffers to accommodate traffic fluctuations, average-case speed seems like an adequate measure. It is hard to come up with reliable average-case scenarios, because they depend heavily on the traffic model and traffic distribution. But the traffic distribution is unknown. In this book, the average case search time is given with the extreme scenario: traffic per prefix is constant.
- Worst-case search time—Unfortunately, it is not known how prefix and traffic distributions will evolve in the future and the Internet so far has rather successfully tried to escape predictability. Therefore, known worst-case bounds are important for designing a system that should work well over the next several years, making worst-case bounds at least as important as knowledge about the average case. In some cases, such as for implementation in hardware, or together with hardware, constant time or constant worst time lookups are a prerequisite.

Table 2.1 Requirement of IP Lookup Speed for Line Cards and Packet Sizes

Line Type in Router	Line Rate (Gbps)	40-Byte Packets (Mpps)	84-Byte Packets (Mpps)	354-Byte Packets (Mpps)
T1	0.0015	0.0468	0.0022	0.00053
OC3c	0.155	0.48	0.23	0.054
OC12c	0.622	1.94	0.92	0.22
OC48c	2.50	7.81	3.72	0.88
OC192c	10.0	31.2	14.9	3.53
OC768c	40.0	125.0	59.5	14.1

Low Storage. Small storage requirements enable the use of fast but expensive memory technologies like synchronous random access memories (SRAMs). A memory-efficient algorithm can benefit from an on-chip cache if implemented in software, and from an on-chip SRAM if implemented in hardware.

Low Preprocessing Time. Preprocessing time is the time taken by an algorithm to compute the initial data structure. An algorithm that supports incremental updates of its data structure is said to be "dynamic." A "static" algorithm requires the whole data structure to be recomputed each time an entry is added or deleted. In general, dynamic algorithms can tolerate larger preprocessing times than static algorithms.

Low Update Time. Forwarding tables have been found to change fairly frequently. As links go down and come back up in various parts of the Internet, routing protocol messages may cause the table to change continuously. Changes include addition and deletion of prefixes, and the modification of next-hop information for existing prefixes. These changes often occur at the peak rate of a few hundred prefixes per second and at the average rate of more than a few prefixes per second [7]. A lookup algorithm should be able to update the data structure at least this fast.

Flexibility in Implementation. The forwarding engine may be implemented either in software or in hardware depending upon the system requirements. Thus, a lookup algorithm should be suitable for implementation in both hardware and software. For the Line Card at higher speeds (e.g., for OC192c), it is necessary to implement the algorithm in hardware. In fact, the lookup algorithms are implemented easier in software than in hardware. The algorithm based on hardware should use a simple data structure, because of the difficulty in managing memory, and can be divided into few operations for the wire-speed.

Scalability to IPv6. IPv6 is the IP of the next generation. Due to its longer IP address, 128 bits instead of 32-bit, IPv6 requires especially suited efficient mechanisms for IP-address lookup. The run-time complexity analysis and lookup time measurement on IPv4 routing tables are necessary for a performance analysis of the algorithms.

2.4 Difficulty of the Longest-Prefix Matching Problem

Before we proceed, it is important understand why Internet address lookup is hard. First, Internet addresses are not specially created addresses to be easy to lookup. Second, the Internet deals with the scaling issue by using address prefixes, which require a more complex lookup. Details are described below.

2.4.1 Comparisons with ATM Address and Phone Number

Looking up an IP address is harder than looking up an asynchronous transfer mode (ATM) address. ATM addresses or virtual circuit indices (VCIs) are carefully chosen to be simple to lookup in a switch table. Unfortunately, ATM addresses must be set up for each conversation, which adds to the delay. By contrast, IP addresses are relatively fixed and there is no additional set up delay per conversation. Secondly, ATM addresses do not currently make much provision for hierarchical networks and so are perceived not to be scalable to truly global networks. IP, through the use of prefixes, has provision for scaling [8].

The IP addresses lookup is harder than looking up a phone number in a phone book, or a word in a dictionary. In those problems, we can search quite fast by first sorting all the words or names. Once sorted, if we are looking up for a word starting with Sea, we simply go to the pages of *S* entries and then look up all entries in a dictionary. In fact, such a lookup is called an exact matching lookup; standard solutions based on hashing and library search provide very fast times for exact matching.

Internet routers store address prefixes in their forwarding tables to reduce the size of their tables. However, the use of such address prefixes makes the lookup problem one of finding the longest matching prefix instead of finding an exact match.

2.4.2 Internet Addressing Architecture

The addressing architecture specifies how the allocation of addresses is performed; that is, it defines how to partition the total IP address space of 2^{32} addresses—specifically, how many prefixes will be allowed and what size each should be. Therefore, before we proceed, it is important to understand addressing architecture.

Class-Based Scheme.   When the Internet was initially designed, the addressing scheme was rather simple, which is known today as the class-based addressing scheme. The IP address space is partitioned into five classes—classes A, B, and C for unicast traffic, class D for multicast traffic, and class E reserved for future use. Network size was determined by the number of bits used to represent the network (Prefix). Prefixes of class A, B, C consisted of an 8, 16, or 24-bit strings, and started with the bit strings "0," "10," "110" respectively, shown in Figure 2.1. Thus the first few most-significant bits of an IP address and prefix determine its class.

Class-based addressing architecture enabled routers to use a relatively simple lookup operation. Typically, the forwarding table had three parts, one for each of the three unicast classes A, B, and C. All prefixes in the same part had the same length. For any IP address D, no more than one prefix in the forwarding table matched with D. The router then searched for an exact match between D and the prefixes in the forwarding table. This exact match search can be performed using, for example, a hashing or a binary search algorithm. Because the class of the IP address D was determined from its most-significant bits, the lookup operation proceeded as in Figure 2.2 [6].

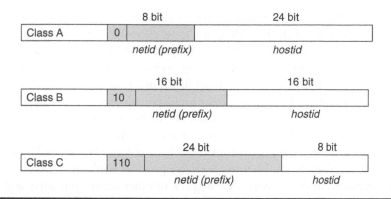

Figure 2.1 Class-based addresses.

With class-based addressing architecture there are 126 ($2^7 - 2$) class A networks with a maximum of ($2^{24} - 2$) hosts or network, 16.384 (2^{14}) class B networks with 65,534 hosts or network, and 2,097,152 (2^{21}) class C networks with 256 hosts or network. The class-based addressing scheme worked well in the early days of the Internet. However, as the Internet grew, two problems emerged: (a) a depletion of the IP address space, and (b) an exponential growth of forwarding tables. Class-based schemes had fixed netid-hostid boundaries. The allocation of IP address was too inflexible and the address space was not used efficiently. For example, there are many organizations in which the number of hosts is more than 254, far less than 65,534. A class B address had to be allocated to these organizations, leading to a large number of wasted addresses and a large number of prefixes to appear in the forwarding table. As a result, the forwarding tables in the backbone IP routers were growing exponentially. The growth rates of forwarding tables were exceeding the growth in the hardware and software capability of Internet routers [6].

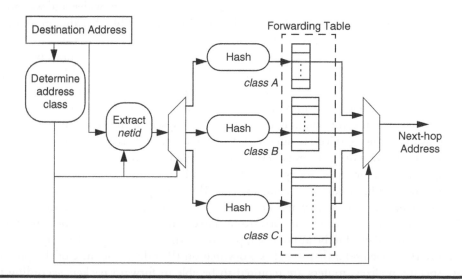

Figure 2.2 Typical implementation of the lookup operation in a class-based addressing scheme. (From Gupta, P., *Algorithms for Routing Lookups and Packet Classification*, 2000. With permission.)

CIDR Addressing Scheme.   In an attempt to slow down the growth of backbone routing tables and allow more efficient use of the IP address space, an alternative addressing and routing scheme called classless inter-domain routing (CIDR) was adopted in 1993. CIDR does away with the class-based partitioning of the IP address space, allowing prefixes in the forwarding table to be of arbitrary length rather than constraining them to be 8, 16, or 24-bits long. The IP address space can therefore be used more efficiently. For example, with CIDR, an organization with 300 hosts can be allocated a prefix of length 23 with a maximum of 512 hosts, not a prefix of length 16 with a maximum of 65,534 hosts, leading to more efficient address allocation.

The adoption of variable-length prefixes enables a hierarchical allocation of IP addresses according to the physical topology of the Internet. A service provider is allocated a short prefix, and can allocate longer prefixes out of its own address space to other service providers and so on. Each service provider can announce the prefix that it has been allocated to the routers that connect to it.

With CIDR, when a service provider has more than one connection to an upstream carrier or other providers, it can announce the longer prefixes out of its own address space to the neighboring routers; the longer prefix is called the more specific prefix. The announcement of the more specific prefixes is to punch a "hole" in the policy of the shorter prefix announcement, creating a different policy to perform some rudimentary form of load balancing and mutual backup for multi-homed networks.

Since 1993, more specific prefixes have appeared in forwarding tables. Broido et al. examined forwarding tables from the University of Oregon Route-Views Project (http://www.antc.uoregon.edu/route-views) and found that the number of the more specific prefixes from the year 2000, is more than half of all prefixes in the forwarding table. Small multi-homed networks with a number of peers and a number of upstream providers who announce more specific prefixes is one of the major drivers of the recent growth in routing tables [10].

To address the problem of the increase in forwarding tables, CIDR allows address aggregation at several levels. For example, in a router, there are two prefixes: $P_1 = 202.114.30/24$ and $P_2 = 202.114.31/24$ that are reachable through the same service provider and we can aggregate the two prefixes into one prefix: $P = 202.114.30/23$. For a prefix $P_1 = 202.114.16/20$ and its more specific prefix $P_2 = 202.114.31/24$ that are reachable through the same service provider, and we can aggregate the two prefixes into one prefix: $P_1 = 202.114.16/20$. It is notable that the precondition of aggregation is that the prefixes are with the same service provider. Because the allocation of an IP address has topological significance, we can recursively aggregate prefixes at various points within the hierarchy of Internet topology. Thus the efficiency of aggregation is related Internet topology. There are two situations that cause a loss of aggregation efficiency. Organizations that are multi-homed must announce their prefix into the routers of their service providers; it is often not feasible to aggregate their routing information into the address space of any one of those providers. Organizations can change service providers but do not renumber. This has the effect of "punching a hole" in the aggregation of the original service provider's advertisement [9].

Although CIDR allows the aggregation of prefixes in the forwarding table, the IP address lookup problem now becomes more complex. In CIDR, the prefixes in the forwarding tables have arbitrary lengths because they are the result of prefix aggregations. Therefore, the destination IP addresses and prefixes in the forwarding table cannot be classified. Because the more specific prefixes are allowed to be announced, prefixes in the forwarding tables may overlap, and there is more than one prefix that matches a destination IP address. Because of the presence of the more specific prefixes, routers are required to be able to forward packets according to the most specific route (the longest matching prefix) present in their forwarding tables. As a result, determining the longest matching prefix involves not only comparing the bit pattern itself, but also finding the appropriate length. In summary, the need to perform LPM has made routing lookups more complicated now than they were before the adoption of CIDR when only one exact match search operation was required.

2.5 Routing Table Characteristics

The Internet can be likened to a loose coalition of thousands of ASs, each of which consists of networks administrated by a single organization with its own policies, prices, services, and customers. Each AS makes independent decisions about where and how to secure the supply of various components that are needed to create the network service. The cement that binds these networks into a cohesive whole is the use of a common address space and a common view of routing. In an AS, routers exchange the routing information using interior gateway protocols, such as RIP, IGRP, and OSPF. They periodically flood an intra-domain network with all known topological information, or link state entries. The interior routing information is limited, while the exterior routing information is represented by an entry-default route.

The scalability of the Internet infrastructure allows more and more ASs to connect to each other. Routers at the border of an AS exchange traffic and routing information with routers at the border of another AS through the BGP. The BGP is an incremental protocol that sends update information only upon changes in network topology or routing policy. Moreover, the BGP uses TCP as its underlying transport mechanism to distribute reachability information to its peer or neighbor routers. The router stores the reachability information as a routing table, in which each entry is a route. A BGP table entry has the format shown in Table 2.2.

Therefore, the BGP routing table not only describes the connectivity between ASs, but also contains the policies concerning how to exchange traffic between ASs. Because each AS runs the BGP to distribute route information (i.e., prefix, AS number) and construct its routing table, a representative set of core BGP routing tables can be used to monitor aspects of the Internet architectural evolution, and the characteristics of BGP. BGP data reflect consumption of vital and finite Internet resources: IP address, AS number, BGP table size limited by router memory, CPU cycles for processing BGP updates, BGP convergence properties, Internet routing stability, and bandwidth consumed by routing update traffic, etc.

Further, analyzing BGP tables can support studies of many questions regarding Internet properties discussed in the Internet community (e.g., NANOG, IETF) [10]:

■ How much IP address space is routable?
■ Are allocated IP addresses actually being routed?
■ Which ASs are the most important?
■ How many ASs are single or multi-homed?

Table 2.2 An Example of a BGP Table

Prefix	Next Hop	Metric	LocPrf	Weight	AS Path
12.0.0.0/8	201.23.239.10	216	100	0	6066 3549 7018
12.0.48.0/20	201.23.239.10	160		2258	6066 3549 209 1742
	210.202.187.254	445	100	0	3257 13646 1742

■ What is the IP hop diameter of an AS?
■ How is Internet topology to be visualized? etc.

In the following, we will discuss the characteristics of the BGP routing table related to IP-address lookup, such as its structure, growth trend, factors that contribute to the growth, and route update.

Because each entry in a BGP routing table has a single prefix, and prefixes are unique, we can define the size of a BGP routing table to be the number of prefixes contained in the routing table.

2.5.1   Routing Table Structure

From Proposition 2.2.2 and 2.2.3, for any two different prefixes, one is a subset (superset) of another, or there is no intersection. We can categorize prefixes in the BGP routing table as [10]:

Definition 2.5.1:
  Stand-alone prefix: a prefix that has no subset or superset in the BGP routing table
  Subroot prefix: a prefix that has a least specific prefix with subsets in the BGP routing table
  More-specific prefix: a prefix that is a subset of some other prefix

Stand-alone prefixes and subroot prefixes are often allocated by Internet Registries, while more-specific prefixes are often announced by Internet service providers. The premise that packets are forwarded according to the route of the most specific announcement implies a degree of cooperation from the owners of the subroot prefix. Therefore, the number of stand-alone prefixes and subroot prefixes are controlled by Internet Registries, and the number of more-specific prefixes is decided by the cooperative relations between the owners of the subroot prefixes.

Definition 2.5.2:   If the prefixes in routing tables are represented as a binary tree, every prefix can have a corresponding node in the tree. If there are n prefixes in the path from root to prefix P, we call the prefix P that is in level n.

For example, there are prefixes 0/1, 101/3, 0110/4, 010/3, and 01001/5 in a routing table. The tree is shown in Figure 2.3a. The prefix 101/3 is a stand-alone prefix, 0/1 is a subroot prefix, and prefixes 010/3, 0110/4, and 01001/5 are more-specific prefixes. 0/1 and 101/3 are in level 0, 010/3 and 0110/4 are in level 1, and 01001/5 is in level 2, as shown in Figure 2.3b.

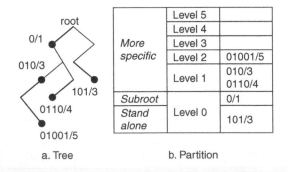

		Level 5	
		Level 4	
More specific		Level 3	
		Level 2	01001/5
		Level 1	010/3 0110/4
Subroot			0/1
Stand alone		Level 0	101/3

a. Tree   b. Partition

Figure 2.3   An example of routing table.

Table 2.3   Distribution of Prefixes

Routing Table	Stand Alone	Subroot	More Specific					
			Level 1	Level 2	Level 3	Level 4	Level 5	Sum
19971108	13751	963	10443	1369	139	4	0	11955
19981101	26970	2208	19736	4786	308	18	2	24850
19991031	29596	3170	31775	6563	487	17	3	38845
20001101	35262	4543	45616	7034	635	20	1	53306
20011101	43580	5823	48758	9201	1042	114	2	59117
20021101	49431	6675	54570	9071	945	46	3	64635

To compute the level of prefixes, we at first sort the prefixes in ascending order and then compare two prefixes if $P_{i+1} \subset P_i$, P_{i+1} is in the next level of P_i. We can partition the BGP routing table by level. Wu et al. analyzed the BGP routing tables from the University of Oregon Route-Views Project (ftp.routeviews.org/bgpdata, 2003), shown in Table 2.3.

The percentage of stand-alone prefixes, subroot prefixes, and more-specific prefixes are 40 percent, 5 percent, and 55 percent respectively. This means all more-specific prefixes are announced by the owners of 5 percent subroot prefixes.

The distribution in levels is shown in Table 2.3. The maximum level is 5. The number of prefixes in level 1 is about ten times that in level 2; the number of prefixes in level 2 is about ten times that in level 3; and so on. About 98 percent of the more-specific prefixes are in level 1 and level 2.

In the tables, at each level, because prefixes are disjoint from each other, there is no more than a matching prefix for any IP address. When we search for an IP address in a routing table using the longest-prefix-matching algorithm from root to leaf, the maximum searching level is 5, and about 98 percent of searching levels are no more than 2.

2.5.2   Routing Table Growth

As more computers connect to the Internet, Internet topology becomes more complex. This results in a large amount of reachability information and larger BGP routing tables that consume vital and finite router resources: memory and CPU cycle. Therefore, to analyze the growth of routing tables is beneficial to the router's design and Internet infrastructural research.

Huston analyzed the growth of BGP routing tables from 1989 to 2000. The size of BGP routing tables are shown in Figure 2.4. At a gross level there appear to be four phases of growth visible as given by Huston [11].

1. First phase—Pre-CIDR growth: From 1988 until April 1994 routing table size shows definite characteristics of exponential growth. Much of this growth can be attributed to the growth in deployment of the historical class C address space (/24 address prefixes). Unchecked, this growth would have lead to saturation of the BGP routing tables in nondefault routers within a small number of years. Estimates of the time at which this would have happened vary somewhat, but the overall observation was that the growth rates were exceeding the growth in hardware and software capability of the deployed network.

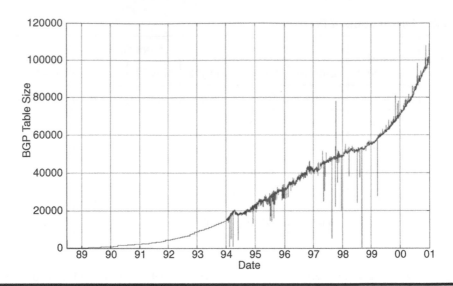

Figure 2.4 BGP routing table growth, 1988–2000. (From Huston, G., *The Internet Protocol Journal*, 4, 2001. With permission.)

2. Second phase—CIDR deployment: The response from the engineering community was the introduction of routing software which dispensed with the requirement for class A, B, and C address delineation, replacing this scheme with a routing system that carried an address prefix and an associated prefix length. A concerted effort was undertaken in 1994 and 1995 to deploy CIDR routing, based on encouraging deployment of the CIDR-capable version of the BGP protocol, BGP4. The effects of this effort are visible in the routing table. Interestingly enough, the efforts of the IETF CIDR deployment Working Group are visible in the table, with downward movements in the size of the routing table following each IETF meeting.

   The intention of CIDR was one of provider address aggregation, where a network provider is allocated an address block from the address registry, and announces this entire block into the exterior routing domain. Customers of the provider use a suballocation from this address block, and these smaller routing elements are aggregated by the provider and not directly passed into the exterior routing domain. During 1994 the size of the routing table remained relatively constant at some 20,000 entries as the growth in the number of providers announcing address blocks was matched by a corresponding reduction in the number of address announcements as a result of CIDR aggregation.

3. Third phase—CIDR growth: For the next four years, until the start of 1998, CIDR proved remarkably effective in damping unconstrained growth in the BGP routing table. Although other metrics of Internet size grew exponentially during this period, the BGP table grew at a linear rate, adding some 10,000 entries per year. Growth in 1997 and 1998 was even lower than this linear rate. Although the reasons behind this are somewhat speculative, it is relevant to note that this period saw intense aggregation within the ISP industry, and in many cases this aggregation was accompanied by large scale renumbering to fit within provider-based aggregated address blocks. Credit for this trend also must be given to Tony Bates, whose weekly reports of the state of the BGP address table, including listings of further potential for route aggregation provided considerable incentive to many providers to improve their levels of route aggregation [12].

A close examination of the table reveals a greater level of stability in the routing system at this time. The short-term (hourly) variation in the number of announced routes reduced, both as a percentage of the number of announced routes, and also in absolute terms. One of the other benefits of using large aggregate address blocks is that the instability at the edge of the network is not immediately propagated into the routing core. The instability at the last hop is absorbed at the point at which an aggregate route is used in place of a collection of more specific routes. This, coupled with widespread adoption of BGP route flap damping, has been every effective in reducing short-term instability in the routing space. It has been observed that although the absolute size of the BGP routing table is one factor in scaling, another is the processing load imposed by continually updating the routing table in response to individual route withdrawals and announcements. The encouraging picture from this table is that the levels of such dynamic instability in the network have been reduced considerably by a combination of route flap damping and CIDR.

4. Fourth phase—The resumption of exponential growth: In late 1998, the trend of growth in the BGP table size changed radically, and the growth from 1999 to 2001 is again showing all the signs of a re-establishment of exponential growth. It appears that CIDR has been unable to keep pace with the levels of growth of the Internet. Once again the concern is that this level of growth, if sustained, will outstrip the capability of hardware.

Is CIDR not effective?—Cengiz [13] analyzed the BGP messages collected by the RIPE-NCC Routing Information Service. The data had been collected for about two years. It is much richer than the daily snapshots often used in analysis and helps us address more detailed questions than simply BGP routing table size growth. For example, we can show the effectiveness of CIDR aggregation, or account for multi-homing and inter-domain traffic engineering more accurately. Cengiz found that CIDR is doing very well, and without CIDR, the BGP routing tables would have been five times larger [13].

The continued growth of the routing table size raises concerns regarding the scalability, stability, management, and increased dynamics of BGP. Therefore, many researchers from academia and industry pay attention to the growth of BGP routing tables. Because the prefixes in routing tables are allocated by the Regional Internet Registries (RIR), announced by its owner, and distributed by the routing protocols, some researchers have explored the extent that these factors contribute to routing table growth. In the next section, we will describe the results from Refs. [14–16] etc.

2.5.3 Impact of Address Allocation on Routing Table

The allocation and management of current IPv4 address space are mainly done by the RIR. IPv4 address blocks are allocated to Internet service providers (ISPs). These ISPs will further assign IP addresses from their allocated address blocks to the end users. All address blocks are represented as address prefixes (for short prefixes). The ISPs run the BGP to announce all the prefixes currently in use on the Internet. Because each allocated address block will eventually be announced, there is potentially an intimate relationship between IP address allocation and the growth of the routing table. Evidently, because it is beneficial to keep the BGP routing table growth in check each ISP announces as few prefixes as possible.

To understand the relationship between the address allocation and routing table growth, Xu et al. analyzed IPv4 address allocation and the routing table over the last 4.5 years [14]. This subsection is partly from [14] (© 2003 IEEE). IPv4 address allocation records (dated up to June 30, 2002) are from three RIRs: ARIN, RIPE, and APINC (ftp://ftp.arin.net/pub/stats/). The BGP routing tables are from two different sources: Oregon Route-Views project (from November 1997 to April 2002) and RIPE NCC (from September 1999 to April 2002).

2.5.3.1 Migration of Address Allocation Policy

Before CIDR deployment, IP address allocation policy is called the class-based addressing scheme, in which the length of the prefixes of each class address have the fixed length. For example, the length of the prefixes of a class A address block is 8 bits, the length of the prefixes of a class B address block is 16 bits, the length of the prefixes of a class C address block is 24 bits, etc.

After CIDR deployment, in theory, the IP address blocks are classless, and the length of the prefixes can be any value from 0 to 32. In practice, the length of the prefixes of the allocated address block was 19 bits before 2000, then 20 bits after 2000. Therefore, the address allocation policy has changed three times until 2002. The total number of the allocation is 50,887. The number of the allocations from 1993 to April 2002 is shown in Figure 2.5.

Figure 2.6 shows the prefix length distributions of the allocated address blocks from 1993 to 2000. From these figures, Xu et al. found the following characteristics.

The first address allocation policy has changed between 1993 and 1996, when CIDR was deployed. Before 1995, the prefixes with lengths 16 and 24 dominated the allocations. From 1995 to 1996, the prefixes with length 19 increased from 5.57 percent to 25.24 percent. This is because RIR recommended the prefixes with length 19. Accordingly, the average length has decreased from 22.38 to 19.27.

The second address allocation policy has changed in 2000. With the explosive growth of the Internet, the unallocated address block becomes fewer and fewer. To utilize the IP address more

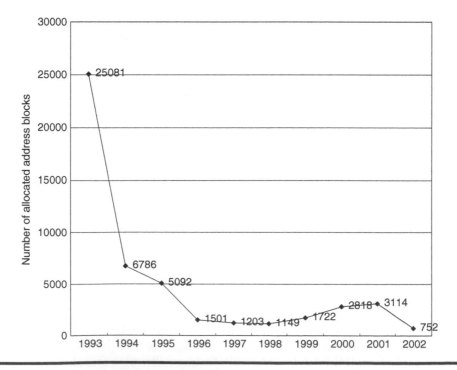

Figure 2.5 Allocated address blocks over time.

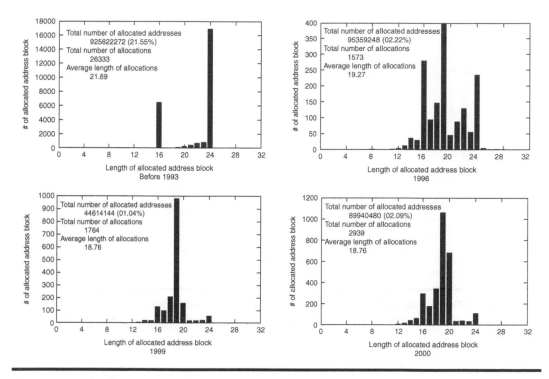

Figure 2.6 **Distribution of address prefix length. (From Xu, Z. et al., *Proceedings of IEEE 18th Computer Communications Workshop*, IEEE Press, 2003. With permission.)**

efficiently, RIR recommended the prefixes with length 21 after 2000. The number of prefixes with length 21 exceeded that of the prefixes with length 19, and this trend continued until 2002. Since the percent of prefixes with lengths 20–22 increases, the average length of prefixes in the routing table reduced from 22.74 in January 1998 to 22.36 in April 2001.

2.5.3.2 Impact of Address Allocations on Routing Table Size

Xu et al. use the following method to measure the impact of address allocations on the static routing table size: for each routing prefix, the addresses it contains must belong to one or multiple allocated blocks $P_1, P_2, \ldots, P_n$. In this case, we say that each of the allocated blocks $P_1, P_2, \ldots, P_n$ contributes $1/n$ to the routing table size. We then check all the routing prefixes and for each involved allocated block its attribution to the routing table size is accumulated. Obviously, the sum of the accumulated contribution of all the involved allocated blocks should be equal to the total number of routing prefixes.

Xu et al. performed the above measurement on the routing table snapshot of April 30, 2002. The routing table had 117,060 routing prefixes, which are generated from 30,705 allocated address blocks. Figure 2.7 plots the cumulative distribution function (CDF) for all the involved allocated blocks' contribution to the routing table size. It shows that 90 percent of the allocated blocks above account for less than 30 percent of the routing prefixes. In other words, 10 percent of the allocated blocks contribute to as high as 70 percent of the routing prefixes. This indicates that the impact on the routing table size is highly skewed among the allocated blocks.

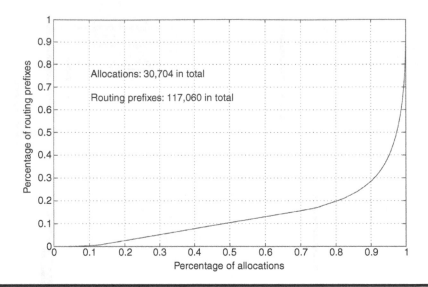

Figure 2.7 CDF of allocated blocks' contribution to the routing table size. (From Xu, Z. et al., *Proceedings of IEEE 18th Computer Communications Workshop,* **IEEE Press, 2003. With permission.)**

2.5.3.3 Impact of Address Allocation on Prefixes with 24-Bit Length

In real routing tables, the percent of prefixes with length 24 is more than 50 percent. For example, in the routing table on April 30, 2002, the number of prefixes with length 24 is 66,148, and the percent is as high as 56.1 percent. About 7409 prefixes are identical allocated blocks, and 58,739 prefixes are fragmented from the larger allocated blocks.

Figure 2.8 shows the distribution of prefixes with length 24. From 1993 to 1996, the number of prefixes with length 24 that are identical to the allocated address blocks decreased sharply. After 1996, the number is less than 80, which is less than 1 percent of all the prefixes with length 24. In 1993 and 1999, the number of prefixes with length 24 that are fragmented for larger-sized allocation blocks increased suddenly, and then decreased in the next year. We can conjecture that the prefixes that are fragmented from larger-sized allocation blocks continue to dominate the prefixes with length 24 in future.

In the routing table on April 30, 2002, there are 50,898 prefixes from the allocated blocks with prefix length ranging from 8 to 23. Figure 2.9 shows the ratio: the numerator is the total number of prefixes with length 24 that are fragmented from the allocation blocks with the corresponding prefix length and the denominator is the number of the matched blocks. About 3964 prefixes with length 24 are fragmented from five allocation blocks with prefix length 10, the ratio is as high as 800. The allocation blocks with prefix length 16 contribute more prefixes with length 24 than the allocation blocks with any other prefix lengths.

2.5.4 Contributions to Routing Table Growth

The introduction of CIDR [9] reduces the routing table size by enabling more aggressive route aggregation in which a single prefix is used to announce the routes to multiple prefixes. Route aggregation, however, might not always be performed. First, an AS can aggregate its prefix with its provider's only

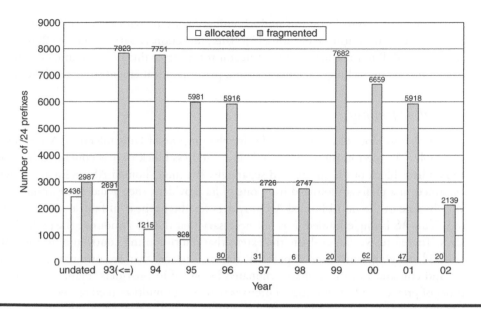

Figure 2.8 Number of routing prefix /24s that are fragmented from and identical to allocated blocks in different time periods. (From Xu, Z. et al., *Proceedings of IEEE 18th Computer Communications Workshop*, IEEE Press, 2003. With permission.)

when the AS is single-homed, that is, the AS has only one provider. For a multi-homed AS, which has multiple providers, its prefix(es) cannot be aggregated by all of its providers. Second, an AS may have to announce several prefixes. One reason is address fragmentation that arises because a set of prefixes originated by the same AS cannot be summarized by one prefix. Another reason is load balancing, where an AS originates several prefixes by announcing different prefixes via different AS paths. The last reason is that an AS may fail to aggregate prefixes that can be aggregated.

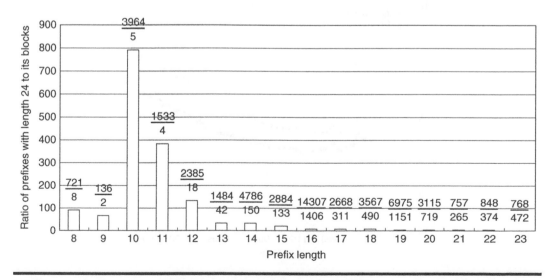

Figure 2.9 Ratio of the number of prefixes with length 24 to the number of its blocks. (From Xu, Z. et al., *Proceedings of IEEE 18th Computer Communications Workshop*, IEEE Press, 2003. With permission.)

Bu et al. explore the contribution of multi-homing, failure to aggregate, load balancing, and address fragmentation to routing table size [15]. This subsection is partly from (© 2002 IEEE). They examine the BGP routing tables of the Oregon route server (http://www.antc.uoregon.edu/route-views/), and present techniques to quantify and perform measurement study on these factors.

2.5.4.1 Multi-Homing

Many ASs connect to more than one provider for the purpose of fault tolerance. Multi-homing may create "holes" in the routing table. A hole is an address block that is contained in another announced address block but is announced separately. If a multi-homed AS originates a prefix p that is contained in a prefix announced from one of its providers, then p has to be announced to the Internet by one of the multi-homed AS providers for the purpose of fault tolerance. On the other hand, if an AS is single-homed, it is not necessary that the AS announce the prefix beyond its providers. Therefore, we can evaluate the extent that multi-homing contributes to the routing table size by identifying multi-homed prefixes, that is, prefixes that are originated by a multi-homed AS and contained in the prefixes originated by one of its providers. Figure 2.10 plots the total number of prefixes and the number of prefixes that are not multi-homed on the Route Viewer router. The number of multi-homed prefixes is shown to be on the rise. Multi-homing introduces approximately 20–30 percent more prefixes.

Although multi-homing increases the routing tale size significantly, it cannot be eliminated. Multi-homing is necessary to ensure the reliability of the Internet connection.

2.5.4.2 Failure to Aggregate

Some AS may fail to aggregate its aggregatable prefixes even though no load balancing is performed among those prefixes. To understand to what extent that failure to aggregate contributes to the routing

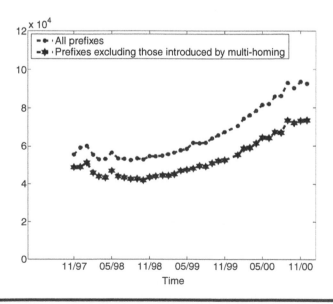

Figure 2.10 Contribution of multi-homing to routing table. (From Bu, T., Gao, L., and Towsley, D., *Proceedings of Global Telecommunications Conference*, IEEE Press, 2002. With permission.)

table size, we aggregate all aggregatable prefixes that are originated by the same AS and are announced identically. First, we classify prefixes into prefix clusters, in each of which prefixes are announced identically. Formally, a prefix cluster is a maximal set of prefixes whose routing table entries are the same in the Route Views server's routing table. Note that although the Route View server has a limited view of the Internet, it does have a good sample of routes because it peers with the many tier-1 ISPs.

Second, we perform aggregation for prefixes from the same prefix cluster iteratively as follows. Initially, we remove all prefixes that are contained in another prefix. In each iteration, we first sort all prefixes in increasing order of their addresses. We aggregate each pair of consecutive prefixes that is aggregatable. For example, a pair prefix 1011010/7 and 1011011/7 can be aggregated by the prefix 101101/6. Figure 2.11 plots the number of all prefixes and the number of prefixes excluding those that are introduced by failure to aggregate. There are approximately 15–20 percent prefixes that could be aggregated beyond what network operators have done.

2.5.4.3   Load Balancing

Failure to aggregate introduces more prefixes because an AS does not aggregate its aggregatable prefixes even though those prefixes are announced identically. Another reason that route aggregation cannot be performed for prefixes originated by the same AS is load balancing. Two aggregatable prefixes might not be aggregated because they are announced differently. To quantify the effect of load balancing on the routing table size, we first compute the number of prefixes resulting from aggregating all aggregatable prefixes originated by the same AS independent of whether those prefixes are announced identically or not. That is, we perform aggregation for prefixes from the same AS iteratively. The total number of prefixes that remain after the aggregation is compared with the number of prefixes excluding those introduced by failure to aggregate. The difference between the

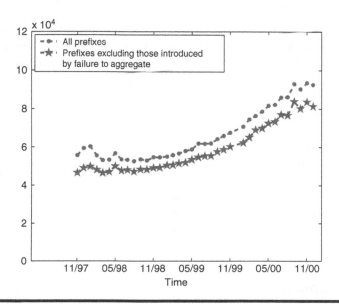

Figure 2.11   Contribution of failure-to-aggregate to routing table size. (From Bu, T., Gao, L., and Towsley, D., *Proceedings of Global Telecommunications Conference*, IEEE Press, 2002. With permission.)

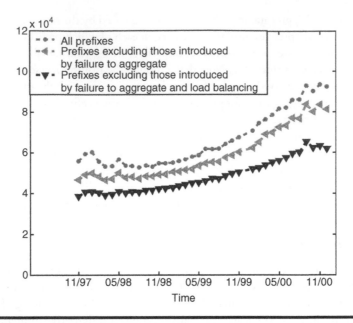

Figure 2.12   Contribution of load balancing to routing table. (From Bu, T., Gao, L., and Towsley, D., *Proceedings of Global Telecommunications Conference,* **IEEE Press, 2002. With permission.)**

two numbers quantifies that load balancing contributes to routing table size, as shown in Figure 2.12. It is observed that the load balancing introduces an additional 20–25 percent more prefixes.

2.5.4.4   Address Fragmentation

Both multi-homing and load balancing are inevitable trends of the Internet. Although it is possible to eliminate the prefixes that are due to failure to aggregate, the reduction of the routing table size is not significant. Bu et al. conjecture that this is due to address fragmentation [15]. Because all of the prefixes within the same prefix cluster are announced identically, a single routing table entry would be sufficient for them if these prefixes could be represented by one prefix cluster. However, the Internet addresses covered by these prefixes may not be summarized by one prefix due to either failure to aggregate or address fragmentation. In this section, we investigate the effect of address fragmentation by comparing the number of prefixes, excluding those contributed by failure to aggregate, with the number of prefix clusters.

We plot the number of prefix clusters in Figure 2.13. The number of prefix clusters is only about 1/5 of the current routing table. The contribution of address fragmentation to the routing table size is the gap between the number of prefixes excluding those introduced by failure to aggregate and the number of prefix clusters. It is suggested by the plot that address fragmentation contributes to more than 75 percent of the routing table size and is the most significant factor.

2.5.5   Route Update

The Internet infrastructure determines whether the Internet routing is unstable. The Internet comprises a large number of interconnected regional and national backbone networks based on contracts, where backbone service providers peer, or exchange traffic and routing information with one another. Any Internet service provider or end user may connect or disconnect with others

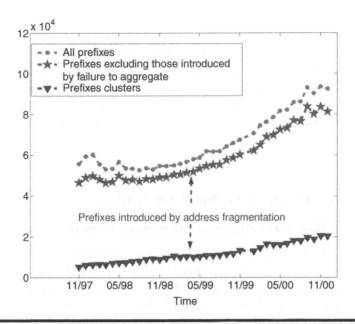

Figure 2.13 Contribution of address fragmentation to routing table. (From Bu, T., Gao, L., and Towsley, D., *Proceedings of Global Telecommunications Conference*, IEEE Press, 2002. With permission.)

according to its requirement. Internet topology is dynamic, and the changes in Internet topology must be recorded in the routing table. Therefore, the routing table is unstable, which is called *routing instability*. Informally, routing instability is defined as the rapid change of network reachability and topology information [7].

Besides Internet topology, there are number of origins of routing instability, such as routing algorithm, routing protocol, route configuration, data link problems, software bugs in router etc. [17].

Although the theoretical properties of routing algorithms have been well studied, the deployed, or actual, behavior of routing protocols has gone virtually without formal analysis. The studies of both widely deployed host protocol implementations, and the behavior of routing protocols in operational networks, have shown that deployed behavior of protocols can often differ drastically from expected, theoretical behavior [7,18]. Moreover, the scale and complexity of large protocol deployments often introduce side-effects not seen in smaller deployments.

Labovitz et al. [7] showed that the majority of Internet information was pathological. Based on discussions with router vendors, Labovitz et al. suggested the vendors modify software in routers. After Internet providers widely deployed the updated routing software across core backbone routers, Labovitz et al. found that the volume of routing-update messages decreased by an order of magnitude. Thus software bugs in routers is one origin of Internet instability.

Routing instability significantly contributes to poor end-to-end network performance and degrades the overall efficiency of the Internet infrastructure. The large volume of updates may consume bandwidth, memory, and CPU cycle, and increase the packet loss to unstable destinations and delays in the time for network convergence. At the extreme, routing instability has led to the transient loss of connectivity for large portions of the Internet.

While the forwarding table is updating, the forwarding engine must block search operations for the IP address until the update operations complete. Consequently, high frequent updating seriously limits the router's lookup performance. A backbone router may experience 100 updates to its forwarding table per second on an average and the number could reach as high as 1000 updates per second [7].

We have analyzed the characteristics of the BGP routing table. Because the routing table is the basic data for an IP-address lookup, the studies of these characteristics is beneficial to the optimizations of IP-address lookup algorithms.

The dramatic growth of routing table decreases the packet forwarding speed and demands more router memory. Thus, reduction of prefix count is typically seen as beneficial to infrastructural integrity, and developing mechanisms to do so is thus important architectural research. In following, we will introduce schemes to construct the optimal routing table.

2.6 Constructing Optimal Routing Tables

2.6.1 Filtering Based on Address Allocation Policies

Network operators configure their routers to apply filters to incoming BGP route advertisements. These filters prevent the router from accepting inappropriate advertisements, such as routes to private addresses. Bellovin et al. [20] investigate the potential for route filtering to help control the growth of BGP routing tables. Three types of filtering rules are discussed. The BGP routing table size can be reduced by aggressively filtering prefixes to enforce the allocation boundaries documented by the numbering authorities.

2.6.1.1 Three Filtering Rules

Martians.   The IPv4 address space includes several "special use" prefixes that have been reserved by the Internet Assigned Numbers Authority (IANA). Network operators should not accept or send advertisements for these addresses (called Martians), as summarized in Table 2.4. The class A

Table 2.4   Martian Address Blocks

Category	Prefix(es)
Default or broadcast	0.0.0.0/8
Loopback	127.0.0.0/8
Private addresses	10.0.0.0/8 172.16.0.0/12 192.168.0.0/16
Class D/E	224.0.0.0/3
Auto-configuration	169.254.0.0/16
Test network	192.0.2.0/24
Exchange points	192.41.177.0/24 192.157.69.0/24 198.32.0.0/16 206.220.243.0/24
IANA reserved	128.0.0.0/16

Source: Bellovin, S., Bush, R., Griffin, T.G., and Rexford, J., Slowing routing table growth by filtering based on address allocation policies. http://citeseer.nj.nec.com/493410.html.

block 0.0.0.0/8 includes the address 0.0.0.0 which is commonly used for default routes. The 127.0.0.0/8 prefix is reserved for loop-back addresses used by a host or router to identify itself. Three prefixes are reserved for private networks that use the IP protocols, as discussed in RFC 1918 [21]. The 224.0.0.0/3 block is devoted to class D (multicast) and class E (reserved) addresses. The 169.254.0.0/16 block is dedicated for auto-configuration of hosts when no dynamic host configuration protocol (DHCP) server is available. The prefix 192.0.2.0/24 is used for example IP addresses in documentation and code fragments. Several prefixes are allocated to the infrastructure at the public Internet exchange points. In addition, prefix 128.0.0.0/16 is reserved by IANA.

Operator Policies.   To limit routing table size, some network operators configure their routers to filter all prefixes with a mask length larger than 24. This protects the routers in the AS from storing and processing routes for a large number of small address blocks. In practice, an AS might apply this filter only to routes learned from a peer or upstream provider although still accepting prefixes with larger mask lengths from customers. That is, an AS may be willing to carry these prefixes on behalf of a paying customer but not for other ASs in the Internet. In addition to removing routes for small address blocks, some operators filter advertisements in the class B space that have a prefix with a mask longer than 16. The RIRs do not allocate small address blocks within the class B space to other institutions. Rather, the RIRs allocate address blocks in the class A and class C regions of the IP address space, as discussed below.

RIR Allocation Policies.   The remainder of the IPv4 address space is allocated by IANA to the three RIRs—Asia-Pacific Network Information Center (APNIC), American Registry for Internet Numbers (ARIN), and Réseaux IP Europeéns (RIPE)—with new registries proposed for Africa and Latin America. Each regional registry allocates address blocks to Local Internet Registries (LIRs) and other organizations within their regions. To reduce the impact on the size of the routing tables, the three regional registries limit the allocation sizes in different parts of the class A and class C portions of the address space, as summarized in Table 2.5. For example, ARIN does not make allocations in the 63.0.0.0/8 space with a mask length longer than 19; similarly, APNIC does not make allocations in the 211.0.0.0/8 space with a mask length longer than 23. These allocation

Table 2.5   RIR A and C Guidelines and Corresponding Counts of Filtered Prefixes for Telstra and Verio, May 16, 2001

RIR	Supernet	Mask Limit	Telstra Filtered	Verio Filtered
A Guidelines				
ARIN	24.0.0.0/8	20	918	19
	63.0.0.0/8	19	2894	310
	64.0.0.0/8	20	2543	318
	65.0.0.0/8	20	863	7
	66.0.0.0/8	20	645	7
RIPE	62.0.0.0/8	19	224	37
APNIC	61.0.0.0/8	22	31	10

continued

Table 2.5   (continued)

C Guidelines				
ARIN	196.0.0.0/8	24	0	0
	198.0.0.0/8	24	0	0
	199.0.0.0/8	24	0	0
	200.0.0.0/8	24	0	0
	204.0.0.0/8	24	0	0
	205.0.0.0/8	24	0	0
	207.0.0.0/8	24	0	0
	208.0.0.0/8	20	4549	4535
	209.0.0.0/8	20	4132	4047
	216.0.0.0/8	20	4541	4526
RIPE	193.0.0.0/8	29	0	0
	194.0.0.0/8	29	0	0
	195.0.0.0/8	29	0	0
	212.0.0.0/8	19	1238	1238
	213.0.0.0/8	19	671	637
	217.0.0.0/8	20	309	293
APNIC	202.0.0.0/8	24	0	0
	203.0.0.0/8	24	0	0
	210.0.0.0/8	22	797	535
	211.0.0.0/8	23	161	160
	218.0.0.0/8	20	0	0

Source: Bellovin, S., Bush, R., Griffin, T.G., and Rexford, J., Slowing routing table growth by filtering based on address allocation policies. http://citeseer.nj.nec.com/493410.html.

policies are publicized to aid the ISP community in filtering and other policy decisions. Users announcing smaller blocks (with longer masks) are warned that network operators throughout the Internet may choose to filter prefixes that exceed the address allocation guidelines.

2.6.1.2   *Performance Evaluation*

To evaluate the influence of filtering policies on routing table size, we applied a collection of filtering rules to the routing table—Telstra (http://www.telstra.net/ops/bgp/bgptable.txt, May 26, 2001).

Table 2.6   Routing Table Size from Telstra on May 16, 2001

Filtering Rules	No. of the Removed Prefixes	No. of the Left Prefixes
Initial		108,400
Martians	33	108,367
Mask length >24	5984	102,383
Class A	8118	94,265
Class B	6975	87,290
Class C	16,398	70,892

Source: Bellovin, S., Bush, R., Griffin, T.G., and Rexford, J., Slowing routing table growth by filtering based on address allocation policies. http://citeseer.nj.nec.com/493410.html.

The Telstra table starts with 108,400 unique prefixes. We first removed any prefixes that fall within the martian address blocks. The table includes 33 martian prefixes. Next, filtering prefixes with a mask length larger than 24 removes 5984 prefixes from the Telstra table; 169 of these prefixes originate directly from Telstra. Then, we removed the prefixes with mask lengths that exceed the RIR allocation policies for the class A portion of the address space, prefixes in class B space that have a mask length larger than 16, and prefixes that exceed the RIR allocation policies for the class C portion of the address space. These allocation rules remove 16,398 prefixes from the Telstra table. Applying the entire set of filters reduced the number of prefixes from 108,400 to 70,892, as summarized in Table 2.6. Experiments with the RouteViews (http://moat.nlanr.net/AS/data) and RIPE data from the vantage point of other ASs reveal similar trends, with a final table size of about 70,000 prefixes after applying the entire collection of filtering rules.

2.6.2   Minimization of Routing Table with Address Reassignments

The classic IP routing strategy is based on a hierarchical approach to assign an address. Where hosts in a subnetwork have a common address prefix, routers must only store prefixes for subnetworks. Nevertheless, the lack of farsighted strategies for assigning IP addresses and the huge and disorganized growth of the Internet from 1990 to 1994 had caused an excessive increase in the number of prefixes in the routing tables and the exhaustion of the 32-bit IP address space. Although CIDR allows multiple contiguous prefixes to be aggregated by a common prefix, which further reduces the routing table size, it is more important to find a new strategy for assigning IP addresses. Bilo and Flammini proposed an algorithm with an asymptotically optimal running time that assigns addresses so as to minimize the size of a single routing table [22]. This section is from [22]. Portions are reprinted with permission (© 2004 IEEE).

For an entry (prefix) e in a routing table, let $mask_e$ be the mask, l_e the length, $next_e$ the next hop. Bilo and Flammini formalized the IP routing table minimization problem (MIN-IP for short) as follows: Let $R = \{r_1, r_2, \ldots, r_n\}$ be a set of n routers, $H = \{h_1, h_2, \ldots, h_m\}$ be a set of $m \leq 2^h$ destination hosts (h is the length of IP address). Each router $r_j \in R$ has a degree δ_j, that is δ_j output interfaces, and a connection requirement $C_j = (H_{j,1}, \ldots, H_{j,\delta_j})$ consisting of a δ_j-tuple of subsets forming a partition of H such that each $H_{j,i}$, $1 \leq i \leq \delta_j$, is the subset of the hosts that must be reached through the i-th output interface of r_j. We must choose the IP addresses of m hosts and construct n IP routing tables so as to minimize the sum of their size, which is the overall number of used entries.

2.6.2.1 Case of a Single IP Routing Table

For the sake of simplicity, we denote by r the router whose table must be minimized, by δ its degree and by $C = (H_1, \ldots, H_\delta)$ its connection requirement. Moreover, let $A = (a_1, \ldots, a_\delta)$ be the δ-tuple of integers such that $a_i = |H_i|$ is the number of hosts reached through the i-th output interface, $1 \leq i \leq \delta$. We first consider the case in which $m = 2^h$ and then show how to generalize the results to any $m \leq 2^h$.

Observe first that if inclusions between entries are not allowed, then because the set of the addresses matching any entry e in the routing table has cardinality 2^{h-l_e}, the minimum number of entries corresponding to the i-th output interface is at least equal to the minimum number of powers of 2 whose sum is equal to a_i. This clearly corresponds to the number of bits equal to 1 in the $h + 1$ bit binary string that encodes a_i. Let $one(a_i)$ denote such a number. Then the minimum size of the table is at least $\sum_{i=1}^{\delta} one(a_i)$. Such a number of entries is always achievable for all the output interfaces as follows. For each i, $1 \leq i \leq \delta$, assign an entry of mask length l_e to each power 2^{h-l_e} decomposing a_i. List all the entries in nonincreasing order of size (or analogously not decreasing order of mask length) and assign in such an order 2^{h-l_e} new addresses to 2^{h-l_e} new hosts reached through $next_e$. At the end, that is, when all the entries have been processed and all the addresses have been assigned, the corresponding routing table is trivially constructed.

Allowing the inclusion of one entry f in one entry e with $next_e \neq next_f$ and $l_e < l_f$ modifies the number of addresses matching the entries corresponding to the output interface $next_e$ from a_{next_e} to $a_{next_e} + 2^{l_f}$. As a consequence, the minimum number of entries corresponding to output interface $next_e$ becomes $one(a_{next_e} + 2^{l_f})$. Concerning the effect of such an inclusion on the entries of the output interface $next_f$, we can charge a cost of one to the inclusion to keep track of the fact that such an entry will be effectively realized inside e, although the number of addresses matching the remaining entries of $next_f$ becomes $a_{next_f} - 2^{l_f}$.

Such a procedure can be extended to handle multiple inclusions. More precisely, if in a table an entry f is included in $s > 1$ entries $e_1, \ldots, e_s$ with $l_1 > \cdots > l_s$, we consider only its direct inclusion in e_1, that is in the smallest size total number of direct inclusions. Clearly, it must also be $e_1 \subset \cdots \subset e_s$, thus making it possible to identify chains of nested containments covering all the direct inclusions. This defines in a natural way a partial order between the direct inclusions, and the above procedure can then be iterated simply by considering direct inclusions according to the partial order, that is, from the smallest to the largest entries.

Let $c_1, \ldots, c_\delta$ be the numbers associated to the δ output interfaces at the end of the process. Then each c_i is the sum of the size of the matching entries of the output interface i, which are not included in any other entry. As a consequence, the minimum number of entries of an IP routing table having such inclusions becomes $\sum_{i=1}^{\delta} one(c_i)$ plus the total number of direct inclusions. In fact, similarly as above, $\sum_{i=1}^{\delta} one(c_i)$ entries are necessary and sufficient for the not included entries, while the total number of included entries is just the total number of direct inclusions. The construction of a table with such a number of entries can be obtained by means of a recursive algorithm that first considers the not included entries and then recurses on the contained ones. We do not give an exact description here because, as we will see in the sequel, it is possible to restrict to proper subclasses of containments that preserve the optimality and allow a much simpler algorithm.

Any IP routing table can be represented by a layered graph $G = (V, E)$, called the representation graph, tracing all the direct inclusions between entries in the following way. G has $h+1$ layers, that is $V = V_0 \cup \cdots \cup V_h$ and $E = E_0 \cup \cdots \cup E_{h-1}$, where $V_k = \{v_i^k \mid 1 \leq i \leq \delta\}$, $1 \leq k \leq h$ and $E_k \subseteq V_k \times V_{k+1}$, $1 \leq k \leq h - 1$. Each node $v_i^k \in V_k$ has a non-negative integer weight v_i^k and every edge $e_{i,j}^k = \{v_i^k, v_j^{k+1}\} \in E_k$ a strictly positive integer weight $w_{i,j}^k$.

Informally speaking, each row of G, that is all the nodes v_i^k for a fixed value of i, corresponds to the i-th output interface. Any edge $e_{i,j}^k \in E_k$ at level k is associated to the direct inclusions of the entries of the output interface i having size 2^k (or mask length $h - k$) in the entries of the output interface j, and $w_{i,j}^k$ is the number of such inclusions. Node weights represent the update of the numbers $a_1, \ldots, a_\delta$ resulting from the direct inclusions as described above, till arriving to $c_1, \ldots, c_\delta$. In particular, $v_i^0 = a_i$, $v_i^h = c_i$, and $v_i^{k+1} = v_i^k + \sum_{e_{j,i}^k \in E_k} 2^k \cdot w_{i,j}^k - \sum_{e_{i,j}^k \in E_k} 2^k \cdot w_{i,j}^k$, $1 \leq i \leq \delta$ and $1 \leq k \leq h - 1$.

Notice that the inclusion of one entry f in another entry of the same output interface is not useful, as f can be simply deleted without violating the connection requirements of the routing table and reducing its size by one. Therefore, in the following we will implicitly assume that edges of the type $e_{i,i}^k$ are not contained in E.

We define the cost of a representation graph $G = (V, E)$ as the minimum number of entries in a table having direct inclusions represented by its weighted edges, that is cos $t(G) = \sum_{i=1}^{\delta} one(v_i^h) + \sum_{e_{i,j}^k \in E} w_{i,j}^k$.

By the above considerations, each routing table T of minimum size among the ones with the same inclusions identifies a representation graph whose cost is equal to its number of entries, although each representation graph represents a family of equivalent minimum routing tables differing only on the assignment of IP addresses but with a same underlying structure of entries.

Let us say that a representation graph G satisfies a connection requirement $C = (H_1, \ldots, H_\delta)$ if $v_i^0 = a_i$, $1 \leq i \leq \delta$. Motivated by the discussion above, we will concentrate on the determination of a representation graph G of minimum cost that satisfies C.

Let us first define a proper subclass of representation graphs that preserves optimality.

Definition 2.6.1: A representation graph G is half-normalized if $\sum_{i=1}^{\delta} one(v_i^h) = 1$.

Given any number a such that $0 \leq a \leq 2^h$, let $b(a)$ be the $h + 1$ bit binary string given by the reverse binary encoding of a, that is, writing the corresponding bits from the least significant to the most significant one. Moreover, let $b_j(a)$ be the $(j + 1)$-st bit of $b(a)$ starting from the least significant bit, $0 \leq j \leq h$.

Notice that, for every layer k of a representation graph G it is $\sum_{i=1}^{\delta} v_i^k = 2^h$. G is half-normalized if and only if in the last layer h all the weights v_i^h are equal to 0, except for one weight equal to 2^h. More precisely, at each level k all the coefficients v_i^k are such that $b(v_i^k) = 0^{k-1}\beta$. In fact, for every fixed i, all the strings $b(v_i^{k'})$ with $k' \geq k$ coincide in the first $k - 1$ bits. This holds because starting from level k all inclusions modify the coefficients of powers of 2 with exponents at least equal to k.

Let us consider the subgraph $G_k = (V_k \cup V_{k+1}, E_k)$ of a representation graph G. Even if in general G can be any weighted bipartite graph, a further normalization step can be performed, allowing a significant simplification.

Definition 2.6.2: G is totally-normalized if it is half-normalized and E_k is a maximum match consisting of edges $e_{i,j}^k$ between nodes $v_i^k \in V_k$ and $v_i^{k+1} \in V_{k+1}$ such that $i \neq j$, $w_{i,j}^k = 1$, $b_k(v_i^k) = 1$, $b_k(v_j^k) = 1$, and no two edges $e_{i,j}^k$ and $e_{j,i}^k$ are contained in E_k. Informally speaking, in the matching E_k if a node v_i^k has an incident edge, then v_i^{k+1} is isolated in E_k and vice versa. Therefore, the matching has a number of edges that is the half of the number of nodes $v_i^k \in V_k$ with $b_k(v_i^k) = 1$. Clearly, because $\sum_{i=1}^{\delta} v_i^k = 2^h$ and each $b(v_i^k) = 0^{k-1}\beta$, for $k < h$ such a number of nodes must be even.

Lemma 2.6.1: Given any representation graph G, there exists a totally-normalized graph $\bar{G}$ with $\cos t(\bar{G}) \leq \cos t(G)$.

Thus, the problem MIN-IP becomes one of determining, at each level k, a maximum match in E_k touching only one of every two nodes v_i^k and v_i^{k+1} with $b_k(v_i^k) = 1$. Notice that once established, the subset T_k of the touched nodes of V_k and the subset T_k' of the touched ones in V_{k+1} are trivially determined because, if $b_k(v_i^k) = 1$, $v_i^{k+1} \in T_k'$ if and only if $v_i^k \notin T_k$. Any matching E_k between T_k and T_k' is clearly equivalent in the sense that it does not influence the cost of G.

We now present a polynomial time algorithm for the determination of a totally normalized graph having minimum cost among the ones satisfying the connection requirements given in the input.

Algorithm Build-Graph

INPUT: The connection requirement C of a router r
    Let the initial graph be $G = (V, E)$ with $E = \phi$ and $v_j^0 = a_j$, $1 \leq j \leq \delta$
    For $k = 0$ to $h - 1$ do
        Let S_k be the set of the nodes of $v_i^k \in V_k$ such that $b_k(v_i^k) = 1$
        Let $T_k \subset S_k$ be the subset of the $\frac{|S_k|}{2}$ nodes $v_i^k \in S_k$ with the lowest $b(v_i^k)$.
        Let E_k be any match of $\frac{|S_k|}{2}$ edges between T_k and T_k'.
        $E = E \cup E_k$.
OUTPUT: $G = (V, E)$

Theorem 2.6.1: The algorithm Build-Graph always returns a (totally normalized) representation graph of minimum cost in time $O(h\delta)$.

Once a totally normalized representation graph G of minimum cost is obtained, the following procedure constructs the corresponding IP routing table. Moreover, it returns δ subsets $D_1 \cdots D_\delta$ such that each D_i is the subset of the addresses that must be assigned to the hosts in H_i, that is, reached through the output interface i. To maintain a low time complexity, the procedure uses a set of entries T that finally will correspond to the returned table and a vector *Small* such that at every step $Small_i$ is the first address of the "free" part of the smallest constructed entry of the output interface i. More precisely, such a remaining part is the one not involved in previous inclusions that must be considered for the new direct inclusions in entries of i. Finally, for the sake of simplicity, each entry e is denoted as the triple $(destt_e, l_e, next_e)$, and given an h bit binary string α encoding an integer n, $\alpha + 2^k$ is the h bit binary string encoding $n + 2^k$.

Algorithm Build-Graph

INPUT: A totally-normalized representation graph $G = (V, E)$
    Let $D_j = \phi$, $1 \leq j \leq \delta$
    Let $T = \{(0^h, 0, i)\}$, $Small_i = 0^h$ and $D_i = [0^h, 1^h]$, where i is the unique output interface such that $v_i^h = 2^h$.
        For $k = h - 1$ down to 0 do
            For each $\{v_i^k, v_i^{k+1}\} \in E_k$ do
            $T = T \cup \{(Small_j, h-k, i)\}$,
            $Small_i = Small_j$,
            $D_i = D_i \cup [Small_j, Small_j + 2^{h-k} - 1]$,

$$D_j = D_j \setminus [Small_j, Small_j + 2^{h-k} - 1],$$
$$Small_j = Small_j + 2^{h-k}$$
OUTPUT: $T, D_1 \ldots D_\delta$

Clearly Build-Table runs in time $O(h\delta)$, as it performs $O(h\delta)$ iterations of time complexity $O(1)$. Added to the time required by Build-Graph, the overall time complexity is $O(h\delta)$ and is asymptotically optimal. In fact, the input size is $\Theta(h\delta)$ because the connection requirement of each output interface is a number with h bits, thus yielding a $\Omega(h\delta)$ lower bound for the problem. Notice also that the above algorithms do not explicitly assign an IP address to every single host. This would clearly require $O(2^h)$ additional steps, still maintaining an asymptotically optimal time complexity.

We finally observe that our algorithm can be easily generalized to the case in which we are given a number of hosts m such that $m < 2^h$. In such a setting, the first $h - \lceil \log m \rceil$ bits of all the addresses are fixed (for instance to 0), and the remaining part of the address is constructed as above considering $h' = \lceil \log m \rceil$ instead of h. Moreover, if m is not a power of 2, it is sufficient to modify the definition of the half-normalized representation graph in the following way.

Definition 2.6.3: A representation graph G is half-normalized if $\sum_{i=1}^{\delta} one(v_i^h) = one(m)$.

The algorithm Build-Graph must slightly modified in such a way that, because the cardinality of each S_k can be odd, T_k is the subset of the $\lceil |S_k|/2 \rceil$ nodes $v_i^k \in S_k$ with the lowest values $b(v_i^k)$. Notice that if $|S_k|$ is odd, then $|T_k'| = |T_k| - 1$ and the matching E_k touches all the nodes in $T_k \cup T_k'$ except one node of T_k. A simple case study shows that, except for little modifications in the algorithm Build-Table, the remaining details and proofs are unaffected.

2.6.2.2 General Case

Definition 2.6.4: Partition by k-sets.
Instance: A universe U of $k \cdot m$ elements, $m > 0$, and a family $S = \{S_1 \ldots S_n\}$ of subsets of U such that for every i, $1 \leq i \leq n$, $|S_i| = k$.
Question: Does S contain a partition of U, that is, m subsets $S_{i_1}, \ldots, S_{i_m}$ such that $\bigcup_{j=1}^{m} S_{ij} = U$?

Lemma 2.6.2: Partition by k-sets is NP-complete.

Proof: We prove Lemma 2.6.2 by means of a polynomial time reduction from partition by 3-sets, a well known NP-complete problem (see [23]).

Given an instance (U, S) of partition by 3-sets, we construct the following instance (U', S') of partition by k-sets: $U' = U \cup \{o_1^1, \ldots o_m^1, \ldots o_1^{k-3}, \ldots, o_m^{k-3}\}$ and $S' = S \times \{o_1^1, \cdots o_m^1\} \times \cdots \times \{o_1^{k-3}, \ldots, o_1^{k-3}\}$ where $o_1^1, \ldots, o_m^{k-3}$ are $(k-3)m$ new elements not contained in U. Clearly this reduction is polynomial because k is constant, that is, it is not part of the input of partition by k-sets. Moreover, if (U, S) is a positive instance of partition by 3-sets and $S_{i_1}, \ldots, S_{i_m}$ form a partition of U, the subsets $S_{i1}', \ldots, S_{i_m}' \in S'$ such that for each j, $1 \leq j \leq m$, $S_j' = S_j \cup \{o_j^1, \ldots, o_j^{k-3}\}$, form a partition of S', that is (U', S') is a positive instance of partition by k-sets. On the other hand, if (U',S') is a positive instance of partition by k-sets with corresponding subsets $S_{i1}', \ldots, S_{im}'$, the subsets $S_{i_1}, \ldots, S_{i_m}$ obtained by pruning from each subset S_{ij}' the elements belonging to $U' \setminus U$ form a partition of U, that is (U, S) is a positive instance of partition by 3-sets.

It is thus possible to prove the following theorem.

Theorem 2.6.2:   MIN-IP is NP-hard.

Proof: Consider an instance (U, S) of partition by 4-sets. Without loss of generality it is possible to assume that $|U|$ is a power of 2. In fact, if such a property does not hold and h is the lowest integer such that $2^h > |U| = 4m$, then by adding $2^h - 4m$ new elements to U and $2^{h-2} - 4m$ subsets to S forming a partition of the new elements, we obtain an equivalent instance such that $|U| = 2^h$.

Let us then consider the following instance of MIN-IP: n routers $R_1, \ldots, R_n$ with $n = |S|$ and $2^h = |U|$ hosts $h_1 \ldots, h_2^h$, where each router r_i has degree 2 and connection requirement $C^i = (H_1^i, H_2^i)$ with $H_1^i = \{h_j | o_j \in S_i\}$ and $H_2^i = H\backslash H_1^i$. Let l be the maximum frequency of a pair of elements of U in the subsets belonging to S. For each pair $\{o_i, o_j\}$ having frequency $l_{i,j}$ in U, add $l - l_{i,j}$ routers of degree 2 for which $H_1^i = \{h_i, h_j\}$ and $H_2^i = H\backslash H_1$. Let n' be the overall number of such added routers (thus in total there are $n + n'$ routers).

We now show that (U, S) is a positive instance of partition by 4-sets if and only if there exists a solution for MIN-IP in which the sum of the table size or analogously their total number of entries is $k = 5n + 3n' - l \cdot 2^{h-1} - 2^{h-2}$.

As can be easily checked, for any possible assignment of addresses to hosts, the minimum number of entries in the table of any router r_i is obtained by including the entries corresponding to the subset H_1^i, that is, to the first output interface in the single entry (matching all the possible addresses) associated to H_2^i. In fact, by construction $|H_1^i| = 2$ or $|H_1^i| = 4$, and not including one or more entries of the first output interface inside the ones of the second interface causes at least $h - 3$ entries for H_2^i. Therefore, under such a table structure, any address assignment affects only the number of entries m_i needed for H_1^i, thus yielding a table of size $m_i + 1$. If all such requirements are covered by entries of size one, the total number of entries is $\sum_{i-1}^{n+n'} (m_i + 1) = 5n + 3n'$. Clearly, two consecutive entries of size one matching two addresses $w0 \in \{0,1\}^h$ and $w1 \in \{0,1\}^h$ can packed to form a single entry of size 2, thus saving one entry. By construction, for every address assignment, the total number of entries of size 2 matching any two consecutive addresses $w0 \in \{0,1\}^h$ and $w1 \in \{0,1\}^h$ is exactly equal to l (thanks to the added m' routers with the corresponding connection requirements). Thus, by also using entries of size 2, the total number of entries becomes $5n + 3n' - l \cdot 2^{h-1}$. Finally, two consecutive entries of size 2 matching respectively the addresses $w00, w01 \in \{0,1\}^h$ and the addresses $w10, w11 \in \{0,1\}^h$ can be packed to form a single entry of size 4, thus saving another entry. According to the above observations, any address assignment can only influence the number of possible entries of size 4, and the theorem follows by observing that such a number can be 2^{h-2}, for a total number of $k = 5n + 3n' - l \cdot 2^{h-1} - 2^{h-2}$ entries, if and only if there exists a partition of U by 4-sets of S. In fact, a one-to-one correspondence between partitions and assignments yielding k entries can be obtained in such a way that the first four addresses are assigned to the hosts corresponding to the first subset $S_{i_1} \in S$ of the partition, the next four addresses to the hosts corresponding to $S_{i_2} \in S$, and so forth, till the last four addresses correspond to 2^{h-2}-th subset of the partition, that is the last one.

We now show that an efficient approximation algorithm exists for the MIN-IP, which exploits a matrix representation of the instances of the problem. In fact, the connection requirements of any router $r_j \in R$ can be represented by means of a $2^h \times \delta_j$ Boolean matrix A^j in which each row i with $1 \leq i \leq m$ is associated to a destination host $h_i \in H$ and each column k to the k-th output interface of r_j. The last $2^h - m$ rows of A have all the entries equal to 0. Then, for $i \leq m$, $A_{i,k}^j = 1$ if at r_j the host h_j must be reached through the output interface k, otherwise $A_{i,k}^j = 0$. Let A be the global matrix given by the horizontal juxtaposition of $A^1, \ldots, A^n$ such that the first δ_1 columns correspond to A^1, columns from $\delta_1 + 1$ to $\delta_1 + \delta_2$ to A^2 and so forth.

Let us define a segment in a column k of A as a maximal vertical sequence of consecutive entries equal to 1 in the column. The approximation algorithm is based on the idea that any permutation π of the rows of A corresponds in a natural way to an assignment of addresses to the hosts in H in such a way that each host $h_{\pi(i)}$, the one that in the permutation corresponds to row i of A, receives an IP address, the h bit binary string encoding $i - 1$. Moreover, each matrix A^j permuted according to π can be seen as the matrix representation of the IP routing table of r_j under such an address assignment in which there is no inclusion of entries. Because each segment corresponds to a limited number of entries in the IP routing table, a permutation causing a low number of segments in A also yields a good solution for MIN-IP.

More precisely, the algorithm exploits a reduction of MIN-IP to the following problem:

Definition 2.6.4: Min consecutive Segments Minimization (MIN-CS).
Instance: An $m \times n$ Boolean matrix A.
Solution: A permutation π of the rows of A.
Goal: Minimize the total number of segments of consecutive 1's in the matrix A permuted according to π, that is, the number of entries such that $A_{\pi(i),\,j} = 1$ and either $A_{\pi(i+1),\,j} = 0$ or $i = m$, where $\pi(i)$ is the index of the i-th row in π.
This problem has been proved NP-hard [23].

A variant of MIN-CS consists of the circular version in which the first and the last row of the matrix A are adjacent, that is a segment can extend from the end to the beginning of the matrix. In this case, the total number of segments corresponds to the number of entries such that $A_{\pi(i),j} = 1$ either $A_{\pi(i+1),j} = 0$ or $i = m$, and $A_{\pi(1),j} = 0$.

A $\frac{3}{2}$ approximation algorithm for this variant can be determined as follows. Let $d_{i,j}$ be the half of the Hamming distance of two rows in A, or analogously the number of columns in which they are different (i.e., such that $A_{i,k} \neq A_{j,k}$), divided by 2. It is easy to see that coefficients $d_{i,j}$ satisfy the triangle inequality ($\forall i,j,k : d_{i,j} \leq d_{i,k} + d_{k,j}$). Moreover, as can be easily checked, for any permutation $\pi : \{1,\ldots,m\} \rightarrow \{1,\ldots,m\}$ of the rows of A, $\sum_{i=1}^{m-1} d_{\pi(i),\,\pi(i+1)} + d_{\pi(m),\,\pi(1)}$ is equal to the number of consecutive segments in the matrix obtained by permuting rows of A according to π. In fact, the Hamming distance of two adjacent rows according to π is the number of segments that start or end between the two rows, and summing up the distances between all the adjacent rows each segment is counted twice.

It is then possible to reduce any instance of the circular version of MIN-CS to an instance of the Min traveling salesman problem (MIN-TSP) in such a way that each node u_i of the graph G corresponds to a row of A and the distance between any two nodes u_i and u_j in G is $d_{i,j}$. By construction, there is a one-to-one mapping between tours in G and permutations of rows of A, and the length of a tour coincides with the number of segments in the matrix A permuted according to corresponding permutation π. As a corollary, optimum tours identify optimal permutations and vice versa.

It is well known that if distances satisfy the triangle inequality, there exists a polynomial algorithm [24], also known as Christofides' heuristic, which returns a tour whose length is at most $\frac{3}{2}$ times the length of the optimal tour. Such a solution clearly corresponds to a permutation of A with a total number of segments being at most $\frac{3}{2}$ times the minimum one.

To solve the basic (noncircular) version of MIN-CS, it suffices to add a dummy row with all entries equal to 1 in A, solve the circular variant exploiting Christofides' heuristic, and then break the circularity deleting the dummy row and shifting the permuted rows in such a way that the row coming right after the dummy one becomes the first in A. This clearly gives a $\frac{3}{2}$ approximation algorithm or terminating at end of A is equal to the number of segments starting or terminating at the dummy row, respectively.

The approximation algorithm then consists in performing the following steps:

Input: Instance of MIN-IP.
Construct the global matrix A given by the horizontal juxtaposition of the matrices $A^1, \ldots, A^n$ representing the connection requirements of the n routers.
Exploit the above reduction to MIN-TSP and Christofides' heuristic to determine a $3/2$ approximate solution for the instance of the noncircular version of MIN-CS consisting of the matrix A.
Assign the address given by the h bit binary representation of the number $i - 1$ to each host $h_{\pi(i)} \in H$ and compute the n routing tables corresponding to the n matrices $A^1, \ldots, A^n$ permuted according to π (producing for each segment a corresponding set of table entries).
Output: The host addresses and the n routing tables.

Given an instance I of MIN-IP, let us denote as $opt_{IP}(I)$ the value of an optimal solution for A. Similarly, given a matrix A instance of MIN-CS let $opt_{CS}(A)$ be the value of an optimal solution for A. Before proving the approximation factor of the algorithm let us prove the following lemma.

Lemma 2.6.3: Let I be an instance of MIN-IP and A be the global matrix representation associated to I. Then $opt_{CS}(A) \leq 2opt_{IP}(I)$.

Proof: Consider an optimal solution for the instance I of MIN-IP, and let $A^1, \ldots, A^n$ be the matrix representations of the n routing tables of $r_1, \ldots, r_n$, respectively.
Even if eliminating containments of entries in the routing table of a router $r_j \in R$ might cause a substantial increment in the number of entries, just a little increase is obtained for the total number of segments in A^j. In fact, each IP entry e corresponds to at most one segment of A^j and in the containments' elimination it is affected only by its direct inclusions, that is, corresponding to entries f such that e is the smallest size entry containing f. The deletion of the addresses matching f in e cuts the corresponding segment in A^j in at most two segments. Then, an increase of at most one segment in A^j, the total number of segments is at most doubled.
Let A be the global matrix obtained by the horizontal juxtaposition of the matrices $A^1, \ldots, A^n$ after the containments deletion. Then A has a total number of segments at most equal to $2\,opt_{IP}(I)$; hence the lemma is true.

Theorem 2.6.3: The above algorithm for MIN-IP has a $3h$-approximation factor.

Proof: Let I be an instance of MIN-IP and A be the global matrix representation associated to I. Let $m_{IP}(I)$ be the total number of entries in n routing tables returned by the algorithm, and $m_{CS}(A)$ the total number of segments in the matrix A permuted according to the permutation π determined by the algorithm. In the permutation of A according to π, each segment of A corresponds to at most h IP entries. Therefore, $m_{IP}(I) \leq h \cdot m_{CS}(A) \leq 3/2h \cdot opt_{CS}(A) \leq 3h \cdot opt_{IP}(I)$.
The approximation factor of the algorithm is asymptotically tight. One fundamental reason is that it produces routing tables without inclusions of entries, and there are cases in which this increases the number of entries of a multiplicative factor proportional to h. A trivial example is that of a single router of degree two in which an output interface corresponds to only one host and the other one to the remaining $2^h - 1$ host. If we allow inclusion we can associate one entry matching only one address to the first interface and one entry matching all the possible addresses to the other interface. If we do not allow inclusions, the second interface requires at least h entries, because $2^h - 1 = 1 + 2 + \cdots + 2^{h-1}$. Another feature causing an approximation factor proportional to h is that

the algorithm does not carefully select the initial positions of the segments. Consider for instance a segment containing $2^b - 1$ entries. If the initial address of the segment is 0^b (or 0^{b-1}) then it corresponds to a single IP entry. On the contrary, if we shift the segment ahead of one address letting it range from $0^{b-1}1$ to $0^{b-2}1$, then h IP entries are needed to represent it in the routing table. Therefore, in order to asymptotically improve the approximation factor h, a substantially different algorithm is needed that allows inclusions and properly selects the initial position of the segments.

Bilo and Flammini have given a polynomial time algorithm for determining the minimum size of an IP routing table in the case of multiple routers. But there are some problems. For example, no one considers the case in which the routing information changes after having assigned the IP addresses.

2.6.3 Optimal Routing Table Constructor

To reduce the number of entries in a routing table, Draves et al. proposed the optimal routing table constructor (ORTC) algorithm that generalizes the subnetting and supernetting techniques [25]. If we use the binary tree to represent a set of prefixes, the idea of ORTC is based on the following observations: The first observation is that the shorter prefixes, close to the root of the tree, should route to the most popular or prevalent next hops; longer prefixes, near the leaves of the tree, should route to less prevalent next hops. Then subnetting will prune the maximal number of routes from the tree. Finally, these operations should be applied recursively, with every subtree. We will describe the three passes from [25] (© 1999 IEEE).

2.6.3.1 Description of the Algorithm

To simplify the description of the ORTC algorithm, Draves et al. proposed two assumptions: first, every routing table has a default route, that is to say, the root of the prefix tree has a next hop; second, every route has only one next hop. An example is shown in Figure 2.14a. In the next subsection, we will remove two assumptions, and apply ORTC algorithm to the real routing table.

Pass 1: The first pass makes preparation for the next two passes. All prefixes in the routing table are represented as a binary tree (Fig. 2.14a), in which each prefix has a corresponding node. If the tree is not a complete binary tree, we enlarge it so that each internal node in tree has two children nodes as follows: for each node (prefix P), if it has the left (right) child node, we create the right (left) child node by adding a bit "0"("1"). The new node inherits the next hop from its parent node. If a node has two children nodes, its next hop can be discarded. At last, all internal nodes have no next hop; each leaf node has a next hop. Figure 2.14b shows the result of Pass 1 processing on the example of a routing table.

An implementation of the first pass might use a preorder traversal of the binary tree or a traversal by levels from the root down. In either case, the traversal pushes next-hop information down to child nodes that do not have a next hop, creating new child nodes when a parent node has only one child.

Pass 2: This pass calculates the most prevalent next hop at each level. From the bottom up towards the root, each internal node is visited, its next hops are calculated from its children. If two children nodes of an internal node (P) have no next hop in common, then the next hop set of the node P has the combination of the next hop sets of two children; otherwise, the next hop set of the node P is the next hop(s) in common. The calculation is shown in Figure 2.14c.

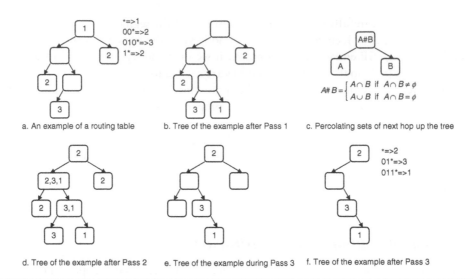

a. An example of a routing table   b. Tree of the example after Pass 1   c. Percolating sets of next hop up the tree

d. Tree of the example after Pass 2   e. Tree of the example during Pass 3   f. Tree of the example after Pass 3

Figure 2.14   An example of ORCT.

When the second pass is complete, every node in the tree is labeled with a set of potential next hops. Figure 2.14d shows the result of pass 2 processing on the example routing table.

Pass 3: From the root down, except for the root node, each node (*P*) visited has a set of potential next hops. If the closest ancestor of the node *P* has a next hop that is a member of the set of next hops of the node *P*, then the next hop set of the node *P* is set NULL; otherwise, the node *P* does only need a next hop, the other next hops are eliminated. The next hop of the node *P* may be chosen randomly from the set of its potential next hops. Thus, this pass may create different trees for different chooses.

Figure 2.14c and f demonstrates this phase of ORTC on the example routing table, using a traversal by level. After Pass 2, the root node has a next hop 2. The two children nodes 0* and 1* of the root node have the next hop 2 in their next hop set. The two children have their next hops moved. The root node becomes the ancestor node of the nodes 00* and 01*, whose next hop set is not NULL. Because the next hop of the root node is a member of the next hop set of the node 00*, the next hop set of the node 00* is set to be NULL. But the next hop of the root node is not a member of the next hop set of the node 01*, and we select a next hop 3 from the next hops set of the node 01*, and the other next hops are removed. The node 01* becomes the ancestor node of the nodes 010* and 011*. The next hop of the node 01* is a member of the next hop set of the node 010*, not a member of the next hop set of the node 011*, and we set the next hop set of the node 010* to be NULL and select the next hop 1 as the next hop of the node 011*. After the next hop sets are chosen for the tree in Figure 2.14d, the result is shown in Figure 2.14e. From the bottom up towards the root, if the next hop set of a leaf node is empty, the leaf node is removed from the tree. Figure 2.14f depicts the final result of ORTC for the tree in Figure 2.14a. The number of prefixes in Figure 2.14a is reduced from 4 to 3.

Three passes can be represented with pseudocode. Let *N* denote a node in the tree, *nexthops*(*N*) denote a set of next hops associated with the node *N*. If the routing table does not assign next hops to *N*, then *nexthops*(*N*) is defined to be the empty set ϕ; we assume *nexthops*(*root*) $\neq \phi$. For nodes with children, *left*(*N*) and *right*(*N*) denote the left and right child nodes. Similarly, we define *parent*(*N*) for all nodes except the root. The operation *choose*(*A*) picks an element from the nonempty set *A*.

As in Figure 2.14c, we define the operation $A\#B$ on two sets of next hops:

$$A\#B = \begin{cases} A\cap B & \text{if } A\cap B \neq \phi \\ A\cup B & \text{if } A\cap B = \phi \end{cases}$$

We define the function *inherited*(N) on nodes other than the root:

$$inherited(N) = \begin{cases} nexthops(parent(N)) & \text{if } \neq \phi \\ inherited(N) & \text{otherwise} \end{cases}$$

The first and third passes perform a traversal from the tree's root down to its leaves. This can be either a preorder traversal or a traversal by levels. Similarly, the second pass performs a traversal from the leaves up to the root, using either a postorder traversal or a traversal by levels. The pseudocode of three phases are shown as follows:

Pass 1
```
for each node N (root to leaves) {
    if N has exactly one child node
        create the missing child node
    if nexthops(N) = φ
        nexthops(N) ← inherited(N)
}
```

Pass 2
```
for each node N (leaves to root) {
    if N is a parent node
        nexthops(N) ← nexthops(left(N))#nexthops(right(N))
}
```

Pass 3
```
for each node N (root to leaves) {
    if N is not the root and inherited(N) ∈ nexthops(N)
        nexthops(N) ← φ
    else
        nexthops(N) ← choose(nexthops(N))
}
```

Although ORTC is very simple in operation, it always yields a routing table that is optimal, in the sense that the output routing table has the smallest number of prefixes possible although still maintaining the same forwarding behavior. In the third pass of the algorithm, the algorithm may choose a next hop from a set of potential next hops. This means that the algorithm may produce many different output routing tables for a given input table. ORTC guarantees that all of these possible output routing tables are optimal, and hence they are all the same size. The complete proof of ORTC's optimality is shown in [25].

2.6.3.2   Improvements

We have described the basic algorithm based on two assumptions. In real routing circumstance, the routing table is dynamic and the routers need more efficient algorithms to construct the optimal routing table. In this subsection, we will propose the schemes to improve the performance of the basic ORTC algorithm and to remove the two assumptions.

Improving Performance.   There are several ways to reduce the number of steps in ORTC and hence improve performance.

It is possible to skip the first pass. When the second pass (now being done first) comes across a parent node with only one child node, then at that time it can create the new child node and assign an inherited next hop to the new child. This is an example of lazy evaluation; the work of the first pass is delayed until it is really necessary.

Another performance improvement saves some work in the third pass by anticipating it in the second pass. In the second pass, when a parent node is assigned the intersection of its child nodes' sets of potential next hops, then the next-hop information for the two child nodes can be deleted. This immediately prunes those prefixes from the routing table. This is a safe optimization because a member of the intersection of two sets is by definition a member of both sets. In the third pass, that parent node will be assigned a next hop from the intersection (or it will inherit such a next hop). If the third pass processed the child nodes, it would see that they inherit a next hop that is a member of their potential set, and prune them at that time.

The complexity of the algorithm is linear in the number of nodes in the tree. The number of tree nodes is $O(wN)$ where w is the maximum number of bits in the prefixes and N is the number of prefixes in the input routing table. For IPv4, $w \leq 32$ and for IPv6, $w \leq 128$. Implementing the algorithm using path-compressed tries [26] would reduce the number of tree nodes to $O(N)$, speeding up the algorithm's operation. ORTC's space complexity is the same as its time complexity.

Improving Stability.   In third pass, the basic ORTC algorithm eliminates some nodes (prefixes) in the tree. If the less original prefixes are eliminated, the faster update can be made. For example, in Figure 2.14, the original prefix 010* is eliminated. If the prefix 010* is deleted from the original routing table, we cannot find it in the optimal table. This needs more operations to delete it. Because the next hop set of the internal node 01* is formed by the union of its two child nodes, we can set its next hop to be NULL and preserve its two child nodes. The output routing table contains prefixes: * → 2, 010* → 3, 011* → 1. The optimal routing table contains the same number of prefixes. If the prefix 010* is in the optimal routing table, it can be directly deleted. Therefore, in the third pass, if we use the original prefixes instead of generating a new prefix, this improves the stability.

Removing the Default Route Assumption. In the first pass, the prefix tree is expanded to a complete binary tree, and every node that has no next hop inherits the next hop from its closet parent node. The default route ensures that every new child node inherits a next hop. In fact, the backbone routers in the Internet use default-free routing tables.

There are two schemes to remove the default route assumption. First, we can remove the nodes (prefixes) that have the same next hop with the root node. In Figure 2.14f, the node 011* contains the next hop 1 (the default route), and it can be removed. Second, the next hop set of the root node is set to be NULL. In the first pass, a new leaf node inherits the next hop from its parent node; if the next hop set of the parent node is NULL, then the next hop set of the new leaf node is NULL too. In the third pass, all leaf nodes which next hop set is NULL are removed.

Table 2.7 ORTC Performance on Internet Backbone Routing Tables

Routing Table	No. of Prefixes		Time (ms)
	Initial	After ORTC	
MaeEast	41,315	23,007 (56%)	400
AADS	24,418	14,964 (61%)	259
MeaWest	18,968	13,750 (73%)	227
Paix	3020	2593 (85%)	51

Source: Draves, R.P., King, C., Venkatachary, S., and Zill, B.D., *Proceedings of Eighteenth Annual Joint Conference of the IEEE Computer and Communications Societies,* 1999. With permission.

Table 2.8 Memory Comparison Using CPE

Number of Levels	Initial Memory (kB)	Memory after ORTC (kB)
2	1028	670
3	496	334
4	402	275
5	379	260

Source: Draves, R.P., King, C., Venkatachary, S., and Zill, B.D., *Proceedings of Eighteenth Annual Joint Conference of the IEEE Computer and Communications Societies,* 1999. With permission.

2.6.3.3 Experiments and Results

Draves et al. analyzed the performance of ORTC using publicly available routing table databases from Merit (http://www.merit.edu, July 7, 1998). Table 2.7 shows the reduction in the number of prefixes. For the 42,315-prefix MaeEast database, ORTC compresses to 32,007 prefixes. For the other database, ORTC reduces to more than 60 percent of their original size. In addition, Table 2.7 presents the ORTC runtime. For the largest routing table, the implementation of ORTC took less than 0.5 second.

Draves et al. also examined ORTC's impact on the size of fast-forwarding data structures built from the routing tables. Table 2.8 shows the reduction in forwarding structure size for the MaeEast database, using the data structure CPE presented in [8]. CPE uses multi-level tries, and allows the number of levels used to be varied. Using more levels decreases the data memory requirement, increases the lookup time because the worst-case lookup time is proportional to the number of levels. For example, using a router configured with 384 KB of fast memory available for forwarding, ORTC allows a 3-level trie to be used instead of a 5-level trie, improving performance.

References

1. Halabi, S. and McPherson, D., *Internet Routing Architectures*, 2nd ed, Cisco Press, 2000.
2. Waldvogel, M., *Fast Longest Prefix Matching: Algorithms, Analysis, and Applications*, Shaker Verlag, GmhH, D-53013 Aachen, Germany, 2000.
3. Knuth, D.E., Morris J., and Pratt, V.R., Fast pattern matching in strings, *SIAM Journal of Computing*, Philadelphia, USA, 1977; 6(1):333–350.
4. Boyer, R.S. and Moore, J.S., A fast string searching algorithm, *Communications of ACM*, 1977; 20(10):762–772.
5. Gusfield, D., *Algorithms on strings, tree, and sequences: Computer science and computational biology*, Cambridge University Press, New York, NY, 1997, pp. 103–106.
6. Gupta, P., *Algorithms for Routing Lookups and Packet Classification*, PhD thesis, Stanford University, Palo Alto, California, December 2000, pp. 7–19.
7. Labovitz, C., Malan, R., and Jahanian, F., Internet routing stability, *IEEE/ACM Transactions. Networking* 1998; 6(5):515–528.
8. Venkatachary, S., *Fast and Efficient Internet Lookups*, Washington University, PhD Thesis, August 1999.
9. Fuller, V., et al., Classless inter-domain routing (CIDR): an address assignment and aggregation strategy, *RFC1519*, September 1993.
10. Broido, A., Claffy, K., and Nemeth, E., Internet expansion, refinement, and churn, *European Transactions on Telecommunications* 2002; 13(1):33–51.
11. Huston, G., Analyzing the Internet's BGP routing table, *The Internet Protocol Journal* 2001; 4(1):2–15.
12. Bates, T., The CIDR Report, update weekly at, http://www.employees.org/~tbates/cidr-report.html.
13. Cengiz, A., RIPE/RIS Project BGP Analysis CIDR at Work, http://www.caida.org/outreach/isma/0112/talks/cengiz /index.pdf, August 2001.
14. Xu, Z. et al., IPv4 address allocation and the evolution of the BGP routing table. Wei, K. (Kevin) Tsai, eds. *Proceedings of IEEE 18th Computer Communications Workshop*, Laguna Niguel, California, 2003, IEEE Press, 2003, pp. 172–178.
15. Bu, T., Gao, L., and Towsley, D., On characterizing routing table growth, *Proceedings of Global Telecommunications Conference*, IEEE Press, 2002, pp. 2185–2189.
16. Bellovin, S., Bush, R., Griffin, T.G., and Rexford, J., Slowing routing table growth by filtering based on address allocation policies. http://citeseer.nj.nec.com/493410.html
17. Labovitz, C., Malan, G.R., and Jahanian, F., Origins of Internet routing instability, *Proceedings of Infocom '99*, New York, NY, March 1999.
18. Dawson, S., Jahanian, F., and Mitton, T., Experiments on six commercial TCP implementations using a software fault injection tool, *Software Practice and Experience* 1997; 27(12):1385–1410.
19. Floyd, S. and Jacobson, V., The synchronization of periodic routing messages, *IEEE/ACM Transactions on Networking* 2(2):122–136.
20. Steve, B. et al., Slowing Routing Table Growth by Filtering Based on Address Allocation Policies, Steve's NANOG talk, Randy's IETF plenary talk, http://www.research.att.com/~jrex/papers/filter.ps, June 2001.
21. Rekhter, Y. et al., Address Allocation for Private Internets, RFC 1918, IETF, February 1996.
22. Bilo, V. and Flammini, M., On the IP routing tables minimization with addresses reassignments, *Proceedings of the 18th International Parallel and Distributed Processing Symposium*, 2004, IEEE Press, 2004, pp. 19–26.
23. Garey, M. and Johnson, D., *Computers and intractability—A guide to the theory of NP-completeness*. W.H. Freeman, San Francisco, 1979.
24. Christofides, N., Worst case analysis of a new heuristic for the traveling salesman problem. Technical Report 388, GSIA Carnegie–Mellon University, Pittsburgh, PA, 1976.
25. Draves, R.P., King, C., Venkatachary, S., and Zill, B.D., Constructing optimal IP routing tables, *Proceedings of Eighteenth Annual Joint Conference of the IEEE Computer and Communications Societies*. 1:88–97, 21–25 March 1999.
26. Knuth, D., *Fundamental algorithms vol. 3: Sorting and searching*. Addison-Wesley, Boston, MA, USA, 1973, pp. 78–114.

Chapter 3

Classic Schemes

With the emergence of the Internet, many researchers from academia and industry have been paying attention to the problem of IP-address lookup. There are many existing schemes. In this chapter, we will discuss the schemes introduced before 1996.

3.1   Linear Search

A linked-list is one of the simplest data structures. It is easy to make a list of all prefixes in the forwarding table. We can traverse the list of prefixes one at a time, and report the longest matching prefix at the end of the traversal. Insertion and deletion algorithms perform trivial linked-list operations. The storage complexity $O(N)$ of this data structure for N prefixes is $O(N)$. The lookup algorithm has time complexity and is thus too slow for practical purposes when N is large. The insertion and deletion algorithms have time complexity (1), assuming the location of the prefix to be deleted is known.

The average lookup time of a linear search algorithm can be made smaller if the prefixes are sorted in order of decreasing length. The lookup algorithm would be modified to simply stop traversal of the linked-list the first time it finds a matching prefix [1].

3.2   Caching

The caching technique remembers the previous results for subsequent reuse. The idea of caching, first used for improving processor performance by keeping frequently accessed data close to the CPU [2], can be applied to routing lookups by keeping recently seen destination addresses and their lookup results in a route-cache. A full lookup (using the longest prefix matching algorithm) is now performed only if the incoming destination address is not already found in the cache. Because the packets destined for the same destination address may arrive again with a certain time interval (called temporal locality), caching the address is one possible accelerating solution. There are some early studies reported by Feldmeier [3], Jain [4,5], etc.

3.2.1 Management Policies

Routing table caching can decrease the average processing time per packet if a locality exists for an IP address. To determine whether IP caching in routers is effective and what type of cache performs best, Feldmeier simulated the performance of several routing-table caches using a trace gathered from a router during normal operation. This trace is then used to drive a cache simulation model. The cache model can be altered to simulate a cache of any type and size, and the simulation can be repeated on the same data set [3]. This subsection is from [3] (© 1988 IEEE).

3.2.1.1 Cache Modeling

A routing table cache differs from a main memory cache in several ways. A main memory cache is bidirectional—it must be able to handle both read and write commands from the processor. A routing table cache is unidirectional, because the packet-forwarding process only reads from the table. A read-only operation eliminates the need for a routing-table update from the cache and simplifies cache operation. Another difference is that on a network with dynamic routing, the routing table changes over time. If the routing table changes, the entries in the cache may become inconsistent with the routing table. The cache model assumes that the cache is never flushed, which approximates true cache operation if the routing table is quasi-static or there exists an inexpensive method of updating the cache. Currently, routing tables are relatively static, so never flushing the cache approximates cache performance on the router.

Different types of caches will behave differently with the same data set, so a type of cache to simulate must be chosen. A routing table cache is defined by its associativity, replacement, and prefetch strategies, as well as size. Each of these must be defined for the simulation model.

The associativity of cache ranges from fully associative to direct mapped. Fully associative caches allow any destination or next-hop pairs to be stored in any cache location. A direct mapped cache allows each destination or next-hop pairs to be stored in only a certain cache location, so multiple destination addresses are mapped into a single cache location. In general, the cache is divided into a disjoint set of locations that may contain the contents of a certain subset of destination or next-hop pairs, and this is a set associative cache. For these measurements, a fully associative cache is used because it provides the best performance for a given number of cache slots. Also, the performance of any set associative cache depends on the value of those items to be stored in the cache and the cache associative mechanism. A fully associative cache avoids the problem of measurement bias because of specific network addresses in the measurements and makes the results general for any set of network addresses with similar characteristics. In addition, broad associative searches are easily implemented in MOS VLSI, so it is reasonable to expect that highly associative caches will be available on a chip.

Whenever there is a cache miss, the missed entry must be entered into the cache and some other item must be discarded. The three most popular strategies are random replacement, first-in first-out (FIFO), and least recent use (LRU). A random replacement strategy simply discards a random location in the cache. The "random" location is most easily chosen based on some counter value (such as a clock) from elsewhere in the system. A FIFO replacement strategy discards the cache entry that is the oldest; a FIFO cache is easily implemented as a circular buffer. The most complex strategy, but also the one that generally provides the best cache performance, is LRU. An LRU cache operates by storing not only the data, but also the time that each datum was last referenced. If a reference is in the cache, its time is updated; otherwise that data that was least recently accessed is discarded and the data for the new reference is cached.

An LRU cache is a good choice for the type of address locality expected and a fully associative, LRU cache forms an upper bound on the cache hit ratio for a fixed number of cache slots. Unfortunately, this type of cache is also the most complex and expensive to implement. A different type of cache may perform worse for a given number of slots, but for a fixed price, a less efficient cache can afford to have more slots and perhaps exceed the performance of a smaller fully associative LRU cache; this is a cost tradeoff that only the router designer can assess.

A strategy for prefetching is suggested by observing that a packet from host A to host B is often followed by a packet from B to A. If the cache prefetches the next hop for the packet source, then a packet traveling through the same router in the reverse direction no longer causes a cache miss. For cache simulation, two situations are analyzed. The first is that only the packet destination is used to update the cache and this corresponds to a case where prefetching is not economical. The second assumes that prefetching is free and should be done any time that the packet source is not in the cache. In reality, the truth is somewhere in-between, and these two sets of curves give the lower and upper bounds for fully associative LRU caches where prefetching occurs.

3.2.1.2 Trace Generation

We would like to estimate the performance of a router with a routing-table cache before the system is built. The performance can be determined by trace-driven simulation, which requires a list of all routing-table accesses on a router. We are interested in the statistics of packet arrivals at various routers. But building a measurement system directly into a router has several disadvantages. The main problem is that each router would need to be reprogrammed to include a monitoring system. Reprogramming is difficult because some of the routers are reluctant to disturb router operation; other routers are unavailable. In addition, any monitoring system installed in a router will use processor time and memory space and this overhead could affect router operation during periods of heavy load, causing packets to be dropped. Because periods of heavy load are of particular interest to determine the efficiency of the caching system, the measurement artifact of such a monitoring program is unacceptable. Also, it is advantageous to be able to try various caching strategies and cache sizes on a single measurement set, so that differences in results are directly attributable to the change in strategy, rather than simply a change in the router traffic. Instead of monitoring the packet arrivals at each router, it is simpler to measure all of the packets on the network that pass through the router. This replaces monitoring programs in each router with a single dedicated monitoring system.

Trace-driven simulation computes the exact cache performance as long as the trace-gathering operation is exact and has no processing cost. Measuring packets on the network with a separate measurement system ensures that the router itself operates as usual. Two types of packets are observed traveling to a router: packets to be forwarded to another network or packets destined to the router itself. Not all packet addresses to the router itself are observed, because only one of the networks attached to the router is monitored, but this is a problem only if a packet to a router causes a routing table reference. Whether a packet to a router causes a routing table reference depends on the router implementation. One possibility is to explicitly check each incoming packet against a table of all of the router's addresses to see if there is a match. This explicit check means that the routing table is never consulted about packets destined to the router. Another possibility is to use the routing table for all packets. The router will discover that it is directly connected to the proper network and will try to send the packet. Before the packet is sent, the router checks if the packet is to its own address on the appropriate network. If the packet is for the router, then it is never transmitted. The explicit check after the routing-table lookup requires checking a smaller number of router addresses at the increased cost of a routing-table lookup. Here, it is assumed that routers do an explicit address check and that the unobserved packets do not cause routing-table accesses.

Another disadvantage of measuring all packets is that the packets in the network to (or from) a router may not be the same as those to be handled by the router software. If the current router drops a packet, some protocols will retransmit that packet. Thus our packet trace now has multiple copies of a single packet, only one of which is actually forwarded by the router. Some of these extraneous packets can be eliminated by filtering the packet trace to remove identical packets with certain inter-arrival times. In any case, these duplicate packets should only appear when the router is overloaded, and this is relatively rare.

The measurement system consists of two computers—a passive monitor and a data analysis machine. The passive monitor receives the packets and timestamps them upon arrival. It then compresses the information of interest from each packet into a large data packet that is sent to the analysis machine. The monitoring system generates less than 2 percent of the network traffic, so the monitoring overhead is acceptable; also, none of packets pass through a router, and so the arrival order of packets at a router remains unchanged. For these measurements, the monitor compresses information only for packets that are to or from nonlocal hosts; in other words, all the packets that pass through the routers. The analysis machine simply stores all of the received data for later analysis.

Only a single trace using complete IP addresses for all packets going through the router is needed for both flat and hierarchical addressing scheme cache simulation. Flat addressing measurements use the entire IP address for simulation; hierarchical addressing uses only the network field of IP addresses.

3.2.1.3 Measurement Results

To avoid cache misses caused by cache initialization, the cache is preloaded so that once the measurement period begins, the cache has many of its most referenced entries. Measurements during the 61-minute period from 11.59 p.m. Sunday, November 29th to 1:00 a.m. on Monday, November 30th 1987 were used to preload the cache to avoid measurements during cache initialization.

The trace is used from the MIT ARPANET router during the 24-hour period from 1:00 a.m. Monday, November 30th to 1:00 a.m. Tuesday, December 1st 1987. The total number of packets processed by the router was 4,546,766. For measurement accuracy, any packets with the same source and destination that arrive within a millisecond or less of each other are discarded as low-level retransmissions.

Results for Flat Addresses.   It is known that a packet from host A to host B has a high probability of being followed by another packet form A to B, which implies time locality of packet addresses. Consider a time-ordered list of previously seen addresses; for a current address that has been previously seen, how far down the list is this address? Figure 3.1 shows the percentage of references versus the time-ordered position of the previous reference for the 400 most recently referenced packet destination addresses. The graph is averaged over all packets received by the router and it is plotted on a logarithmic scale so that the relative popularity of the slots is more easily seen. The graph for destination/source caching is similar.

The decrease in probability for both graphs is nearly monotonic until about 50 most recently referenced addresses. This monotonically decreasing function implies that the LRU cache

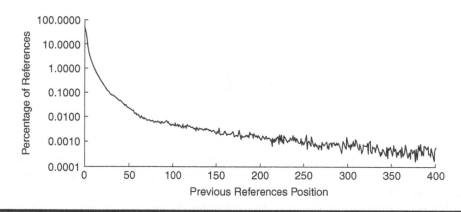

Figure 3.1   Percentage of references versus time of last references. (From Feldmeier, D.C., ***Proceedings IEEE Infocom'88,*** **1988. With permission.)**

management strategy is optimal for caches of less than 50 slots. After reference 50, the probability continues to decrease generally, but there is more variance. One reason for the increased variance is that the probability of reference decreases with increasing age, so that the amount of data that falls into the farther positions is too small to properly approximate the probability distribution curve at these points. At these outer parts of the curve, the average number of packets per slot is about 15, as compared to between 500 and 876,000 in the first 50 positions. Another explanation is that the relationship between probability of reference and time of last reference changes at some point on the curve. This change in relationship might be caused by two separate mechanisms that determine whether the current packet address is in the cache.

The first mechanism that produces cache hits is a current packet that is part of a train of packets. The first packet of the train initializes the cache and as long as the cache is not too small, all following packets are cache hits. The second mechanism is that the first packet of a train is to a destination already in the cache. If N is the number of packets in a train, the number of packets that arrive due to the first mechanism must be greater than $N-1$ times the number of packets that arrive due to the second mechanism. This means that the destination addresses of packets in an active train are most likely to be near the first slot of an LRU cache and that addresses of inactive but popular train destinations are likely to be between slot i, where i is the number of simultaneously active trains, and the last slot of the cache. The relationship among train arrivals at a destination is unknown, but undoubtedly it is weaker than the relationship among packets in a train, and this could explain the higher variance in the later part of the probability distribution. Additional information is necessary to decide whether an insufficient number of samples or two cache hit mechanisms more correctly explains the increased variance in the later part of the probability distribution above.

If the probability distribution does change with age after a certain point, the most cost-effective strategy may be to have an LRU cache with 50 slots, followed by a FIFO or random management cache with a few hundred slots. If the probability of reference does not decrease monotonically with increasing last-reference age, then an LRU cache decreases in efficiency relative to other cache types and its higher cost may not be justified because other cache types allow a larger cache for the same price and perhaps a higher cache hit ratio for a given cost.

Given the data above, an LRU cache should be effective for reducing the number of routing-table references. The probability of reference versus time of previous reference data allows simple calculation of an LRU cache hit ratio for a given number of cache slots. The relationship between cache hit ratio and probability of access is:

$$f_h = \sum_{j=1}^{i} p_j$$

where $f_h(i)$ is the cache hit ratio as a function of i, the number of cache slots; and p_j is the probability of a packet address being the jth previous reference. Figure 3.2 shows that the ratio of the percentage of cache hits to cache size is high. Table 3.1 shows just how quickly the probability of a cache hit climbs even for small cache sizes. With as few as nine slots, the hit ratio is already above 0.9.

Another way of looking at the ratio data is to plot the average number of packets passing through the router between cache misses. The relation between cache hit ratio and packets between cache misses is:

$$f_m(i) = \frac{1}{1 - f_h(i)}$$

where $f_m(i)$ is the number of packets between cache misses as a function of i, the number of cache slots; and $f_h(i)$ is the probability of a cache hit as a function of the number of cache slots.

The effectiveness of the cache depends on the ratio of the packets-between-misses values before and after the cache, not an absolute number, so the graph in Figure 3.3 is plotted on a logarithmic scale so that equal ratios appear as equal vertical intervals. This graph shows the incremental efficiency of each additional slot in the cache. In addition to curves for destination LRU, destination source FIFO, also shown are the destination only and destination source curves for a two-cache system consisting of a 64-slot LRU cache followed by a FIFO cache.

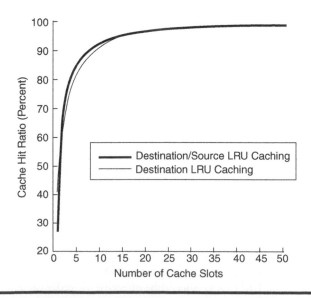

Figure 3.2   Percentage of cache hits versus cache size. (From Feldmeier, D.C., *Proceedings IEEE Infocom'88*, 1988. With permission.)

Table 3.1 Cache Hit Ratio Percentage versus Cache Size

Number of Cache Slots	Destination Only	Destination Source
1	40.78	27.05
2	63.53	67.83
3	72.34	76.30
4	78.19	82.12
5	81.96	85.34
6	84.78	87.44
7	86.95	89.15
8	88.72	90.48
9	90.17	91.57
10	91.37	92.52
11	92.38	93.30
12	93.22	93.97
13	93.92	94.54
14	94.52	95.05
15	95.02	95.48
16	95.45	95.87
17	95.82	96.19
18	96.13	96.47
19	96.40	96.73
20	96.64	96.96

Source: Feldmeier, D.C., *Proceedings IEEE Infocom'88,* 1988. With permission.

As expected, the destination source LRU cache performed best, followed closely by the destination LRU cache. The destination source cache outperforms the destination only cache, which confirms that much of the traffic through the router is bidirectional. Performance of both FIFO caches is relatively poor until the destination FIFO cache size exceeds 1250. It turns out that 1035 is the number of distinct destination addresses handled by the router in a 24-hour period, in addition to 230 destinations in the preload, giving a total of 1265. For destination source caching, the FIFO buffer size does not help until 1370 slots, which is about the number of distinct destination and source addresses processed by the router in a 24-hour period: 1130, plus 254 during preload, giving a total of 1384.

The incremental efficiency of each slot in the cache is the slope of the curve at slot number. A sharp decrease in the slope indicates a point of diminishing returns for larger cache sizes. Particularly notice the drop in LRU cache efficiency around cache slot 50.

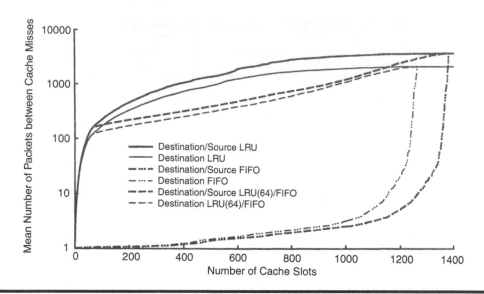

Figure 3.3 Number of packets between cache misses versus cache size (log-linear). (From Feldmeier, D.C., *Proceedings IEEE Infocom'88*, 1988. With permission.)

The dual cache systems perform moderately well. For example, consider a 128-slot destination LRU cache (this can be thought of as two 64-slot LRU caches back to back). The same performance can be obtained with a 64-slot destination LRU cache followed by a 241-slot destination FIFO cache. A FIFO cache is simpler to implement than an LRU cache and if a 241-slot FIFO cache is cheaper than a 64-slot LRU cache, then the dual cache combination provides a better hit ratio for a given cost than a single LRU cache.

If the interest is in determining how many cache slots are necessary for a given number of packets between cache misses, the above graph is better plotted on a linear scale. The graph in Figure 3.4 is interesting because each of the curves seems to be composed of two or three linear regions. At least for the LRU curves, the number of packets between misses seems proportional to the number of cache slots, until the curve approaches its maximum, where the incremental slot efficiency drops.

Results for Hierarchical Addressing.   Currently, routing is done based on the network address (prefix) and because the number of networks is smaller than the number of hosts, the cache hit ratio for a network cache is higher than the hit ratio of a flat address cache. To cache network addresses, the complete address must pass through an address decoder that extracts the network address before CIDR deployment. The tradeoff is that the hit ratio of the cache is higher than for flat addresses, but each address must be decoded.

Hierarchical addressing is complex because the separation of the network and host fields of the IP address is determined by the type of address. To avoid this complication, the separation of fields for the purposes of this research is that the first three bytes of the full address refer to the network and the last byte refers to the host. This procedure never removes part of the network address as long as the address is not a Class C address with subnetting (which is not known to be used by anyone). In some cases, some of the host's address will be left with the network address, which means that the cache hit ratio will be lower than if the address were completely separated.

Because the purpose of a cache is to speed up routing table access, it is probably advantageous to place the cache before the address separation, rather than after it, so that each cache hit saves

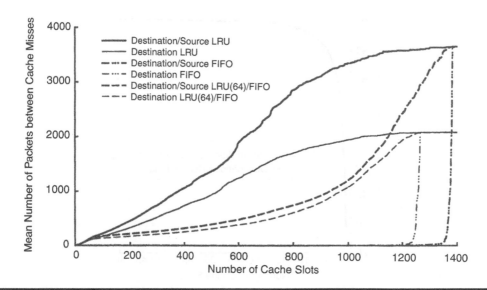

Figure 3.4 **Number of packets between cache misses versus cache size (linear-linear). (From Feldmeier, D.C.,** *Proceedings IEEE Infocom'88,* **1988. With permission.)**

not only a table lookup, but also the trouble of dissecting the full address. It may be advantageous to use a simple address separation procedure before the cache, such as the suggested one of removing the last byte of the address. This preprocessing step increases the cache hit ratio by eliminating most, if not all, of the host address from the full address, and yet avoids the complexity of complete address decoding.

Figure 3.5 shows the percentage of references versus the time-ordered position of the previous reference for the 80 most recently referenced packet destination addresses. The graph is averaged over all packets received by the router and it is plotted on a logarithmic scale so that the relative popularity of the slot is more easily seen. The graph for destination source caching is similar.

Figure 3.6 shows the percentage of cache hits versus cache size for both destination and destination source caches. Notice that even a relatively small cache has a high cache hit ratio.

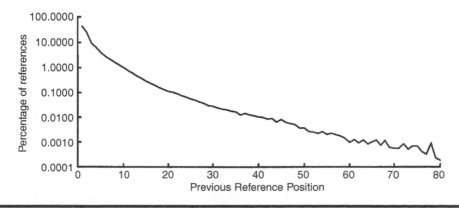

Figure 3.5 **Percentage of reference versus time of last reference. (From Feldmeier, D.C.,** *Proceedings IEEE Infocom'88,* **1988. With permission.)**

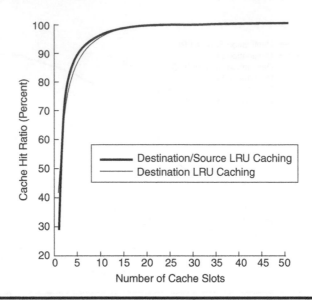

Figure 3.6 Percentage of cache hits versus cache size. (From Feldmeier, D.C., *Proceedings IEEE Infocom'88***, 1988. With permission.)**

Table 3.2 shows just how quickly the probability of a cache hit climbs even for small cache sizes. With as few as seven slots, the hit ratio is already above 0.9.

Another way of looking at this data is to plot the average number of packets going through the router between cache misses. The graph in Figure 3.7 shows the incremental efficiency of each additional slot in the cache. Curves are shown for destination LRU, destination or source LRU, destination FIFO, and destination source FIFO.

Table 3.2 Cache Hit Ratio Percentage versus Cache Size

Number of Cache Slots	Destination Only	Destination Source
1	42.31	29.94
2	66.54	71.28
3	76.03	80.81
4	82.52	86.32
5	86.61	89.61
6	89.53	91.81
7	91.74	93.48
8	93.44	94.76
9	94.75	95.77
10	95.77	96.56

Source: Feldmeier, D.C., *Proceedings IEEE Infocom'88,* 1988. With permission.

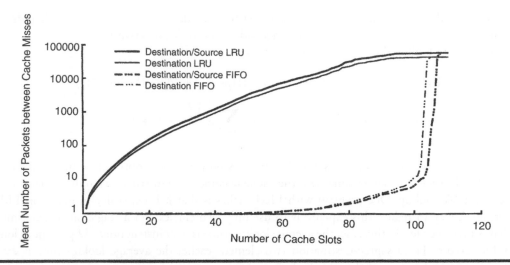

Figure 3.7 Number of packets between cache misses versus cache size (log-linear). (From Feldmeier, D.C., *Proceedings IEEE Infocom'88*, 1988. With permission.)

As expected, the destination or source LRU cache performed best, followed closely by the destination LRU cache. Performance of both FIFO caches is relatively poor until the destination FIFO cache size exceeds 102. It turns out that 52 is the number of distinct destination addresses handled by the router in a 24-hour period, in addition to 52 destinations in the preload, giving a total of 104. For destination source caching, the FIFO buffer size does not help until 104 slots, which is about the number of distinct destination and source addresses processed by the router in a 24-hour period, 54 plus 53 during preload, giving a total of 107.

3.2.1.4 Caching Cost Analysis

Although the hit ratios have been given for various caches, the important question is what is the performance improvement that is gained for the router. The following analysis is meant to be a simple worst-case analysis. The cache used for this analysis is a fully associative cache with an LRU replacement policy. The cache is implemented completely in software as a doubly-linked linear list. The list is searched linearly from the most recently used to the least recently used and when the list is rearranged to preserve the LRU ordering, this is done by swapping pointers of the doubly linked list. Only destination addresses are cached.

Estimates of the number of instructions are based on a load-store machine architecture. Loading the pointer to the cache requires one instruction. A matching entry is recognized in three instructions; in addition, seven instructions are needed for each previously checked cache slot. After each cache search, the cache must be reordered to preserve the LRU ordering (except if the entry was found in the first slot) and this takes 21 instructions (except for the last cache slot, which needs 11 instructions).

An estimate of the minimum-time for address decoding and routing table lookup for a packet is 80 instructions, a figure obtained by estimating the number of load-store instructions generated by the C code at MIT. The code would take even longer; this estimate is conservative because it does not include function call overhead, error handling, or the additional routing table lookup necessary for subnet routing and assumes that there are no hashing collisions. This estimate also does not include the time necessary to map a next hop IP address into a local network address, which is at least two instructions, but varies depending on the local network.

The average lookup time with the cache is now:

$$1 + 3p_1 + (7 + 3 + 21)p_2 + (14 + 3 + 21)p_3 + \cdots + (7[k-1] + 3 + 21)p_k$$

$$+ \cdots + (7[n-1] + 3 + 11)p_n + \left(1 - \sum_{i=1}^{n} p_i\right)(7n + 21 + 80)$$

If the cache stores flat addresses and destinations only, then the optimum number of cache slots is 16. The average lookup time with the cache becomes 37.9 instructions, only 47 percent as long as a table lookup. One problem with LRU caches is that it is expensive to maintain LRU ordering. A way to reduce this expense is to use only two entries and a single bit to determine which is first in order. In this case, the average lookup time is 33.9 instructions, 42 percent as long as table lookup. For a source-destination two-element cache, the average lookup time is 29.6 instructions or 37 percent of the table lookup.

With hierarchical addressing, the performance is even better. To obtain a network address from a flat address, the last byte is dropped from the address. Although this does not necessarily result in a network address, the number of distinct addresses that the cache must handle is reduced and the overhead is only a single AND instruction. The optimal cache size is 51 slots and the average lookup time is 31.5 instructions, 39 percent as long as table lookup. A two-element simple cache as described above averages 33.3 instructions, 42 percent of a table lookup. A source-destination cache requires 27.7 instructions, or 35 percent of a table lookup.

Although the figures above minimize average lookup time, it may be that a lower maximum lookup time is desired at the cost of a slightly higher average lookup time.

Obviously, better cache implementation would increase performance. Cache lookup could use a hash table rather than a linear search, the cache slots need not be fully associative, a replacement strategy approximating LRU may be implemented less expensively, the cache fetch strategy could use both destination and source addresses, or a hardware cache could be built into the system. But even with this simple cache implementation and optimistic assumptions about routing-table lookup time, the best cache above reduces lookup time 65 percent.

3.2.2 Characteristics of Destination Address Locality

It is known that there is locality behavior of destination addresses on computer networks [4]. An understanding of this behavior will help us design a caching scheme for efficiently looking through large tables with a high probability. Jain [5] explained various locality concepts and analyzed the applicability of different locality models and the performance of various cache replacement algorithms.

3.2.2.1 Locality: Concepts

The locality of a reference pattern may be temporal or spatial. Temporal locality implies a high probability of reuse. For example, the reference string {3, 3, 3, 3, 3, ...} has a high temporal locality, because the address 3 is used repeatedly once it is referenced. Spatial locality implies a high probability of reference to neighboring addresses. For example, the string {1, 2, 3, 4, 5, ...} has high spatial locality because after a reference to address k, the probability of reference to $k+1$ is very high. Spatial locality is useful in designing prefetching algorithms because the information likely to be used in the near future is fetched before its first reference, thereby avoiding a cache miss.

The terms persistence and concentration have also been used to characterize locality behavior [6]. Persistence refers to the tendency to repeat the use of a single address. This is, therefore, similar to temporal locality. Concentration, on the other hand, refers to the tendency of the references to be limited (concentrated) to a small subset of the whole address space. A reference string with high concentration is good in that a small cache would produce large performance gains. Persistence can be measured by counting consecutive references to the same address, although concentration can be measured by computing the fraction of address space used for a large fraction of the reference string. For example, in a reference string with high persistence, the probability of the same address being referenced consecutively may be 60 percent. Similarly, in a string with high concentration, 99 percent of the references may be to 1 percent of the address space.

Virtual memory is one of the first applications of locality concepts in computer systems design, and was first applied in 1970 [7]. From this time, some applications appeared, such as file caching and destination address caching. A very large cache can give better performance (low miss rate), but is too expensive. How can we optimize the cache size for a given acceptable miss rate?

3.2.2.2 Cache Replacement Algorithms

The traditional metric for performance of a cache is the number of misses. A miss is said to occur when an address is not found in the cache. On a cache miss, one of the entries in the cache must be replaced to bring in the missed entry. Several replacement algorithms can be found in the literature on processor design and virtual memory. We chose four popular algorithms for comparison: LRU, FIFO, random (RAND), and a theoretically optimal algorithm called MIN [8]. Given a reference trace and a fixed-size cache, it has been proven that the MIN algorithm would cause less faults than any other algorithm. MIN chooses the address that will be referenced farthest in future. It, therefore, requires looking ahead in the reference string. Obviously, it cannot be implemented in a real system. Nonetheless, it provides a measure of how far a particular algorithm is from the theoretical optimal.

We used the following three metrics to compare the replacement algorithms: miss probability, interfault distance, normalized search time.

To compare various replacement algorithms, Jain et al. used a trace of destination addresses observed on an extended local area network in use at Digital's King Street, Littleton facility [5]. The trace with 200,000 destination addresses was a result of approximately one hour of monitoring.

Miss Probability.   It is defined as the probability of not finding an address in the cache. For a given trace, it is simply the ratio of the number of faults to the total number of references in the trace. The lower the miss probability, the better the replacement algorithm.

The miss probability for various cache sizes for the four replacement algorithms are presented in Figure 3.8. From the figure we see that for small caches, LRU, FIFO, and RAND are not very different for this trace. The miss probability for MIN is better by approximately a factor of two. Thus, there is sufficient room for improvement by designing another replacement algorithm.

For large cache sizes, the miss probability curves are too close to make any inferences. The interfault distance curves provide better discrimination for such sizes.

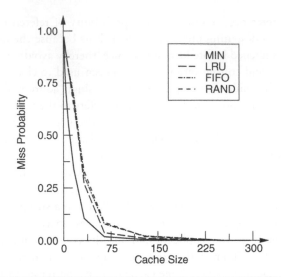

Figure 3.8   Cache miss probability for various cache replacement algorithms. (From Jain, R., *Computer Networks and ISDN Systems*, **1990. With permission.)**

Interfault Distance.   It is defined as the number of references between successive cache misses. For a given trace, the average interfault distance can be computed by dividing the total number of references by the number of faults. Thus, the average interfault distance is the reciprocal of the miss probability.

Average interfault distances for our trace using the four replacement algorithms are shown in Figure 3.9.

From Figure 3.9, we see that for large caches, LRU is close to optimal. FIFO and RAND are equally bad for this trace. Thus, unless one discovers a better replacement algorithm, we can use the LRU replacement algorithm with large caches.

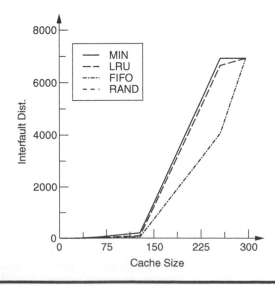

Figure 3.9   Interfault distance for various cache replacement algorithms. (From Jain, R., *Computer Networks and ISDN Systems*, 1990. With permission.)

This leads us to wonder what the optimal cache size is. If a cache is too small, we have a high miss rate. If the cache is too large, we do not gain much even if the miss rate is small because we have to search through a large table. The question of optimal cache size is answered by our third metric, normalized search time.

Normalized Search Time.   Caches are useful for several reasons. First, they may have a faster access time then the main database. This is particularly true if the main database is remotely located and the cache is local. Second, they may have a faster access method. For example, caches may be implemented using associated memories. Third, the references have a locality property so that entries in the cache are more likely to be referenced than other entries.

We need to separate the effect of locality and find out if there is sufficient locality in the address reference patterns to warrant the use of caches. If there were enough locality, one would want to use a cache even if the access time to the cache was the same as that of the main database and the cache used the same access method (say binary search) that would be used for the main database.

Assuming that the access time and the access method for the cache are the same, we can compute the average access time with and without cache and use the ratio of the two as the metric of contribution to performance due to locality alone.

Assuming that a full database of n entries would generally require a search time proportional to $1 + \log_2(n)$, we have:

$$\text{Time to search without cache} = 1 + \log_2(n)$$

With a cache, if p is the miss probability, we need to search through both the cache and the full table with probability p, and the normalized search time is defined as the ratio:

$$\text{Normalized Search Time} = \frac{\text{Search time with cache}}{\text{Search time without cache}}$$

$$= \frac{(1-p)[1+\log_2(c)] + p[1+\log_2(c)+1+\log_2(n)]}{1+\log_2(n)}$$

The normalized search time for the four replacement algorithms considered is shown in Figure 3.10. From the figure, we see that with a cache using the MIN replacement algorithm, we could achieve up to 33 percent less search time than that without caching. The payoff with other replacement algorithms is much less. It is more important to observe, however, that with LRU, FIFO, and RAND, the total search time may be more with a small cache size of eight: these three algorithms would require 20 percent more search time than without a cache. This trace, therefore, shows a reference pattern in which caching can be harmful.

With a very large cache, the cache does reduce the search time, but the gain decreases as the cache size increases. The optimal cache size for this trace is approximately 64, which produces 20–25 percent reduction in search time.

3.2.2.3 *Stack Reference Frequency*

Jain et al. showed the cumulative probability distribution function using a stack model [5]. If the addresses are arranged in a stack so that the address referenced is always taken out of its current position in the stack and pushed to the top of the stack, the probability p_i of the ith stack position (counting from the top toward the bottom of the stack) being referenced is a decreasing function of i. If, instead of adding the probability for successive stack positions, we plot the

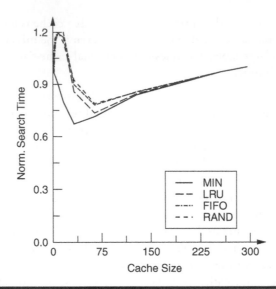

Figure 3.10 Normalized search time for various caches. (From Jain, R., *Computer Networks and ISDN Systems*, 1990. With permission.)

probability for individual stack positions, we get the probability density function (pdf) curve as shown in Figure 3.11. In this figure, we can see that the pdf is not a continuously decreasing function. Instead, there is a hump around stack position 30. For this environment, the most likely stack position to be referenced is the 30th position and not the stack top.

LRU is not the best replacement strategy for such a reference string. In general, it is better to replace the address least likely to be referenced again, that is, the address with minimum probability. For the stack reference probabilities shown in Figure 3.11, the minimum probability does not

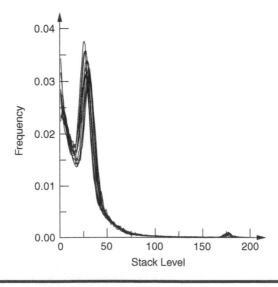

Figure 3.11 Stack reference frequency. (From Jain, R., *Computer Networks and ISDN Systems*, 1990. With permission.)

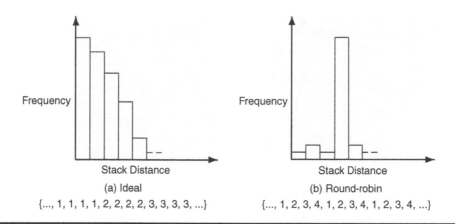

Figure 3.12 shows two frequency versus stack distance plots:

(a) Ideal — {..., 1, 1, 1, 1, 2, 2, 2, 2, 3, 3, 3, 3, ...}

(b) Round-robin — {..., 1, 2, 3, 4, 1, 2, 3, 4, 1, 2, 3, 4, ...}

Figure 3.12 A round-robin reference pattern results in a hump in the stack reference frequency. (From Jain, R., *Computer Networks and ISDN Systems,* 1990. With permission.)

always occur at the highest possible stack distance. For example, if the cache size is 30, the address at stack position 15 has a lower probability of reference than at position 30 and is, therefore, a better candidate for replacement.

One possible cause of the hump could be a round-robin behavior in our reference pattern. To understand this we consider the two hypothetical reference patterns shown in Figure 3.12. The first pattern shows high persistence. Once an address is referenced, it is referenced again several times. Such a reference string would result in a continuously decreasing stack pdf of the type shown in Figure 3.12a. The second pattern shows a round-robin reference string consisting of k addresses, for instance, repeated over and over again {1, 2, 3, ..., k, 1, 2, 3, ..., k, 1, ...}. The stack pdf for this string would be an impulse (or Dirac delta) function at k, that is, all references would be to stack position k.

A mixture of round-robin behavior and persistent traffic would result in a curve with a hump similar to the one observed in Figure 3.11. This round-robin behavior could be caused by the periodic nature of some of the protocols such as local area terminals (LAT) [9]. Each LAT server is connected to a number of terminals and provides a virtual connection to several hosts on the extended LAN. To avoid sending several small frames, the terminal input is accumulated for 80 milliseconds and all traffic going to one host is sent as a single frame. This considerably reduces the number of frames and improves the performance of the terminal communication. A large number of LAT servers transmitting at regular intervals of 80 milliseconds could very well be responsible for the round-robin behavior observed in the reference pattern.

To verify the above hypothesis, we divided our trace into two subtraces: one consisting entirely of interactive (LAT) frames, and the other consisting of the remaining noninteractive traffic. The stack pdf for these two subtraces are shown in Figures 3.13 and 3.14. Notice that the interactive traffic exhibits a hump, although the noninteractive traffic does not. Thus, the interactive traffic does seem to be responsible for the hump, leading to the conclusion that for environments dominated by LAT and similar protocols one would need either a cache size equal to the number of LAT servers or to develop a cache prefetch policy that would bring the right address into the cache just before it is referenced.

The observation that the noninteractive traffic has a continuously decreasing stack pdf is an interesting one. Because the LAT traffic is limited to a single extended LAN, it does not go through routers, which are used to connect several extended LANs to wide area networks. The reference

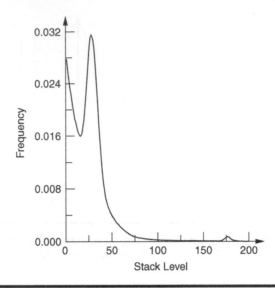

Figure 3.13   Stack distance density function for LAT traffic. (From Jain, R., *Computer Networks and ISDN Systems,* **1990. With permission.)**

pattern seen at routers is expected to be similar to that of the noninteractive traffic, though we have not yet verified this observation. If this is so, it would be interesting to see if caching would pay off for noninteracrive traffic alone. We, therefore, analyze noninteractive traffic in the next section.

3.2.2.4   Analysis of Noninteractive Traffic

We will present the graphs for miss probability, interfault distance, and normalized search time for noninteractive traffic alone. There are two reasons: noninteractive traffic may give us some

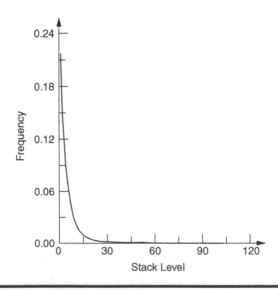

Figure 3.14   Stack distance density function for noninteractive traffic. (From Jain, R., *Computer Networks and ISDN Systems,* **1990. With permission.)**

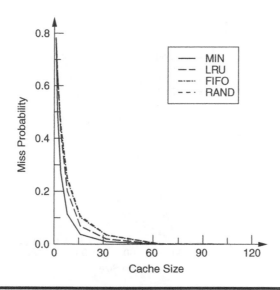

Figure 3.15 Cache miss probability for noninteractive frames. (From Jain, R., *Computer Networks and ISDN Systems,* **1990. With permission.)**

indication of the behavior of references in routers and helps us illustrate how some of the conclusions reached earlier would be different in a different environment.

Figure 3.15 shows the miss probability for the four replacement algorithms. Notice that even for small caches, LRU is significantly better than FIFO and RAND. This is not surprising considering the fact that for any reference trace with nondecreasing stack pdf, LRU is the optimal cache replacement algorithm [10]. LRU is optimal in the sense that no other practical algorithm can give a lower number of faults for any given cache size. MIN does give a lower number of faults and, hence, a lower miss probability, but that is due to its knowledge of future references. For reference patterns similar to noninteractive traffic, therefore, we do not need to look for other replacement algorithms. Of course, if LRU is too complex to implement, which is often the case, one would go for simpler algorithms, but that would always come at the cost of increased faults.

Figure 3.16 shows the interfault distances for the four replacement algorithms. We see that for a large cache size also, LRU is far superior to FIFO and RAND for this subtrace.

The normalized search time for noninteractive traffic is shown in Figure 3.17. Notice that for small caches, we now have a valley where we had a peak in Figure 3.10. Thus, not only are the small caches helpful they are also optimal. The optimal cache size with LRU is about eight entries. This reduces the search time by about 40 percent.

3.2.2.5 Cache Design Issues

There are many cache design issues that remain to be addressed before caching of network addresses can become a reality. The issues can be classified as cache management, cache structuring, and multicache issues.

Cache management issues relate to algorithms for replacement, fetching, lookup, and deletion. Several replacement algorithms have been compared above. We assumed demand fetching where the address is brought into the cache when it is actually referenced. Prefetching, such as the most

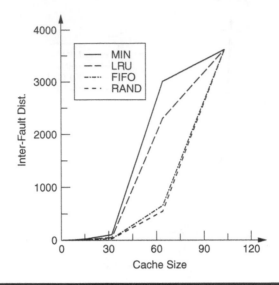

Figure 3.16 Interfault distances for noninteractive frames. (From Jain, R., *Computer Networks and ISDN Systems*, 1990. With permission.)

significant octet first or the least significant octet first may produce different performances. Finally, the issue of deleting addresses periodically needs to be studied.

Processor caches are generally structured as sets [11]. Each set consists of several entries. A given address is first mapped to a set and the replacement, lookup, etc. is then confined to the set. Two extreme cache structures are direct mapped, in which each set consists of only one entry, and fully associative, in which all entries are part of the same set and there is no mapping.

Another issue related to cache structuring is that of organizing separate caches for different types of addresses. For example, in many computer systems, instruction and data caches are

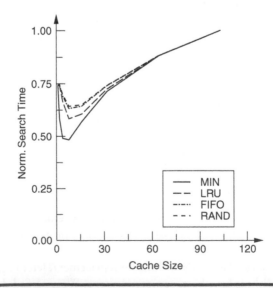

Figure 3.17 Normalized search time for noninteractive frames. (From Jain, R., *Computer Networks and ISDN Systems*, 1990. With permission.)

organized separately because their reference patterns are so different [12]. In computer networks, one may want to study the effect of organizing separate caches for interactive and noninteractive traffic, or a separate cache for each protocol type.

Multicache consistency [13] is also an interesting issue, particularly in multiport intermediate systems in which each port has a separate cache of addresses.

3.2.3 Discussions

An early solution proposed by Cisco [14] in 1993 consists in putting a three-level system of hierarchical caching. Later, the router vendors did not use cache for packet forwarding.

Partridge [15] reports a hit rate of 97 percent with a cache of size 10,000 entries, and 77 percent with a cache of size 2000 entries. Partridge [15] suggests that the cache size should scale linearly with the increase in the number of hosts or the amount of Internet traffic. This implies the need for exponentially growing cache sizes. Cache hit rates are expected to decrease with the growth of Internet traffic because of decreasing temporal locality [16]. The temporal locality of traffic is decreasing because of an increasing number of concurrent flows at high-speed aggregation points and decreasing duration of a flow, probably because of an increasing number of short web transfers on the Internet.

A cache management scheme must decide which cache entry to replace upon addition of a new entry. For a route cache, there is an additional overhead of flushing the cache on route updates. Hence, low hit rates, together with cache search and management overheads, may even degrade the overall lookup performance. Furthermore, variability in lookup times of different packets in a caching scheme is undesirable for the purpose of hardware implementation. Because of these reasons, caching has generally fallen out of favor with router vendors in the industry (see Cisco [17], Juniper [18], and Lucent [19]) who tout fast hardware lookup engines that do not use caching. But the research of the caching technique for IP lookup never stops. We will introduce the last results in Chapter 5.

3.3 Binary Trie

A binary trie, or simply a trie, is the simplest species in this tree-like family. The name is apparently based on "retrieval" [20,21]. A binary trie has labeled branches, and is traversed during a search operation using individual bits of the search key. The left branch of a node is labeled "0" and the right branch is labeled "1." A node in the binary trie, v, represents a bit-string formed by concatenating the labels of all branches in the path from the root node to v. This section is partly from [22] (© 1988 IEEE).

A natural way to represent prefixes is using a binary trie. A prefix, P, is stored in the node that represents the bit-string P. Figure 3.18 shows a binary trie representing a set of prefixes of a forwarding table. A trie for W-bit prefixes has a maximum depth of W nodes.

In a trie, a node on level 1 represents the set of all addresses that begin with the sequence of l bits consisting of the string of bits labeling the path from the root to that node. For example, node c in Figure 3.18 is at level 3 and represents all addresses beginning with the sequence 011. The nodes that correspond to prefixes are shown in a darker shade; these nodes will contain the forwarding information or a pointer to it. Note also that prefixes are not only located at leaves but also at some internal nodes. This situation arises because of exceptions in the aggregation process. For example, in Figure 3.18 the prefixes b and c represent exceptions to prefix a. The address spaces covered by prefixes b and c overlap with the address space covered by prefix a. Thus, prefixes b and c represent exceptions to prefix a, and refer to specific subintervals of the address interval covered

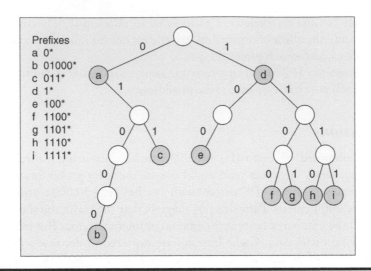

Figure 3.18   A binary trie for a set of prefixes. (From Ruiz-Sanchez, M.A., Biersack, E.W., and Dabbous, W., *IEEE Network,* 2001. With permission.)

by prefix *a*. In the trie in Figure 3.18, this is reflected by the fact that prefixes *b* and *c* are descendants of prefix *a*; in other words, prefix *a* is itself a prefix of *b* and *c*. As a result, some addresses will match several prefixes. For example, addresses beginning with 011 will match both prefixes *c* and *a*. Nevertheless, prefix *c* must be preferred because it is more specific (longest match rule).

Tries allow us to find, in a straightforward way, the longest prefix that matches a given destination IP address. The search in a trie is guided by the bits of the destination address. At each node, the search proceeds to the left or right according to the sequential inspection of the address bits. Although traversing the trie, every time we visit a node marked as a prefix (i.e., a dark node) we remember this prefix as the longest match found so far. The search ends when there are no more branches to take, and the longest or best matching prefix will be the last prefix remembered. For instance, if we search the best matching prefix (BMP) for an address beginning with the bit pattern 10110 we start at the root in Figure 3.18. Because the first bit of the address is 1 we move to the right, to the node marked as prefix *d*, and we remember *d* as the BMP found so far. Then we move to the left because the second address bit is 0; this time the node is not marked as a prefix, so *d* is still the BMP found so far. Next, the third address bit is 1, but at this point there is no branch labeled 1, so the search ends and the last remembered BMP (prefix *d*) is the longest matching prefix.

In fact, what we are doing is a sequential prefix search by length, trying at each step to find a better match. We begin by looking in the set of length-1 prefixes, which are located at the first level in the trie, then in the set of length-2, located at the second level, and so on. Moreover, using a trie has the advantage that although stepping through the trie, the search space is reduced hierarchically. At each step, the set of potential prefixes is reduced, and the search ends when this set is reduced to 1.

Therefore, a trie for *W*-bit prefix needs $O(NW)$ memory (where *N* is the number of prefixes), finding the longest matching prefix takes *W* memory accesses in the worst-case, that is, it has time complexity $O(W)$ [22].

Update operations are also straightforward to implement in binary tries. Inserting a prefix begins with a search. When arriving at a node with no branch to take, we can insert the necessary nodes. Deleting a prefix starts again with a search, unmarking the node as a prefix and, if necessary, deleting unused nodes (i.e., leaf nodes not marked as prefixes). Note finally that because the bit strings of prefixes are represented by the structure of the trie, the nodes marked as prefixes do not need to store the bit strings themselves. Hence, insertion of a new prefix can lead to the addition of at most W other trie nodes; deletion of a prefix can lead to the removal of at most W other trie nodes.

3.4 Path-Compressed Trie

While binary tries allow the representation of arbitrary-length prefixes, they have the characteristic that long sequences of one-child nodes may exist (see prefix b in Fig. 3.18). Because these bits need to be inspected, even though no actual branching decision is made, search time can be longer than necessary in some cases. Also, one-child nodes consume additional memory. In an attempt to improve time and space performance, a technique called path compression can be used. Path compression consists of collapsing one-way branch nodes. When one-way branch nodes are removed from a trie, additional information must be kept in the remaining nodes so that a search operation can be performed correctly. This section is partly from [22] (© 2001 IEEE).

There are many ways to exploit the path compression technique; perhaps the simplest to explain is illustrated in Figure 3.19, corresponding to the binary trie in Figure 3.18. Note that the two nodes preceding b have now been removed. Note also that because prefix a was located at a one-child node, it has been moved to the nearest descendant that is not a one-child node. In a path to be compressed, because several one-child nodes may contain prefixes; in general a list of prefixes must be maintained in some of the nodes, and as one-way branch nodes are now removed, we can jump directly to the bit where a significant decision has to be made, bypassing the bit inspection of some bits. As a result, a bit number field must be kept now to indicate which bit is the next bit to inspect. In Figure 3.19, these bit numbers are shown next to the nodes. Moreover, the bit strings of prefixes must be explicitly stored. A search in this kind of path-compressed trie is as follows. The algorithm performs, as usual, a descent in the trie under the guidance of the address bits, but this time only inspecting bit positions indicated by the bit-number field in the nodes traversed. When a node marked as a prefix is encountered, a comparison with the actual prefix value is performed. This is necessary because

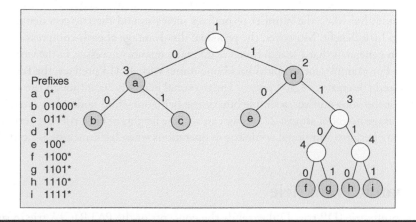

Figure 3.19   A path-compressed trie. (From Ruiz-Sanchez, M.A., Biersack, E.W., and Dabbous, W., *IEEE Network*, 2001. With permission.)

during the descent in the trie we may skip some bits. If a match is found, we proceed traversing the trie and keep the prefix as the BMP so far. The search ends when a leaf is encountered or a mismatch found. As usual, the BMP will be the last matching prefix encountered. For instance, if we look for the BMP of an address beginning with the bit pattern 010110 in the path-compressed trie shown in Figure 3.19, we proceed as follows. We start at the root node and, because its bit number is 1, we inspect the first bit of the address. The first bit is 0, so we go to the left. Because the node is marked as a prefix, we compare prefix *a* with the corresponding part of the address (0). Because they match, we proceed and keep *a* as the BMP so far. Because the node's bit number is 3, we skip the second bit of the address and inspect the third one. This bit is 0, so we go to the left. Again, we check whether the prefix *b* matches the corresponding part of the address (01011). Because they do not match, the search stops, and the last remembered BMP (prefix *a*) is the correct BMP.

Path compression reduces the height of a sparse binary trie, but when the prefix distribution in a trie gets denser, height reduction is less effective. Hence, the complexity of search and update operations in path-compressed tries is the same as in classical binary tries. Path-compressed tries are full binary tries. Full binary tries with N leaves have $N - 1$ internal nodes. Hence, space complexity for path compressed tries is $O(N)$ [22].

Path compression was first proposed in a scheme called practical algorithm to retrieve information coded in alphanumeric (PATRICIA) [23], but this scheme does not support longest-prefix matching. Sklower proposed a scheme with modifications for longest-prefix matching in [24]. In fact, this variant was originally designed to support not only prefixes but also more general non-contiguous masks. Because this feature was really never used, current implementations differ somewhat from Sklower's original scheme. For example, the BSD version of the path-compressed trie (referred to as a BSD trie) is essentially the same as that just described. The basic difference is that in the BSD scheme, the trie is first traversed without checking the prefixes at internal nodes. Once at a leaf, the traversed path is backtracked in search of the longest matching prefix. At each node with a prefix or list of prefixes, a comparison is performed to check for a match. The search ends when a match is found. Comparison operations are not made on the downward path in the hope that not many exception prefixes exist. Note that with this scheme, in the worst-case the path is completely traversed twice. In the case of Sklower's original scheme, the backtracking phase also needs to do recursive descents of the trie because noncontiguous masks are allowed.

The longest matching prefix problem was addressed by using data structures based on path-compressed tries such as the BSD trie. Path compression makes a lot of sense when the binary trie is sparsely populated; but when the number of prefixes increases and the trie gets denser, using path compression has little benefit. Moreover, the principal disadvantage of path-compressed tries, as well as binary tries in general, is that a search needs to do many memory accesses, in the worst-case 32 for IPv4 addresses. For example, for a typical backbone router with 47,113 prefixes, the BSD version for a path-compressed trie creates 93,304 nodes. The maximal height is 26, although the average height is almost 20. For the same prefixes, a simple binary trie (with one-child nodes) has a maximal height of 30 and an average height of almost 22. As we can see, the heights of both tries are very similar, and the BSD trie may perform additional comparison operations when backtracking is needed.

3.5 Dynamic Prefix Trie

The dynamic prefix trie (DP-trie) [25] extends the concepts of the PATRICIA trie to support the storage of prefixes and to guarantee retrieval times at most linear in the length of the input key irrespective of the trie size, even when searching for longest-matching prefixes. DP-trie permits insertion and deletion operation. This section is from [25]. Portions are reprinted with permission (© 1996 IEEE).

3.5.1 Definition and Data Structure

The algorithms presented here operate on (binary) keys, defined as nonempty bit sequences of arbitrary but finite size. Keys are represented as $k = b_0 b_1 \ldots b_{l-1}$, with elements $k[i] = b_i \in \{0,1\}$, $0 \le i \le |k|$, with $|k| = l - 1$ representing the width of k and $\|k\| = l$ its length. Following standard terminology, a key k' is said to be a (strict) prefix of key k, denoted by $k' < k$, iff $|k'| \le |k|$ holds, and $k'[i] = k[i]$, $0 \le i \le |k'|$. In particular, two keys are identical if they are prefixes of each other. Keys are stored in nodes. Each comprises an index, at most two keys, and up to three links to other nodes. That is, a given node n has the following components (see Fig. 3.20a):

Index(n): The relevant bit position for trie construction and retrieval—the prefix to this bit position is common among all the keys in a subtrie that has node n as its root—the index allows the nonrelevant bits to be skipped, thus obviating one-way branches.

Leftkey(n): A key with $LeftKey[Index(n)] = 0$ or NIL.

Rightkey(n): A key with $RightKey[Index(n)] = 1$ or NIL.

Parent(n): A link to the parent node of node n, NIL for the root node.

LeftSubTrie(n): A link to the root node of that subtrie of node n with keys k such that $k[Index(n)] = 0$ or NIL.

RightSubTrie(n): A link to the root node of that subtrie of node n with keys k such that $k[Index(n)] = 1$ or NIL.

To illustrate the data structures and particular relation between nodes, we will now construct a DP-Trie for a sample set {1000100, 1001, 10, 11111, 11} using graphical representations for the nodes depicted in Figure 3.20a. When the first key, 1000100, is inserted into the trie, we obtain Figure 3.20b. The trie consists of only a root node, a. As key 1000100 is the only key of node a, the index takes its maximum value, that is, the width of the key. The key has 0 at bit position and hence becomes the left key of the root. Figure 3.20c depicts the DP-trie after key 1001 is inserted. The index of node a becomes three, the first bit position at which the two keys differ, and the

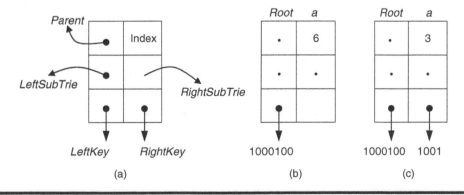

(a)   (b)   (c)

Figure 3.20 Node structure and insertions: (a) node structure; (b) insertion of 1000100; (c) insertion of 1001. (From Doeringer, W., Karjoth, G., and Nassehi, M., *IEEE/ACM Transactions on Networking*, 1996. With permission.)

common prefix 100 is thus ignored. As the new key has 1 in position three, it becomes the right key of node *a*. (If the new key had 0 at this bit position and the old key had 1, the old key would have been moved to become the right key and the new key would have become the left key.) In this trie, a search is guided to either stored key by the bit value at position three [*Index*(*a*)] of an input key; the bit positions 0, 1, and 2 are skipped. For example, a search with an input 1001001111 returns 1001 as the longest matching prefix.

Adding key 10 results in Figure 3.21a. As key 10 is a prefix to both keys of node *a*, a new node, *b* is created and made the parent of node *a*. The index of node *b* is one, the width of key 10. Key 10 and node *a* become, respectively, the left key and the left subtrie of node *b*, because all the stored keys have zero at position 1. Adding key 11111, we obtain Figure 3.21b. This key differs from key 10 in the bit position 1. So without a change in the index value, node *b* can accommodate the new key as its right key. Figure 3.21c shows the DP-trie after key 11 is added. Key 11 is a prefix of key 11111. Therefore, key 11111 is replaced by key 11 and pushed lower to the new node *c*, that becomes the right subtrie of node *b*.

Now, let us consider some more search operations. A search for *k* = 10011 at node *b* is first guided to node *a* because *k*[1] = 0, and then to 1001; bit positions 0 and 2 are skipped. The latter key is identified as the longest matching prefix after a total of two bits and one key comparison, which are very basic and easily implemented operations in suitable hardware. In contrast, a search for 100011 proceeds to node *a* only to find that 1000100 is not a prefix of the input key. Using the pointer reversal provided by the parent links, key 10 is identified as the longest matching prefix. Similarly, a search for key 001 branches to node *a* and terminates unsuccessfully at node *b*.

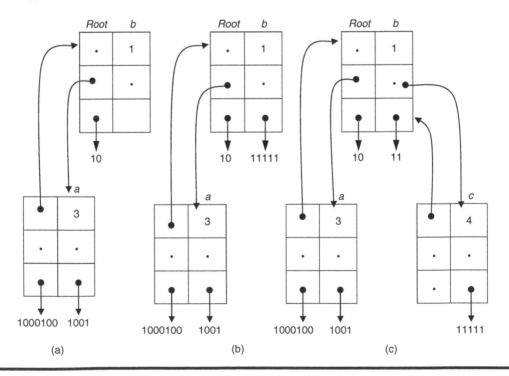

Figure 3.21 More insertions: (a) insertion of key 10; (b) insertion of key 11111; (c) insertion of key 11. (From Doeringer, W., Karjoth, G., and Nassehi, M., *IEEE/ACM Transactions on Networking,* 1996. With permission.)

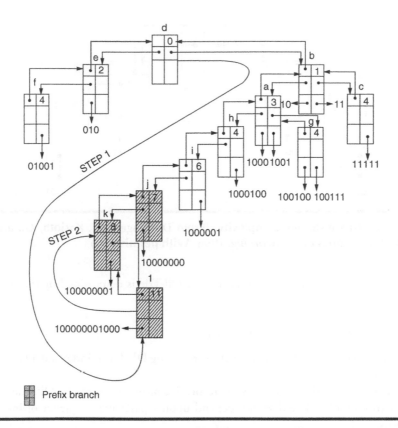

Figure 3.22   Example of a DP-Trie. (From Doeringer, W., Karjoth, G., and Nassehi, M., *IEEE/ACM Transactions on Networking,* 1996. With permission.)

Further insertion of the keys depicted in Figure 3.22 produces a sample DP-Trie that covers all structure aspects of a general DP-Trie and will be used henceforth as a reference. This figure can also be used to examine search operations involving more than a few nodes. For example, a search for key 1000011 is guided through nodes *d, b, a, h,* and *i*. Skipping over bit position 2 and 5, through pointer reversal 1000 is identified as the maximum length matching prefix.

3.5.2   Properties of DP-Tries

Let us call a Prefix Branch any sequence of nodes $n_1, n_2, ..., n_k$, such that n_{i+1} is the root of the only subtrie of n_i for $1 \leq i \leq k - 1$, and all nodes store exactly one key as strict prefixes of each other. Then we can derive the following result that shows that the algorithms presented preserve the storage efficiency of PATRICIA tries in a naturally generalized way.

Lemma 3.5.1:   Storage Complexity of DP-Tries: Let a^+ denote $Max(0, a)$ for a number a. The algorithms of a DP-Trie guarantee a minimal storage complexity described by the following relation, which is an optimal estimate for a general DP-Trie: $\neq nodes \leq \neq keys + (\neq PrefixBranches - 1)^+$.

The example in Figure 3.23 shows that the upper bound is optimal for a DP-Trie, and the established results for PATRICIA tries follow from observing that in those tries, no prefix branches may exit.

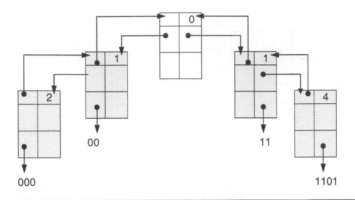

Figure 3.23 Worst-case storage complexity. (From Doeringer, W., Karjoth, G., and Nassehi, M., *IEEE/ACM Transactions on Networking*, 1996. With permission.)

The next result confirms that the structure of a DP-Trie is determined by the set of inserted keys only.

Lemma 3.5.2: Invariance Properties of DP-Tries

1) For any given set of keys, the structure of the resulting DP-Trie is independent of the sequence of their insertion.
2) The delete operation always restores the respective prior structure of a DP-Trie.
3) The structure of a DP-Trie does not depend upon a particular sequence of insert or delete operations.

Lemma 3.5.3: General Properties of DP-Tries

N1 For a given node n, let $Keys(n)$ denote the set of keys stored in n and its subtries. All such keys have a width of at least $Index(n)$ and hence a length of at last $Index(n) - 1$.

N2 $Keys(n)$ denote the subset of keys that share the respective common prefix of $Index(n)$ bits in length, denoted by $Prefix(n)$. That is, all keys in $Keys(n)$ match in bit positions up to and including $Index(n) - 1$. If $Index(n) + 1$ is distinguishing, $Prefix(n)$ is of maximum length.

N3 The keys in any subtrie of node n share a common prefix of at least $Index(n) + 1$ bits and are at least of such width.

N4 The index of a node n is maximal in the following sense: In a leaf node having only one key k, we have $Index(n) = |k|$, and in all other nodes there exist keys in $Keys(n)$ that differ at position $Index(n)$, or one of the keys stored in n has a width if $Index(n)$.

N5 Every node contains at least one key or two nonempty subtries. That is, a DP-Trie contains no chain nodes.

N6 If the left (right) key as well as the left (right) subtrie of node n is nonempty, then $LeftKey(n) = Index(n)$ ($RightKey(n) = Index(n)$).

N7 If the left (right) key of a node n is empty and the left (right) subtrie is not, then the subtrie contains at least two keys.

N8 For a given node n, the left (right) key and all keys in the left (right) subtrie of b have a 0 (1) bit at position $Index(n)$.

P1 $Prefix(n)$ increases strictly as n moves down a DP-Trie from the root to a leaf node, as does $Index(n)$. Hence, there are no loops and there always exists a single root node.

P2 Every full path terminates at a leaf node that contains at least one key, and there exists only one path from the root to each leaf node.

The proof of Lemma 3.5.1, 3.5.2, and 3.5.3 is seen in [25].

3.5.3 Algorithms for DP-Tries

This section provides a description of the algorithms for DP-Tries. They are represented in a notation that avoids distracting details, but is kept at a level that allows straightforward translations into real programming languages, such as C. The correctness of the algorithms follows directly from the properties of DP-Tries stated in Lemma 3.5.3 and the observation that these properties are invariant under the insertion and deletion operations. We preface the description with the definition of a small set of commonly used operators, predicates, and functions. For convenience of representation, we shall generally denote a trie or subtrie by a pointer to its root node.

The first bit position where two keys k and k' differ is called their distinguishing position, and is defined as $DistPos(k, k') = Min\{i \mid k'[i] \neq k[i]\}$, with the convention that $DistPos(k, k') = |k'| + 1$ for $k' < k$. Furthermore, a node is said to be a (single key) leaf node if it has no subtries (and only stress one key). For example, in Figure 3.21c, node c is a single-key leaf-node, whereas node a is only a leaf-node, and $DistPos(101, 111000) = 1$.

The two operations that select the pertinent part of a node relative to an input key are defined below based on the convention that nonexisting subtries or keys are represented by the special value NIL:

```
Key (node, key) {
    If (|key| < Index(node)) then return (NIL)
    If (key[Index(node)]) = 0) then return (LeftKey(node))
     else return (RightKey(node))
}

SubTrie(node, key) {
    If (|key| < Index(node)) then return (NIL)
    If (key[Index(node)] = 0) then return (LeftSubTrie(node))
     else return (RightSubTrie(node))
}
```

As an example, consider the nodes in Figure 3.21c. For $k = 0001$ we get as $SubTrie(b, k)$ the trie starting at node a, and $Key(b, k) = 10$.

3.5.3.1 Insertion

The insertion of a new key proceeds in three steps. First, the length of the longest prefix of the new key common to any of the keys stored in leaf nodes is defined (STEP1). On the basis of this information, the node around which the new key is to be inserted is located (STEP2). Finally, the key is added to the trie in the appropriate place (STEP3).

To accomplish the first step, we descend to a leaf node guided by the input key and make default choices where required. That is, as long as the width of the new key is greater than or equal to the respective nodal index, we follow the pertinent subtrie, if it exists. Otherwise we proceed along any subtrie, possibly with some implementation-specific (probabilistic) bias, until we reach a leaf node. The keys in that node may then be used to calculate the longest common prefix (see Fig. 3.22 and Lemma 3.5.4). The node for insertion is identified by backtracking downward along the path followed in the first step until a node is reached whose index is less than or equal to the length of the common prefix and to the width of the input key, or the root node is reached (see Fig. 3.22). Depending on the prefix relations between the input key and those in the insertion node, the new key may then be inserted above (Fig. 3.21c), in Figure 3.20b, or below the selected node (Fig. 3.21b).

We now proceed to the description of the insertion algorithm, making use of the following auxiliary operations that attempt to select subtries and keys of a node as closely related to an input key as possible. The procedures will be used to formalize the best effort descent of the first step.

```
ClosestKey(node, key) {
    if (|key ≥ Index(node)|) then /* A correct selection may be
possible. */
            if (Key(node, key) ≠ NIL) then
                    return (Key(node, key))
            else
                    return (Key(node, BitComplement(key))) /*
Return the 'other' key.*/
    else
        return (any key of node or NIL if none exists)
}
```

```
ClosestSubTrie(node, key) {
    if (|key ≥ Index(node)|) then /* A correct selection may be
possible. */
            if (Subtrie(node, key) ≠ NIL) then
                    return (SubTrie(node, key))
            else
                    return (SubTrie(node, BitComplement(key))) /*
Return the 'other' key.*/
    else
        return (any key of node or NIL if none exists)
}
```

The operation that allocates and initializes the new nodes of a DP-Trie is formalized as

```
Allocate Node (index, key) {
    local node
    NEWNODE(node)/* allocate space for a new node*/
    LeftKey(node) := RightKey(node) := NIL
    Parent(node) := LeftSubTrie(node) := RightSubTrie(node) := NIL
    Index(node) := index;
    Key(node, key) := key
    return(node)
}
```

Given the above definitions, the algorithm for a key insertion is defined in the following procedure.

```
Insert(key) {
   local node, distpos, index

   /*Empty trie: just add a node. Root is the global pointer to
the root node. */
   if ( Root = NIL ) then Root := AllocateNode(|key|, key)

   /* NonEmpty trie: insertion proceeds in three steps.*/
   else
     /*STEP1: Descend to a leaf node, identify the longest
common prefix, and the index of the node to store the new
key. */
   node := Root /* start at the root */
   while (Not LeafNode(node))
    do node := ClosestSubTrie(node, key)
   distpos := DistPos(key, ClosetKey(node, key))
   index := Min(|key|, distpos) /* of the node to store the
new key */
   /* STEP2: Ascend toward the root node, identify the insertion
node.*/
   while ( index < Index(node) and node ≠ Root)
    do node := Parent(node)

   /* STEP3: Branch to the appropriate insert operation. */
   if (node = Root) then
       InsertInOrAbove(node, key, distpos)
   elseif(SubTrie(node, key) = NIL) then
       InsertWithEmptySubTrie(node, key, distpos)
   else
       InsertWithNonEmptySubTrie(node, key, distpos)
}
```

The relevant properties of the local variable distpos and of the insertion node identified in STEP2 are summarized in the following Lemma.

Lemma 3.5.4: Properties of the Insertion Node: Let n denote the insertion node of STEP2 of the insertion algorithm, then:

a) The new key does not belong to any subtrie of n.
b) If Min(distpos, $|key| \geq Index(n)$), then the key belongs to the trie starting at n.
c) If Min(distpos, $|key| \leq Index(n)$), then n is the root and the new key needs to be inserted in or above n.
d) Distpos is the length of the longest prefix of the insert key common to any key stored in a leaf node of the trie.

InsertInOrAbove. This procedure deals with insertions above the root node. If the new key is not a prefix of a single key in the root node, it is simply added and the node index is adjusted accordingly. Otherwise, the key is stored in a new node that becomes the root, and the current trie is added as its only subtrie.

```
InsertInOrAbove(node, key, distpos) {
    local newnode, index := Min(|key|, distpos)

 /* InorAbove-1: add the new key to the root node. */
 if(|key| ≥ distpos and SinngleKeyLeafNode(node)) then
    Index(node) := index
    Key(node, BitComplement(key)) := ClosestKey(node, key)
    Key(node, key) := key

 /* InorAbove-2: Add the new key in a new node above the root
and the current trie as its subtrie at the appropriate side of
the new node. */
 else
 newnode := AllocateNode(index, key)
 Parent(node) := newnode
 if(|key| > distpos) then /* InorAbove-2.1: No prefix */
    SubTrie(newnode, BitComplement(key)) := node
 else /* InorAbove-2.2: The new key is a prefix of the stored
keys!*/
    SubTrie(newnode, key) := node
 Root := newnode /* A new Root has been created. */
}
```

InsertWithNonEmptySubTrie.   This procedure covers the cases of nonempty subtries at the insertion node. The new key is added if it fits exactly [see Lemma 3.5.3 (N6)], otherwise a new node with this key is inserted below the located node and above its subtrie. To improve the readability of the main algorithm, we introduce an additional subfunction that links two nodes as required for a given input key.

```
LinkNodes(node, subnode, key) {
  SubTrie(node, key) := subnode
  Parent(subnode) := node
}
```

```
InsertWithNonEmptySubTrie(node, key, dispos) {
  local newnode, subnode, index := Min(|key|, distpos)
  /* NonEmpty-1: The new key fits exactly. */
if (|key| = Index(node)) then Key(node, key) := key
/* NonEmpty-2: A new node is inserted below node and above the
subtrie */
else
  subnode := SubTrie(node, key) /*Save pointer. */
  newnode := AllocateNode(Index, key)
  Parent(SubTrie(node, key)) := newnode
  LinkNodes(node, newnode, key)
```

```
    /* Add the old subtrie at the appropriate side */
    if(|key| ≥ distpos) then
    /* Nonempty-2.1: no prefix */
    if(SingleKeyLeafNode(subnode)) then
        /* Garbage collect this node */
        Key(newnode, BitComplement(key)) := ClosestKey(subnode,
key)
        DeallocateNode(subnode)
      else
      SubTrie(newnode, BitComplement(Key)) := subnode
    else /* Nonempty-2.2: New key is a prefix */
      SubTrie(newnode, key) := subnode
}
```

InsertWithEmptySubTrie.   This procedure performs the additions of keys in or below nodes whose pertinent subtrie is empty. It is the most complex of all insertion functions in that it must distinguish between five cases. The insertion node may not have a respective key stored, the new key and the stored key may be equal, the new or the stored key fits exactly and one of them needs to be stored in a new node below the current one, or, the most involved case, one key is a strict prefix of the other and both keys need to be placed into two separate nodes.

```
InsertWithEmptySubTrie(node. Key, distpos) {
  local storedkey, dpos, newnode, newnewnode, index := Min(|key|,
distpos)

  /* NonEmpty-1: The new key may just be added to an existing
node. */
if (Key(node, key)= NIL) then Key(node, key) := key

/* NonEmpty-2: The new is a duplicate. */
elseif(Key(node, key) = key) then
      Key(node, key) = key
/* Empty-3: The stored key fits, and the new one needs to be
added below node.*/
elseif(|Key(node, key)| = Index(node)) then
  newnode := AllocateNode(|key|, key)
  LinkNodes(node, newnode, key)
  /* Empty-4: The new key fits, and the stored key needs to be
pushed down */
  elseif(|key| = Index(node) then
    newnode := AloocateNode(|Key(node, key)|), Key(node, key))
    Key(node, key) := key
    LinkNodes(node, newnode, key)

  /* Empty-5: the stored key and the new key are longer than
the node index and need to be stored below node */
else
```

```
    storedkey := Key(node, key) /* Save the stored key. */
    Key(node, key) := NIL /* will be moved. */
    dpos := DistPos(key, storedkey) /* Distinguishing position */

/* Empty-5.1: The keys are not prefixes of each other and may
hence be stored in one node. */
  if(dpos ≤ Min(|key|, |storedkey|)) then
    newnode := AllocateNode(dpos, key)
    Key(newnode, storedkey) := storedkey
    LinkNodes(node, newnode, storedkey)

  /* Empty-5.2: The stored key is a strict prefix of the new
key: Each key is stored in a separate new node.*/
  elseif(dpos > |storedkey|) then
    newnode := AllocateNode(|storedkey|, storedkey)
    LinkNodes(node, newnode, storedkey)
    newnewnode := AllocateNode(|key|, key)
    LinkNodes(newnode, newnewnode, key)

  /* Empty-5.3: The new key is a strict prefix of the stored
key: Each key is stored in a separate new node.*/
  else
    newnode := AllocateNode(|key|, key)
    LinkNodes(node, newnode, storedkey)
    newnewnode := AllocateNode(|storedkey|, storedkey)
    LinkNodes(newnode, newnewnode, storedkey)
}
```

Algorithmic Complexity. STEP3 of the algorithm has a complexity of $O(1)$, and the first two steps are linear in the depth of the trie. In the absence of prefixes, the depth depends logarithmically on the number of keys (see Lemma 3.5.1), whereas the general case may degenerate to a linear list of prefixes. However, by storing at each node the respective common prefix of all keys stored in that node and its subtries [see Lemma 3.5.3 (N2)] the first two steps may be combined and terminated, at the latest, when the insert key is exhausted. That is, the insert operation can be performed independent of the size of the database as a tradeoff against a small increase in storage complexity. In all cases, STEP3 affects, at most, the insertion node, its parent or the root node of one of its subtries.

3.5.3.2 Deletion

To delete a key, we simply erase it from its node. However, the storage and computational efficiency of the DP-Trie hinges on the fact that the structure of a DP-Trie is determined solely by its keys and not by a respective sequence of insert and delete operations. Hence, the deletion algorithm needs to run a garbage-collection function in nodes from which keys have been deleted to restore the respective prior trie structure. If a node becomes empty after a key deletion, it is simply removed. Nodes that no longer store keys and have only one subtrie, so-called Chain nodes, are also removed by linking their parent node and subtrie directly. When a node becomes a single-key leaf-node, its index is maximized by appropriately swapping its key. As a last step of the garbage-collection function, an attempt is made to move the key from a new single-key leaf node into its parent node.

```
DeleteKey(key) {
      local node, collnode := NIL, storedkey := NIL

      /* STEP1: Search for the key to be deleted */
      Node := Root
      if(node ≠ NIL and |key| ≥ Index(node)) then
          while (SubTrie(node, key) ≠ NIL and |key| ≥ Index
(SubTrie(node, key) )
          do node := SubTrie(node, key)

          /* if the node or key cannot be found return error */
          if( node = NIL or Key(node, key) ≠ key ) return (NotFound)
          /* STEP2: Delete the key and garbage collection nodes */
          else
              Key(node, key) := NIL
      /* DelEmpty: Delete an empty node */
      if ( Empty(node)) then
        /* DelEmpy-1: The root is not deleted. */
        if(node ≠ Root) then
          SubTrie(Parent(node), key) := NIL
          collnode := Parent(node)
        /* DelEmpty-2: the root is deleted. The trie is now
empty.*/
        else
            Root := NIL
            DeallocateNode(node)
 /* DelChain; Delete Chain nodes */
 elseif (ChainNode(node)) then
     if(node ≠ Root)
         SubTrie(Parent(node), key) := ClosestSubTrie(node, key)
         Parent(ClosestSubTrie(node, key)) := Parent(node)
         collnode := ClosestSubTrie(node, key)
     else
         Root := ClosestSubTrie(node, key)
         Parent(Root) := NIL
         DeallocateNode(node)

      /* DeMax: Maximize the index of single-key leaf-nodes */
      elseif(SingleKeyLeafNode(node)) then
         storedkey := Key(node, BitComplement(key))
         LeftKey(node) := RightKey(node) := NIL
         Index(node) := |storedkey|
         Key(node, storedkey) := storedkey
```

```
          Collnode := node

     /* DelSingle: Handle a single-key subtrie */
     elseif( SubTrie(node, key) ≠ NIL and SingleKeyLeafNode(Sub
Trie(node, key)) then
          collnode := SubTrie(node, key)

     /* SETP3: Last step of garbage collection */
     /* DelGC: Attempt to move keys from single-key leaf-node
to the parent node*/
     if(collnode ≠ NIL and SingleKeyLeafNode(collnode)) then
          storedkey := ClosestKey(collnode, key)

     if(Parent(conllnode) ≠ NIL and Key(Parent(collnode),
storedkey) = NIL ) then
          Key(Parent(collnode), storedkey) := storedkey
          SubTrie(Parent(collnode), stoedkey) := NIL
          DeallocateNode(collnode)
}
```

Algorithm Complexity. The search down the trie is linear in the length of the input key, and the removal of the key and the garbage collection of nodes has a complexity of $O(1)$. Hence, the complexity of the delete operation does not depend on the size of the trie. The deletion has only a local impact on the trie structure in that it affects, at most, the node that stores the pertinent key and the node below and above it.

3.5.3.3 Search

The search algorithm performs a descent of the trie under strict guidance by the input key, inspecting each of its bit positions, at most, once. One this first step has terminated, the traversed path is backtracked in search of the longest prefix. The decision not to perform key comparisons on the downward path resulted from a bias toward successful searches. In a networking environment, such as when performing routing decisions, negative results typically resemble error situations that cause data packets to be discarded and error messages to be sent with low priority. Notification or recovery of such errors is not deemed be overly time critical [26,27].

```
SearchKey(key) {
 local node := Root

 /* Check for empty trie and short keys */
if( node = NIL or |key|<Index(node) ) then
 return(NIL)

/* STEP1: Downward path */
while(SubTrie(node, key) ≠ NIL) and (|key| ≥
Index(SubTrie(node, key))
```

```
    do node := SubTrie(node, key)

/* STEP2: Backtracking to find the longest prefix */
while ((node ≠ NIL) and
        ((Key(node, key) = NIL)
    or  (Key(node, key) = ≰ key)))
    do node := Parent(node)

/* if a node was found, then it stores the longest prefix */
if(node ≠ NIL) then return (Key(node, key))
else return(NIL)
}
```

Algorithm Complexity. The complexity of the downward and upward searches is linear in the size of the input key independent of the size of the trie. In most implementations, the operations on the downward path will require only very little processing because only pointers need to be moved and single bits to be compared, on the upward path, the test for prefix relationships may be optimized for a given environment, such as by taking advantage of processor word sizes and instruction sets.

3.5.4 *Performance*

The DP-Trie is a variant of the PATRICIA data structure that supports nonrecursive search-and-update operations. In the case of fixed length keys, a comparison is made with the performance of the AVL trie [28]. For all three operations, the DP-Trie outperforms the AVL trie [25]. In the case where keys may be prefixes of each other, the average search time increases linearly as a function of $\log(N)$, (N is the number of keys in the tree). A lookup operation requires two traversals along the tree, the first traversal descends to a leaf node and the second backtracks to find the longest prefix matching the given address. The shorter the matching prefix is, the longer it takes for the search operation to find it. The insertion and deletion algorithms as reported in [25] need to handle a number of special cases and seem difficult to implement in hardware.

References

1. Gupta, P., Algorithms for Routing Lookups and Packet Classification, PhD thesis, Stanford University, 2000.
2. Hennessey, J. and Patterson, D., *Computer Architecture: A Quantitative Approach*, Morgan Kaufmann Publishers, San Francisco, 2nd edition, 1996.
3. Feldmeier, D.C., Improving gateway performance with a routing-table cache. *Proceedings IEEE Infocom'88*, Mar. 1988, pp. 298–307.
4. Jain, R. and Routhier, S., Packet train: measurements and new model for computer network traffic. *IEEE Journal on Special Areas in Communications* 1986; 4(6):986–994.
5. Jain, R., Characteristics of destination address locality in computer networks: a comparison of caching schemes. *Computer Networks and ISDN Systems* 1990; 18(4):243–254.
6. Bunt, R.B. and Murphy, J.M., The measurement of locality and the behavior of programs. *The Computer Journal* 1984; 27(3):238–245.

7. Mattson, R.L. et al., Evaluation techniques for storage hierarchies. *IBM System Journal* 1970; 9(2): 78–117.

8. Belady, L.A., A study of replacement algorithms for a virtual-storage computer. *IBM Systems Journal* 1996; 5(2):78–101.

9. Mann, B., Stutt, C., and Kempf, M., Terminal severs on ethernet local area networks. *Digital Technical Journal* 1986; 3(9):73–87.

10. Spirn, J.R., *Program Behavior: Models and Measurements*, New York: Elsevier, 1977, pp. 210–230.

11. Smith, A.J., Cache memories. *Computing Survey* 1982; 14(8):473–530.

12. Smith, A.J., Problems, directions and issues in memory hierarchies. *Proc. 18th Annual Hawaii Intl. Conf. on System Sciences*, 1985, pp. 468–476.

13. Karlin, A.R. et al., Compatitive Snoopy Caching, Carnegie-Mellon University, Report CMU-CS-1986, 86–164.

14. Bashinski, J., Cisco's AGS+ router: the architecture that keeps evolving. *The Packet, Cisco Systems User Magazine*, 5, 1993.

15. Partridge, C., Locality and route caches, NSF workshop on Internet Statistics Measurement and Analysis, San Diego, February 1996. Also see http://moat.nlanr.net/ISMA/Positions/partridge.html.

16. Naldi, M., Size estimation and growth forecast of the Internet, Centro Vito Volterra preprints, University of Rome, October 1997.

17. Cisco 12000 series GSR, at http://www.cisco.com/warp/public/cc/pd/rt/12000/index.shtml.

18. Juniper Networks Inc., Internet Processor II: Performance without compromise, at http://www.juniper.net/products/brochures/150006.html.

19. Lucent NX64000 multi-terabit switch/router, at http://www.lucent.com/ins/products/nx64000.

20. Birandais, R., File searching using variable length keys. *Proceedings of the Western Joint Computer Conference*, 1959, pp. 295–298.

21. Fredkin, E., Trie memory. *Communication of the ACM*, 1960; 3:490–499.

22. Ruiz-Sanchez, M.A., Biersack, E.W., and Dabbous, W., Survey and taxonomy of IP address lookup algorithms, *IEEE Network* 2001; 15(2):8–23.

23. Morrison, D.R., PATRICIA—Practical algorithm to retrieve information coded in alphanumeric, *Journal of ACM* 1968; 15(4):514–534.

24. Sklower, K., A tree-based packet routing table for Berkeley Unix, *Proc. 1991 Winter Usenix Conf.*, 1991, pp. 93–99.

25. Doeringer, W., Karjoth, G., and Nassehi, M., Routing on longest-matching prefixes. *IEEE/ACM Transactions on Networking* 1996; 4(1):86–97.

26. Comer, D., *Internetworking with TCP/IP. Principle, Protocols and Architecture*, Englewood Cliffs, NJ: Prentice-Hall, 1989.

27. Perlman, R., *Interconnections: Bridges and Routers*. Reading, MA: Addison-Wesley, 1992, pp. 183–192.

28. Knuth, D.E., *The Art of Computer Programming*, Vol. 3 Sorting and Searching. Reading, MA: Addison-Wesley, 1991.

Chapter 4

Multibit Tries

Binary tries provide an easy way to handle arbitrary length prefixes. Lookup and update operations are straightforward. Nevertheless, the search in a binary trie can be rather slow because we inspect one bit at a time and in the worst-case 32 memory accesses are needed for an IPv4 address. One way to speedup the search operation is to inspect not just one bit a time but several bits simultaneously. For instance, if we inspect 4 bits at a time we would need only 8 memory accesses in the worst-case for an IPv4 address. The number of bits to be inspected per step is called stride and can be constant or variable.

A trie structure that allows the inspection of bits in strides of several bits is called a multibit trie. Thus, a multibit trie is a trie where each node has 2^k children, where k is the stride [1]. There are a lot of data structures based on multibit tries for IP-address lookup. In this chapter, we will introduce some of them.

4.1 Level Compression Trie

4.1.1 Level Compression

The binary strings in Figure 4.1 correspond to the trie in Figure 4.2a. The trie is not very efficient, because the number of nodes is large and the average depth (the average length of a path from the root to a leaf) is long. The traditional technique to overcome this problem is to use path compression: each internal node with only one child is removed. Of course, we have to somehow record which nodes are missing. In each node, we use the skip value to indicate how many bits have been skipped on the path. The path-compressed version of the trie in Figure 4.2a is shown in Figure 4.2b. The number of internal nodes in Figure 4.2b is less than that in Figure 4.2a. Therefore, path compression gives a significant overall size reduction. In fact, it compresses those parts of the trie that are sparsely populated.

Level compression [2] is a technique for compressing parts of the trie that are densely populated. The idea is to replace the i highest complete levels of the binary trie with a single node of degree 2^i; this replacement is performed recursively on each subtrie. The level-compressed version, the *LC-trie* [3], of the trie in Figure 4.2b is shown in Figure 4.2c. This section is partly from [3] (© 1999 IEEE).

107

No	string
0	0000
1	0001
2	00101
3	010
4	0110
5	0111
6	100
7	101000
8	101001
9	10101
10	10110
11	10111
12	110
13	11101000
14	11101001

Figure 4.1 Binary strings to be stored in a trie structure. (From Nilsson, S. and Karlsson, G., *IEEE Journal on Selected Areas in Communication*, 17, 6, 1999. With permission.)

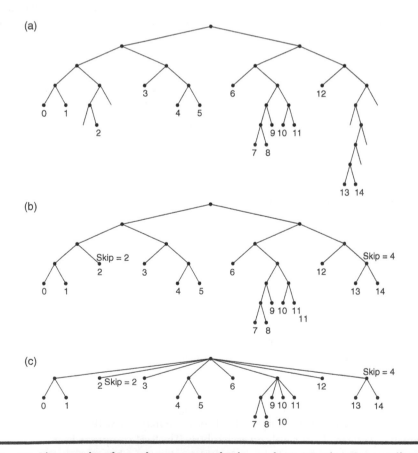

Figure 4.2 (a) Binary trie, (b) path-compressed trie, and (c) LC-trie. (From Nilsson, S. and Karlsson, G., *IEEE Journal on Selected Areas in Communication*, 17, 6, 1999. With permission.)

For an independent random sample with a density function that is bounded from above and below, the expected average depth of an LC-trie is $\Theta(\log^* n)$, where $\log^* n$ is the iterated logarithm function, $\log^* n = 1 + \log^* (\log n)$, if $n > 1$, and $\log^* n = 0$ otherwise. For data from a Bernoulli-type process with character probabilities not all equal, the expected average depth is $\Theta(\log \log n)$ [4]. Uncompressed tries and path-compressed tries both have expected average depth $\Theta(\log n)$ for these distributions.

4.1.2 Representation of LC-Tries

The standard implementation of a trie, where a set of children pointers are stored at each internal node is not a good solution, because it has a large space overhead. If we want to achieve the efficiency promised by the theoretical bounds, it is of course important to represent the trie efficiently.

A space efficient alternative is to store the children of a node in consecutive memory locations. In this way, only a pointer to the leftmost child is needed. In fact, the nodes may be stored in an array and each node can be represented by a single word. In our implementation, the first 5 bits represent the branching factor, the number of descendants of the node. This number is always a power of 2 and hence, using 5 bits, the maximum branching factor that can be represented is 2^{31}. The next 7 bits represent the skip value. In this way, we can represent values in the range from 0 to 127, which is sufficient for IPv6 addresses. This leaves 20 bits for the pointer to the leftmost child and hence, using this very compact 32-bit representation, we can store at least $2^{19} = 524,288$ strings. Figure 4.3 shows the array representation of the LC-trie in Figure 4.2c; each entry represents a node. The nodes are numbered in breadth-first order starting at the root. The number in the branch column indicates the number of bits used for branching at each node. A value $k \geq 1$ indicates that the node is a leaf. The next column contains the skip value, the number of bits that can be shipped during a search operation. The value in the pointer column has two different interpretations. For an internal node, it is used as a pointer to the leftmost child; for a leaf it is used as a pointer to a base vector containing the complete strings.

The search algorithm can be implemented very efficiently, as shown in Figure 4.4. Let s be the string searched for and let EXTRACT(p, b, s) be a function that returns the number given by the b bits starting at position p in the string s. We denote the array representing the tree by T. The root is stored in T[0]. Note that the address returned only indicates a possible hit; the bits that have been skipped during the search may not match. Therefore we need to store the values of the strings separately and perform one additional comparison to check whether the search was actually successful.

As an example we search for the string 10110111. We start at the root, node number 0. The branching value is 3 and skip value is 0 and therefore we extract the first three bits from the search string. These three bits have the value 5, which is added to the pointer, leading to position 6 in the array. At this node the branching value is 3 and the skip value is 0 and therefore we extract the next 2 bits. They have the value 2. Adding 2 to the pointer we arrive at position 15. At this node the branching value is 0, which implies that it is a leaf. The pointer value 5 gives the position of the string in the base vector. Observe that it is necessary to check whether this constitutes a true hit. We need to compare the first 5 bits of the search string with the first 5 bits of a value stored in the base vector in the position indicated by the pointer (10) in the leaf. In fact, Figure 4.1 contains a prefix 10110 matching the string and the search was successful.

For IP lookup, the data structure consists of four parts: an LC-trie, a base vector, a next-hop table, and a prefix table. The LC-trie is the heart of the data structure. The leaves of the LC-trie contain pointers into a base vector, which contains the complete binary strings and two pointers: one pointer

No	branch	skip	pointer
0	3	0	1
1	1	0	9
2	0	2	2
3	0	0	3
4	1	0	11
5	0	0	6
6	2	0	13
7	0	0	12
8	1	4	17
9	0	0	0
10	0	0	1
11	0	0	4
12	0	0	5
13	1	0	19
14	0	0	9
15	0	0	10
16	0	0	11
17	0	0	13
18	0	0	14
19	0	0	7
20	0	0	8

Figure 4.3 Array representation of the LC-trie. (From Nilsson, S. and Karlsson, G., *IEEE Journal on Selected Areas in Communication*, 17, 6, 1999. With permission.)

into the next-hop table and one pointer into the prefix table. The next-hop table is an array containing all possible next-hops addresses. Each entry of the prefix table contains a number that indicates the length of the prefix and a pointer into an other entry of the prefix table. The pointer has the special value −1 if no prefix of the string is present. The actual value of the prefix need not be explicitly stored, because it is always a proper prefix of the corresponding value in the base vector.

The search routine follows the next-hop pointer if the search is successful. If not, the search routine tries to match a prefix of the string with the entries in the prefix table, until there is a match or the pointer of an entry in the prefix table has the special value −1.

```
node = trie[0];
pos = node.skip;
branch = node.branch;
adr = node.adr;
while (branch !=0 ) {
    node = trie[adr + EXTRACT(pos, branch, s)];
    pos = pos + branch + node.skip;
    branch = node.branch;
    adr = node.adr;
}
return adr;
```

Figure 4.4 Pseudocode for find operation in an LC-trie. (From Nilsson, S. and Karlsson, G., *IEEE Journal on Selected Areas in Communication*, 17, 6, 1999. With permission.)

4.1.3 Building LC-Tries

Building the LC-trie is straightforward. The first step is to sort the base vector. Then we make a top-down construction of the LC-trie as demonstrated by the pseudocode in Figure 4.5. This recursive procedure builds an LC-trie covering a subinterval of the base vector. This subinterval is identified by its first member (first), the number of strings (n), and the length of a common prefix (pre). Furthermore, the first free position (pos) in the vector that holds the representation of the trie is passed as an argument to the procedure. There are two cases. If the interval contains only one string, we simply create a leaf; otherwise we compute the skip and branch values (the details are discussed below), create an internal node and, finally, the procedure is called recursively to build the subtries.

To compute the skip value, the longest common prefix of the strings in the interval, we only need to inspect the first and last string. If they have a common prefix of length k, all the strings in between must also have the same prefix, because the strings are sorted.

The branching factor can also be computed in a straightforward manner. Disregarding the common prefix, start by checking if all four prefixes of length two are present (there are at least two branches, because the number of the strings are at least two and the entries are unique). If these prefixes are present we continue the search, by examining if all eight prefixes of length three are present. Continuing this search we will eventually find a prefix that is missing and the branching factor has been established.

Allocating memory in the vector representing the trie can be done by simply keeping track of the first free position in this vector. (In the pseudocode it is assumed that memory has been allocated for the root node before the procedure is invoked.) A more elaborate memory allocation scheme may be substituted. For example, in some cases it may be worthwhile to use a memory layout that optimizes the cashing behavior of the search algorithm for particular machine architecture.

```
build (int first, int n, int pre, int pos)
{
    if (n = 1) {
        trie[pos] = {0, 0, first};
        return;
    }
    skip = computeSkip (pre, first, n);
    branch = computeBranch (pre, first, n, skip);
    adr = allocateMemory (2^branch);
    trie[pos] = {branch, skip, adr};

    p = first;
    for bitpat = 0 to 2^branch - 1 {
        k = 0;
        while ( EXTRACT(pre + skip, branch, base[p + k]) = bitpat)
            k = k + 1;
        build (p, k, pre + skip, adr + bitpat);
        p = p + k;
    }
}
```

Figure 4.5 Pseudocode for building an LC-trie given a sorted base vector. (From Nilsson, S. and Karlsson, G., *IEEE Journal on Selected Areas in Communication*, 17, 6, 1999. With permission.)

To estimate the time complexity we observe that the recursive procedure is invoked once for each node of the trie. Computing the skip value takes constant time. The branching factor of an internal node is computed in the time proportional to the number of strings in the interval. Similarly, the increment statement of the inner loop is executed once for every string in the interval. This means that at each level, the trie can be built in $O(n)$ time, where n is the total number of strings. Hence the worst-case time complexity is $O(nh) + S$, where h is the height of the trie, and S is the time to sort the strings. Using a comparison based sorting algorithm and assuming that the strings are of constant length $S = O(n \log n)$.

It is important to select the branching factor to build an LC-trie. A node can have a branching factor 2^k only if all prefixes of length k are present. This means that a few missing prefixes might have a considerable negative influence on the efficiency of the level compression. To avoid this, it is natural to require only that a fraction of the prefixes are present. In this way we get a tradeoff between space and time. Using larger branching factors will decrease the depth of the trie, but it will also introduce superfluous empty leaves into the trie. When computing the branching factor for a node covering k strings, we use the highest branching that produces at most $\lceil k(1-x) \rceil$ empty leaves, where x is a fill factor, $0 < x \leq 1$. Because it is particularly advantageous to use a large branching factor for the root, it is fixed independently of the fill factor.

4.1.4 Experiments

To evaluate the performance of the LC-trie, experiments were performed on two different machines: a SUN Ultra Sparc with two 296-MHz processors and 512 MB RAM, and a personal computer with a 133-MHz Pentium processor and 32 MB RAM. The forwarding tables were from the IPMA project (http://www.merit.edu/ipma, 1997) and the router of the Finnish University and Research Network (FUNET). The traffic is a random permutation of all entries in a forwarding table, in which the entries were extended to 32-bit numbers by adding zeroes. Table 4.1 shows some measurements for an LC-trie with fill factor 0.50 and using 16-bits for branching at the root. The size of the LC-trie is less than 500 kB. The speed is over half a million complete lookup operations per second on a 133 MHz Pentium PC, and more than two million per second on a more powerful SUN Sparc Ultra II workstation.

Table 4.1 Results for a Structure with 16-Bits Used for Branching at Root and a Fill Factor of 0.5

Site	Routing Entries	Next-Hop	Memory of LC-trie	Average Depth (Max. Depth)	Lookups	
					Sparc	PC
Mae-East	38,367	59	457,276	1.66 (5)	2.0	0.6
Mae-West	15,022	57	327,268	1.29 (5)	3.8	0.8
AADS	20,299	19	364,596	1.42 (5)	3.2	0.7
Pac-Bell	20,611	3	367,484	1.43 (5)	2.6	0.7
FUNET	41,578	20	515,460	1.73 (5)	5.0	1.2

Note: The speed is measured in million lookups per second.

Source: Nilsson, S. and Karlsson, G., *IEEE Journal on Selected Areas in Communication*, 17, 6, 1999. With permission.

4.1.5   Modified LC-Tries

An LC-trie is used for IP lookup with the additional data structure like base, prefix and next-hop vectors. Ravikumar et al. proposed a modified LC-trie to avoid additional storage and backtracking [5]. The modified LC-trie stores all prefixes either in the internal nodes or the leaf nodes. The building algorithm does not change, but each entry in the data structure has five fields, shown in Figure 4.6.

Branching factor: [0:3]: This indicates the number of descendents of a node;
Skip [4:10]: This indicates the number of bits that can be skipped in an operation;
Port [11:15]: This represents the output port for the current node in case of a match;
Pointer [16:31]: This a pointer to the leftmost child in case of an internal node and NULL in case of a leaf node. This is a 16-bit value and can represent a maximum of 65,536 prefixes;
String [32:63]: This represents the actual value of the prefix the node represents. It can be extended to 128-bits for IPv6.

The search algorithm is the same as that of the LC-trie. But at each node, we check if the IP address matches the prefix that the node represents. If it matches, the output is stored in the register, and the search continues. Otherwise, the search stops and, we use the previous value of the output port from the register to route the packet.

Because this approach checks for the match in the string at every step, it is necessary to traverse the entire trie in case the IP address is not present. In the LC-trie approach, after we traverse through the trie we perform a check for the string match in the base vector, consuming at least one memory fetch. If there is a mismatch a check is done again on the prefix table and this requires checking for a prefix match, which again requires another memory fetch. Thus, compared with the LC-trie, the modified LC-trie can save at least two memory cycles for every routing lookup. Also, the approach avoids storage for the base, prefix, and next-hop vectors and, hence occupies less storage. The simulation results show that the modified LC-trie works [5] 4.11 times better than the LC-trie, and that the storage is 2.38 times less.

4.2   Controlled Prefix Expansion

Before 1993, Internet addressing architecture partitioned the IP address space into five classes—A, B, C, D, and E. Typically, the forwarding table had three parts: A, B, C. All prefixes in the same part had fixed length: 8, 16, and 24. The fact that restricted prefix lengths lead to faster search is well known. After the deployment of Classes Inter-Domain Routing in 1993, prefixes could be of arbitrary length from 8 to 32 rather than being constrained to 8, 16, or 24 bits. The idea of the controlled prefix expansion (CPE) was to reduce a set of arbitrary length prefixes to a set of the predefined length prefixes [6].

Branching factor	Skip	Port	Pointer	String
4bit	7bit	5bit	16bit	32bit

Figure 4.6   Data structure for each entry. (From Ravikumar, V.C., Mahapatra, R., and Liu, J.C., *IEEE/ACM Proceedings on MASCOTS*, 2002. With permission.)

4.2.1 Prefix Expansion

An example of a forwarding table is shown in Figure 4.7. The prefixes range from length 1 all the way to length 7. Suppose we want to reduce the forwarding table to an equivalent forwarding table with prefixes of length 2, 5, and 7 (3 distinct lengths). All prefixes are expanded to the closest admissible length. The prefix P1 = 0* is expanded to two prefixes 00* and 01*, which inherit the next-hop of P1. The prefix P2 = 1* is expanded two prefixes 10* and 11*. As the prefix 10* is already an existing prefix P3, it cannot inherit the next-hop of P2, otherwise the expanded table is not an equivalent of the original. Prefix expansion can reduce any set of arbitrary length prefixes into an expanded set of prefixes of any prespecified sequence of lengths $L_1, L_2, ..., L_k$ [6]. The expanded table is shown in Figure 4.7.

Expansion can be performed in time proportional to the number of expanded prefixes. The algorithm uses an array A and maintains an invariant that $A[i]$ contains a pointer to the set of current prefixes of length i. Initially, we load the original prefixes into A to preserve the invariant. We also take as input the sequence of expansion levels $L_1, L_2, ..., L_k$. The main loop scans the lengths in A in increasing order of length. If we are currently at length i and length i is in the target sequence of levels, we skip this length. Otherwise, we expand each prefix in set $A[i]$ to length $i + 1$, while getting rid of the expanded prefixes which are the same as the existing prefixes.

Original			Expanded(3 levels)		
Prefix	Next hop	Length	Prefix	Next hop	Length
P1 = 0*	A	1	00*	A	
			01*		
P2 = 1*	B	1	11*	B	2
P3 = 10*	C	2	10*	C	
P4 = 111*	D	3	11100*	D	
			11101*		
			11110*		
			11111*		5
P5 = 11001*	E	5	11001*	E	
P6 = 1000*	F	4	10000*	F	
			10001*		
P7 = 100000*	G	6	1000001*	G	7
P8 = 1000000*	H	7	1000000*	H	

(a) Expanded prefixes

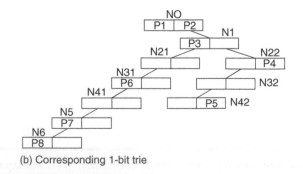

(b) Corresponding 1-bit trie

Figure 4.7 **An example of prefix expansion.**

4.2.2 Constructing Multibit Tries

Figure 4.8 is a multibit trie of the expanded prefixes in Figure 4.7. Because we have expanded to length 2, 5, and 7, the first level of the trie uses 2-bits, the second uses 3-bits (5 − 2) and the third uses 2 = bits (7 − 2). If a trie level uses m bits, then each trie node at that level is an array of 2^m locations, we called m the stride of the trie level. Each array element contains a pointer Q (possibly null) and an expanded prefix P. Prefix P is stored along with pointer Q if the path from the root to Q (including Q itself) is equal to P.

Thus in comparing Figures 4.7 and 4.8, notice that the root trie node contains all the expanded prefixes of length 2 (first four expanded prefixes in Fig. 4.7). For example, the first expanded prefix of length 5 in Figure 4.7 is 11100*. This is stored in a second level trie node pointed to by the pointer corresponding to the 11 entries in the first trie level, and stored at location 100.

To search for a destination IP address, we break the destination address into chunks corresponding the strides at each level (e.g., 2, 3, and 2 in Fig. 4.8) and use these chunks to follow a path through the trie until we reach a nil pointer. As we follow the path, we keep track of the last prefix that was alongside a pointer we followed. The last prefix encountered is the best matching prefix when we terminate. The search algorithm is shown in Figure 4.9. It is easy to see that the search time is $O(k)$, where k is the maximum path through the multibit trie, which is also the maximum number of expanded lengths.

In controlled expansion, at first we pick the target prefix lengths: $L_1, L_2, ..., L_k$ or the stride of each level: $m1 = L_1, m2 = L_2, - L_1, m3 = L_3 - L_1, L_2, ..., m_k = L_{k-1} - L_k$, then we expand the original prefixes into n expanded prefixes, assuming the storage is measured by the total number of the expanded prefixes. Clearly, we wish to make k and n as small as possible. For example, if we wanted to guarantee a worst-case trie path of four nodes, we would choose $k = 4$. But which four target lengths do we choose to minimize the total storage? A naïve algorithm would be to search through all possible $k = 4$ target expansion lengths. For IPv4, there are 32!/(32−4)!4! possible choices. A second naïve technique is to use heuristics like picking the highest density lengths. It works reasonably, but is not optimal [6].

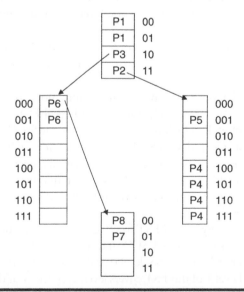

Figure 4.8   The multibit trie of the prefixes. (From Srinivasan, V. and Varghese, G., *ACM Transaction on Computer Systems*, 17, 1, 1999. With permission.)

```
SEARCH(D)  {  // D is a destination address
        BMP = default; // the best matching prefix encountered on path so far
        N = root; // current trie node we are visiting is root
    While (N ≠ nil) do
                C = chunk of D of length N.stride starting from bit N.startBit
        If N[bmp, C] ≠ nil then
                        BMP = N[bmp, C]; // find a longer prefix
            Endif;
                N = N[ptr, C];
    Endwhile
    Return (BMP);
}
```

Figure 4.9 Pseudocode of the search algorithm in a multibit trie. (From Srinivasan, V. and Varghese, G., *ACM Transaction on Computer Systems*, 17, 1, 1999. With permission.)

Srinivasan and Varghese used dynamic programming to find the optimal set of k target lengths (levels) [6]. The algorithm for optimal fixed-stride tries (FST) has complexity $O(k * W^2)$, where W is the width of the destination address. The algorithm for optimal variable-stride has a complexity of $O(n * k * W^2)$, where n is the number of original prefixes in forwarding table. Sahni and Kim developed a new dynamic programming formulation [7], which is more efficient than that in [6]. We will describe the algorithm from [7] (© 2003 IEEE) in the next sections.

4.2.3 Efficient Fixed-Stride Tries

In a multibit trie, the stride of a node is the number of bits used at that node to determine which branch to take. A node whose stride is s has 2^s child fields (corresponding to the 2^s possible values for the s-bits that are used) and 2^s data fields. Such a node requires 2^s memory units. The fixed-stride trie (FST) is the trie in which all nodes at the same level have the same stride; nodes at different levels may have different strides.

For a set P of prefixes and an integer k, we will find the strides for a k-level FST in such a manner that the k-level FST for the given prefix uses the smallest amount of memory, called the fixed-stride optimization (FSTO) problem.

For a set of prefixes P, a k-level FST may actually require more space than a $(k-1)$-level FST. For example, when $P = \{00*, 01*, 10*, 11*\}$, the unique 1-level FST for P requires four memory units while the unique 2-level FST (which is actually the 1-bit trie for P) requires six memory units. Because the search time for a $(k-1)$-level FST is less than that for a k-level-tree, we would actually prefer $(k-1)$-level FSTs that take less (or even equal) memory over k-level FSTs. Therefore, in practice, we are really interested in determining the best FST that uses at most k levels (rather than exactly k levels). The modified FSTO problem (MFSTO) is to determine the best FST that uses at most k levels for the given prefix set P.

Let O be the 1-bit trie, F be any k-level FST for P. Let $s_0, s_1, ..., s_{k-1}$ be the strides for F. We shall say that level 0 of F covers levels 0, ..., $s_0 - 1$ of O, and that level j, $0 < j < k$ of F covers levels a, ..., b of O, where $a = \sum_0^{j-1} s_q$ and $b = \sum_0^j s_q - 1$. So, level 0 of the FST of Figure 4.8 covers levels 0 and 1 of the 1-bit trie of Figure 4.7b. Level 1 of this FST covers levels 2, 3, and 4 of the 1-bit trie of Figure 4.7b; and level 2 of this FST covers levels 5 and 6 of the 1-bit trie. We shall refer to levels $e_u = \sum_0^u s_{q'}$, $0 \leq u < k$ as the expansion levels of O. The expansion levels defined by the FST of Figure 4.8 are 0, 2, and 5.

Let nodes (*i*) be the number of nodes at level *i* of the 1-bit trie *O*. For the 1-bit trie in Figure 4.7b, nodes(0:6) = [1, 1, 2, 2, 2, 1, 1]. The memory required by *F* is $\sum_0^{k-1}$ nodes(e_q) * 2^{s_q}. For example, the memory required by the FST of Figure 4.8 is nodes(0) * 2^2 + nodes(2) * 2^3 + nodes(5) * $2^2 = 24$.

Let T(*j*, *r*), *r* ≤ *j* + 1 be the cost (i.e., memory requirement) of the best way to cover levels 0 through *j* of *O* using exactly *r* expansion levels. When the maximum prefix length is *W*, *T*(*W* − 1, *k*) is the cost of the best *k*-level FST for the given set of prefixes. Sirnivasan et al. have obtained the following dynamic programming recurrence for *T*:

$$T(j, r) = \min_{m \in \{r-2..j-1\}} \{T(m, r-1) + nodes(m+1) * 2^{j-m}\}, r > 1 \tag{4.1}$$

$$T(j, r) = 2^{j+1} \tag{4.2}$$

The rationale for Equation 4.1 is that the best way to cover levels 0 through *j* of *O* using exactly *r* expansion levels, *r* > 1, is to have its last expansion level at level *m* + 1 of *O*, where *m* must be at least *r* − 2 (as otherwise, we do not have enough levels between levels 0 and m of *O* to select the remaining *r* − 1 expansion levels) and at most *j* − 1 (because the last expansion level is ≤ *j*). When the least expansion level is *nodes*(*m* + 1), For optimality, levels 0 through *m* of *O* must be covered in the best possible way using exactly *r* − 1 expansion levels.

Using Equation 4.1, we may determine *T*(*W* − 1, *k*) in $O(kW^2)$ time [excluding the time needed to compute *O* from the given prefix set and determine *nodes*()]. The stride for the optimal *k*-level FST can be obtained in additional *O*(*k*) time. Because, Equation 4.1 may also be used to compute *T*(*W* − 1, *q*) for all *q* ≤ *k* in $O(kW^2)$ time, we can actually solve the MFSTO problem in the same asymptotic complexity as required for the FSTO problem.

We can reduce the time needed to solve the MFSTO problem by modifying the definition of *T*. The modified function is *C*, where *C*(*j*, *r*) is the cost of the best FST that uses at most *r* expansion levels. It is easy to see that *C*(*j*, *r*) ≤ *C*(*j*, *r* − 1), *r* > 1. A simple dynamic programming recurrence for *C* is:

$$C(j, r) = \min_{m \in \{-1..j-1\}} \{C(m, r-1) + nodes(m+1) * 2^{j-m}\}, \quad j \geq 0, \ r > 1 \tag{4.3}$$

$$C(-1, r) = 0 \text{ and } C(j, 1) = 2^{j+1}, \quad j \geq 0 \tag{4.4}$$

To see the correctness of Equations 4.3 and 4.4, note that when *j* ≥ 0, there must be at least one expansion level. If *r* = 1, then there is exactly one expansion level and the cost is 2^{j+1}. If *r* > 1, the last expansion level in the best FST could be at any of the levels 0 through *j*. Let *m* + 1 be this last expansion level. The cost of the covering is *C*(*m*, *r* − 1) + nodes(*m* + 1) * 2^{j-m}. When *j* = −1, no levels of the 1-bit trie remain to be covered. Therefore, *C*(−1, *r*) = 0.

We may obtain an alternative recurrence for *C*(*j*, *r*) in which the range of *m* on the right side is *r* − 2 … *j* − 1 rather than −1 … *j* − 1. First, we obtain the following dynamic programming recurrence for *C*:

$$C(j, r) = \min\{C(j, r-1), T(j, r)\}, r > 1 \tag{4.5}$$

$$C(j, 1) = 2^{j+1} \tag{4.6}$$

The rationale for Equation 4.5 is that the best FST that uses at most *r* expansion levels either uses at most *r* − 1 levels or uses exactly *r* levels. When at most *r* − 1 levels are used, the cost is *C*(*j*, *r* − 1) and when exactly *r* levels are used, the cost is *T*(*j*, *r*), which is defined by Equation 4.1.

Let $U(j, r)$ be as defined below:

$$U(j, r) = \min_{m \in \{r-2 \ldots j-1\}} \{C(m, r-1) + nodes(m+1) * 2^{j-m}\}$$

From Equations 4.1 and 4.5 we obtain:

$$C(j, r) = \min\{C(j, r-1), U(j, r)\} \tag{4.7}$$

To see the correctness of Equation 4.7, note that for all j and r such that $r \leq j + 1$, $T(j, r) \geq C(j, r)$.
Furthermore,

$$\min_{m \in \{r-2 \ldots j-1\}} \{T(m, r-1) + nodes(m+1) * 2^{j-m}\}$$

$$\geq \min_{m \in \{r-2 \ldots j-1\}} \{C(m, r-1) + nodes(m+1) * 2^{j-m}\} = U(j, r) \tag{4.8}$$

Therefore, when $C(j, r-1) \leq U(j, r)$, Equations 4.5 and 4.7 compute the same value for $C(j, r)$. When $C(j, r-1) > U(j, r)$, it appears from Equation 4.8 that Equation 4.7 may compute a smaller $C(j, r)$ than is computed by Equation 4.5. However, this is impossible, because

$$C(j, r) = \min_{m \in \{-1 \ldots j-1\}} \{C(m, r-1) + nodes(m+1) * 2^{j-m}\}$$

$$\leq \min_{m \in \{r-2 \ldots j-1\}} \{C(m, r-1) + nodes(m+1) * 2^{j-m}\}$$

where $C(-1, r) = 0$. Therefore, the $C(j, r)$ computed by Equations 4.5 and 4.7 are equal.

In the following, we use Equations 4.3 and 4.4 for C. The range for m (in Equation 4.3) may be restricted to a range that is (often) considerably smaller than $r-2 \ldots j-1$. To obtain this narrower search range, Sahni and Kim establish a few properties of 1-bit tries and their corresponding optimal FSTs [7].

The algorithm *FixedStrides* is shown in Figure 4.10. The complexity of this algorithm is $O(kW^2)$. Although *FixedStrides* has the same asymptotic complexity as does the CPE algorithm in [7], experiments in [7] indicate that *FixedStrides* runs two to four times as fast, shown in Table 4.2.

4.2.4   Variable-Stride Tries

In a variable-stride trie (VST) [6], nodes at the same level may have different strides. FSTs are a special case of VSTs. The memory required by the best VST for a given prefix set P and the number of expansion levels k is less than or equal to that required by the best FST for P and k. Despite this, FSTs many be preferred in certain router applications because of their simplicity and slightly faster search time [6].

Let r-VST be a VST that has at most r levels. Let $Opt(N, r)$ be the cost (i.e., memory requirement) of the best r-VST for a 1-bit trie whose root is N. Srinivasan and Varghese have obtained the following dynamic programming recurrence for $Opt(N, r)$.

```
Algorithm FixedStrides(W, k)
// W is length of longest prefix.
// k is maximum number of expansion levels desired
// Return C(W–1,k) and compute M(*, *).
{
    for (j = 0; j < W; j++) {
        C(j, 1): = 2^{j+1};
        M(j, 1):= –1;}
    for (r = 1; r < k; r++)
        C(-1, r): = 0;
    for (r = 2; r < k; r++)
        for ( j = r-1; j < W; j + +) {
        // Compute C(j, r).
            minJ:= max(M(j–1, r), M(j, r–1));
            minCost:= C(j, r–1);
            minL:= M(j, r–1);
            for(m = minJ; m < j; m++) {
                cost:= C(m, j–1) + nodes(m + 1)*2^{j–m};
                if (cost < minCost) then
                            {minCost:= cost; minL:= m;}
        }
        C(j, r):= minCost;
        M(j, r):= minL;
        }
    return C(W–1, k);
}
```

Figure 4.10   Algorithm for fixed-stride tries. (From Sahni, S. and Kim, K., *IEEE/ACM Transactions on Networking*, 11, 4, 2003. With permission.)

Table 4.2   Execution Time (in μ sec) for FST Algorithms

	Paix		Pb		Mae-West		Aads		Mae-East	
k	CPE	FST	CPE	FST	CPE	FST	CPE	FST	CPE	FST
2	39	21	41	21	39	21	37	20	37	21
3	85	30	81	30	84	31	74	31	96	31
4	123	39	124	40	128	38	122	40	130	40
5	174	46	174	48	147	47	161	45	164	46
6	194	53	201	54	190	55	194	54	190	53
7	246	62	241	63	221	63	264	62	220	62

Note: The experiments run on a SUN Ultra Enterprise 4000/5000 computer. The data is from IPMA.

Source: Sahni, S. and Kim, K., *IEEE/ACM Transactions on Networking*, 11, 4, 2003. With permission.

$$Opt(N,r) = \min_{s \in \{1...1+height(N)\}} \{2^s + \sum_{M \in D_s(N)} Opt(M, r-1)\}, \quad r > 1 \tag{4.9}$$

where $D_s(N)$ is the set of all descendents of N that are at level s of N. For example, $D_1(N)$ is the set of children of N and $D_2(N)$ is the set of grandchildren of N. $height(N)$ is the maximum level at which the trie rooted at N has a node. When $r = 1$,

$$Opt(N, 1) = 2^{1+height(N)} \tag{4.10}$$

Equation 4.10 is equivalent to Equation 4.2 (in Section 4.2.3); the cost of covering all levels of N using at most one expansion level is $2^{1 + height(N)}$. When more than one expansion level is permissible, the stride of the first expansion level may be any number s that is between 1 and $1 + height(N)$. For any such selection of s, the next expansion level is level s of the 1-bit trie whose root is N. The sum in Equation 4.9 gives the cost of the best way to cover all subtrees whose roots are at this next expansion level. Each such subtree is covered using at most $r - 1$ expansion levels. It easy to see that $Opt(R, k)$, where R is the root of the overall 1-bit trie for the given prefix set P, is the cost of the best k-VST for P. Srinivasan and Varghese describe a way to determine $Opt(R, k)$, using Equations 4.9 and 4.10. The complexity of their algorithm is $O(n * W^2 * k)$, where n is the number of prefixes in P and W is the length of the longest prefix.

By modifying the equations of Srinivasan and Varghese slightly, we are able to compute $Opt(R, k)$ in $O(mWk)$ time, where m is the number of nodes in the 1-bit trie. Because $m = O(n)$ for realistic router prefix sets, the complexity of our algorithm is $O(mWk)$. Let

$$Opt(N, s, r) = \sum_{M \in D_s(N)} Opt(M, r), \quad s > 0, \quad r > 1,$$

and let $Opt(N, 0, k) = Opt(N, r)$. From Equations 4.9 and 4.10, we obtain:

$$Opt(N, 0, r) = \min_{s \in \{1...1+height(N)\}} \{2^s + Opt(N, s, r-1)\}, \quad r > 1 \tag{4.11}$$

and

$$Opt(N, 0, 1) = 2^{1+height(N)} \tag{4.12}$$

For $s > 0$ and $r > 1$, we get

$$Opt(N, s, r) = \sum_{M \in D_s(N)} Opt(M, r) = Opt(LeftChild(N), s-1, r) + Opt(RightChild(N), s-1, r) \tag{4.13}$$

For Equation 4.13, we need the following initial condition:

$$Opt(null, *, *) = 0 \tag{4.14}$$

With the assumption that the number of nodes in the 1-bit trie is $O(n)$, we see that the number of $Opt(*, *, *)$ values is $O(nWk)$. Each $Opt(*, *, *)$ value may be computed in $O(1)$ time using Equations 4.11 through 4.17 provided the Opt values are computed in postorder. Therefore, we may compute $Opt(R, k) = Opt(R, 0, k)$ in $O(nWk)$ time. The algorithm requires $O(W^2k)$

memory for $Opt(*, *, *)$ values. Notice that there can be at most $W + 1$ nodes N whose $Opt(N, *, *)$ values must be retained at any given time, and for each of these at most $W + 1$ nodes, $O(W^2k)$ $Opt(N, *, *)$ values must be retained. To determine the optimal strides, each node of the 1-bit trie must store the stride s that minimizes the right side of Equation 4.11 for each value of r. For this purpose, each 1-bit trie node needs $O(k)$ space. The total memory required is, therefore, $O(nk + W^2k)$.

In practice, we may prefer an implementation that uses considerably more memory. If we associate a cost array with each of the $O(n)$ nodes of the 1-bit trie, the memory requirement increases to $O(nWk)$. The advantage of this increased memory implementation is that the optimal strides can be recomputed in $O(W^2k)$ time [rather than $O(nWk)$] following each insertion or deletion of a prefix. This is so because $Opt(N, *, *)$ values need be recomputed only for nodes along the insert/delete path of the 1-bit trie. There are $O(W)$ such nodes. For the optimal 2(3)-VST, Sahni et al. propose a faster algorithm [7].

To explain the process of finding an optimal VST, Sahni et al. give an example [7] with a prefix set P as shown in Figure 4.11a. Figure 4.12 shows the 1-bit trie.

To determine the cost, $Opt(N0, 0, 4)$, of the best 4-VST, we must compute all the Opt values shown in Figure 4.13. In this figure, $Opt1$, for example, refers to $Opt(N1, *, *)$ and Opt_{42} refers to $Opt(N42, *, *)$. The Opt arrays shown in Figure 4.13 are computed in postorder, that is, in the order N41, N31, N21, N6, N5, N42, N32, N22, N1, N0. The Opt values shown in Figure 4.13 were computed using Equations 4.11 through 4.14.

From Figure 4.13, we determine that the cost of the best 4-VST for the given prefix set is $Opt(N0, 0, 4) = 18$. To construct this best 4-VST, we must determine the strides for all nodes in the best 4-VST. These strides are easily determined if, with each $Opt(*, 0, *)$, we store the s value that minimizes the right side of Equation 4.11. For $Opt(N0, 0, 4)$, this minimizing s value is 1. This means that the stride for the root of the best 4-VST is 1, its left subtree is empty (because N0 has an empty left subtree), and its right subtree is the best 3-VST for the subtree rooted at N1. The minimizing s value for $Opt(N1, 0, 3)$ is 2 (actually, there is a tie between $s = 2$ and $s = 3$; ties may be broken arbitrarily). Therefore, the right child of the root of the best 4-VST has a stride of 2. Its first subtree is the best 2-VST for N31; its second subtree is empty; its third subtree is the best 2-VST for N32; and its fourth subtree is empty. Continuing in this manner, we obtain the 4-VST of Figure 4.14. The cost of this 4-VST is 18.

P1 = 0*	0*(P1)
P2 = 1*	1*(P2)
P3 = 11*	101*(P4)
P4 = 101*	110*(P3)
P5 = 10001*	111*(P3)
P6 = 1100*	10001*(P5)
P7 = 110000*	11000*(P6)
P8 = 1100000*	11001*(P6)
	1100000*(P8)
	1100001*(P7)
(a) Original prefixes	(b) Expanded prefixes

Figure 4.11 A prefix set and its expansion to four lengths. (From Sahni, S. and Kim, K., *IEEE/ACM Transactions on Networking*, 11, 4, 2003. With permission.)

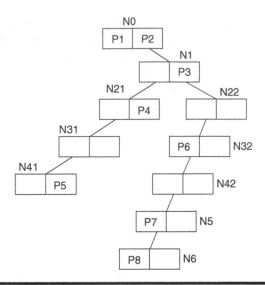

Figure 4.12 1-bit trie for prefixes. (From Sahni, S. and Kim, K., *IEEE/ACM Transactions on Networking*, 11, 4, 2003. With permission.)

Opt_0	$k = 1$	2	3	4		Opt_1	$k = 1$	2	3	4
$s = 0$	128	26	20	18		$s = 0$	64	18	16	16
1	64	18	16	16		1	40	18	16	16
2	40	18	16	16		2	20	12	12	12
3	20	12	12	12		3	10	8	8	8
4	10	8	8	8		4	4	4	4	4
5	4	4	4	4		5	2	2	2	2
6	2	2	2	2						

Opt_{22}	$k = 1$	2	3	4
$s = 0$	32	12	10	10
1	16	18	18	18
2	8	6	6	6
3	4	4	4	4
4	2	2	2	2

Opt_{21}	$k = 1$	2	3	4
$s = 0$	8	6	6	6
1	4	4	4	4
2	2	2	2	2

Opt_{31}	$k = 1$	2	3	4
$s = 0$	4	4	4	4
1	2	2	2	2

Opt_{32}	$k = 1$	2	3	4
$s = 0$	16	8	8	8
1	8	6	6	6
2	4	4	4	4
3	2	2	2	2

Opt_{41}	$k = 1$	2	3	4
$s = 0$	2	2	2	2

Opt_5	$k = 1$	2	3	4
$s = 0$	4	4	4	4
1	2	2	2	2

Opt_{42}	$k = 1$	2	3	4
$s = 0$	8	6	6	6
1	4	4	4	4
2	2	2	2	2

Opt_6	$k = 1$	2	3	4
$s = 0$	2	2	2	2

Figure 4.13 Opt values in the computation of Opt(N0, 0, 4). (From Sahni, S. and Kim, K., *IEEE/ ACM Transactions on Networking*, 11, 4, 2003. With permission.)

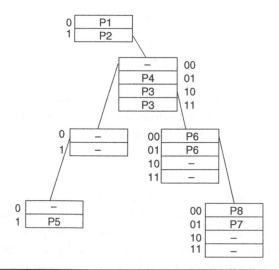

Figure 4.14 Optimal 4-VST for prefixes. (From Sahni, S. and Kim, K., *IEEE/ACM Transactions on Networking*, 11, 4, 2003. With permission.)

4.3 Lulea Algorithms

The Lulea algorithms [8] are designed for quick IP routing lookup. Its data structure is small enough to fit in the cache of a conventional general-purpose processor and it is feasible to do a full routing lookup for each IP packet at gigabit speeds without special hardware.

The forwarding table is a representation of the binary tree spanned by all routing entries. This is called the prefix tree. We require that the prefix tree is complete, that is, that each node in the tree has either two or no children. Nodes with a single child must be expanded to have two children; the children added in this way are always leaves, and their next-hop information is the same as the next-hop of the closet ancestor with next-hop information, or the "undefined" next-hop if no such ancestor exists. This procedure, illustrated in Figure 4.15, increases the number of nodes in the prefix tree, but allows building of a small forwarding table. Note that it is not needed to actually build the prefix tree to simplify our explanation. The forwarding table can be built during a single pass over all routing entries.

A set of routing entries partitions the IP address space into sets of IP addresses. Each set (range) of IP address has a length that is a power of two.

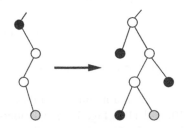

Figure 4.15 Expanding the prefix tree to be complete. (From Degermark, M. et al., *Proceedings ACM SIGCOMM '97*, Cannes, 1997. With permission.)

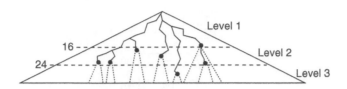

Figure 4.16   The three levels of the data structure. (From Degermark, M. et al., *Proceedings ACM SIGCOMM '97*, Cannes, 1997. With permission.)

As shown in Figure 4.16, level one of the data structure covers the prefix tree down to depth 25–32. Wherever a part of the prefix tree extends below level 16, a level two chunk describes that part of the tree. Similarly, chunks at level 3 describe parts of the prefix tree that are deeper than 24. The result of searching one level of the data structure is either an index into the next-hop table or an index into an array of chunks for the next level.

The core result of the Lulea algorithm is that we can represent a complete binary tree of height h using only one bit per possible leaf at depth h, plus one base index per 64 possible leafs, plus the information stored in the leaves. For $h > 6$, the size in bytes of a tree with l leaves holding information of size d is

$$2^{h-3} + b \times 2^{h-6} + l \times d$$

where b is the size of a base index. With two-byte base indices, a tree of height 8 (a chunk) requires 40 bytes plus leaf information, and a tree of height 16 requires 10 kBytes plus leaf information.

4.3.1   Level 1 of the Data Structure

The first level is essentially a tree node with 1-64K children. It covers the prefix tree down to depth 16. Imagine a cut through the prefix tree at depth 16. The cut is represented by a bit-vector, with one bit per possible node at depth 16. 216 bits = 64 Kbits = 8 kBytes are required for this. To find the bit corresponding to the initial part of an IP address, the upper 16-bits of the address is used as an index into the bit-vector.

Heads.   When there is a node in the prefix tree at depth 16, the corresponding bit in the vector is set. Also, when the tree has a leaf at a depth less than 16, the lowest bit in the interval covered by that leaf is set. All other bits are zero. A bit in the bit-vector can thus be

- A one indicating that the prefix tree continues below the cut; a root head (bits 6, 12, and 13 in Fig. 4.17), or
- A one representing a leaf at depth 16 or less; a genuine head (bits 0, 4, 7, 8, 14, and 15 in Fig. 4.17)
- A zero which means that this value is a member of a range covered by a leaf at a depth less than 16 (bits 1, 2, 3, 5, 9, 10, and 11 in Fig. 4.17). Members have the same next-hop as the largest head smaller than the member

The bit-vector is divided into bit-masks of length 16. There are $2^{12} = 4096$ of those.

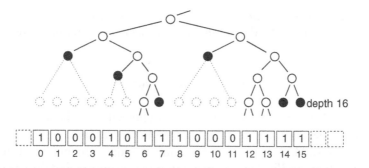

Figure 4.17 Part of cut with corresponding bit-vector. (From Degermark, M. et al., *Proceedings ACM SIGCOMM '97*, Cannes, 1997. With permission.)

Head Information. For genuine heads we need to store an index into the next-hop as the largest head smaller than the member. For root heads, we need to store an index at the level two chunks that represent the corresponding subtree.

The head information is encoded in 16-bit pointers stored consecutively in an array. Two bits of each pointer encode what kind of pointer it is, and the 14 remaining bits either form an index into the next-hop table or an index into an array containing level two chunks. Note that there are as many pointers associated with a bit-mask as its number of set bits.

Finding Pointer Groups. Figure 4.18 is an illustration of how the data structure for finding pointers corresponds to the bit-mask. The data structure consists of an array of code words, as many as there are bit-masks plus an array of base indices, one per four code words. The code words consists of a 10-bit value (r1, r2, ...) and a 6-bit offset (0, 3, 10, 11, ...).

The first bit-mask in Figure 4.18 has three-set bits. The second code word thus has an offset of three because three pointers must be skipped over to find the first pointer associated with that bit-mask. The second bit-mask has seven-set bits and consequently the offset in the third code word is $3 + 7 = 10$.

After four code words, the offset value might be too large to represent with 6-bits. Therefore, a base index is used together with the offset to find a group of pointers. There can be at most 64K pointers in level 1 of the data structure, so the base indices need to be at most 16-bits ($2^{16} = 64K$). In Figure 4.18, the second base index is 13 because there are 13 set bits in the first four bit-masks.

This explains how a group of pointers is located. The first 12-bits of the IP address are an index to the proper code word, and the first 10-bits are index to the array of base indices.

Maptable. It remains to explain how to find the correct pointer in the group of pointers. This is what the 10-bit value is for (r1, r2, ... in Fig. 4.18). The value in an index into a table that maps bit-numbers is 16-bits long. One might think that the table needs 64K entries. However, bit-masks are generated from a complete prefix tree, so not all combinations of the 16-bits are possible.

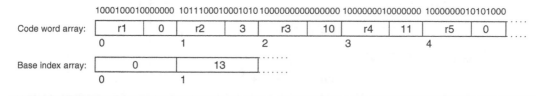

Figure 4.18 Bit-masks versus code words and base indices. (From Degermark, M. et al., *Proceedings ACM SIGCOMM '97*, Cannes, 1997. With permission.)

A nonzero bit-mask of length $2n$ can be any combination of two bit-masks of length n or the bit-mask with value 1. Let $a(n)$ be the number of possible nonzero bit-masks of length 2^n; $a(n)$ is defined by the recurrence

$$a(0) = 1, a(n) = 1 + a(n-1)^2$$

$16 = 2^4$ are a(4) + 1 = 678 = 2^9 + 166 (<2^{10}), the additional one is because the bit mask can be zero. Therefore, the indexes ($r1$, $r2$, ... in Fig. 4.18) only need 10 bits.

We keep such a table, Maptable, to map bit numbers within a bit-mask to 4-bit offsets, shown in Figure 4.19. The offset specifies how many pointers to skip over to find the one required, so it is equal to the number of set bits smaller than the bit index. These offsets are the same for all forwarding tables, regardless of what values the pointers happen to have. Maptable is constant, and is generated once and for all.

Searching. The steps in Figure 4.20 are required to search the first level of the data structure; the array of code words is called code, and the array of base indices is called base. Figure 4.19 illustrates the procedure. The index of the code word, ix, the index of the base index, bix, and the bit number, bit, are first extracted from the IP address. Then the code word is retrieved and its two parts are extracted into ten and six. The pointer index, pix, is then obtained by adding the base index, the 6-bit offset six, and the pointer offset obtained by retrieving the column bit from row ten of the maptable. After the pointer is retrieved from the pointer array, it will be examined to determine if the next-hop has been found or if the search should continue on the next level.

The code is extremely simple. A few bit extractions, array references, and additions are all that are needed. No multiplication or division instructions are required except for the implicit multiplications when indexing an array.

A total of 7 bytes need to be accessed to search in the first level: a two-byte code word, a two-byte base address, one byte (4-bits, really) in the maptable, and finally a two-byte pointer. The size of the first level is 8 kBytes for the code word array, 2 kBytes for the array of base indices, plus a number of pointers. The 5.3 kBytes required by the maptable are shared among all three levels.

When the bit-mask is zero or has a single bit set, the pointer must be an index into the next-hop table. Such pointers can be encoded directly into the codeword, and thus maptable need not contain entries for bit-masks one and zero. The number of maptable entries is thus reduced to 676

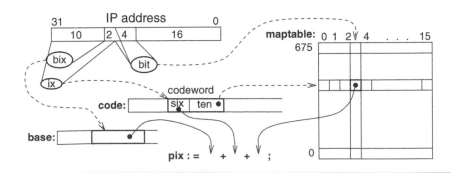

Figure 4.19 Finding the pointer index. (From Degermark, M. et al., *Proceedings ACM SIGCOMM '97*, Cannes, 1997. With permission.)

```
ix := high 12 bits of IP address
bix := high 10 bits of IP address
bit := low 4 of high 16 bits of IP address
codeword := code[ix]
ten := ten bits from codeword
six := six bits from codeword
pix := base[bix] + six + maptable[ten][bit]
pointer := level1_pointers[pix]
```

Figure 4.20   Steps to search the first level. (From Degermark, M. et al., *Proceedings ACM SIGCOMM '97*, Cannes, 1997. With permission.)

(indices 0 through 675). When the ten bits in the codeword (ten above) are larger than 675, the code word represents a direct index into the next-hop table. The six bits from the code word are used as the lowest 6-bits in the index, and (ten—676) are the upper bits of the index. This encoding allows at most $(1024 - 676) \times 2^6 = 22272$ next-hop indices, which is more than the 16K we are designing for. This optimization eliminates three memory references when a routing entry is located at depth 12 or more, and reduces the number of pointers in the pointer array considerably. The cost is a comparison and a conditional branch.

4.3.2   Levels 2 and 3 of the Data Structure

Levels 2 and 3 of the data structure consist of chunks. A chunk covers a subtree of height 8 and can contain at most $2^8 = 256$ heads. A root head in level $n-1$ points to a chunk in level n.

There are three varieties of chunks depending on how many heads the imaginary bit-vector contains. When there are

- 1–8 heads, the chunk is sparse and is represented by an array of the 8-bit indices of the heads, plus eight 16-bit pointers; a total of 24 bytes.
- 9–64 heads, the chunk is dense. It is represented analogously with level 1, expect for the number of base indices. The difference is that only one base index is needed for all 16 code words, because 6-bit offsets can cover all 64 pointers. A total of 34 bytes are needed, plus 18–128 bytes for the pointer.
- 65–256 heads, the chunk is very dense. It is represented analogously with level 1. Sixteen code words and four base indices give a total of 40 bytes. In addition, the 65–256 pointers require 130–512 bytes.

Dense and very dense chunks are searched analogously with the first level. For sparse chunks, the 1–8 values are placed in decreasing order. To avoid a bad worst-case when searching, the fourth value is examined to determine if the desired element is among the first four or last four elements. After that, a linear scan determines the index of the desired element, and the pointer with that index can be extracted. The first element less than or equal to the search key is the desired element. At most 7 bytes need to be accessed to search a sparse chunk.

Dense and very dense chunks are optimized analogously with level 1. In sparse chunks, consecutive heads can be merged and represented by the smallest if their next-hops are identical. When deciding whether a chunk is sparse or dense, this merging is taken into account so that the chunk is deemed sparse when the number of merged heads is eight or less. Many of the leaves added to make the tree complete will occur in order and have identical next-hops. Heads corresponding to such leaves will be merged into sparse chunks.

This optimization shifts the chunk distribution from the larger dense chunks toward the smaller sparse chunks. For large tables, the size of the forwarding table is typically decreased by 5–15 percent.

4.3.3 Growth Limitations in the Current Design

The data structure can accommodate considerable growth in the number of routing entries. There are three limits in the current design.

1. The number of chunks of each kind is limited to $2^{14} = 16,384$ per level. If the limit is ever exceeded, the data structure can be modified so that pointers are encoded differently to give more room for indices, or so that the pointer size is increased.
2. The number of pointers in levels 2 and 3 is limited by the size of the base indices. The current implementation uses 16-bit base indices and can accommodate a growth factor of 3–5. If the limit is exceeded it is straightforward to increase the size of base pointers to 3 bytes. The chunk size is then increased by 3 percent for dense chunks and 10 percent for very dense chunks. Sparse chunks are not affected.
3. The number of distinct next-hops is limited to $2^{14} = 16,384$. If this limit is exceeded all next-hop indices cannot be encoded directly into code words. It is possible to avoid storing a pointer when the bit-mask is zero. When the bit-mask has one head, however, it is necessary that a pointer is stored. Consequently, the size of the data structure will increase because there needs to be one pointer per interval and pointers are larger.

To conclude, with small modifications the data structure can accommodate a large increase in the number of routing entries.

4.3.4 Performance

Forwarding tables are constructed from routing tables of various sizes. For the largest routing tables with 40,000 routing entries, the data structure is 150–160 kBytes, which is small enough to fit in the cache of a conventional general-purpose processor. With the table in the cache, a 200-MHz Pentium Pro or a 333-MHz Alpha 21,164 can perform a few million IP lookups per second without special hardware, and no traffic locality is assumed.

4.4 Elevator Algorithm

The PATRICIA trie is a binary trie as a result of compressing each maximal nonbranching path. Searching for the longest matching prefix for an IP address using the PATRICIA trie takes $O(W)$ time in the worst-case regardless of the number and length of prefixes, where W is the length of the IP address. To improve the time complexity, Sangireddy et al. propose a data structure—the kth-level tree, in which the search algorithm can jump k levels of the PATRICIA trie, where k is an integer between 1 and W, and present two algorithms—Elevator-Stairs and logW-Elevators [9]. This section is from [9]. Portions are reprinted with permission (© 2005 IEEE).

4.4.1 Elevator-Stairs Algorithm

The Elevator-Stairs algorithm is similar to a scenario wherein a tall building has an elevator that stops only at certain intermediate floors. A passenger desiring to reach any floor from the topmost floor of the building can take the elevator up to the nearest possible upper floor and then reach the destination floor by taking the stairs. To implement the ideal, Elevator-Stairs algorithm uses hash tables with multiple bits as the key to skip multiple levels of the PATRICIA trie.

An example of the PATRICIA trie is shown in Figure 4.21. A PATRICIA trie edge may represent a sequence of bits, called its *edge label*. Let the *path label* of a path in a PATRICIA trie be the representation of edge labels along the path, and let the *string depth* of a node be the length of the path label from root to node. We define level l in a PATRICIA trie to be the set of all the nodes that are at string depth l, and the edges between two nodes such that one is at a string depth $< l$ and the other is at a string depth $> l$. For IP address p, $p[i]$ represents the ith of p, and $p[i \dots j]$ represents the substring of p between the ith and jth positions. Let the k-level of a PATRICIA trie denote a level (string depth) ik for some integer i such that $0 \leq i \leq \lfloor W/k \rfloor$.

The PATRICIA trie is first modified to facilitate a search with the kth-level tree. For all the edges that cross a k-level, a nonbranching node is created at the k-level so that the node can be indexed in the kth-level tree. The kth-level tree consists of nodes corresponding to the nodes at all the k-levels in the PATRICIA trie. Each node u in the kth-level tree at string depth has a pointer to the corresponding node in the PATRICIA trie, and a hash table $H(u)$ that stores pointers to all the nodes at string depth $(i + 1)k$ in the subtrie of the PATRICIA trie with u as the root node. The key for a pointer stored in the hash table is obtained by the concatenation of edge labels between u and the corresponding node. Therefore, the length of each key is k. Each node in the kth-level tree at string depth ik keeps a copy of the next-hop port (NHP) of the corresponding node in the PATRICIA trie. If the corresponding node in the PATRICIA trie does not carry a port number, the kth-level tree at string depth ik keeps a copy of the NHP of the nearest ancestor in the PATRICIA trie between levels $(i - 1)k + 1$ and ik. The root of the kth-level tree carries the default port number. The algorithm to build the kth-level tree is shown in Figure 4.22.

The search for longest matching prefix on an IP address p using the kth-level tree, shown in Figure 4.23, starts at its root and sets the variable *current_port_number* to be the default port number. The variable current_port_number stores the port number assigned to the longest matching prefix of length less than the string depth of the current node. At a string depth of ik in the kth-level

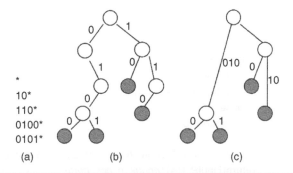

(a) (b) (c)

Figure 4.21 A binary trie and a PATRICIA trie representation for a simple set of prefixes. (a) A small set of prefixes; (b) binary trie representation; (c) PATRICIA trie representation. (From Sangireddy, R., Futamura, N., Aluru, S., and Somani, A., *IEEE/ACM Transactions on Networking,* **13, 4, 2005. With permission.)**

procedure Build k^{th}-level-tree(v)
    Create an empty hash table $H(v)$
    For each edge that crosses level k from v
        Add a non-branching node at level k
    For each node u at level k
        path = path label between v and u
        p = Build k^{th}-level-tree(u)
        Insert p into $H(v)$ with key = *path*

Figure 4.22 Algorithm to build kth-level-tree. (From Sangireddy, R., Futamura, N., Aluru, S., and Somani, A., *IEEE/ACM Transactions on Networking*, 13, 4, 2005. With permission.)

tree, the search algorithm updates the current_port_number to the port number stored in the current node in the kth-level tree, if the node has a copy of the port number. Subsequently, the search algorithm checks the hash table to see if a node with the key $p[ik + 1 \ldots (i + 1)k]$ exists. If such a node exists, the lookup mechanism follows the pointer to the node at level $(i + 1)k$ and continues the search. If such a node does not exist, it indicates that the search for longest matching prefix must end in the PATRICIA trie between levels ik and $(i + 1)k$. Hence, the lookup follows the pointer to the corresponding node in the PATRICIA trie and performs regular PATRICIA trie traversal, as shown in Figure 4.24, which gets completed within k node traversals.

Figure 4.25 shows two cases of LPM search, one at level $3k$ and the other at level $2k + 2$ from the root node. In the former case, the search finds successive matches in the hash tables at levels k, $2k$, and $3k$, but fails to find a match at level $4k$ and also fails to traverse the PATRICIA trie. Thus, an inference is drawn that the LPM for the IP address is in level $3k$ and subsequently the

procedure FindIP(*node*, *p*, *port*)
    port = copy of NHP at *node*
    if *node* represents a leaf in PATRICIA trie
        return(*port*)
    key = $p[pos + 1 \ldots pos + k]$
    if *key* is in $H(node)$
        v = node corresponding to *key* in $H(node)$
        if v is no more then k level away from node
            return(FindIP(v, p, $pos + k$, *port*))
        else
            e = edge between *node* and v
            if p matches the edge label of e
                l = length of e
                return(FindIP(v, p, $pos + l$, k, *port*))
            else
                return(*port*)
    else
        pnode = node in PATRICIA tree corresponding to *node*
        return(FindIP_PAT(*pnode*, p, *pos*, *port*))

Figure 4.23 IP address search algorithm. When function is called initially, node is the root of kth-level-tree, p is the IP address, pos = 0, port is the default NHP. (From Sangireddy, R., Futamura, N., Aluru, S., and Somani, A., *IEEE/ACM Transactions on Networking*, 13, 4, 2005. With permission.)

```
procedure FindIP_PAT(node, p, pos, port)
    if a NHP is assigned to node
        port = NHP assigned to the node
    if the node is leaf
        return(port)
    e = edge of node that starts with p[pos + 1]
    if no such edge exits
        return(port)
    if edge label of e matches p
        child = node at the end of e
        l = length of e
        return(FindIP_PAT(child, IP, pos + l, port))
    else
        return(port)
```

Figure 4.24 **IP address search for Elevator-Stairs algorithm. This function is called from FindIP, and performs search in PATRICIA tree. (From Sangireddy, R., Futamura, N., Aluru, S., and Somani, A., *IEEE/ACM Transactions on Networking*, 13, 4, 2005. With permission.)**

NHP information is retrieved from the corresponding node in the hash table. In the latter case, the search finds a match in the hash tables at levels k and $3k$, but fails to find a match in level $3k$. Subsequently the search traverses the PATRICIA trie starting from the corresponding node in level $2k$, and finds a match at a node in level $2k + 2$. Thus, in the worst case, the search goes through $\lfloor W/k \rfloor$ levels of the kth-level tree and $k - 1$ node traversals in the PATRICIA trie.

Optimal Complexity of Multiple Metrics. As discussed in the above cases, going through each level takes constant time provided hash table lookup takes constant time, and hence the total search time is $O((W/k) + k)$, which is minimized when $k = O(\sqrt{W})$ and achieves a search time of $O(\sqrt{W})$.

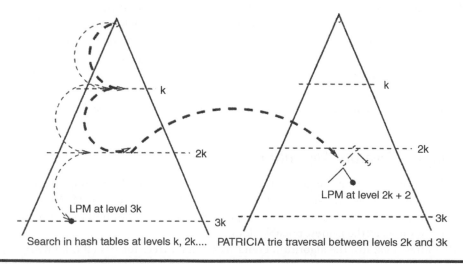

Figure 4.25 **Multi-level hash table lookup using Elevator-Stairs algorithm. Thick dashed lines depict the lookup hops for LPM at level $2k + 2$ and thin dashed lines show the lookup hops for LPM at level $3k$. (From Sangireddy, R., Futamura, N., Aluru, S., and Somani, A., *IEEE/ACM Transactions on Networking*, 13, 4, 2005. With permission.)**

For a routing table with N prefixes, the memory required for storing the kth-level tree is $O(NW/k)$, because $O(N)$ edges may cross multiple k-levels and many nonbranching nodes may be created in the PATRICIA trie. The memory space can be reduced by restricting the edges to have up to two nonbranching nodes: one at the topmost k-level and the other at the lowest k-level that the edge goes through. Search along a long edge must be treated in a special way: when the search algorithm reaches the topmost nonbranching node, the algorithm just compares the edge label between the two nonbranching nodes with the corresponding substring of the query. If the comparison is successful, the search proceeds in a regular way from the lowest nonbranching edge. If the comparison is a mismatch, the longest prefix match is found. Construction of an $O(N)$ space kth-level tree takes $O(N)$ time as well. The optimal complexity of the algorithm for updating the routing tables is next discussed.

Update of Address Lookup Data Structure. Insertion of a prefix in the routing table starts with the search for the longest match to the prefix being inserted. It is the same as searching for the longest matching prefix of the IP address except that the prefix may be shorter than bits. When the search is completed, the nodes of kth-level and PATRICIA trees that need to be modified are known. There are two cases of insertion:

a) The prefix that needs to be inserted forms a new leaf in the PATRICIA trie, or
b) The inserted prefix does not form a new leaf and in that case the structures of the PATRICIA tree and the kth-level tree do not change.

In case (a), a new leaf is attached to the trie at the point where the search ends. It may require the creation of a new internal node to attach an edge to the new leaf. If the edge to the leaf crosses a k-level, a new node is created on this edge at that level in the PATRICIA trie. A node in the kth-level tree is also created to represent the new node in the PATRICIA trie, and a pointer to the new node is added to the hash table in its ancestor in the kth-level tree at $k(i-1)$-level with the $p[ik+1...(i+1)\,k]$ key, where p is the newly inserted prefix. In case the hash table becomes full (exceeds a predefined loading factor), the hash table size is doubled and all the elements are rehashed. This results in an amortized constant time per insertion [10]. Thus, the update takes $search\_time + O(1) = O((W/k) + k)$ amortized time.

In case (b), there are no changes in the structure of the PATRICIA trie, but the new port number must be copied to the nearest kth-level below the updated node in kth-level tree. This insertion takes $search\_time + O(2^k) = O((W/k) + k + 2^k)$ time. The sequence of operations for insertion of a prefix in the routing data structure is shown in Figure 4.26. The procedure for deletion of a prefix occurs in the same way as insertion, and is depicted in Figure 4.27, and the update algorithm in case of a change in the port number of a prefix is shown in Figure 4.28.

4.4.2   *logW-Elevators Algorithm*

The log$W$-Elevators algorithm is equivalent to a scenario wherein a tall building has elevators and each elevator halts at designated floors with regular intervals of 2, 4, 8, and 16, respectively for a 32-floor building and 2, 4, 8, 16, 32, and 64, respectively for a 128-floor building. A passenger can select one of elevators to the nearest possible floor.

For IP address lookups, the algorithm contains kth-level trees for $k = (W/2), (W/4), ..., 2$, in addition to the PATRICIA trie. Thus, the total space required is $O(N\log W)$ because each kth-level tree consumes $O(N)$ space. Building of kth-level trees for a *logW-Elevators* algorithm is shown in Figure 4.29. If the root node of $W/2$-level tree has a pointer to a node key $p[1 ... (W/2)]$, such a

procedure Insert(*p*)

    Search prefix *p* on the same way as FindIP

    if the search finishes inside an edge

        Create a node three

    pnode = node where the search ends

    if *p* = path-label(*pnode*)

        pn = NHP assigned to *p*

        Copy *pn* to appropriate nodes in *k*th-level-tree

    else

        Attach a leaf for *p* at *pnode*

        e = the new edge between *p* and *pnode*

        if *e* crosses a *k*-level

            Create a node at the *k*-level on *e*

            parent = *pnode's* parent in *k*th-level-tree

            knode = node in the *k*th-level-tree to represent *pnode*

            Insert a pointer to *knode* in *H*(*parent*)

            if *H*(*parent*) becomes full

                Double the size of *H*(*parent*)

                Rehash the elements in *H*(*parent*)

Figure 4.26 **Insertion of prefix *p* for Elevator-Stairs algorithm. (From Sangireddy, R., Futamura, N., Aluru, S., and Somani, A.,** *IEEE/ACM Transactions on Networking,* **13, 4, 2005. With permission.)**

procedure Delete(*p*)

    Search prefix *p* on the same way as FindIP

    pnode = node matching *p*

    if *p* = path-label(*pnode*)

        pn = NHP assigned to *p's* parent

        Copy *pn* to appropriate nodes in*k*th-level-tree

    else

        e = the edge to *p*

        delete the leaf for *p*

        if *e* crosses a *k*-level

            knode = the node on *e* at *k*-level

            klnode = node in *k*th-level-tree representing *knode*

            parent = *pnode's* parent in *k*th-level-tree

            Delete *knode*

            Delete the pointer to *kpnode* in *H*(*parent*)

            Delete *Klnode*

            Delete *pnode*

            if *H*(*parent*) has few entries

                Half the size of*H*(*parent*)

                Rehash the element in *H*(*parent*)

Figure 4.27 **Deletion of prefix *p* for Elevator-Stairs algorithm. (From Sangireddy, R., Futamura, N., Aluru, S., and Somani, A.,** *IEEE/ACM Transactions on Networking,* **13, 4, 2005. With permission.)**

```
procedure Change(p)
    Search prefix p on the same way as FindIP
    pnode = node matching p
    Change the NHP for pnode
    if pnode is an intenal node
        Copy new NHP to appropriate nodes in kth-level-tree
```

Figure 4.28 Change of a NHP assigned to prefix *p* for Elevator-Stairs algorithm. (From Sangireddy, R., Futamura, N., Aluru, S., and Somani, A., *IEEE/ACM Transactions on Networking*, 13, 4, 2005. With permission.).

path exists in the PATRICIA trie and the search reaches the node that is indicated by the pointer. It is also possible to determine if the longest prefix match is already found at level $W/2$ by examining if $((W/2) + 1)$th bit is a match in the PATRICIA tree:

a) If the node at level $W/2$ does not have an edge starting with $((W/2) + 1)$th bit of p, the longest match in the PATRICIA trie is found, or

b) If the node at level $W/2$ has an edge starting with $((W/2) + 1)$th bit of p, the search should finish between level $((W/2) + 1)$ and level N, or

c) If there is no hash table entry with $p[1...(W/2)]$ as the key, the search must finish between levels 1 and $W/2$.

In cases (b) and (c), where the longest match in the PATRICIA trie is not found, the range of the string depth where the search has to be performed is halved, and the node from which the search has to continue is known. Thus, the search range keeps halving by using the $(W/4)$-level, $(W/8)$-level, $(W/16)$-level, ..., 2-level trees and eventually ends by searching the PATRICIA trie for at most one level unless the longest match in it is found by case two in some tree along the way. The procedure for the lookup operation using $\log W$-Elevators algorithm is shown in Figure 4.30. All the nodes in the PATRICIA trie store a port number of the longest matching prefix in this algorithm, and the port number can be read from the PATRICIA trie directly. This strategy is chosen to facilitate an update time of $O(N)$. Note that about half of the levels in a kth-level tree are never used. For example, a node in level eight in the 8th-level tree has pointers to nodes in level 16. But these pointers are never used because 16th-level tree must be used to jump to level 16. Hence, those levels that carry pointers to jump to a level that are already covered by another kth-level tree with a larger k can be discarded to save memory.

Figure 4.31 shows two cases of an LPM search, one at level $(W/4) + (W/8)$ (12-bit prefix for IPv4) and the other at level $(W/4) + (W/8) + 1$ (21-bit prefix for IPv4) from the root node. In the former case, the search fails to find a match in the hash table at level $W/2$ and subsequently finds successive matches in hash tables at levels $(W/4)$ and $(W/4) + (W/8)$. At each of these hash table

```
procedure Build_fast
    Make k-level trees with k = W/2, ..., 2
    Link nodes that represent same PATRICIA  trie node
    for each node u in PATRICIA trie in pre-order traversal order
    if u does not have NHP
        Copy NHP from parent
```

Figure 4.29 Algorithm to build *k*th-level-trees for log*W* Elevators algorithm. (From Sangireddy, R., Futamura, N., Aluru, S., and Somani, A., *IEEE/ACM Transactions on Networking*, 13, 4, 2005. With permission.)

```
procedure FindIP_fast(node, p, pos, L, port)
    if L = 1
        return(FindIP_PAT(node, p, pos, port))
    else
        key = p[pos + 1ε...pos + L]
        look up H(node) with key
        if the key is found
            v = node corresponding to key in H(node)
            return(FindIP_fast(v, p, pos + L, L/2, port))
        else
            v = node in L/2 tree corresponding to node
            return (FindIP_fast(v, p, pos, L/2, port))
```

Figure 4.30 log W-Elevators lookup algorithm. When function is called initially node is the root of $(W/2)$th-level-tree, p is the IP address, $pos = 0$, $L = W/2$, current_port is the default NHP. (From Sangireddy, R., Futamura, N., Aluru, S., and Somani, A., *IEEE/ACM Transactions on Networking*, 13, 4, 2005. With permission.)

hits, a single-bit check is made for the existence of further traversals. At level $(W/4) + (W/8)$, it fails the check and thus it is concluded that the LPM for the IP address is at level $(W/4) + (W/8)$ and, subsequently, the NHP information is retrieved from the corresponding node in the hash table. In the latter case, the search consecutively finds a match in the hash table at level $W/2$, fails to find it in level $(W/2) + (W/4)$, finds it in level $(W/2) + (W/8)$, fails to find it in level $(W/2) + (W/8) + (W/16)$, and finally jumps to the PATRICIA trie to find a match at level $(W/2) + (W/8) + 1$.

Optimal Complexity of Multiple Metrics. As discussed in the above cases, with regard to the number of steps it takes for a successful search for a longest prefix match using this algorithm, the IP address lookup can be performed in $O(\log W)$ time using a memory space of $O(N \log W)$. This is achieved with the help of kth-level trees for $k = (W/2), (W/4), \ldots, 2$ in addition to the PATRICIA trie. Thus, the total space required is $O(N \log W)$ because each kth-level tree consumes $O(N)$ space. The complexity of the algorithm for updating the routing table is next discussed.

Update of Address Lookup Data Structure. Update (insertion and deletion) of a prefix in the routing table is more complex than the Elevator-Stairs algorithm because there are multiple

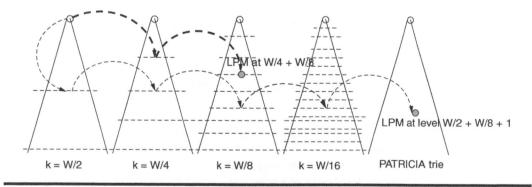

Figure 4.31 Multi-level hash table lookup using log W-Elevators algorithm. Thick dashed lines depict the lookup hops for LMP at level $((W/4) + (W/8))$ and thin dashed lines show the lookup hops for LPM at level $((W/4) + (W/8)) + 1$. (From Sangireddy, R., Futamura, N., Aluru, S., and Somani, A., *IEEE/ACM Transactions on Networking*, 13, 4, 2005. With permission.)

*k*th-level trees and can be as large as *W*/2. If we choose to update *k*th-level trees in the same way as the Elevator-Stairs algorithm, we may have $O(N)$ time to do so, resulting in a total update time of $O(N \log W)$. The port number for the longest matching prefixes can be assigned to each of the nodes in PATRICIA trie by a top-down traversal of the PATRICIA trie. There exist, again, two cases for insertion:

a) The prefix that needs to be inserted forms a new leaf in the PATRICIA trie, or
b) The inserted prefix does not form a new leaf in the PATRICIA trie and in that case the structures of the PATRICIA tree and the *k*th-level tree do not change

In case (a), a new edge is created in the PATRICIA trie and all the *k*th-level trees in which a level crosses the edge to the newly created leaf may be modified. Modification of each *k*th-level tree takes amortized constant time, considering the occasional doubling of the hash tables. Total modification consumes *search_time* + $O(\log W) = O(\log W)$ time. In case (b), a modification of a port number has to be copied to the descendant nodes that do not have a port number assigned. This may take *search_time* + $O(N) = O(N)$ time.

The sequence of operations for insertion of a prefix in the routing data structure is shown in Figure 4.32. The procedure for deletion of a prefix occurs in the same way as an insertion, and is depicted in Figure 4.33. Deletion starts with the searching of the prefix to be deleted. In case (a), where the leaf of the PATRICIA trie is deleted, a deletion from the hash table may occur. In case (b), where the deletion occurs in an internal node in the PATRICIA trie, a port number stored in the parent node is copied to the deleting node and also to the nodes that have a copy of the port number of the deleting node. The update operation in case of a change in the port number for a prefix is shown in Figure 4.34.

4.4.3 Experiments

The experiment platform is a Sun Ultra-Sparc III-based computer with a clock speed of 750 MHz and a 64 KB L1 data cache. The routing tables are extracted from the IPMA project (http://www. merit.edu/, 2004).

For the Elevator-Stairs algorithm, because the lookup complexity is $O((W/k) + k)$, the experiments were done with different values of *k*, from 2 to 12, and selected a value of *k* that optimizes lookup time as shown in Table 4.3. This value of *k* does not necessarily optimize update time or memory usage. The Elevator-Stairs algorithm gives an average throughput of 15.08 million

```
procedure Insert_fast(p)
    Search prefix p on the same way as FindIP_fast
    pnode = node where search ends in PATRICIA trie
    if p = path-label(pnode)
        pn = NHP assigned to p
        Copy pn to appropriate nodes in PATRICIA trie
    else
        Attach a leaf for p at pnode
        Insert (p) for all k-level-trees
```

Figure 4.32 Insertion of prefix *p* for log *W*-Elevators algorithm. (From Sangireddy, R., Futamura, N., Aluru, S., and Somani, A., *IEEE/ACM Transactions on Networking*, 13, 4, 2005. With permission.)

```
procedure Delete_fast(p)
    Search prefix p on the same way as FindIP_fast
    pnode = node matching p in PATRICIA trie
    if p = path-label(pnode)
        pn = NHP assigned to the parent of p
        Copy pn to appropriate nodes in PATRICIA trie
    else
        Delete the pnode for PATRICIA trie
        Delete(p) for all kth-level-tree
```

Figure 4.33   Deletion of prefix *p* for log *W*-Elevators algorithm. (From Sangireddy, R., Futamura, N., Aluru, S., and Somani, A., *IEEE/ACM Transactions on Networking*, 13, 4, 2005. With permission.)

```
procedure Change_fast(p)
    Search prefix p on the same way as FindIP_fast
    pnode = node matching p
    pn = NHP assigned to p
    if pnode is an internal node
        Copy pn to appropriate nodes in PATRICIA trie
    else
        Change the NHP of pnode to pn
```

Figure 4.34   Change of a NHP assigned to prefix *p* for log *W*-Elevators algorithm. (From Sangireddy, R., Futamura, N., Aluru, S., and Somani, A., *IEEE/ACM Transactions on Networking*, 13, 4, 2005. With permission.)

lookups per second (Mlps) with a memory consumption of 517 KB for an AADS router with 33,796 prefixes, and gives an average throughput of 15.74 Mlps with a memory consumption of 459 KB for a MAE-West router with 29,487 prefixes.

Table 4.4 gives the performance of log *W*-Elevators algorithm for the different routers. The number of average lookups represents the average number of *k*th-level trees visited per lookup, and in the worst-case it would be log *W*. The log *W*-Elevators algorithm gives an average throughput of 20.5 Mlps with a memory consumption of 1413 KB for the AADS router, and gives an average throughput of 21.41 Mlps with a memory consumption of 1259 KB for the Mae-West router.

Table 4.3   Lookup Time Performance Analysis for Elevator-Stairs Algorithm

| Routing Table | Prefix Count | k Value for Best Lookup | Lookup Time (μs) | | Memory (kBytes) Consumed |
			Average	Worst-Case	
Mae-west	29,487	8	0.0635 (47.6 cycles)	1.3217	459
Mae-east	24,792	12	0.0860 (64.5 cycles)	1.4334	344
Aads	33,796	8	0.0663 (49.7 cycles)	1.3801	517
PacBell	6822	9	0.0869 (65.0 cycles)	1.4671	88

Source: Sangireddy, R., Futamura, N., Aluru, S., and Somani, A., *IEEE/ACM Transactions on Networking*, 13, 4, 2005. With permission.

Table 4.4   Lookup Time Performance Analysis for log *W*-Elevators Algorithm

Routing Table	Prefix Count	Lookup Time (µs)		Average Lookups	Memory Consumed (kBytes)
		Average	Worst Case		
Mae-west	29487	0.0476 (35.0 cycles)	0.6403	1.3495	1259
Mae-east	24792	0.0864 (64.8 cycles)	0.6334	4.4639	1037
Aads	33796	0.0488 (36.6 cycles)	0.6438	1.3231	1413
PacBell	6822	0.0711 (53.0 cycles)	0.5096	4.5949	305

Source: Sangireddy, R., Futamura, N., Aluru, S., and Somani, A., *IEEE/ACM Transactions on Networking*, 13, 4, 2005. With permission.

The algorithms implemented in a general-purpose processor have given significant throughput rates of packet processing. These, when implemented in hardware, are bound to enhance the packet processing rates.

4.5   Block Trees

A prefix in the routing table is a range in the IP number line. The starts and ends of ranges partitions all IP addresses into intervals. Some researchers construct forwarding table data structure-based interval endpoints. For example, Srinivasan et al. used a two level 16–8 variable-stride trie [6] on top of a set of tree structures, called block trees, shown in Figure 4.35. To optimize multiple performance metrics simultaneously, Sundstrom and Larzon [12] proposed a block tree-based algorithm supporting high lookup performance, fast incremental updates, and a guaranteed memory compression ratio. This section is from [12]. Portions are reprinted with permission (© 2005 IEEE).

4.5.1   Construction of Block Trees

A block tree, or more precisely a (t, w) block tree, is an $O(n)$ space *implicit* tree structure that represents a partition of a set of w bits non-negative integers that supports search operations using

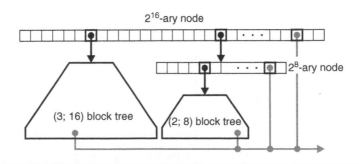

Figure 4.35   Outline of the data structure. The first level is implemented as an array of pointers, referring to a (3, 16) block tree, a 2^8-ary node or a next-hop index (*shaded arrows*). At the second level, each pointer contains either a next-hop index or a reference to a (2, 8) block tree. (From Mikael Sundstrom and Lars-Ake Larzon, *Proceedings—24th Annual Joint Conference of the IEEE Computer and Communications Societies: INFOCOM 2005.* IEEE, 2005. With permission.)

at most t memory accesses. For given values of t and w there are a maximum number of intervals $n_{max}(t, w)$ that can be represented. If the number of intervals at the second level is $\leq n_{max}(3, 16)$ we can use a (3, 16) block tree as an alternative to the 2^8-ary trie node. We can reduce the worst-case amortized space per interval and achieve a better-guaranteed compression ratio for the whole data structure.

After the prefixes in the routing table are transferred into intervals, the sorted list of interval endpoints and next-hop indices is reorganized into *nodes* and *leaves* stored in b bits blocks. To obtain the best compression ratio and lookup performance for a given input size the subtree pointers in the nodes are skipped altogether. Instead, we store the subtrees of a node in order in the blocks immediately after the node. Moreover, we make sure that each subtree, except possibly the last, is *complete* or *full* so that we know its size in advance. We can then search the node to obtain an *index*. The index is then multiplied with the size of the subtree to obtain an offset ahead, from the node to the subtree where the search is continued.

A general block tree is characterized by block size b and the maximum number of blocks t needed to be accessed during lookup. We will sometimes refer to t as the number of *levels* or the *height* of the block tree. Besides b and t, a block tree is also characterized by the size of the data d and the number of bits required to represent a range boundary w. For given values of t, w, b, and d the maximum number of intervals that can be represented by a (t, w, b, d) block tree is

$$n_{max}(t, w, b, d) = \left(\left\lfloor \frac{b}{w} \right\rfloor + 1 \right)^{t-1} \cdot \left\lfloor \frac{b+w}{d+w} \right\rfloor$$

For fixed values of $b = 256$, $d = 13$, and $(t, w) = (3, 16)$ and (2, 8), respectively, we get $n_{max}(3, 16) = 17^2 \cdot 9 = 2601$, $n_{max}(2, 8) = 33 \cdot 12 = 396$.

Observe that $n_{max}(2, 8) > 2^8$. This means that a (2, 8) block tree can always handle the intervals below two levels of trie nodes. By definition, a block tree of height one occupies one block. A complete block tree of height two requires one block for the node and one for each possible leaf. For a (2, 16) block tree we get a total of 18 blocks because we can fit 16 interval endpoints into 256-bits to support up to 17 leafs. Storing a complete block tree in general requires $S(t, w, b, d)$ blocks. A more thorough description of block trees and their properties can be found in [13].

$$S(t, w, b, d) = \frac{\left(\left\lfloor \frac{b}{w} \right\rfloor + 1 \right)^t - 1}{\left(\left\lfloor \frac{b}{w} \right\rfloor + 1 \right) - 1}$$

Constructing a complete $(t, 16)$ block tree from a sorted list of interval endpoints $r_1, r_2,..., r_{n-1}$ is achieved by the following recursive procedure. Interval $R_i = \{ r_{i-1}, r_{i-1} + 1,..., r_i - 1 \}$ is associated with next-hop index D_i and T represents a pointer to the first block available for storing the resulting block tree. The procedure is expressed as a base case (leaf) and inductive step (node).

```
build(1, 16)  (T,  r1...r8,  D1...D9)  =
     store r1...r8 and D1...D9 in block T
build(t, 16)  (T,  r1...rn-1,  D1...Dn)   =
     m ← nmax  (t-1, 16)
     store rm, r2m,  ...,  r16m in block T
     T ← T + 1
```

```
for i ← 0... 16 do
    R ← r_{im+1} ... r_{(i+1) · m-1}
    D ← D_{im} ... D_{(i+1) · m}
    build_{(t-1, 16)} (T, R, D)
    T ← T + S(t - 1, 16)
end
```

Observe that in each round in the loop the block tree pointer is moved forward according to the size of the subtree constructed recursively. To construct a *partial* $(t, 16)$ block tree we use essentially the same procedure except that we bail out when running out of intervals.

Looking up a query key q in a $(3, 16)$ block tree starting at block T is achieved by the following simple procedure. (Endpoints between parentheses are stored implicitly.)

```
lookup_{(3, 16)} (T, q) =
    find min r_i > q in r_1, r_2, ..., r_{16}, (r_{17})
    T ← T + 1 + 18(i - 1)
    find min r_i > q in r_1, r_2, ..., r_{16}, (r_{17})
    T ← T + 1 + (i - 1)
    find min r_i > q in r_1, r_2, ..., r_8, (r_9)
    return D_i
```

There are corresponding but slightly simpler procedures for construction of and lookup in $(2, 8)$ block trees.

4.5.2 Lookup

An overview of the complete forwarding table data structure—how it is organized in memory and how a lookup is performed—is shown in Figure 4.36. The memory is organized into four major areas: *level 1* containing *one* array of 2^{16} pointers, the *next-hop* table which contains up to 2^{13} next-hop

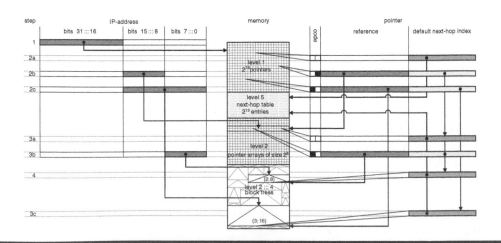

Figure 4.36 Overview of the complete forwarding table lookup. (From Mikael Sundstrom and Lars-Ake Larzon, *Proceedings—24th Annual Joint Conference of the IEEE Computer and Communications Societies: INFOCOM 2005*. IEEE, 2005. With permission.)

entries, *level 2* containing an array of arrays of 2^8 pointers each, and *level 2 ... 4* which contains (3, 16) block trees and (2, 8) block trees of various sizes. We refer to the areas as levels because the LPM data structure consists of four levels where one memory access is spent in each level to compute the next-hop index into the fifth level—the next-hop table.

Looking up the next-hop index and retrieving the next-hop information is described by the procedure below:

```
lookup(a) =                                            step   line
    p ← L1[a₃₁...₁₆]                                     1      1
    def ← p.def                                                2
    if p.code = 00_bin then return L5[def]             2a      3
    elsif p.code = 10_bin then                          2c      4
    ix ← lookup(3, 16) (L2...4[p.ref], a₁₅...₀)         2c      5
    if ix ≠ 0 then def ← ix end                                6
    return L5[def]                                      3c      7
    end                                                        8
    p ← L2[p.ref][a₁₅...₈]                              2b      9
    if p.def ≠ 0 then def ← p.def end                         10
    if p.code = 00_bin then                             3a     11
    return L5[def]                                      3a     12
    end                                                       13
    ix ← lookup(2, 8) (L2...4[p.ref], a₇...₀)           3b     14
    if ix ≠ 0 then def ← index end                            15
    return L5[def]                                      4      16
```

The steps in the lookup procedure refer to the steps in Figure 4.36. There are however a number of steps that are not obvious and we will now go through these. We have reserved next-hop index zero to represent lookup failure and this is used throughout the data structure. Initially during lookup, no prefix (range) has yet matched the address. In line 2, we record the default next-hop index field from the pointer in variable *def*. If the value is zero, it means that there is no prefix of length 0...16 that matches the IP-address. On the other hand, if the value is non-zero, *def* will contain the next-hop index of the longest matching prefix P of a among the prefixes of length 0...16. Observe that among the prefixes of length 0 ... 16, P is the longest matching prefix of any IP-address a' where $a'_{31...16} = a_{31...16}$. In fact, *def* will contain the *default next-hop index* for subuniverse $a_{31...16}$. After performing the block tree lookup in line 5 (14), we store the index in variable *ix* and in line 6 (15) we check if *ix* is the non-zero update *def*. If *ix* is zero, it means that a does not match any prefix of length 17...32 and that the longest matching prefix of a is either P or that a does not have any matching prefix.

Slightly similar to the unconditional assignment of *def* in line 2, we update *def* in line 10 if and only if the default next-hop index field in the pointer extracted from level two is non-zero, that is, if and only if a has a matching prefix of length 17... 24. This technique for storing and managing default next-hop indices for subuniverses is one of the key components in achieving high performance incremental updates while simultaneously maintaining a high compression ratio without affecting the lookup performance. In effect, we create a seal between classes of prefixes of length 0...16 and prefixes of length 17... 32 which allows us to insert or delete a prefix from one class without affecting the other class. In the high density case, when we use 8-bit trie nodes and (2, 8) block trees, there is a corresponding seal (locally) between prefixes of lengths 17 ... 24 and 25 ... 32.

4.5.3 Updates

When a new entry is inserted, the operations performed on the forwarding and next-hop table depend on the length of the prefix and therefore prefixes are classified according to their prefix length.

There are four different classes α, β, γ, and δ containing prefixes of length 0…8 bits, 9…16-bits, 17…24 bits, and 17…32 bits, respectively. Each prefix class has a custom insertion procedure and we will now go through these in detail.

Class α Prefixes. Insert the prefix P into a separate table T_a containing only class α prefixes. For each IP-address of the form a.0.0.0 where P is the longest matching prefix in T_a, store the next-hop information associated with P in slot $a + 1$ in the next-hop table.

Class β Prefixes. Insert the next-hop information in the next-hop table and obtain a next-hop index *index*. Insert the prefix P into a separate table T_β containing only class β prefixes. For each IP-address of the form $a.b.0.0$ where P is the longest matching prefix in T_β, store *index* in the default next-hop index field of pointer $2^8 \cdot a + b$ in the first-level trie node.

Class γ and δ Prefixes (low density). This is applicable if the current subuniverse is empty or is represented by a block tree in the forwarding table. Insert the next-hop information in the next-hop table and obtain a next-hop index *index*. Let u be the 16 most significant bits of the prefix P. (u is referred to as the subuniverse index of P.) Insert P in a table $T_{\gamma \text{ and } \delta}[u]$ containing only class γ and δ prefixes with the same subtree index as P. Compute the *density n* of subuniverse u, that is, the number of basic intervals in the partition of the subuniverse defined by the prefixes in $T_{\gamma \text{ and } \delta}[u]$. If $n > 2295$ the current subuniverse must be handled as a high-density case. Otherwise, we allocate space for constructing a new block tree (or reuse the space for the current one if possible) and then construct a new block tree from the list of intervals in the partition. Each interval that does not correspond to a matching class γ or δ prefix will be associated with next-hop index zero.

Class γ Prefixes (high-density). This is applicable if a 2^8-ary trie node that corresponds to the current subuniverse is in use in the forwarding table or if the density is too high for a low-density case. Insert the next-hop information in the next-hop table and obtain a next-hop index index. As above, let u be the subuniverse index of P. Insert P in a table $T_\gamma[u]$ containing only class γ with the same subtree index as P. For each IP-address of the form a.b.c.0 where P is the longest matching prefix in $T_\gamma[u]$, store the index in the next-hop index field of pointer c in the second-level trie node associated with subuniverse u.

Class δ Prefixes (high-density). Insert the next-hop information in the next-hop table and obtain a next-hop index *index*. Let u be the 24 most significant bits of the prefix P. (u is referred to as the subuniverse index of P.) Insert P in a table $T_\delta[u]$ containing only class δ prefixes with the same subtree index as P. Compute the *density n* of subuniverse u, that is, the number of basic intervals in the partition of the subuniverse defined by the prefixes in $T_\delta[u]$. The density cannot exceed 2^8 and we can always use a (2, 8) block tree to represent the partition. Therefore, allocate space for constructing a new block tree (or reuse the space of the current one if possible) and then construct a new block tree from the list of intervals in the partition. As above, uncovered intervals are associated with next-hop index zero.

For class α prefixes, we use a technique not mentioned in the previous lookup section. The reason for treating these short class α prefixes as a special case instead of treating all prefixes of length 0 … 16 the same is to avoid updates of large portions of the first-level trie node in the forwarding table. If the result from the lookup described in the previous section is zero, the query IP-address does not match any prefix of length 9 ... 32 but there is still a possibility that a class α prefix is matched. By reserving slots 1 ... 256 of the next-hop table for class α prefixes and using this technique, a copy of the next-hop information associated with the longest such prefix is stored in the slot with the same index as the eight most significant bits of the query IP-address plus one. The corresponding procedures for deleting prefixes of different classes are essentially the inverses of the insert procedures.

What then is the cost for incremental updates of the forwarding table? Because we consider *only* memory accesses in the forwarding table and the next-hop table, inserting or deleting a class α prefix may require that the first 256-slots in the next-hop table are accessed. Assuming that each piece of the next-hop information occupies eight bytes, we then need to access 64 memory blocks. For class β prefixes, the updates take place in the first-level trie node. We only update pointers that correspond to the inserted or deleted prefix and because the prefix is at least 9-bits long, at most 128 pointers need to be updated. Because each pointer occupies four bytes, the total number of blocks accessed for class β updates is 16. For class γ and δ prefixes, the cost for incremental updates is directly related to the size of the block trees and the second-level trie nodes constructed, because they must be filled with information. However, even more important is the cost of *allocating* and *deallocating* space. In the following subsection, we present a memory management algorithm which allows us to perform allocation and deallocation at virtually the same cost as the actual construction cost while maintaining zero fragmentation and maximum compression ratio.

4.5.4 Stockpiling

Consider the general problem of allocating and deallocating memory areas of different sizes from a *heap* while maintaining zero fragmentation. In general, allocating a *contiguous* memory area of size s blocks is straightforward—we simply let the heap grow by s blocks. Deallocation is however not so straightforward. Typically, we end up with a hole somewhere in the middle of the heap and a substantial reorganization effort is required to fill the hole. An alternative would be to relax the requirement that memory areas need to be contiguous. It would then be easier to create patches for the holes, but nearly impossible to use the memory areas for storing data structures and so on.

We need a memory management algorithm, which is something between these two extremes. The key to achieving this is the following observation: In the block tree lookup, the leftmost block in the block tree is always accessed first followed by accessing one or two additional blocks beyond the first block. It follows that a block tree can be stored in two parts where information for locating the second part and computing the size of the respective parts is available after accessing the first block.

A *stockling* is a *managed memory area* of s blocks (i.e., b-bit blocks) that can be moved and stored in two parts to prevent fragmentation. It is associated with information about its size s, whether or not the area is divided in two parts, and the location and size of the respective parts. Moreover, each stockling must be associated with the *address to the pointer* to the data structure stored in it so that it can be updated when moved. Finally, it is associated with a (possibly empty) procedure for encoding the location and size of the second part and the size of the first part in the first block.

Let n_s be the number of stocklings of size s. These stocklings are stored in, or actually constitute, a *stockpile*, which is a contiguous $s \cdot n_s$ block memory area. A stockpile can be moved one block to the left by moving one block from its left side to its right side (the information stored in the block in the leftmost block is moved to a free block at the right of the rightmost block). Moving a stockpile one block to the right is achieved by moving the rightmost block to the left side of the stockpile. The rightmost stockling in a stockpile is possibly stored in two parts while all other stocklings are contiguous. If it is stored in two parts, the left part of the stockling is stored in the right end of the stockpile and the right end of the stockling at the left end of the stockpile.

Assume that we have c different sizes of stocklings $s_1, s_2, \ldots, s_c$ where $s_i > s_{i+1}$. We organize the memory so that the stockpiles are stored in sorted order by increasing size in the growth direction. Furthermore, we assume without loss of generality that the growth direction is to the right. Allocating and deallocating a stockling of size s_i from stockpile i is achieved as follows:

Allocate s_i. Repeatedly move each of the stockpiles $1, 2, \ldots, i-1$ one block to the right until all stockpiles to the right of stockpile i have moved s_i blocks. We now have a free area of s_i blocks to the right of stockpile i. If the rightmost stockling of stockpile i is stored in one piece, return the free area. Otherwise, move the left part of the rightmost stockling to the end of the free area (without changing the order between the blocks). Then return the contiguous s_i block area, beginning where the rightmost stockling began before its leftmost part was moved.

Deallocate s_i. Locate the rightmost stockling that is stored in one piece (it is either the rightmost stockling itself or the stockling to the left of the rightmost stockling) and move it to the location of the stockling to be deallocated. Then reverse the allocation procedure.

Figure 4.37 illustrates the stockpiling technique in the context of insertion and deletion of structures of size 2 and 3 in a managed memory area with stockling sizes 2, 3, and 5. Each structure consists of a number of blocks, and these are illustrated by squares with a shade of gray and a symbol. The shades are used to distinguish between blocks within a structure and the symbol is used to distinguish between blocks from different structures. We start with a 5-structure, and then in Figure 4.37a we insert a 2-structure after allocating a 2-stockling. Observe that the 5-structure is stored in two parts with the left part starting at the sixth block and the right part at the third block. In Figure 4.37b we allocate and insert three blocks, and, as a result, the 5-structure is restored into one piece. A straightforward deletion of the 2-structure is performed in Figure 4.37c resulting in both remaining structures being stored in two parts. Finally, in Figure 4.37d a new 3-structure is inserted. This requires that we first move the 5-structure three blocks to the right. Then, the left part (only the white block in this case) of the old 3-structure is moved next to the 5-structure and finally the new 3-structure can be inserted.

The cost for allocating an s_i stockling and inserting a corresponding structure is computed as follows. First, we have to spend $(i-1) \cdot s_i$ memory accesses for moving the other stockpiles to create free space at the end of the stockpile. We then have two cases: (*i*) Insert the data structure directly into the free area. The cost for this is *zero memory* accesses because we have already accessed the free area when moving the stockpiles (insertion can be done simultaneously while moving the stockpiles). (*ii*) We need to move the leftmost part of the rightmost stockling. However, it occupies an area that will be overwritten when inserting the data structure. Therefore, we get an additional s_i memory access for inserting the data structure. For deallocation, we get an additional cost of s_i memory access because we may need to overwrite the deleted stockling somewhere in the middle of the stockpile. We also need to account for the cost for updating pointers to the data structures that are moved. Because the stockpiles are organized by increasing size, at most one pointer needs to be updated for each stockpile moved plus two extra pointer updates in the current stockpile. It follows that

Lemma 4.5.1: The cost for inserting a s_i block data structure when using stockpile memory management is $i \cdot s_i + (i-1) + 2 = i \cdot s_i + i + 1$ memory accesses and the cost for deletion is $(i+1) \cdot s_i + (i-1) + 2 = (i+1) \cdot s_i + i + 1$ memory accesses.

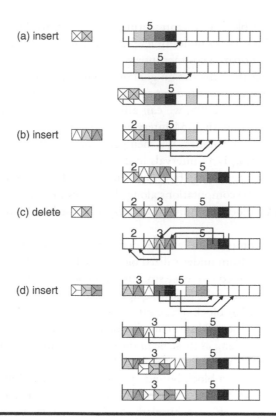

Figure 4.37 Examples of stockpiling. (From Mikael Sundstrom and Lars-Ake Larzon, *Proceedings—24th Annual Joint Conference of the IEEE Computer and Communications Societies: INFOCOM 2005.* IEEE, 2005. With permission.)

Stockpiling can also be used if it is not possible to store data structures in two parts. In each stockpile, we have a dummy stockling and ensure that after each reorganization, it is always the dummy stocklings that are stored in two parts. The extra cost for this is $\sum s_i$ space and, in the worst-case, $\sum_{j=1}^{i} s_j$ memory accesses for swapping the dummy with another stockling that happened to be split up after a reorganization.

4.5.5 Worst-Case Performance

In Section 4.5.1, we have already fixed the lookup costs to four memory accesses for the longest prefix match operation and one additional memory access for retrieving the next-hop information. Here, we analyze the worst-case amortized space and update costs.

We begin by computing the space required for the forwarding table. The cost for the first level is fixed to 2^{16} times the size of a pointer, which is 4 bytes. This gives a total of 2^{18} bytes. For each subuniverse of density ≤2295, a (3, 16) block tree is used. If the density is 2296 or larger, a 2^8-ary trie node of size $4 \cdot 2^8 = 1024$ bytes is used and below the node there will be a number of (2, 8) block trees. The cost for the trie node is amortized over at least 2296 intervals. Hence, its contribution to the amortized cost per interval is 1024/2296 bytes. Our goal is to achieve a compression ratio of 10 bytes per prefix (not counting the first level). This corresponds to 5 bytes per interval in (3, 16) block trees and 5 − 1024/2296 ≈ 4.554 bytes per interval in (2, 8) block trees.

Now consider a (3, 16) block tree representing 2295 intervals. Storing such a tree requires $\lceil 2295/9 \rceil = 255$ leaves, $\lceil 255/17 \rceil = 15$ nodes in level 2, and 1 node in level 1 giving a total of $255 + 15 + 1 = 271$ blocks. The amortized cost per interval is $271 \cdot 32/2295 \approx 3.7786$ bytes per interval, which is considerably better than the design goal of 5 bytes. In fact, the number of intervals can be reduced to 1734 before it becomes too expensive to use 271 blocks $(271 \cdot 32/1734 = 5.0012)$. Hence, if the density is 1735 ... 2295, we can afford to allocate 271 blocks for the block tree. Similarly, storing a (3, 16) tree with 1734 intervals requires $\lceil 1734/9 \rceil = 193$ leaves, $\lceil 193/17 \rceil = 12$ levels of 2 nodes and 1 level of 3 nodes. We then get a total of $193 + 12 + 1 = 206$ blocks. The amortized cost for representing 1734 intervals in 206 blocks is $206 \cdot 32/1734 \approx 3.8016$ bytes per interval and the number of intervals can be reduced to 1318 before it becomes too expensive to use 206 blocks. By continuing the computations along these lines we obtain a mapping between densities and a minimal set of allocation units or *stockling sizes* used in the reference implementation according to Table 4.5.

To achieve the desired compression ratio for densities of 37 intervals or 18 prefixes or less, the quantization effects resulting from underutilized blocks need to be reduced. In the process, we will slightly deviate from the presented reference stockling sizes but not in a way that affects the analysis of the update costs. We use one of the following approaches or a combination of the two.

Quarter Block Trees. (3, 16)-block trees containing 10...37 intervals are represented using one node and a number of leaves. The maximum number of interval endpoints in the node is $\lceil 37/9 \rceil - 1 = 4$, occupying 8 bytes, and as long as these lie in the same block we can search among them in one memory access. We then have two memory accesses left for the lookup and, therefore, it does not matter if leaves are stored across a block boundary. Hence, for 10...37 intervals, we can store the block trees in a special memory area as 8-byte blocks instead of 32-byte blocks and this is sufficient to achieve the desired compression ratio. Similarly, for 1...9 intervals, we use a single leaf stored in a special memory area as 2-byte blocks. These smaller block sizes require that we increase the reference fields in the pointers to 21-bits instead of 17 and this requires that 4 of the 13-bits from the default next-hop index field are stored in the block tree itself. We use code 11_{bin} to distinguish this from the three standard cases.

Prefix Lists. Each (16-bits) prefix and the corresponding next-hop index require $(16 - x)$ bits for the prefix itself, $(x + 1)$ bits for the prefix length (represented in base 1), and 13 bits for the next-hop index. The total number of bits for storing up to 18 prefixes and next-hop indices including the number of prefixes is $18 \cdot (16 - x + x + 1 + 13) + 5 = 545$. We can skip block trees and store the prefixes directly as a list. By storing prefix lists of length 18 in an 8-byte aligned contiguous memory area and smaller prefix lists in a 4-byte aligned memory area, we can guarantee that the prefix list does not straddle more than two block boundaries and can thus be searched in at most three memory accesses as required. The storage requirement is 10 bytes per prefix and with this representation we use only 4 bytes per prefix for 1...18 prefixes.

For (2, 8) block trees, we use similar techniques to map their densities to stocklings of size 23 and smaller and the same approach to reduce quantization effects. Recall that the level 3 node is not fully utilized if the density is 2295 and the reason for this is that we need to use four bytes to encode stockling information (if the stockling is stored in two parts). The same goes for level 2 nodes. Therefore, when the density is below 135, we first step down to the 2-level block trees. This is the reason for the jump from 21 blocks to 16 blocks. Of the four bytes, we use 17-bits to reference to the right part of the stockling, 5-bits for the size (as there are 23 different sizes), and 9 bits for the size of the left part. At the cost of some additional pointer updates during incremental updates, we can even skip the encoding of the size of the left part to save 9 bits. This is achieved as follows: Let i be the index of the leftmost block of an s-stockling relative to the base of the memory. Make sure that $i \equiv 0 \pmod{s}$ when allocating an s-stockling. For stocklings stored in two parts,

Table 4.5 Relation between Densities and Stockling Sizes

Number of Basic Intervals	Stockling Size
1735 … 2295	271
1319 … 1734	206
1005 … 1318	157
768 … 1004	120
596 … 767	93
461 … 595	72
365 … 460	57
288 … 364	45
224 … 287	35
180 … 223	28
148 … 179	23
135 … 147	21
103 … 134	16
84 … 102	13
71 … 83	11
58 … 70	9
52 … 57	8
45 … 51	7
36 … 44	6
27 … 35	5
18 … 26	4
10 … 17	3
1 … 9	1

Source: Mikael Sundstrom and Lars-Ake Larzon, *Proceedings— 24th Annual Joint Conference of the IEEE Computer and Communications Societies: INFOCOM 2005.* IEEE, 2005. With permission.

store i plus the size of the left part in the pointer. Computing the size of the left part during lookup is then achieved by taking the pointer value modulo s.

Conceptually, we allocate a big chunk of memory for the forwarding table from the beginning. Because we have a guaranteed compression ratio, we can guarantee that for a given maximum number of prefixes N_{max} we will not run out of memory. At one end of the chunk, we place the first level trie node and the next-hop table as in Figure 4.36. The rest of the chunk is the dynamic memory. A set of stockpiles is implemented in each end of the dynamic memory. They are called sp_{lo} and sp_{hi}, respectively and grow, in opposite directions, toward each other. By the worst-case amortized space boundary, we can guarantee that they will not grow into each other as long as the number of prefixes is less than N_{max}. The stockpile sizes in the respective stockpile set are given by

$$sp_{lo} = 206, 157, 93, 72, 5, 4, 3, 1, 32, \text{ and}$$
$$sp_{hi} = 271, 120, 57, 45, 35, 28, 23, 21, 16, 13, 11, 9, 8, 7, 6.$$

Observe that we added stocklings of size 32 that will contain 2^8-ary trie nodes. Moreover, we have configured the stockpile so that this is the innermost size class. This means that the trie nodes will be located beside the next-hop table, but more importantly, trie nodes will be stored in one part. Thus it is no problem to perform direct indexing.

By Lemma 4.5.1, we can compute the cost of allocation and deallocation from each stockpile. To compute the worst-case update cost, we only need consider transitions that can occur such that deallocating one size and allocating the next smaller or larger size (assuming that we can perform updates without worrying about simultaneous lookups). By an exhaustive search among the possible transitions, we have found that the worst-case transition is when we go from 271 blocks, causing a deallocation cost of 544 memory accesses to 206 blocks at an allocation cost of 208 memory accesses. The total worst-case transition cost between block trees is then $208 + 544 = 752$ memory accesses. Will it be more expensive than this when we go from 2295 intervals in a block tree to 2296 intervals, which requires a trie node and possibly some (2, 8) block trees? Handling this in a straightforward fashion will result in an explosive transition at considerable expense. To achieve a smooth transition, we allocate a trie node (32-stockling) and then construct the initial set of (2, 8) block trees inside the 271-stockling. Moreover, we keep the 271-stockling during the course of the lifetime of the trie node (i.e., as long as the density is larger than 2295) and use it as the preferred choice for storing (2, 8) block trees. In this way, we avoid explosive and implosive transitions and obtain smooth transitions also on the verge of using block trees. Hence, the worst-case cost for incremental updates is 752 memory accesses.

4.5.6 Experiments

The experiment platform is a standard Pentium 4 PC with a 3.0 GHz CPU. A dataset (including a routing table and updates) was collected in London from the RIS project (http://www.ripe.net/ripencc/pub-services/np/ris/index.html, on January 31, 2004). The other datasets are from the IPMA project (http://www.merit.edu/). The measured results are shown in Table 4.6. The average lookup time is measured by looking up 10,000,000 random IP addresses on each dataset. Each address is looked up twice to get measurements of the uncached and cached entries respectively.

In Table 4.6, we observe that the compression ratio increases with the dataset size. This is because a larger dataset increases the possibility of aggregating prefixes and the cost of the root array is amortized over a larger number of prefixes. The average lookup speeds are more than 27 million packets per second, which is equivalent to wire speeds exceeding 10 Gbit/s. The caches can speedup more than 40 percent.

When measuring the incremental update performance, we insert all entries in each dataset into the data structure, and then randomly choose one entry from the data set. If the entry is already present in the data structure, it is removed—otherwise it is inserted. This procedure is repeated 1,000,000 times. The average updates for each of the datasets are more than 200,000 updates per second, while the largest observed update rates in the Internet is of the order of 400 updates per second [14].

Table 4.6 Dataset Information and Measured Performance

Dataset		London	Mae-East	Pac-Bell	Mae-West
No. of unique prefixes		131,227	32,061	17,431	13,294
Total size (byte)		948,196	414,432	344,156	319,004
Average bytes per routing entries	Total	7.23	10.77	16.68	21.20
	Exc. root array	5.23	3.96	3.97	3.78
The average updates per second		219,988	300,384	327,465	414,948
Uncached lookups	CPU cycles/lookup	107	65	39	36
	Lookups/second	27,799,411	45,516,566	76,514,811	82,204,341
	Supported wire speed(Gbit/s)	10.2	16.75	28.1	30.25
Cached lookups	CPU cycles/lookup	44	28	18	16
	Lookups/second	67,673,468	103,558,663	165,602,475	179,201,603
	Supported wire speed(Gbit/s)	24.9	38.1	60.9	65.9

Source: Mikael Sundstrom and Lars-Ake Larzon, *Proceedings—24th Annual Joint Conference of the IEEE Computer and Communications Societies: INFOCOM 2005.* IEEF, 2005. With permission.)

4.6 Multibit Tries in Hardware

4.6.1 Stanford Hardware Trie

The Stanford hardware trie [11] is based on the following two key observations: On backbone routers there are very few routes with prefixes longer than 24-bits and memory is getting cheaper and cheaper. This observation provides the motivation for trading off large amounts of memory for lookup speed. This subsection is from [11] (© 1998 IEEE).

The basic scheme *DIR-24-8-BASIC* uses a two-level multibit trie with fixed-strides: the first level corresponds to a stride of 24-bits (called TBL24) and the second level to a stride of 8-bits (called TBLlong), shown in Figure 4.38.

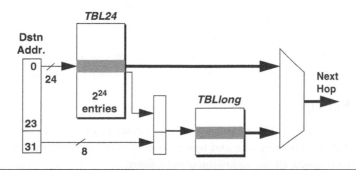

Figure 4.38 DIR-24-8-BASIC architecture. (From Gupta, P., Lin, S., and McKeown, N., *Proceedings IEEE INFOCOM '98*, 1998. With permission.)

If longest prefix with this 24-bit prefix is <25 bits long:

0	Next-hop
1 bit	15 bits

If longest prefix with this 24-bit prefix is >25 bits long:

0	Index into 2nd table TBLlong
1 bit	15 bits

Figure 4.39 TBL24 entry format. (From Gupta, P., Lin, S., and McKeown, N., *Proceedings IEEE INFOCOM '98*, 1998. With permission.)

TBL24 stores all possible prefixes that are up to, and including, 24-bit long. This table has 2^{24} entries, addresses from 0.0.0 to 255.255.255. Each entry in TBL24 has the format shown in Figure 4.39. TBLlong stores all route-prefixes in the routing table that are longer than 24-bits.

A prefix, X, is stored in the following manner: if X is less than or 24-bits long, it need only be stored in TBL24: the first bit of the entry is set to zero to indicate that the remaining 15-bits designate the next-hop. If, on the other hand, the prefix X is longer than 24-bits, then we use the entry in TBL24 addressed by the first 24-bits of X. We set the first bit of the entry to one indicate that the remaining 15-bits contain a pointer to a set of entries in TBLlong. In effect, prefixes shorter than 24-bits are expanded. For example, the prefix 128.23./16 will have 256 entries associated with it in TBL24, ranging from the memory address 128.230.0 through 128.23.255. All 256 entries will have exactly the same content (the next-hop corresponding to the prefix 128.23/16). By using memory inefficiently, we can find the next-hop information within one memory access.

TBLlong contains all prefixes that are longer than 24-bits. Each 24-bit prefix that has at least one prefix longer than 24-bits is allocated $2^8 = 256$ entries in TBLlong. Each entry in TBLlong corresponds to one of the 256 possible longer prefixes that share the single 24-bit prefix in TBL24. Because we are simply storing the next-hop in each entry of TBLlong, it need be only 1 byte wide.

When a destination address is presented to the route lookup mechanism, the following steps are taken.

1. Using the first 24-bits of the address as an index into the first table TBL24, we perform a single memory read, yielding 2 bytes.
2. If the first bit equals zero, then the remaining 15-bits describe the next hop.
3. Otherwise (if the bit equal one), we multiply the remaining 15-bits by 256, add the product to the last 8-bits of the original destination address (achieved by shifting and concatenation), and use this value as a direct index into TBLong, which contains the next hop.

Entries in TBL24 need 2 bytes to store a pointer; hence, a memory bank of 32 MBytes is used to store 2^{24} entries. The size of TBLlong depends on the expected worst-case prefix length distribution because there are very few prefixes longer than 24-bits; in practice, prefix length is smaller.

This scheme requires a maximum of two memory accesses for a lookup. Nevertheless, because the first stride is 24-bits and leaf pushing [6] is used, updates may take a long time in some cases.

4.6.2 Tree Bitmap

Tree Bitmap [15] is a multibit trie algorithm that allows fast searches (one memory reference per trie node) and allows faster update and fewer memory storage requirements. The Tree Bitmap design and analysis is based on the following observations:

■ A multibit node (representing multiple levels of unibit nodes) has two functions: to point at children multibit nodes, and to produce the next-hop pointer for searches in which the longest matching prefix exists within the multibit node. It is important to keep these purposes distinct from each other.

■ With burst-based memory technologies, the size of a given random memory access can be very large (e.g., 32 bytes for SDRAM). This is because while the random access rate for core DRAMs have improved very slowly, high speed synchronous interfaces have evolved to make the most of each random access. Thus the trie node stride sizes can be determined based on the optimal memory burst sizes.

■ Hardware can process complex bitmap representations of up to 256-bits in a single cycle. Additionally, the mismatch between processor speeds and memory access speeds have become so high that even software can do extensive processing on bitmaps in the time required for a memory reference.

■ To keep update times bounded it is best not to use large trie nodes (e.g., 16-bit trie nodes used in the Lulea algorithm). Instead, we use smaller trie nodes (at most 8 bits). Any small speed loss due to the smaller strides used is offset by the reduced memory access time per node (one memory access per trie node versus three in the Lulea algorithm).

■ To ensure that a single node is always retrieved in a single page access, nodes should always be power of two in size and properly aligned (8 byte nodes on 8-byte boundaries etc.) on page boundaries corresponding to the underlying memory technology.

Based on these observations, the Tree Bitmap algorithm is based on three key ideas. The first is that all child nodes of a given trie node are stored contiguously. This allows us to use just one pointer for all children (the pointer points to the start of the child node block) because each child node can be calculated as an offset from the single pointer. This can reduce the number of required pointers by a factor of two compared with standard multibit tries. More importantly it cuts down the size of trie nodes, as we see below. The only disadvantage is that the memory allocator must, of course, now deal with larger and variable sized allocation chunks. Using this idea, the same 3-bit stride trie of Figure 4.40 is redrawn as Figure 4.41.

The second idea is that there are two bitmaps per trie node, one for all the internally stored prefixes and one for the external pointers. See Figure 4.42 for an example of the internal and external bitmaps for the root node. The use of two bitmaps allows us to avoid leaf pushing [6]. The internal bitmap is very different from Lulea encoding, and has a 1-bit set for every prefix stored within this node. Thus for an r bit trie node, there are $2^{(r-1)}$ possible prefixes of lengths $< r$ and thus we use a $2^r - 1$ bitmap.

For the root trie node of Figure 4.41, we see that we have three internally stored prefixes: P1 = *, P2 = 1*, and P3 = 00*. Suppose our internal bitmap has one bit for prefixes of length 0,

P1	*
P2	1*
P3	00*
P4	101*
P5	111*
P6	1000*
P7	11101*
P8	111001*
P9	1000011*

Figure 4.40   An example of forwarding table. (From Eatherton, W., Varghese, G., and Dittia, Z., *ACM SIGCOMM Computer Communication Review Archive*, 34, 2, 2004. With permission.)

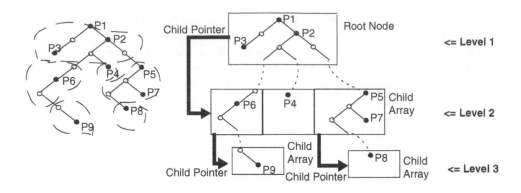

Figure 4.41   Sample database with tree bitmap. (From Eatherton, W., Varghese, G., and Dittia, Z., *ACM SIGCOMM Computer Communication Review Archive*, 34, 2, 2004. With permission.)

two following bits for prefixes of length 1, 4 following bits for prefixes of length 2 and so on. Then for 3-bits the root internal bitmap becomes 1011000. The first 1 corresponds to P1, the second to P2, the third to P3. This is shown in Figure 4.42.

The external bitmap contains a bit for all possible 2^r child pointers. Thus in Figure 4.41, we have eight possible leaves of the 3-bit subtrie. Only the fifth, sixth, and eighth leaves have pointers to children. Thus the extending path (or external) bitmap shown in Figure 4.42 is 00011001. As in the case of the Lulea algorithm, we need to handle the case where the original pointer position contains a pointer and a stored prefix (e.g., location 111, which corresponds to P5 and also needs a pointer to prefixes like P7 and P8). The trick we use is to push all length 3 prefixes to be stored along with the zero length prefixes in the next node down. For example, in Figure 4.41, we push P5 to be in the rightmost trie node in Level 2. In the case of P4, we actually have to create a trie node just to store this single zero length prefix.

The third idea is to keep the trie nodes as small as possible to reduce the required memory access size for a given stride. Thus a trie node is of fixed size and only contains an external pointer bitmap, an internal next-hop information bitmap, and a single pointer to the block of child nodes. But what about the next-hop information associated with any stored prefixes?

The trick is to store the next-hops associated with the internal prefixes stored within each trie node in a *separate* array associated with this trie node. For memory allocation purposes, result arrays are normally an even multiple of the common node size (e.g., with 16-bit next-hop pointers, and 8-byte nodes, one result node is needed for up to 4 next-hop pointers, two result nodes are

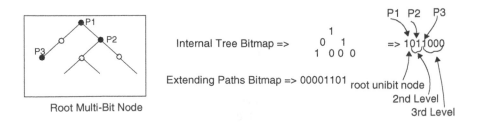

Figure 4.42   Multibit node compression with tree bitmap. (From Eatherton, W., Varghese, G., and Dittia, Z., *ACM SIGCOMM Computer Communication Review Archive*, 34, 2, 2004. With permission.)

needed for up to 8, etc.) Putting next-hop pointers in a separate result array potentially requires two memory accesses per trie node (one for the trie node and one to fetch the result node for stored prefixes). However, we use a simple lazy strategy to not access the result nodes until the search terminates. We then access the result node corresponding to the last trie node encountered in the path that contained a valid prefix. This adds only a single memory reference at the end besides the one memory reference required per trie node.

The search algorithm is now quite simple. We start with the root node and use the first bits of the destination address (corresponding to the stride of the root node, 3 in our example) to index into the external bitmap at the root node at say position P. If we get 1 in this position there is a valid child pointer. We count the number of 1s to the left of this 1 (including this 1) as say I. Because we know the pointer to the start position of the child block (say C) and the size of each trie node (say S), we can easily compute the pointer to the child node as C + (I * S).

Before we move on to the child, we also check the internal bitmap to see if there is a stored prefix corresponding to position P. This requires a completely different calculation from the Lulea style bit calculation. To do so, we can imagine that we successively remove bits of P starting from the right and index into the corresponding position of the internal bitmap looking for the first 1 encountered. For example, suppose P is 101 and we are using a 3-bit stride at the root node bitmap of Figure 4.42. We first remove the rightmost bit, which results in the prefix 10*. Because 10* corresponds to the sixth bit position in the internal bitmap, we check if there is a 1 in that position (there is not in Fig. 4.42). If not, we need to remove the rightmost two bits (resulting in the prefix 1*). Because 1* corresponds to the third position in the internal bitmap, we check for a 1 there. In the example of Figure 4.42, there is a 1 in this position, so our search ends. (If we do not find a 1, however, we simply remove the first 3-bits and search for the entry corresponding to * in the first entry of the internal bitmap.)

This search algorithm appears to require a number of iterations proportional to the logarithm of the internal bitmap length. However, in hardware for bitmaps of up to 512-bits or so, this is just a matter of simple combinational logic (which intuitively does all iterations in parallel and uses a priority encoder to return the longest matching stored prefix). In software this can be implemented using table lookup. Thus while this processing appears more complex than the Lulea bitmap processing, it is actually not an issue in practice.

Once we know we have a matching stored prefix within a trie node, we do not immediately retrieve the corresponding next-hop information from the result node associated with the trie node. We only count the number of bits before the prefix position (more combinational logic!) to indicate its position in the result array. Accessing the result array would take an extra memory reference per trie node. Instead, we move to the child node while remembering the stored prefix position and the corresponding parent trie node. The intent is to remember the last trie node T in the search path that contained a stored prefix, and the corresponding prefix position. When the search terminates (because we encounter a trie node with a 0 set in the corresponding position of the external bitmap), we have to make one more memory access. We simply access the result array corresponding to T at the position we have already computed to read off the next-hop information.

Figure 4.43 gives the pseudocode for a full tree bitmap search. It assumes a function tree-Function that can find the position of the longest matching prefix, if any, within a given node by consulting the internal bitmap (see description above). "LongestMatch" keeps track of a pointer to the longest match seen so far. The loop terminates when there is no child pointer (i.e., no bit set in the external bitmap of a node) upon which we still have to do our lazy access of the result node pointed to by LongestMatch to get the final next-hop. We assume that the address being searched is already broken into strides and *stride*[i] contains the bits corresponding to the *i*th stride.

```
node:= root; (* node is the current trie node being examined; so we start with root as the first trie
node *)
i:= 1; (* i is the index into the stride array; so we start with the first stride *)
do forever
    if (treeFunction(node.internalBitmap,stride[i]) is not equal to null) then
                                                    (* there is a longest matching
prefix, update pointer *)
        LongestMatch:= node.ResultsPointer + CountOnes(node.internalBitmap,
                        treeFunction(node.internalBitmap, stride[i]));
    if (externalBitmap[stride[i]] = 0) then (* no extending path through this trie node for this search
*)
        NextHop:= Result[LongestMatch]; (* lazy access of longest match pointer to get next hop
pointer *)
        break; (* terminate search)
    else (* there is an extending path, move to child node *)
        node:= node.childPointer + CountOnes(node.externalBitmap, stride[i]);
        i=i+1; (* move on to next stride *)
    end do;
```

Figure 4.43 Tree bitmap search algorithm for destination address whose bits are in an array called stride. (From Eatherton, W., Varghese, G., and Dittia, Z., *ACM SIGCOMM Computer Communication Review Archive*, 34, 2, 2004. With permission.)

So far we have implicitly assumed that processing a trie node takes one memory access. This is valid if the size of a trie node corresponds to the memory burst size (in hardware) or the cache line size (in software). That is why we have tried to reduce the trie node sizes as much as possible to limit the number of bits accessed for a given stride length. In what follows, we describe some optimizations that reduce the trie node size even further.

4.6.3 Tree Bitmap Optimizations

For a stride of 8, the data structure for the core algorithm would require 255-bits for the Internal Bitmap, 256-bits for the External Bitmap, 20-bits for a pointer to children, and 20-bits for a result pointer, which is 551. The next larger power of 2 node size is 1024-bits or 128-bytes. This is a much larger burst size than any that technology optimally supports. Thus we can seek optimizations that can reduce the access size. Additionally, the desire for node sizes that are powers of two means that in this case only 54 percent of the 128-byte node is being used. So, we also seek optimizations that make more efficient use of space.

Initial Array Optimization.   Almost every IP lookup algorithm can be speeded up by using an initial array (e.g., [8, 16–18]). Array sizes of 13-bits or higher could, however, have poor update times. An example of initial array usage would be an implementation that uses a stride of 4, and an initial array of 8-bits. Then the first 8-bits of the destination IP address would be used to index into a 256 entry array. Each entry is a dedicated node possibly 8-bytes in size, which results in a dedicated initial array of 2 kBytes. This is a reasonable price in bytes to pay for the savings in memory accesses. In hardware implementation, this initial array can be placed in on-chip memory.

End Node Optimization.   We have already seen an irritating feature of the basic algorithm in Figure 4.41. Prefixes like P4 will require a separate trie node to be created (with bitmaps that are

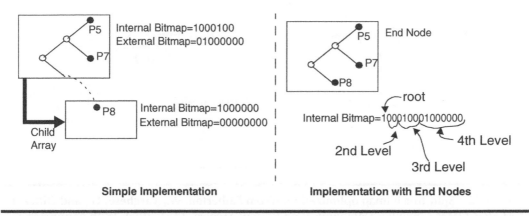

Figure 4.44   End node optimization. (From Eatherton, W., Varghese, G., and Dittia, Z., *ACM SIGCOMM Computer Communication Review Archive*, 34, 2, 2004. With permission.)

almost completely unused). Let us call such nodes "null nodes." While this cannot be avoided in general, it can be mitigated by picking strides carefully. In a special case, we can also avoid this waste completely. Suppose we have a trie node that only has external pointers that point to null nodes. Consider Figure 4.44 as an example that has only one child, which is a null node. In that case, we can simply make a special type of trie node called an endnode (this type of node can be encoded with a single extra bit per node) in which the external bitmap is eliminated and substituted with an internal bitmap of twice the length. The endnode now has room in its internal bitmap to indicate prefixes that were previously stored in null nodes. The null nodes can then be eliminated and we store the prefixes corresponding to the null node in the endnode itself. Thus in Figure 4.44, P8 is moved up to be stored in the upper trie node, which has been modified to be an endnode with a larger bitmap.

Split Tree Bitmaps.   Keeping the stride constant, one method of reducing the size of each random access is to split the internal and external bitmaps. This is done by placing only the external bitmap in each "Trie" node. If there is no memory segmentation, the children and the internal nodes from the same parent can be placed contiguously in memory. If memory segmentation exists, it is bad design to have the internal nodes scattered across multiple memory banks. In the case of segmented memory, one option is to have the trie node point at the internal node, and the internal node point at the results array. Or the trie node can have three pointers: to the child array, to the internal node, and to the results array. Figure 4.45 shows both options for pointing to the results array and implementing a split tree bitmap.

To make this optimization work, each child must have a bit indicating if the parent node contains a prefix that is a longest match so far. If there is a prefix in the path, the lookup engine records the location of the internal node (calculated from the data structure of the last node) as containing the longest matching prefix thus far. Then when the search terminates, we first have to access the corresponding internal node and then the results node corresponding to the internal node. Notice that the core algorithm accesses the next-hop information lazily; the split tree algorithm accesses even the internal bitmap lazily. What makes this work is that any time a prefix P is stored in a node X, all children of X that match P can store a bit saying that the parent has a stored prefix. The software reference implementation uses this optimization to save internal bitmap processing; the hardware implementations use it only to reduce the access width size (because bitmap processing is not an issue in hardware).

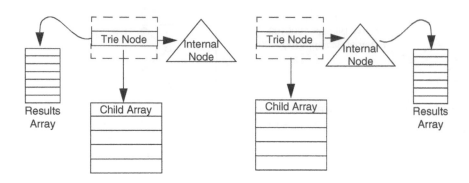

Figure 4.45 Split tree bitmap optimization. (From Eatherton, W., Varghese, G., and Dittia, Z., *ACM SIGCOMM Computer Communication Review Archive*, 34, 2, 2004. With permission.)

A nice benefit of split tree bitmaps is that if a node contained only paths and no internal prefixes, a null internal node pointer can be used and no space will be wasted on the internal bitmap. For the simple tree bitmap scheme with an initial stride of eight and a further stride of four, 50 percent of the 2971 trie nodes in the MaeEast database do not have internally terminating prefixes.

Segmented Bitmaps.   After splitting the internal and external bitmaps into separate nodes, the size of the nodes may still be too large for the optimal burst size. The next step for reducing the node sizes is to segment the multibit trie represented by a multibit node. The goal behind segmented bitmaps is to maintain the desired stride, keep the storage space per node constant, but reduce the fetch size per node. The simplest case of segmented bitmaps is shown in Figure 4.46 with a total initial stride of 3. The subtrie corresponding to the trie node is split into its two child subtries, and the initial root node duplicated in both child subtries. Notice that the bitmap for each child subtrie is half the length (with one more bit for the duplicated root). Each child subtrie is also given a separate child pointer as well as a bitmap and is stored, as a separate "segment." Thus each trie node contains two contiguously stored segments.

Because each segment of a trie node has its pointers, the children and result pointers of other segmented nodes are independent. While the segments are stored contiguously, we can use the high order bits of the bits that would normally have been used to access the trie node to access only the required segment. Thus we need to roughly access only half the bitmap size. For example, using 8-bit strides, this could reduce the bitmap accessed from 256-bits to 128-bits.

When this simple approach is extended to multiple segmentations, the complexity introduced by the segmented bitmap optimization is that if k is the initial stride, and 2^j is the number of final segmented nodes, the internal prefixes for the top k-j rows of prefixes are shared across multiple segmented nodes. The simplest answer is to simply do CPE and push down to each segmented node the longest matching result from the top k-j rows.

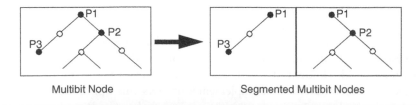

Figure 4.46 Segmented tree bitmap. (From Eatherton, W., Varghese, G., and Dittia, Z., *ACM SIGCOMM Computer Communication Review Archive*, 34, 2, 2004. With permission.)

CAM Nodes. Empirical results show that a significant number of multibit nodes have only a few internal prefixes. In these cases, the space normally occupied by internal bitmaps and a pointer to a results arrays can be replaced by simple CAM-type entries that have match bits, match length, and next-hop pointers. The gain is that the next-hop pointers are in the CAM nodes and not in a separate results array taking up space. For end nodes and internal nodes, even single entry CAM nodes was found to typically result in over half of the next-hop pointers moving from results arrays to inside CAM end nodes. There are quickly diminishing returns however.

4.6.4 Hardware Reference Design

In this section, we investigate the design of a hardware lookup engine based on the Tree Bitmap algorithm. The goal here is not to present comprehensive design details, but to discuss design and illustrate the interesting features that the tree bitmap algorithm provides to hardware implementation. The issues discussed include the base pipelined design, hardware-based MM, incremental updates, and finally an analysis of the design. An area not explored here due to space limitations is the ability to use tree bitmaps, coupled with the presented memory management scheme and path compression to give impressive worst-case storage bounds.

Design Overview. The reference design presented is a single chip implementation of an IP lookup engine using embedded SRAM. The first target of this design is deterministic lookup rates supporting the worst-case TCP/IP over OC-192c link rates (25 million lookups per second). Another target is high speed and deterministic update rates transparent to the lookup processes and easily handled in hardware. A final goal is memory management integrated with updates (so easily handled in hardware) and able to provide high memory utilization.

Figure 4.47 is a block diagram of the IP lookup engine. The left side of the engine has the lookup interface with a 32-bit search address, and a return next-hop pointer. The lookup engine is pipelined so multiple addresses will be simultaneously evaluated. On the right side of the lookup engine is an update interface. Updates take the form: type {insert, change, delete}, prefix, prefix length, and for insertions or table changes, a next-hop pointer. When an update is completed the lookup engine returns an acknowledgment signal to conclude the update.

Looking inside the block diagram of Figure 4.47, there are two SRAM interface blocks. The right SRAM interface block connects to an SRAM containing the initial array. The initial array optimization trades off a permanently allocated memory for a reduction in the number of memory accesses in the main memory. For the reference design the initial array SRAM is 512 entries and is 54-bits wide (this contains a trie node and a 12-bit next-hop pointer). The left SRAM interface

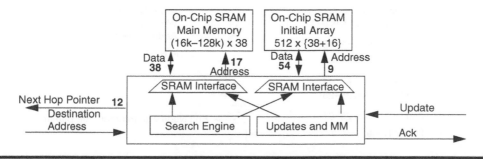

Figure 4.47 Block diagram of IP lookup engine core. (From Eatherton, W., Varghese, G., and Dittia, Z., *ACM SIGCOMM Computer Communication Review Archive*, 34, 2, 2004. With permission.)

block connects to the main memory of the lookup engine. The reference design has a 38-bit wide interface to the main memory and supports implementations of up to 128k memory entries in depth. It is important to note that all addressing in the main memory is done relative to the node size of 38-bits.

The design parameters are: a 9-bit initial stride (so the first 9-bits of the search address are used to index into the initial array), a 4-bit regular stride, Split Tree Bitmap Optimization, End nodes (note the lowest level of the trie must use end nodes and will encompass the last 3 search address bits), and CAM optimization for end nodes and internal nodes.

Each node first has a type (trie, end, or cam), and for the basic trie node there is an extending bitmap and a child address pointer. Additionally, there are a couple of extra bits for various optimizations like skip_length for path compression and parent_has_match to flag a match in the parent.

Figure 4.48 illustrates an example longest node access pattern for the reference design. In this Figure, the first 9-bits of the search address (bits 0–8) are used to index into the initial array memory. At the indexed location a node is stored (in this illustration it is a trie node) which represents bits 9–12 of the search address. After the first trie node is fetched from the initial array, 4 trie nodes and then an end node are fetched from the main memory. The end node points at a result node, which contains the next-hop pointer for this example search.

In the worst-case there are seven memory accesses per lookup slot for search accesses. Devoting one out of every eight memory accesses to control operations (updates and memory management), there are eight memory accesses required per lookup. At 200 Mhz, with full pipelining the lookup rate will be deterministically 25 million per second.

Memory Management. A very important part of an IP lookup engine design is the handling of memory management. A complex memory management algorithm requires processor management of updates, which is expensive and hard to maintain. Poor memory utilization due to fragmentation can negate any gains made by optimization of the lookup algorithm itself for space savings. Memory management is therefore a very important part of a lookup engine and requires careful consideration.

The memory management problem is more difficult for variable-length allocation blocks (which most compression trie schemes require) than for fixed-sized allocation blocks. The simplest method of handling a small fixed number of possible lengths is to segment memory such that there is a separate memory space for each allocation block size. Within each memory space a simple list-based memory management technique is used. The problem with this approach is that it is possible that one memory space will fill while another memory space goes very underutilized.

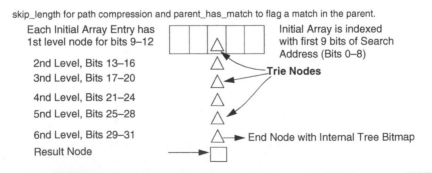

Figure 4.48 **Node sequence for a full 32-bit search. (From Eatherton, W., Varghese, G., and Dittia, Z.,** *ACM SIGCOMM Computer Communication Review Archive,* **34, 2, 2004. With permission.)**

This means that the critical point at which a filter insertion fails can happen with total memory utilization being very low. A possible way to avoid underutilization is to employ programmable pointers that divide the different memory spaces. A requirement of this technique is an associated compaction process that keeps nodes for each bin packed to allow pointer movement. With perfect compaction, the result is perfect memory management in that a filter insertion will only fail when memory is 100 percent utilized.

Compaction without strict rules is difficult to analyze, difficult to implement, and does not provide any guarantees with regard to update rates or memory utilization at the time of route insertion failure. What is needed is compaction operations that are tied to each update such that guarantees can be made. Presented here is a new and novel memory management algorithm that uses programmable pointers, and only does compaction operations in response to an incremental update (*reactive compaction*). This approach is most effective with a limited number of allocation blocks, which would result from small strides.

For the reference design, there are 17 different possible allocation block sizes. The minimum allocation block is 2 nodes because each allocation block must have an allocation block header. The maximum allocation block size is 18 nodes. This would occur when you have a 16-node child array, an allocation header node, and an internal node of the parent. The memory management algorithm presented here first requires that all blocks of each allocation block size be kept in a separate contiguous array in memory. A begin and end pointer bounds the 2 edges of a given memory space. Memory spaces for different block sizes never intermingle. Therefore, memory as a whole will contain 17 memory spaces, normally with free space between the memory spaces (it is possible for memory spaces to abut). To fully utilize memory we need to keep the memory space for each allocation block size tightly compacted with no free space. For the reference design, there are 34 pointers in total: 17 α pointers representing the "top" of a memory space and 17 β pointers representing the "bottom" of a memory space. The first rule for allocating a new free block is that when an allocation block of a given size is needed, check if there is room to extend the size of the memory space for that block size either up or down. After allocating the new node, adjust the appropriate end pointer (either α or β) for the memory space. Second, if the memory space cannot be extended either up or down, a linear search is done in both directions until a gap is found between two memory spaces large enough for the desired allocation size. For, whatever distance from the free space to the memory space needing an allocation, the free space will have to be passed by copying the necessary number of allocation blocks in each memory space from one end to the other. For each memory space the free space must be passed through, this might require copying one or more allocation blocks from one end to the other and adjusting its pointers.

Figure 4.49 shows an example of a portion of memory space near the top of the main memory with memory spaces for allocation block sizes of 2, 3, and 4 nodes.

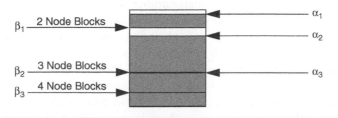

Figure 4.49 Programmable pointer based memory management. (From Eatherton, W., Varghese, G., and Dittia, Z., *ACM SIGCOMM Computer Communication Review Archive*, 34, 2, 2004. With permission.)

For the example shown in Figure 4.49, if an allocation is needed of a block that is 4 nodes in size, it will need the free space between α_2 and β_1 (we will assume the free space is at least six nodes in size). To do this allocation, two blocks from the 3-node memory space will be moved from the bottom of the 3-node memory space to the top of that space. Then α_2 and β_2 will be shifted UP 6 memory locations and α_3 will be shifted UP 4 memory locations, making room for the newly allocated block. Note that 2 nodes are left between the memory space for 3 and 4 nodes.

For the reference design, the worst-case allocation scenario is that the 18-node memory space runs out of adjacent free space, and all the free space remaining is at the other end of the memory, between the 2-node memory space and the "top." With the assumption that memory spaces are organized in order (allocation block size of 2 nodes, 3 nodes, ..., 18 nodes) then for this worst-case scenario, 34 nodes of free space must be passed from the far side of the 2-node memory space to the 18-node memory space. The reason that, 34 free nodes must be passed from end to end rather then the exact 18-nodes needed is that as the free space propagates through memory, the larger nodes (9–17) must still copy 2 nodes from one end of their memory space to the other to pass at least 18-nodes of free space. A very simple upper bound (overtly conservative but easy to calculate) on the number of memory accesses to allocate a block can be found by counting the number of memory accesses to move 34 nodes across all of memory, and adding six memory accesses for every block moved to account for the search and modify operation for the parent of each block moved (explained in more detail below). This results in 1852 memory accesses, which means that if dedicating 1 out of 8 memory accesses to updates, the worst-case memory allocation time is 74 microseconds.

One of the problems with compaction mentioned above is that when an allocation block is moved in memory, the parent of the allocation block must have its pointer modified to reflect the new position. One method for finding a parent is to put with every allocation block the search value that would be used to get to this node. To find the parent of a block, a search is performed until the parent of a node being moved is found and modified. A secondary use of this header block is for the efficient implementation of path compression. There can be an arbitrary skip between parent and child because at the bottom of the search, a complete address for the block containing the final node is checked.

Deallocation of blocks is very simple. For the majority of the cases the deallocated block will be inside the memory space and not on the edge. For this typical case, simply copy one of the blocks at the edge of the memory space into the now vacant block. Then adjust the appropriate end pointer to reflect the compressed memory space. If the deallocated block is on the edge of the memory space then all that needs to be done is pointer manipulation.

Updates. For all update scenarios, the allocation time will dominate. In the worst case, a block of the maximum number of nodes and a block with 2 nodes need to be allocated. In the discussion of memory management we said that 74 microseconds was the worst-case allocation time for a maximum size node. It can similarly be shown that 16 microseconds is the maximum for a 2-node allocation block. Given the small number of memory accesses for the update itself, 100 microseconds is a conservative upper bound for a worst-case prefix insertion. This means that a deterministic update rate of 10,000 per second could be supported with this design. Also note that because the updates are incremental, hardware can be designed to handle the updates from a very high-level filter specification. This lowers the bandwidth necessary for updates and lends itself to atomic updates from the lookup engine perspective; hardware can play a useful role in bypassing lookups from fetching data structures that are currently being altered.

Empirical Result. The experiments were done by creating a C program that models the update and memory management algorithms to be implemented in hardware. The results are presented in Table 4.7.

Table 4.7   Analysis of Empirical Results

Data	MaeEast	MaeWest	PacBell	AADS	PAIX
No. of prefixes	40902	19229	22246	23373	2994
Total size	1.4 Mbits	702 Kbits	748 Kbits	828 Kbits	139 Kbits
Storage per next-hop Pointer bit	2.85	3.04	2.8	2.95	3.87
Memory Mgmt. (MM) Overhead	20%	21%	13.5%	19.4%	21%
Total size without MM	1.1 Mbits	551 Kbits	636 Kbits	668 Kbits	109 Kbits
Storage per next-hop Pointer bit (without MM)	2.24	2.38	2.38	2.38	3.03

Source: Eatherton, W., Varghese, G., and Dittia, Z., *ACM SIGCOMM Computer Communication Review Archive*, 34, 2, 2004. With permission.

The first row in Table 4.7 represents the prefix count for each database, the second row gives the total size of the table for each database. The third row gives the total number of bits stored for every bit of next-hop pointer. This number can be useful for comparing storage requirements with other lookup schemes because the size of the next-hop pointer varies in the published literature for other schemes. The next row summarizes the percent overhead of the allocation headers used for memory management. The last two rows of Table 4.7 are the table size and "bits per next-hop pointer bit" without the allocation headers counted. The reason the table size is recalculated without the memory management overhead is that results for other IP address lookup schemes traditionally do not contain any memory management overhead.

The total storage per next-hop pointer bit without memory management for MaeEast is 2.24 bits per prefix. For the Lulea algorithm [8] with the MaeEast database (dated January 1997 with 32k prefixes) 160 kB are required, which indicates 39 bytes per prefix or 2.8 bits per next-hop pointer bit (they use 14-bit next-hop pointers). This analysis suggests that the reference design of a Tree Bitmap is similar in storage to the Lulea algorithm but without requiring a complete table compression to achieve results.

The reference design has assumed a range of possible main memory sizes from 16k nodes to 128k nodes. From the empirical evidence of 2.24 bits per prefix, it can be extrapolated that a 128k node memory system would be able store 143k prefixes. A 128k node memory system would require 4.8 Mbits of SRAM, and ignoring routing channels would occupy approximately 50 percent of a 12mm × 12mm die in the IBM.25 micron process (SA-12). The logic to implement the reference lookup engine described here would require a very small amount of area compared to the SRAM.

A Tree Bitmap has small memory requirements, fast lookup and update times, and is tunable over a wide range of architectures. The algorithm can provide a complete suite of solutions for the IP lookup problem from the low end to the high. It has the desired features of CAM solutions while offering a much higher number of supported prefixes.

There are other schemes based on the Multibit trie. The full expansion compression scheme [19] proposed by Crescenzi et al. runs length encoding to efficiently compress the forwarding table. The pipelined architecture based on the Multibit trie is proposed by Ting Zhao et al. [20]. Taylor et al. presented the Fast Internet Protocol Lookup (FIPL) architecture [21], which utilizes Eatherton's Tree Bitmap algorithm. Striking a favorable balance between lookup and update performance, memory efficiency, and hardware resource usage, each FIPL engine supports over

500 Gb/s of link traffic while consuming less than 1 percent of available logic resources and approximately 10 bytes of memory per entry. Song et al. have presented a novel data structure—*shape shifting* trie and an IP lookup algorithm that uses it [22]. The algorithm outperforms the well-known and highly successful tree bitmap algorithm, and can be used in high performance routers to perform IP route lookup.

References

1. Ruiz-Sanchez, M., Biersack, E.W.A., and Dabbous, W., Survey and taxonomy of IP address lookup algorithms. *IEEE Network* 2001; 15(2):8–23.
2. Andersson, A. and Nilsson, S., Improved behavior of tries by adaptive branching. *Information Processing Letters* 1993; 46(6):295–300.
3. Nisson, S. and Karlsson, G., IP-address lookup using LC-trie. *IEEE Journal on Selected Areas in Communication* 1999; 17(6):1083–1092.
4. Anderson, A. and Nilsson, S., Faster searching in tries and quadtrees—an analysis of level compression. *Algorithms ESA'94 proceedings of the second annual European symposium on algorithm*, pp. 82–93, 1994. LNCS 855.
5. Ravikumar, V.C., Mahapatra, R., and Liu, J.C., Modified LC-trie based efficient routing look up. *IEEE/ACM Proceedings on MASCOTS*, October 2002, pp. 177–182.
6. Srinivasan, V. and Varghese, G., Fast address lookups using controlled prefix expansion. *ACM Transaction on Computer Systems* 1999; 17(1):1–40.
7. Sahni, S. and Kim, K., Efficient construction of multibit tries for IP lookup. *IEEE/ACM Transactions on Networking* 2003; 11(4):650–662.
8. Degermark, M. et al., Small forwarding tables for fast routing lookups, *Proceedings ACM SIGCOMM '97*, Cannes, 14–18 September 1997, pp. 3–14.
9. Sangireddy, R., Futamura, N., Aluru, S., and Somani, A., Scalable, memory efficient, high-speed IP lookup algorithms. *IEEE/ACM Transactions on Networking* 2005; 13(4):802–812.
10. Cormen, T.H., Leiserson, C.E., and Rivest, R.L., *Introduction to Algorithms*. Cambridge, MA: MIT Press, 2000. 24th printing.
11. Gupta, P., Lin, S., and McKeown, N., Routing lookups in hardware at memory access speeds, *Proceedings IEEE INFOCOM '98*, April 1998, pp. 1240–1247.
12. Mikael Sundstrom and Lars-Ake Larzon, High-performance longest prefix matching supporting high-speed incremental updates and guaranteed compression. *Proceedings—4th Annual Joint Conference of the IEEE Computer and Communications Societies: INFOCOM 2005*. IEEE, 2005, pp. 1641–1652.
13. Mikael Sundstrom, Block trees—an implicit data structure supporting efficient searching among interval endpoints. *Tech. Rep.*, Lulea University of Technology, 2005.
14. Lan Wang, Xiaoliang Zhao, Dan Pei, and Randy Bush et al., Observation and analysis of BGP behavior under stress. *Proceedings of the second ACM SIGCOMM Workshop on Internet Measurement*, 2002, pp. 183–195, ACM Press.
15. Eatherton, W., Varghese, G., and Dittia, Z., Tree bitmap: hardware/software IP lookups with incremental updates. *ACM SIGCOMM Computer Communication Review Archive* April 2004; 34(2):97–122.
16. McKeown, N., Fast switched backplane for a gigabit switched router. *Business Communications Review* 1997; 27(12):125–158.
17. Chiueh, T. and Pradhan, P., High performance IP routing table lookup using CPU caching. *IEEE INFOCOMM* 1999; pp. 1421–1428.
18. Waldvogel, M. et al., Scalable high speed IP routing lookups. *Proc. ACM SIGCOMM '97*, pp. 25–37, Cannes (14–18 September 1997) pp. 25–36.

19. Crescenzi, P., Dardini, L., and Grossi, R., IP address lookup made fast and simple, *7th Annual Euro. Symp. Algorithm.*

20. Zhao, T., Lea, C.T., and Huang, H.C., Pipelined Architecture for fast IP lookup. *IEEE High Performance Switching and Routing*, 2005. On 12–14 May 2005 pp. 118–122.

21. Taylor, D., Rurner, J., Lockwood, J., Scalable IP lookup for internet router. *IEEE Journal on Selected Areas in Communication* 2003; 21(4):523–33.

22. Song, H., Turner, J., Lockwood, J., Shape shifting tries for faster IP route lookup. *Proc. ICNP*, Boston, MA, November 6, 2005, pp. 358–367.

Chapter 5

Pipelined Multibit Tries

Application-specific integrated circuit (ASIC)-based architectures usually implement a trie data structure using some sort of high-speed memory such as static RAMs (SRAMs). If a single SRAM memory block is used to store the entire trie, multiple accesses (one per trie level) are required to forward a single packet. This can slow down the lookups considerably, and the forwarding engine may not be able to process the incoming packets at the line rate. A number of researchers have pointed out that forwarding speeds can be significantly increased if pipelining is used in ASIC-based forwarding engines [1–3]—with multiple stages in the pipeline (e.g., one stage per trie level), one packet can be forwarded during every memory access time period.

In addition, pipelined ASICs provide a general and flexible architecture for a wide variety of forwarding tasks. This flexibility is a major advantage in today's high-end routers, which have to provide IPv6 and multicast routing in addition to IPv4 routing. Therefore, the pipelined ASIC architecture can produce significant savings in cost, complexity, and space for the high-end router.

Despite the various advantages of pipelined ASIC architectures, it is difficult to design an efficient-pipelined ASIC architecture for the forwarding task. For example, managing routing trie during route updates in such architectures is difficult. Basu and Narlikar proposed an efficient data structure to support the fast incremental updates [4]. To minimize the total memory used in [4], Kim and Sahni proposed the optimal algorithms that guarantee not only to minimize the maximum per-stage memory but also to use the least total memory [5]. The algorithms from [4,5] are based on the optimal fixed-stride tries. Lu and Sahni proposed a heuristic for the construction of pipelined variable-stride tries (PVSTs) that require significantly less per-stage memory than that required by the optimal pipelined fixed-stride tries [6], etc. We will describe these algorithms subsequently.

5.1 Fast Incremental Updates for the Pipelined Fixed-Stride Tries

5.1.1 Pipelined Lookups Using Tries

Tries are a natural candidate for pipelined lookups; each trie level can be stored in a different pipeline stage. In a pipelined hardware architecture, each stage of the pipeline consists of its own fast

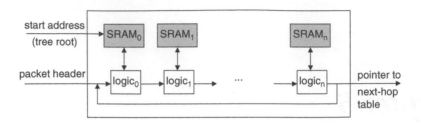

Figure 5.1   A typical *n*-stage forwarding pipeline. In general, a packet header can travel through the pipeline multiple times. (From Basu, A. and Narlikar, G., *IEEE or ACM Transactions on Networking*, 13, 3, 2005. With permission.)

memory (typically SRAMs) and some hardware to extract the appropriate bits from a packet's destination address. Figure 5.1 shows a typical *n*-stage forwarding pipeline. These bits are concatenated with the lookup result from the previous stage to form an index into the memory for the current stage. A different packet can be processed independently in each stage of the pipeline. It is easy to see that if each packet traverses the pipeline once, the forwarding result for one packet can be output every memory access cycle.

Using an optimization called leaf-pushing [3], the trie memory and the bandwidth required between the SRAM and the logic can be halved. Here, the prefixes at the nonleaf nodes are pushed down to all the leaf nodes under it that do not already contain a more specific prefix. In this manner, each node need only contain one field—a prefix pointer or a pointer to an array of child nodes. Thus, each trie node can now fit into one word instead of two. In a leaf-pushed trie, the longest-matching prefix is always found in the leaf at the end of the traversed path (see Fig. 5.2c). We will only consider the leaf-pushed tries. This section is from [4]. Portions are reprinted with permission (© 2005 IEEE).

Because the pipeline is typically used for forwarding, its memories are shared by multiple tables. Therefore, evenly distributing the memory requirement of each table across the pipeline

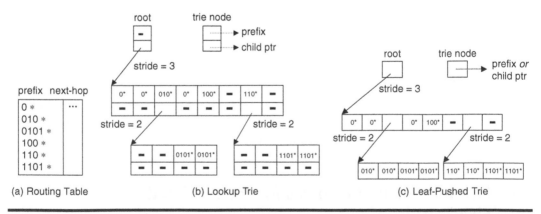

Figure 5.2   (a) A routing table, the * in the prefixes represent the do not care bits, (b) the corresponding forwarding trie, and (c) the corresponding trie with leaf pushing. a "–" represents a null pointer. In the trie without leaf pushing, each node has two fields: (1) a prefix and (2) a pointer to an array of child nodes. In the leaf-pushed trie, each node has only one field, a prefix, or a pointer to an array of child nodes. (From Basu, A. and Narlikar, G., *IEEE or ACM Transactions on Networking*, 13, 3, 2005. With permission.)

memories simplifies the task of memory allocation to the different tables. It also reduces the likelihood of any one-memory stage overflowing due to route additions.

The updates to the forwarding table go through the same pipeline as the lookups. A single route update can cause several write messages to be sent through the pipeline. For example, the insertion of the route 1001* in Figure 5.2c will cause 1 write in the level 2 node (linking the new node to the trie) and 4 writes in the level 3 node (2 writes for 1001* and 2 writes for pushing down 100*).

These software-controlled write messages from one or more route updates are packed into special write packets and sent down the pipeline, similar to the reads performed during a lookup. We call each write packet a *bubble*—each bubble consists of a sequence of (stage, location, and value) triples, with at most one triple for each stage. The pipeline logic at each stage issues the appropriate write command to its associated memory. Minimizing the number of write bubbles introduced by the route updates reduces the disruption to the lookup process. Finally, care should be taken to keep the trie in a consistent state between consecutive write bubbles, as route lookups may be interleaved with write bubbles.

5.1.2 Forwarding Engine Model and Assumption

To address the issues described in the previous section, Basu and Narlikar developed a series of optimizations using a combination of simulation and analysis [4]. To this end, they first developed a simulation model of the packet forwarding components in a typical (high-end) router line card. They then used certain well-known characteristics of the IPv4 address allocation process and the BGP routing protocol to develop the optimizations for balancing the memory allocations and reducing the occurrence of "write bubbles" in the pipeline. Finally, they validated the findings by incorporating these optimizations in the simulator and running a set of route update traces through the simulator. We now briefly describe the simulator for the packet forwarding engine and the assumptions.

Forwarding Engine Model. The distributed router architecture described in this section is similar to that of commercially available core routers (e.g., Cisco GSR12000, Juniper M160): the router has a central processor that processes BGP route updates from neighboring routers and communicates the resulting forwarding table changes to the individual line cards. Each line card has a local processor that controls the pipelined forwarding engine on the line card. Using a software shadow of the pipelined routing trie, the local processor computes the changes to be made to each stage of the forwarding pipeline for each route update. It also typically performs all the memory management for the pipeline memories. Here, we focus on optimizations that will minimize the cost of route updates and not on memory management issues.

The simulation model consists of three components—first, we have the trie component that constructs and updates the routing trie (Fig. 5.3). It processes one route update at a time and

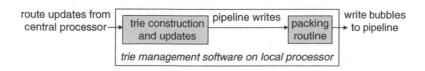

Figure 5.3 The software system on the line card that constructs and maintains the forwarding trie. (From Basu, A. and Narlikar, G., *IEEE or ACM Transactions on Networking*, 13, 3, 2005. With permission.)

generates the corresponding writes to the pipeline memories. The second component is the packing component that a pack writes from a batch of consecutive route updates into write bubbles that are sent down the pipeline. When a new subtree is added to the trie (due to a route add), the pipeline write that adds the root of the subtree is tagged by the trie component. The packing component ensures that this write is not packed into a write bubble before any of the writes to the new sun-tree (to prevent any dangling pointers in the pipelined trie). Finally, we have a pipeline component that actually simulates the traversal of these write bubbles through a multi-stage pipeline.

Assumptions.   We now describe the assumptions about the forwarding engine setup that we made when designing the simulation model (which therefore affect the nature of the optimizations that we have developed).

- The initial trie construction takes as input a snapshot of the entire table. After this construction phase, it is only updated incrementally for the several million updates in our data sets.
- The pipeline in the same order processes bubbles as they are generated by the packing component. A bubble is not interrupted in its progression from the first to the last stage of the pipeline. However, consecutive bubbles may be interspersed with lookups. Therefore, the trie structure should always be consistent between consecutive write bubbles.
- Only tries with fixed strides are considered. Although variable-stride tries could be more memory-efficient, they are difficult to maintain during incremental updates. Because each node can have a different stride, there is no straightforward way to determine the strides for trie nodes that are created when the new prefixes are added.
- We focus on the leaf-pushed tries. Although the updates to leaf-pushed tries can result in more pipeline writes, leaf pushing allows for a higher pipeline throughput and a more efficient use of the pipeline memory. All optimizations, however, are also applicable to nonleaf-pushed tries.
- Writes to different pipeline stages can be combined into a single write bubble—each bubble can contain at most one write to each stage of the pipeline. We use the number of bubbles created as a measure of how disruptive incremental updates are to the search process. Note that it is possible to have more than one write to each pipeline stage in a single bubble. For example, this would be the case when the pipeline memory can service multiple back-to-back write requests. Assumptions (of at most one write per stage) therefore provide an upper bound on the number of bubbles generated (worst case).
- The packing component is permitted to pack pipeline writes from multiple route updates into a single write bubble, as long as the route updates arrive with the same timestamp (granularity—one second). To further limit the delay in updating the pipelined trie, we combine writes from batches of at most 100 such concurrent route updates.
- The experiments focus on IPv4 lookups, because no extensive data are currently available for IPv6 tables or updates. We focus on a pipeline with eight stages and assume that only one pass is required through the pipeline to look up an IPv4 packet. Section 5.1.5 also shows the results for pipelines with fewer stages.
- The next hop information is stored in a separate next hop table that is distinct from the pipelined trie.
- Memory management is done by a separate component. In particular, we assume the existence of malloc and free-like primitives.

5.1.3 Routing Table and Route Update Characteristics

In Chapter 2, we have described the characteristics of the IPv4 address allocation process and the BGP routing table. Here, we will analyze the routing table and the update trace used in the situations.

O–I: A majority of the prefixes in the routing tables of today are 24-bit prefixes—consequently most routing updates affect 24-bit prefixes. For example, two routing tables (RRC03 and RRC04) are from RIS project (http://www.ripe.net/ripencc/pub-services/np/ris/index.html); 57.7 percent of the RRC03 prefixes and 58.8 percent of the RRC04 prefixes were 24-bit prefixes, and 64.2 percent of RRC03 route updates and 61.6 percent of RRC04 route updates were to 24-bit prefixes.

O–II: The number of very small (≤8-bit) prefixes is very low, and very few updates affect them. For example, about 0.02 percent of the prefixes in the RRC03 and RRC04 prefixes were less than 8-bit long, and 0.04 percent of the RRC03 updates and 0.03 percent of the RRC04 updates were to such prefixes. However, because a short route is typically replicated a number of times in a trie, each update to it may result in modifications to a large number of memory locations.

O–III: Prefixes corresponding to the customers of a given ISP are typically the neighboring 24-bit prefixes. Hence, prefixes close together and differing in only a few low order bits (but with possibly different next hops) often fully populate a range covered by a single, shorter prefix.

O–IV: A link failure (recovery) in an ISP network disconnects (reconnects) some or all of its customer networks (represented by neighboring prefixes in the routing trie). In turn, this can cause updates to the corresponding neighboring routes to occur simultaneously. For example, in the RRC03 and RRC04 traces, we observed that about 80 percent of the routes that are withdrawn get added back within 20 minutes.

O–V: Finally, recent studies of BGP dynamics [8,9] indicate the following. First, the proportion of route updates corresponding to network failure and recovery is fairly high: about 40 percent of all updates are route withdrawals and additions [9]. Second, once a network failure occurs, the mean time to repair is of the order of several minutes [6]. Thus, a large proportion of routes that are withdrawn get added back a few minutes later. For example, our analysis of the RRC03 and RRC04 traces indicate that about 80 percent of the routes that are withdrawn get added back within 20 minutes.

The optimizations for constructing and maintaining the pipelined lookup trie are based on the above observations. To test the effectiveness of the optimizations, we use seven datasets. These data sets will be referenced in the following sections by the labels listed in Table 5.1. The RRC** data sets were collected from the Routing Information Service (http://www.ripe.net/ripencc/pub-services/np/ris/index.html), while MaeEast and MaeWest databases were obtained from the internet performance measurement and analysis (IPMA) project (http://www.merit.edu/). Note that the data sets have widely distributed spatial (location of collection) and temporal (time of collection) characteristics—this ensures that optimizations are not specific to a particular router interface or a specific time interval.

The update traces used in our experiments were collected starting from the time of the table snapshot. Each BGP update, which has a timestamp granularity of one second, is either a route addition or a route withdrawal. Updates from several consecutive days were combined, and some were filtered out as follows. Adds to prefixes already in the table can be treated as a change of the next-hop entry and simply involve rewriting an entry in the next-hop table (thus not affecting the forwarding trie). Such adds were filtered out; similarly, withdraws for prefixes not currently in the table were also filtered out. In our experiments, the filtered update sequence was applied incrementally to the initial routing table snapshot.

Table 5.1   Datasets Used in Experiments

		No. of Prefixes	No. of Next Hops	No. of Updates	Updates per Second	
Label	Site				Average	Maximum
RRC01	RRC01	103,555	37	4,719,521	12.52	30,060
RRC03-a	RRC03	89,974	74	3,769,903	16.70	8867
RRC03-b	RRC03	108,267	80	4,722,493	12.45	30,060
RRC04	RRC04	109,600	29	3,555,839	13.80	36,326
me-a	MaeEast	50,374	59	3,685,469	10.63	8509
me-b	MaeEast	46,827	59	4,350,898	9.66	10,301
Mw	MaeWest	46,732	53	2,856,116	10.71	14,024

Source: Basu, A. and Narlikar, G., *IEEE or ACM Transactions on Networking*, 13, 3, 2005. With permission.

5.1.4 Constructing Pipelined Fixed-Stride Tries

We now present a trie construction algorithm that takes as input the number of stages in the pipeline and a set of routing table prefixes and finds the trie that minimizes the size of the largest trie level (pipeline stage). If there are multiple such tries, the algorithm finds the most compact (least total size) trie among them. Such an algorithm makes it easier to pack multiple different protocol tables into the set of available pipelined memories and to avoid memory overflows as routing tables grow. We also provide upper bounds on the worst-case performance of this algorithm; this enables hardware designers to decide how big each pipeline stage should be.

Tries present a tradeoff between space (memory requirement) and packet lookup time (number of trie levels, assuming one lookup operation per level). Large strides reduce the number of levels (and hence, the lookup time), but may cause a large amount of replication of prefixes (e.g., Fig. 5.4a).

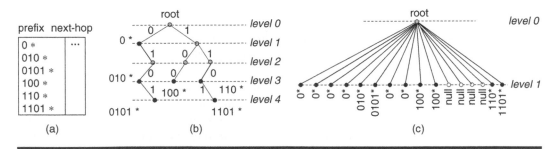

Figure 5.4   (a) A sample routing table, (b) the corresponding 1-bit trie, and (c) the corresponding 4-bit trie. The *dotted lines* show the nodes at each level in the tries; of these, only the black nodes contain a prefix. The 4-bit trie has a smaller depth (1 versus 4 memory accesses) but a larger number of nodes compared with the 1-bit trie. (From Basu, A. and Narlikar, G., *IEEE or ACM Transactions on Networking*, 13, 3, 2005. With permission.)

To balance the space–time tradeoff in trie construction, Srinivasan and Varghese [3] use controlled prefix expansion to construct memory-efficient tries for the set of prefixes in a routing table. Given the maximum number of memory accesses allowed for looking up any IP address (i.e., the maximum number of trie levels), they use dynamic programming (DP) to find the FST with the minimum total memory requirement.

The problem of constructing an FST reduces to finding the stride size at each level, that is, finding the bit positions at which to terminate each level. The DP technique in controlled prefix expansion works as follows. First, a 1-bit auxiliary trie is constructed from all the prefixes (e.g., Fig. 5.4b). Let $nodes(i)$ be the number of nodes in the 1-bit trie at level i. If we terminate some trie level at bit position i and the next trie level at some bit position $j > i$, then each node in $nodes(i + 1)$ gets expanded out to 2^{j-i} nodes in the multi-bit trie. Let $T[j, r]$ be the optimal memory requirement (in terms of the number of trie nodes) for covering bit position 0 through j using r trie levels (assuming that the leftmost bit position is 0). Then $T[j, r]$ can be computed using DP as (Fig. 5.5a):

$$T[j,r] = \min_{m \in \{r-1,\dots,j-1\}} (T[m, r-1] + nodes(m+1) \times 2^{j-m}) \tag{5.1}$$

$$T[j,1] = 2^{j+1} \tag{5.2}$$

Here, we choose to terminate the $(r-1)$th trie level at bit position m, such that it minimizes the total memory requirement. For prefixes with at most W bits, we need to compute $T[W-1, k]$, where k is the number of levels in the trie being constructed. This algorithm takes $O(k \times W^2)$ time; for IPv4, $W = 32$.

The controlled prefix-expansion algorithm finds the FST with the minimum total memory. It can easily be applied to pipelined lookup architecture by fixing the number of trie levels to be the number of pipeline stages (or some multiples of it). However, the algorithm does not attempt to distribute the memory equally across the different pipeline stages. As a consequence, some stages may be heavily loaded, whereas others may be very sparse. Figure 5.6 shows the amount of memory allocated in each stage of an eighth-stage pipeline for the two routing tables: RRC03 and RRC04.

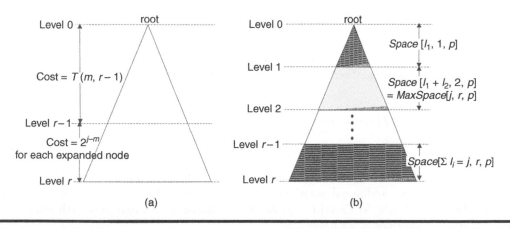

Figure 5.5 Optimizing memory (a) using the controlled prefix-expansion algorithm by Srinivasan and Varghese, and (b) using our MinMax algorithm. Here, the second level occupies the most memory, hence *maxspace[j, r, p] = space[l₁ + l₂, 2, p]* for the partition *p*. (From Basu, A. and Narlikar, G., *IEEE or ACM Transactions on Networking*, 13, 3, 2005. With permission.)

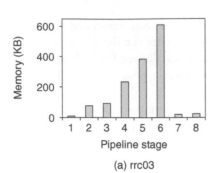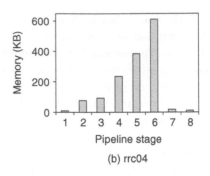

Figure 5.6   Memory allocation for the forwarding trie in different stages of the pipeline using controlled prefix expansion for the two tables: (a) RRC03 and (b) RRC04. (From Basu, A. and Narlikar, G., *IEEE or ACM Transactions on Networking*, 13, 3, 2005. With permission.)

Each FST was constructed with controlled prefix expansion and uses leaf pushing. The memory allocations are highly variable across the different stages. In particular, the stage that contains 24-bit prefixes has the highest memory requirement. Besides increasing the chance of memory overflows in this stage, this imbalance can negatively impact the update performance: the overloaded stage contains the 24-bit prefixes and therefore typically gets many more writes than the other stages. This makes it difficult to pack the pipeline writes into a small number of bubbles.

Based on controlled prefix expansion, Basu and Narlikar have developed an algorithm that attempts to evenly allocate memory across the different stages. The algorithm assumes that each pipeline stage contains exactly one level in the lookup trie, and constructs a trie that satisfies the following constraints:

- Each level in the FST must fit in a single pipeline stage.
- The maximum memory allocated to a stage (over all stages) is minimized. This ensures that the memory allocation is balanced across all the pipelined stages.
- The total memory used is minimized subject to the first two constraints (we explain why this is important later in this section).

More formally, as before, let $T[j, r]$ denote the total memory required for covering bit positions 0 through j using r trie levels, when the above conditions are satisfied. Furthermore, let N denote the size of each pipeline stage. Then, the first and the third constraints are satisfied by the following equations:

$$T[j,r] = \min_{m \in S(j,r)} (T[m, r-1] + nodes(m+1) \times 2^{j-m}) \tag{5.3}$$

$$T[j,1] = 2^{j+1} \tag{5.4}$$

where $S(j,r) = \{m \mid r - 1 \leq m \leq j - 1 \text{ and } nodes(m+1) \times 2^{j-m} \leq P\}$. To satisfy the second constraint, we first introduce some additional notations. For a partition p of bit position 0 through j into r levels in the multi-bit trie, let $Space[j, r, p]$ denote the memory allocated to the rth level in the multi-bit trie. In other words, we have:

$$Space[j,r,p] = nodes(m + 1) \times 2^{j-m} \tag{5.5}$$

where bit positions 0 through m are covered by $r - 1$ levels in the trie. (Note that this implies that the rth level in the trie covers bit position $m + 1$ through j.) We then define $MaxSpace[j, r, p]$ as the maximum memory allocated to any trie level when partition p is used to split bit positions 0 through j among r levels in the multi-bit trie (Fig. 5.5b). More formally, we have:

$$MaxSpace[j,r,p] = \max_{1 \le m \le r} Space\left[\sum_{i=1}^{m} l_i, m, p\right]$$

where l_i denotes the stride size of the ith level in the trie and $\sum_{i=1}^{r} l_i = j$. Now let $MinMaxSpace[j, r]$ be the minimum value of $MaxSpace[j, r, p]$ for all possible partitions p of bit positions 0 through j into r levels. Then $MinMaxSpace[j, r] = \min_{p \in Part} MaxSpace[j, r, p]$, where $Part$ is the set of all possible partitions. Then, in addition to Equations 5.3 and 5.4, following equation must also be satisfied by the variable m:

$$MinMaxSpace[j,r] = \min_{m \in S(j,r)} (\max(nodes(m+1) \times 2^{j-m}, MinMaxSpace[m, r-1])) \qquad (5.6)$$

$$MinMaxSpace[j, 1] = 2^{j+1}$$

Equation 5.6 selects the bit position m to terminate the $(r-1)$th level of the trie such that it minimizes the size of the largest trie level. Here the largest level (in terms of memory requirement) is the bigger of the last (rth) level and the largest level of the first $r-1$ levels.

We give Equation 5.6 precedence over Equation 5.3 when choosing m. When multiple values of m yield the same value of $MinMaxSpace[j, r]$, Equation 5.3 is used to choose between these values of m. In other words, the primary goal is to reduce the maximum memory allocation across the pipeline stages, and the secondary goal is to minimize the total memory allocation. We found that it was important to maintain this secondary goal of memory efficiency to produce tries with low update overheads. A memory-efficient trie typically has smaller strides and hence less replication of routes in the trie; a lower degree of route replication results in fewer trie nodes that need to be modified for a given route update.

For a set of prefixes where the longest prefix length is W bits and the maximum number of lookups is k (i.e., k trie levels), this algorithm takes $O(k^2 \times W)$ operations (same as the controlled prefix-expansion algorithm). We refer to this algorithm as the *MinMax* algorithm.

The *MinMax* algorithm is intended for use in hardware design—an important measure of performance in such cases is the worst-case memory usage (in addition, to lookup and update times). In other words, given a k-stage pipeline and a prefix table of size N with a maximum prefix length of W, we would like to compute the worst-case memory size for a pipeline stage. Let the worst possible input of N prefixes to the *MinMax* algorithm be WC, and let max_{WC} be the size of the maximum pipeline stage as computed by the MinMax algorithm.

We shall now construct the 1-bit auxiliary trie that corresponds to the input WC and prove that WC is indeed the worst possible input. The construction of the 1-bit auxiliary trie is shown in Figure 5.7. The trie that is constructed has two halves. In the upper half, which consists of levels 1 through $\lceil \log N \rceil$, each node has two children; in other words, there is a maximum fan out in the upper half. In lower half, which spans levels $\lceil \log N \rceil + 1$ through W, each node has exactly one child. Note that each prefix represented by this trie has length W, which is the maximum allowed.

To prove that WC is the worst possible input, we introduce some more notations. Let D denote any input of N prefixes to the MinMax algorithm. We denote by max_D, the size of the maximum pipeline stage as computed by the MinMax algorithm for the input D. Furthermore, let $nodes_{WC}(m)$ be the number of nodes in level m of the 1-bit auxiliary trie for input WC. The quantity $nodes_D(m)$ is similarly defined. We now prove the following lemma.

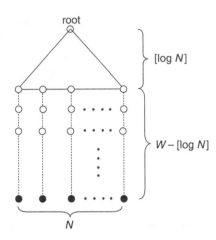

Figure 5.7 **The 1-bit auxiliary trie corresponding to the worst possible input of *n* prefixes for MinMax algorithm; the longest length is *w* bits. (From Basu, A. and Narlikar, G., *IEEE or ACM Transactions on Networking*, 13, 3, 2005. With permission.)**

Lemma 5.1.1: For any level m in the auxiliary trie, $nodes_D(m) \leq nodes_{WC}(m)$

Proof: We first observe that the auxiliary trie for input has the maximum possible fan out for each node in the first $\lceil \log N \rceil$ levels. Therefore, at any level $m \leq \lceil \log N \rceil$, the number of nodes in the auxiliary trie for input WC must be the maximum possible. Now consider any level $m > \lceil \log N \rceil$. The number of nodes in the auxiliary trie for input WC is N. Suppose there are some inputs consisting of N prefixes where the auxiliary trie has $L > N$ nodes at level m. If we traverse down this trie, each node at level m must lead to a leaf node which contains a prefix. This implies that the auxiliary trie has $L > N$ prefixes, which is a contradiction.

Now let p_D be the partition produced by the MinMax algorithm for the input D, and let p_{WC} be the partition produced for the input WC. Consider the 1-bit trie for input D with the partition p_{WC} superposed on it. Let $\max_D(WC)$ be the size of the largest pipeline stage in this case; note that $\max_D$ is the largest pipeline stage corresponding to the partition p_D which is optimal for the input D. Therefore, we conclude that

$$\max_D \leq \max_D(WC) \tag{5.7}$$

We now consider the levels in the 1-bit auxiliary trie (for input D) that are spanned by the pipeline stage with size $\max_D(WC)$. Let the start level be $max(m)$ and the number of levels be x. Then, we have (using Equation 5.5),

$$\max_D(WC) = nodes_D(\max(m)) \times 2^x \tag{5.8}$$

Using Lemma 5.1.1 and Equation 5.8, we assert that

$$\max_D(WC) = nodes_D(\max(m)) \times 2^x \leq nodes_{WC}(\max(m)) \times 2^x \leq \max_{WC} \tag{5.9}$$

Using Equations 5.7 and 5.9, we conclude that

$$\max_D \leq \max_{WC} \tag{5.10}$$

This leads to the following memory bound.

Theorem 5.1.1:   For any given set of N prefixes of maximum length W, the maximum memory per pipeline stage required to build a k-level trie using the MinMax algorithm is bounded by $2 \left\lceil \frac{W + (k-1) \lceil \log N \rceil}{k} \right\rceil$ trie nodes.

Proof: Let N' be the smallest power of 2 that is greater than or equal to N. Then $\log N' = \lceil \log N \rceil$. We will compute the memory bound for $N' \geq N$ prefixes; the same upper bound should hold for N prefixes.

Using Equation 5.10, we conclude that the worst possible input with N' prefixes for the MinMax algorithm is shown by the construction in Figure 5.7. Now, we need to compute the bit positions at which each of the k levels will be terminated by the MainMax algorithm.

We can show that here is at least one level in the resulting trie with over N' nodes in it. Therefore, there is no benefit from terminating the first level at less than $\log N'$ bits. There may however, be some benefit from terminating the first level at a bit value grater than $\log N'$, so that the sizes of the lower levels can be reduced. Let the optimal bit value at which the first level terminates be $\log N' + x$ (where $x > 0$ is an integer). Then the first level has $2^{\log N' + x}$ nodes.

Now, the worst-case 1-bit trie (Fig. 5.7) has exactly N' nodes in every level at bit position greater than $\log N'$. Therefore, MinMax will simply attempt to split them at equal intervals to create the remaining $k - 1$ levels. Thus, the remaining $(W - \log N' - x)$ bit positions are split evenly into $k - 1$ levels. The largest of these levels will cover $\left\lceil \frac{W - \log N' - x}{k-1} \right\rceil$ bits. This level will contain $N' \cdot 2^{\left\lceil \frac{W - \log N' - x}{k-1} \right\rceil}$ nodes.

MinMax will select the value of x such that the size of the first partition equals the largest of the lower partitions. Therefore, $2^{\log N' + x} = N' \cdot 2^{\left\lceil \frac{W - \log N' - x}{k-1} \right\rceil} = 2^{\log N' + \left\lceil \frac{W - \log N' - x}{k-1} \right\rceil}$ or $\log N' + x = \log N' + \left\lceil \frac{W - \log N' - x}{k-1} \right\rceil$.

Solving for x gives us

$$x = \left\lceil \frac{W - \log N'}{k} \right\rceil \tag{5.11}$$

Therefore, the number of nodes in the largest partition is bounded by

$$2^{\left\lceil \frac{W + (k-1)\log N'}{k} \right\rceil} = 2^{\left\lceil \frac{W + (k-1)\lceil \log N \rceil}{k} \right\rceil} \tag{5.12}$$

For $N = 2^{20}$ (1M) prefixes, $k = 8$, and $W = 32$, this bound amounts to 2^{22} (4M) trie nodes per stage. Assuming a pointer size of 22 bits (to address each of the 4M entries in a stage), we get a maximum memory requirement of 11 MBytes gets per stage. For $N = 128\,\text{K}$ (a typical size for a current core routing table), a similar calculation yields 1.19 MBytes gets per stage.

The performance of the MinMax algorithm is shown in Figure 5.8; it reduces the maximum memory allocation across the pipeline stages (by over 40 percent) at the cost of a slightly higher (13–18 percent) total memory overhead compared with the controlled prefix expansion (Table 5.2). We also point out that in each of the graphs in Figure 5.8, some of the levels show disproportionately low memory usage even after the MinMax algorithm is applied (for e.g., levels 7 and 8 in

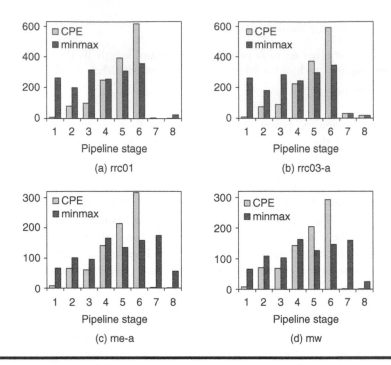

Figure 5.8 **Memory allocation (in kBytes) in the different pipeline stages using the MinMax algorithm for trie construction, compared with using controlled prefix expansion. (From Basu, A. and Narlikar, G., *IEEE or ACM Transactions on Networking*, 13, 3, 2005. With permission.)**

Fig. 5.8a). In each of these cases, the levels in question are the ones that are assigned to bit positions 24–31 (recall that bit positions are numbered from 0). Because there are very few prefixes of length more than 24, the number of prefixes terminating in these levels is low, hence the low memory usage. The MinMax algorithm assigns one level solely to the 24-bit prefixes (which typically form the largest level in the trie). Therefore, it cannot further balance out the trie levels by combining the 24-bit prefixes with the sparsely populated prefixes of length greater than 24.

Finally, we note that instead of minimizing the maximum, minimizing some other metrics could also balance out the memory allocation across the multiple stages. We have experimented with three other strategies, namely, (a) minimizing the standard deviation of all the $Space[j, r, p]$'s, (b) minimizing the sum of squares of all the $Space[j, r, p]$'s, and (c) minimizing the difference between the maximum $Space[j, r, p]$ and the minimum $Space[j, r, p]$. However, the quality of the

Table 5.2 **Additional (Total) Memory Overhead of the MinMax Algorithm and the Reduction in the Memory Requirement to Corresponding Values for CPE**

Table	RRC01	RRC03a	RRC03b	RRC004	me-b	me-a	mw
Overhead (%)	17.2	17.6	16.4	16.5	16.5	17.8	13.8
Max reduce (%)	42.2	41.7	42.0	41.8	44.3	44.5	44.2

Source: Basu, A. and Narlikar, G., *IEEE or ACM Transactions on Networking*, 13, 3, 2005. With permission.

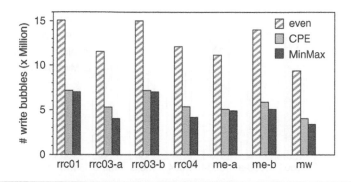

Figure 5.9 **Number of write bubbles sent to the pipeline when different trie construction algorithms are used: "even" denotes using equal (4-bit) strides in each of the eight levels, "CPE" denotes using controlled prefix expansion, and "MinMax" denotes using MinMax. (From Basu, A. and Narlikar, G., *IEEE or ACM Transactions on Networking*, 13, 3, 2005. With permission.)**

results for the MinMax algorithm proves to be just better than the other (computationally more intensive) algorithms. We will use the MinMax algorithm to calculate stride lengths.

As shown in Figure 5.9, both controlled prefix expansion (CPE) and MinMax are fairly effective in reducing the number of write bubbles generated by the packing component, when compared with the baseline case of using a trie with equal strides (4 bits) at each level. We also point out that the CPE algorithm was not explicitly designed to solve the pipelined architecture problem. However, we will omit the baseline case (of tries with equal strides) and the CPE algorithm because it is the most competitive algorithm we could find.

5.1.5 Reducing Write Bubbles

We now describe a series of optimizations aimed at reducing the number of write bubbles that are sent to the pipeline when routes are added to or withdrawn from the forwarding trie. These optimizations are implemented in either the trie component or the packing component (Fig. 5.3). In some cases, the packing component applies an optimization based on writes that are specially tagged by the trie component. As we describe each optimization, we show its benefits when applied incrementally along with the previous optimizations.

5.1.5.1 Separating Out Updates to Short Routes

The number of short prefixes (≤8 bits) in all the tables is very small. However, even a small number of updates to these prefixes can cause a big disruption to the pipeline. For example, if the trie root has a stride of 16, the addition of a 7-bit prefix can cause up to $2^{16-7} = 512$ writes to the first stage of the pipeline. These writes cannot be packed into a smaller number of bubbles because they all target the same pipeline stage. Because the trie construction algorithms do not take into account such update effects when determining the strides into in the trie, we instead suggest storing all short prefixes of up to 8 bits in length, in a separate table with $2^8 = 256$ entries. Assuming 32 bits per entry, this only requires an additional 1 kByte of fast memory. The IP lookup process now searches the pipeline first. If no prefix is found, an additional lookup is performed in this table using the first 8 bits of the destination address. This lookup can also be pipelined similar to the trie lookups.

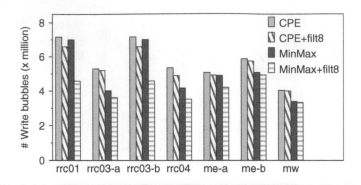

Figure 5.10 Reduction in the number of write bubbles when short prefixes (≤8 bits) are filtered out and stored in a separate 265-entry table. Bubbles include the cost of updating this separate table. CPE is controlled prefix expansion, and MinMax is the new tree-building algorithm. "filt8" denotes the runs when the short prefixes are stored separately. (From Basu, A. and Narlikar, G., *IEEE or ACM Transactions on Networking*, 13, 3, 2005. With permission.)

Figure 5.10 shows the benefit of using this simple optimization. The figure includes the cost of writing to the new table. For example, an update to a 7-bit prefix now causes 2 writes to the new table (which cannot be packed into the same bubble). The benefit of this simple optimization ranges between 1.6 percent and 35 percent when using MinMax to construct the trie. The routing tables for RRC01 and RRC03b benefit more than the rest because they have a larger number of updates to very short prefixes, including 1-bit routes.

An alternative to adding a separate table for short prefixes would be to simply add another stage (with a stride of 8 bits) at the beginning of the pipeline. However, with such a pipeline stage, many copies of the short prefixes would be pushed down into lower parts of the trie (due to leaf pushing)—adding a separate table at the end of the pipeline avoids this overhead.

5.1.5.2 Node Pullups

In FSTs, all 24-bit prefixes lie in a single level of the trie. Because most updates are to 24-bit prefixes, a large fraction of the pipeline writes are directed to that level, which resides in a single stage of the pipeline. This makes it harder to pack the writes efficiently into bubbles. Node pullups are an optimization aimed at spreading out the 24-bit prefixes in the trie. Given that there are many groups of neighboring 24-bit prefixes in the trie, we can move entire such groups above the level that contains the 24-bit prefixes. This pullup can be performed by increasing the stride of the node that is the lowest common ancestor of all the neighboring prefixes in the group. Let l be the level that contains the 24-bit prefixes. Consider a node in a level k above level l; say k terminates at bit position n (<24). For some node in level k, if all of the 2^{24-n} possible 24-bit prefixes that can be descendants of this node are present in the trie, we pull all of them up into level k. The stride of the parent of the pulled-up node is increased by $24-n$. We start examining the nodes to pull up in a top-down manner, so that the 24-bit prefixes are pulled up as far as possible.

The node pullup optimization ensures that the memory requirement of the transformed trie can possibly reduce, but not increase (Fig. 5.11). Thus, the MinMax algorithm constructs a strictly FST; node pullups subsequently modify the strides of some nodes in a controlled manner. This optimization is similar to the level compression scheme described in [10]. However, the motivation

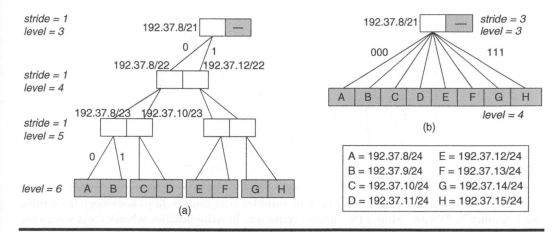

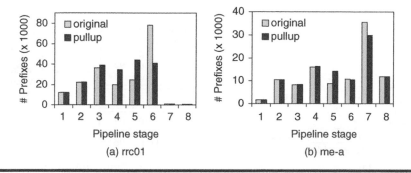

Figure 5.11 A subtree (a) before and (b) after node pullup has been performed. The 24-bit prefixes A, B, ..., H can pulled up two levels. The number of trie nodes can at best decrease (in this case from 16 to 10) after node pullups. (From Basu, A. and Narlikar, G., *IEEE or ACM Transactions on Networking*, 13, 3, 2005. With permission.)

here is different (we use it to reduce the update overheads) and we have also developed a modification to enhance the performance of this optimization.

State Tries.   Figure 5.12 shows the distribution of prefixes in different levels of an 8-level trie before and after node pullups have been performed for three of the data sets. Here we used the MinMax algorithm to construct the trie. Node pullups are successful in spreading out the prefixes across the pipeline stages. They also helped reduce the size of the largest pipeline stage by an average of 6.5 percent compared with MinMax. However, simply using the node pullup optimization is not sufficient. The pullup information (in the form of a changed stride length) is stored in the node where the pullup has occurred. Therefore, if the node itself is deleted, and then reinserted (due to a route withdrawal, followed by an insertion), this information cannot be reconstructed. Instead, the only information available is the trie-level information that has been calculated by the MinMax algorithm. To remedy this shortcoming, we use a state trie in software when pullups are applied. The

Figure 5.12   Number of prefixes in each level of the pipelined trie before and after node pullup. For both the data sets, the tallest bar in the original trie represents the stage with the 24-bit prefixes. (From Basu, A. and Narlikar, G., *IEEE or ACM Transactions on Networking*, 13, 3, 2005. With permission.)

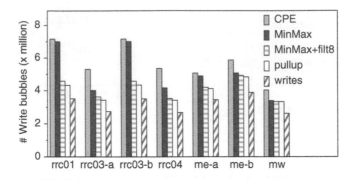

Figure 5.13 **Reduction in the number of write bubbles after node pullups have been performed (label "pullup"). "Writes" shows the further reduction in write bubbles when excess writes are eliminated. For both these optimizations, the trie was constructed using the MinMax algorithm, and shorter prefixes were filtered out. (From Basu, A. and Narlikar, G., *IEEE or ACM Transactions on Networking*, 13, 3, 2005. With permission.)**

state trie stores the pullup information at each node. When there is a deletion followed by an insertion, the stride size of the inserted node is obtained from the corresponding node in the state trie.

Figure 5.13 shows the benefit of node pullups (with a state trie) for reducing the total number of write bubbles. The benefits were lower than that we expected: updates to neighboring routes often appear together, and all the neighboring routes are typically in the same level, whether or not they have been pulled up. This makes it difficult to pack the resulting writes to that level into write bubbles.

5.1.5.3 Eliminating Excess Writes

Because neighboring routes are often added in the same time step, the same trie node can be over-written multiple times before all the nodes are added. We eliminate these extra writes (by eliminating all except the last write to the same trie node). At the same time, we take care not to create any dangling pointers between consecutive write bubbles. For example (Fig. 5.14), when neighboring 24-bit

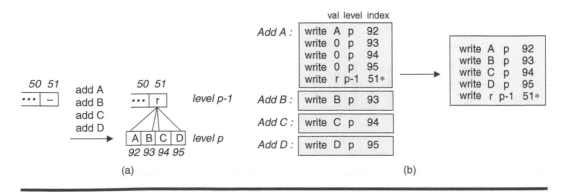

Figure 5.14 **(a) The portion of the trie before and after four neighboring prefixes A, B, C, and D are added. (From Basu, A. and Narlikar, G., *IEEE or ACM Transactions on Networking*, 13, 3, 2005. With permission.)**

prefixes A, B, C, and D are added in the same timestamp, the first route (say *A*) may cause all four new nodes to be created. A pointer to *A* will be written in one node and a null pointer will be written to its three neighboring nodes (a total of 4 write bubbles). When B, C, and D are added, pointers to them are written in these neighboring nodes. Thus, the first of the two writes to each neighboring node (*B*, *C*, and *D*) can be eliminated.

Note that care must be taken to ensure that the trie is not left in an inconsistent state. Consider the case when the pointer to the node in level p is written before the new node is written in. In this case, if a search operation is inserted between the two write operations, it may return erroneous results if it tries to access the *dangling pointer* at level $p - 1$. To eliminate these kinds of update dependency, the trie update algorithm tag writes that link a whole new subtree to the existing trie. The packing component then makes sure that these tagged writes are executed after the subtree that they point to has been written in.

Excess writes can also be eliminated when neighboring routes are withdrawn. Often an entire subtree is deleted when neighboring routes are withdrawn. The trie component from Figure 5.3 tags any pipeline write that deletes the root of an entire subtree. The packing component then eliminates all the writes in that subtree that occur with the same timestamp before the tagged write. Figure 5.15 shows how writes are eliminated when the four neighboring routes added, as shown in Figure 5.14, are withdrawn.

The effectiveness of both optimizations increases when a large number of neighboring routes are added or withdrawn in a single timestamp. Figure 5.13 shows the benefit of eliminating the excess writes. Combined with node pullups, it results in an incremental reduction of 18–25 percent in the number of write bubbles by using MinMax and filtering out short prefixes.

5.1.5.4 Caching Deleted SubTrees

A route is often withdrawn and added back a little later with possibly a different next hop. Because the withdrawal and the add often do not appear with the same timestamp, we cannot simply update the next hop table in this case. Instead, when a route withdrawal causes a subtree to be deleted, the trie component caches the subtree in software and remembers the location of the cached trie in the pipeline memory. The deleted subtree contains pointers to the withdrawn route as well as (possibly) pointers to a shorter routing prefix that was pushed down into the subtree due to leaf pushing. Therefore, the only information that must be stored with the cached subtree is the prefix that was pushed down and the last route in the subtree that was withdrawn.

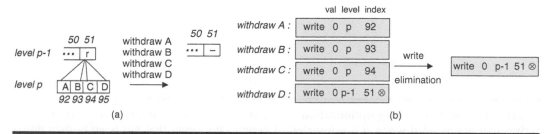

Figure 5.15 (a) The portion of the trie before and after four neighboring prefixes A, B, C, and D are withdrawn. (b) Three excess writes are eliminated by the packing component; the write that deleted the entire subtree is tagged by the trie component to aid this optimization. (From Basu, A. and Narlikar, G., *IEEE or ACM Transactions on Networking*, 13, 3, 2005. With permission.)

If the same route is added before any other neighboring route gets added, and the prefix that is pushed down remains unchanged, we can simply add back a pointer to the deleted subtree in the pipeline memory. All this checking is performed in software by the trie component, and the pipeline sees only one write instead of a number of writes. The caching optimization is equivalent to allowing a fast "undo" of one route withdrawal. Figure 5.16 shows how this optimization works.

When multiple routes withdrawn in the same timestep result in the deletion of a subtree, the caching optimization can conflict with the optimization of eliminating excess writes. For example, consider Figure 5.15 after routes *A*, *B*, and *C* are withdrawn. If route *D* is withdrawn now and we cache the deleted subtree, the cached subtree would still contain routes A, B, and C—the pipeline writes that deleted these routes were eliminated by the packing component as an optimization. If route D is now added back, and if the cached subtree is reinserted into the trie, routes *A*, *B*, and *C* would also (incorrectly) be added back to the trie. To avoid this error, the trie component caches a subtree only if a single route is deleted from it in one timestamp. By storing timestamps in the shadow trie nodes when they are modified, the trie component checks for this condition before caching a subtree. Figure 5.17 shows that caching the subtrees reduces the number of write bubbles by an additional 13–16 percent.

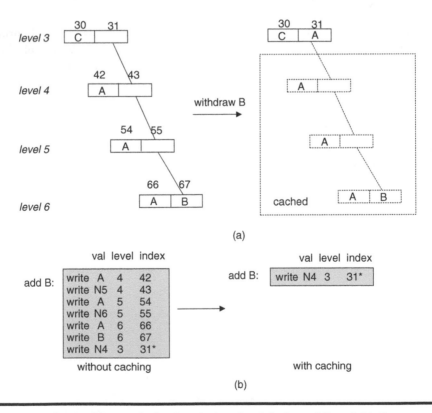

Figure 5.16 (a) The caching optimization during the deletion of *B* and (b) the corresponding writes that are avoided when *B* is reinserted. *A* denotes the route that has been pushed down the subtree containing *B*. *N4*, *N5*, and *N6* denote the pointers to nodes at levels 4, 5, and 6, respectively, and the number denotes the memory location. (From Basu, A. and Narlikar, G., *IEEE or ACM Transactions on Networking*, 13, 3, 2005. With permission.)

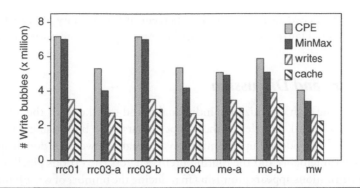

Figure 5.17 **Effect of caching the subtrees that are withdrawn and added back soon after. The label "cache" denotes using the caching optimization (along with all the previous optimizations). (From Basu, A. and Narlikar, G.,** *IEEE or ACM Transactions on Networking,* **13, 3, 2005. With permission.)**

Memory Requirements. Not all withdrawn routes are added back soon after. Therefore, caching the subtrees can consume precious pipeline memory. However, because a route withdraw is often closely followed by an add, we get most of the benefit of caching the subtrees by incurring small memory overheads. Therefore, we limited the amount of caching memory to a fixed size. We observed that nodes withdrawn several timestamps ago are less likely to be added back; hence, we maintained an FIFO list of cached nodes. When the caching memory in use went over the fixed size limit, the oldest cached nodes were deleted. We experimented with the amount of memory required for caching to get good performance from the caching optimization. The results are shown in Figure 5.18. The *X*-axis shows the caching memory overhead per pipeline stage as a percentage of the trie memory usage (actual allocation, not worst case) in that stage. In the *Y*-axis, we show how effective the optimization was when compared with the case where we could use unlimited caching memory. Thus, the 100 percent reduction number (on the *Y*-axis) refers to the case where we can cache as many deleted nodes as we want. We find that as much as 80 percent of the optimization benefits can be obtained

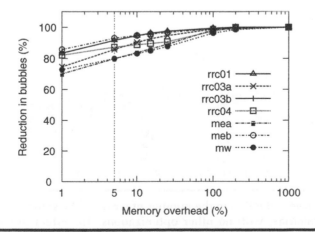

Figure 5.18 **Memory requirements for caching deleted nodes. (From Basu, A. and Narlikar, G.,** *IEEE or ACM Transactions on Networking,* **13, 3, 2005. With permission.)**

even if we restrict the cached memory to only 5 percent of the memory allocated to the trie. The caching results shown in Figure 5.17 were obtained using a 5 percent memory overhead threshold.

5.1.6 Summary and Discussion

The benefits of applying each optimization individually, and together with the other optimizations is shown in Figure 5.19. All the tries in these experiments were built using MinMax. As noted before, the benefits of the node pullup optimization are small. However, it does help to further balance the memory requirements and the routing prefixes across the various pipeline stages. Each of the remaining optimizations appears promising in reducing the number of write bubbles. Compared with using the controlled prefix expansion to construct the trie, the optimizations (along with MinMax for trie construction) result in (on average) a factor of two fewer bubbles (Fig. 5.20).

We also tested the effectiveness of our optimizations with fewer pipeline stages. Figure 5.20 shows the results of applying all the optimizations when the pipeline has four or six stages (and therefore, four or six trie levels, respectively) instead of eight. The graphs show that we get similar improvements for both the four- and the six-stage pipeline; hence, the optimizations are not specific to a given number of pipeline stages.

Prefix Table Dynamics.   The MinMax algorithm attempts to balance the memory allocations across the pipeline stages. One of the motivations behind this was to avoid frequent rebuilding of the trie after incremental updates. However, performing a large number of incremental updates may cause the trie to gradually become unbalanced. Indeed, it is possible that because of this unbalanced growth, the memory requirements for some of the pipeline stages may exceed the capacity of the pipeline stage when updates occur, and MinMax would need to be reapplied to balance out the memory allocations. Hence, it can be argued that optimizing the memory allocation based on an initial prefix table snapshot may not be the right approach.

To explore this idea further, Basu and Narlikar did some measurements on how much the structure of the optimal trie (as chosen by MinMax) changes when large update sequences are

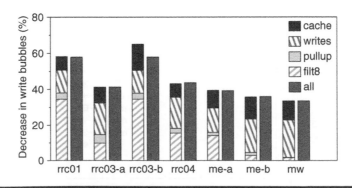

Figure 5.19   Percentage reduction in the number of write bubbles; the base case here is a trie constructed using MinMax, with no other optimizations. The effect of each optimization is shown here in isolation (with no other optimizations turned on); "all" shows the effect of turning on all the optimizations. (From Basu, A. and Narlikar, G., *IEEE or ACM Transactions on Networking*, 13, 3, 2005. With permission.)

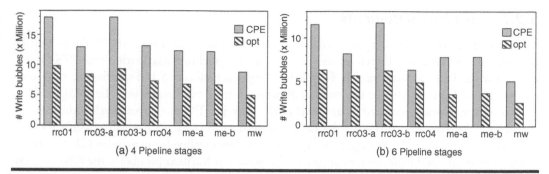

Figure 5.20 **Number of write bubbles when the lookup pipeline has fewer (four or six) stages. Here "CPE" denotes the number of bubbles generated when the trie is constructed using controlled prefix expansion (with no additional optimizations), and "opt" is when the trie is constructed using MinMax and all optimizations listed in this section are applied. (From Basu, A. and Narlikar, G., *IEEE or ACM Transactions on Networking*, 13, 3, 2005. With permission.)**

applied to an initial trie [4]. The update traces are the same as that have been used so far. The percentage difference in the size of the largest stage after all the updates are applied and the size of the largest stage if the MinMax algorithm were applied at the end of the updates to rebalance the trie once more is 3.4, 1.5, 0.09, 0.05, 7.9, 1.8, and 0.38 percent for RRC01, RRC03a, RRC03b, RRC04, me-b, me-a, and mw, respectively. For all the tables, the basic structure of the trie (in terms of strides at each level) chosen by MinMax remained the same before and after the updates. The small difference in the sizes of the largest stage is due to some additional nodes being pulled up when the trie is rebuilt at the end of the updates. Note that the optimal trie structure at the end of the updates remained unchanged even for the me-a and me-b tables, which grew by 34 percent and 61 percent, respectively. In conclusion, Basu and Narlikar's approach based on optimal memory allocation, using an initial prefix table snapshot, works reasonably well.

5.2 Two-Phase Algorithm

In Section 5.1, Basu and Narlikar propose a DP alogrithm to construct the optimal FSTs subject to the following constraints.

C1: Each level in the FST must fit in a single pipeline stage.
C2: The maximum memory allocated to a stage (over all stages) is minimum.
C3: The total memory used is minimized subject to the first two constraints.

Constraint 3 is important because studies conducted by Basu and Narlikar [4] using real Internet traces indicate that the pipeline disruptions due to update operations are reduced when the total memory used by the trie is reduced. But the algorithm of Basu and Narlikar does not minimize the total memory subject to constraints C1 and C2. Kim and Sahni developed a two-phase algorithm that constructs FSTs that satisfy the three constraints C1–C3 and a partitioning algorithm to reduce the maximum per-stage memory. In the following, we will give the problem statements and introduce the two-phase algorithms which is from [5]. Portions are reprinted with permission (© 2005 IEEE).

5.2.1 Problem Statements

When mapping an FST onto a pipelined architecture, we place each level of the FST into a different stage of the pipeline. Therefore, the problem of the pipelined fixed-stride trie (PFST) is to determine the FST that uses exactly k levels for the given prefix set S and satisfies the three constraints C1–C3. Let $nodes(i)$ be the number of nodes in the 1-bit trie at level i. Let $MS(j, r, R)$ be the maximum memory required by any level of the r-level FST R that covers levels 0 through j of the given 1-bit trie. Let $MMS(j, r)$ be the minimum of $MS(j, r, R)$ over all possible FSTs R. $MMS(j, r)$ (called $MinMaxSpace[j, r]$ in [4]) is the minimum value for constraint C2. Let $T(j, r)$ be the minimum total memory for constraint C3. Basu and Narlikar [4] obtain the following DP recurrence equations for MMS and T:

$$MMS(j,r) = \min_{m \in [r-2,\dots,j-1]} \{\max\{MMS(m,r-1), nodes(m+1) * 2^{j-m}\}\}, r-1 \qquad (5.13)$$

$$MMS(j,1) = 2^{j-m} \qquad (5.14)$$

$$T(j,r) = \min_{m \in [r-2,\dots,j-1]} \{T(m,r-1) + nodes(m+1) * 2^{j-m}\}, r > 1 \qquad (5.15)$$

$$T(j,1) = 2^{j+1} \qquad (5.16)$$

where $nodes(m + 1) * 2^{j-m} \le P$, P is the memory capacity of a pipeline stage.

Kim and Sahni found that although the algorithm of Basu and Narlikar computes $MMS(j, r)$ correctly, the values of $T(j, r)$ are not computed the best possible. This is because, in their algorithm, the computation of Ts is overconstrained. The computation of $T(j, r)$ is contained by $MMS(j, r)$ rather than by the potentially larger $MMS(W - 1, k)$ value. To correctly compute $T(W - 1, k)$ under constraints C1 and C2, Kim and Sahni proposed a two-phase algorithm that computes $MMS(W - 1, k)$ in the first phase and then uses the computed $MMS(W - 1, k)$ in the second phase to obtain $T(W - 1, k)$. To describe the two-phase algorithm, we use the following lemmas.

5.2.2 Computing MMS(W − 1, k)

We will describe three algorithms to compute $MMS(W - 1, k)$: DP, binary search, and parametric search.

Lemma 5.2.1: For every 1 bit, (a) $nodes(i) \le 2^i$, $i \ge 0$, and (b) $nodes(i) * 2^j \ge nodes(i + j)$, $i, j \ge 0$.

Proof: The proof follows from the fact that a 1-bit trie is a binary tree.

Lemma 5.2.2: Let q be an integer in the range $[r - 2, j - 2]$ and let

$$minMMS = \min_{m \in \{q+1\cdots j-1\}} \{\max\{MMS(m, r-1), nodes(m+1) * 2^{j-m}\}\}, r > 1.$$

If $nodes(q + 1) * 2^{j-q} > minMMS$, then for every s in the range $[r - 2, q]$,

$\max\{MMS(s, r-1), nodes(s + 1) * 2^{j-s}\} > minMMS$. Further, $MMS(j, r) = minMMS$.

Proof: From Lemma 5.2.1(b), it follows that $nodes(s + 1) * 2^{q-s} \ge nodes(q + 1)$. So,

$nodes(s + 1) * 2^{j-s} \ge nodes(q + 1) * 2^{j-q} > minMMS$. Hence,

$\max\{MMS(s, r-1), nodes(s + 1) * 2^{j-s}\} > minMMS$

From this, the definition of min*MMS*, and Equation 5.13, it follows that $MMS(j, r)$ = min*MMS*.

Using Lemma 5.2.2 and Equations 5.13 through 5.16, we may develop a new single-phase algorithm, called the reduced-range 1-phase algorithm, to compute the same values for *MMS* $(W-1, k)$ and $T(W-1, k)$ as the algorithm of Basu and Narlikar.

$MMS(W-1; k)$ may be computed from Equations 5.13 and 5.14 using the standard methods for DP recurrences. The time required is $O(kW^2)$. Although we cannot reduce the asymptotic complexity of the DP approach, we can reduce the observed complexity on actual prefix databases by a constant factor by employing certain properties of *MMS*.

Lemma 5.2.3:   When $r \leq l \leq W$, $MMS(l, r+1) \leq MMS(l, r)$.

Proof: When $l \geq r$, at least one level of the FST F that achieves $MMS(l, r)$ has a stride ≥ 2 (F has r levels, whereas the number of levels of the 1-bit trie that are covered by F is $l + 1$). Consider any level v of F whose stride is two or more. Let s be the stride for level v and let w be the expansion level represented by level v of F. Split level v into two to obtain an FST G that has $r + 1$ levels. The first of these split levels has a stride of 1 and the stride for the second of these split levels is $s - 1$. The remaining levels have the same strides as in F. From Lemma 5.2.1, it follows that *nodes* $(w + 1) \leq 2 * nodes(w)$. Therefore,

$$\max\{nodes(w) * 2, nodes(w+1) * 2^{s-1}\} \leq nodes(w) * 2^s$$

So, the maximum per-stage memory required by G is no more than that required by F. Because G is an $(r + 1)$-level FST, it follows that $MMS(l, r+1) \leq MMS(l, r)$.

Lemma 5.2.4:   When $l < W$ and $\lceil l/2 \rceil \geq r - 1 \geq 1$, $MMS(l, r) \leq nodes\left(\lceil l/2 \rceil\right) * 2^{l - \lceil l/2 \rceil + 1}$.

Proof: From Equation 5.13, it follows that

$$MMS(l, r) \leq \max\{MMS(\lceil l/2 \rceil - 1, r-1), nodes(\lceil l/2 \rceil) * 2^{l-\lceil l/2 \rceil+1}\}$$

From Lemma 5.2.3 and Equation 5.14, we obtain

$$MMS(\lceil l/2 \rceil - 1, r - 1) \leq MMS(\lceil l/2 \rceil - 1, 1) = 2^{\lceil l/2 \rceil}$$

The lemma now follows from the observation that $nodes(\lceil l/2 \rceil) \geq 1$ and $2^{l-\lceil l/2 \rceil+1} \geq 2^{\lceil l/2 \rceil}$.

Lemma 5.2.5:   Let $M(j, r)$, $r > 1$, be the largest m that minimizes $\max\{MMS(m, r - 1), nodes(m + 1) * 2^{j-m}\}$ in Equation 5.13. For $l < W$ and $r \geq 2$, $M(l,r) \geq \lceil l/2 \rceil - 1$.

Proof: From Equation 5.13, $M(l, r) \geq r - 2$. So, if $r - 2 \geq \lceil l/2 \rceil - 1$, the lemma is proved. Assume that $r - 2 < \lceil l/2 \rceil - 1$. Now, Lemma 5.2.4 applies and $MMS(l,r) \leq nodes(\lceil l/2 \rceil) * 2^{l-\lceil l/2 \rceil+1}$. From the proof of Lemma 4, we know that this upper bound is attained, for example, when $m = \lceil l/2 \rceil - 1$ in Equation 5.13. From the proof of Lemma 5.2.4 and from Lemma 5.2.1, it follows that for any smaller m value, say i,

$$\max\{MMS(i, r-1), nodes(i+1) * 2^{l-i}\} = nodes(i+1) * 2^{l-i}$$

$$\geq nodes\left(\lceil l/2 \rceil\right) * 2^{i-\lceil l/2 \rceil+1} * 2^{l-i} = nodes\left(\lceil l/2 \rceil\right) * 2^{l-\lceil l/2 \rceil+1}$$

So a smaller value of m cannot get us below the upper bound of Lemma 5.2.4. Hence, $M(l,r) \geq \lceil l/2 \rceil - 1$.

Theorem 5.2.1: From Lemma 5.2.5 and Equation 5.13, for $l < W$ and $r \geq 2$, $M(l,r) \geq$ max $\{r - 2, \lceil l/2 \rceil - 1\}$

Theorem 5.2.1 and a minor extension of Lemma 5.2.2 lead to DynamicProgramming*MMS*, which computes $MMS(W - 1, k)$, shown in Figure 5.21. The complexity of this algorithm is $O(kW^2)$.

Binary Search. It is easy to see that for $k > 1$, $0 < MMS(W - 1, k) < 2^W$. The range for $MMS(W - 1, k)$ may be narrowed by making the following observations.

LB: From Lemma 5.2.3, $MMS(W - 1, W) \leq MMS(W - 1, k)$. Because,

$$MMS(W - 1, W) = \max_{0 \leq i \leq W} \{nodes(i) * 2\} \tag{5.17}$$

We obtain a better lower bound on the search range for $MMS(W - 1, k)$.

Algorithm DynamicProgramming*MMS*(*W, k*)

// *W* is length of longest prefix.

// *k* is maximum number of expansion levels desired.

// Return *MMS*(*W* – 1, *k*).

{

    for ($j = 0$; $j < W$; $j + +$)

        $MMS(j, 1) = 2^{j+1}$;

        for ($r = 2$; $r \leq k$; $r + +$)

            for ($j = r - 1$; $j < W$; $j + +$){

        // Compute MMS(j, r).

            $minJ = max(r - 2, [j / 2] - 1)$;

            min*MMS* = ∞ ;

            for ($m = j - 1$; $m \geq minJ$; m – –){

                if ($nodes(m + 1) * 2^{j-m} \geq$ min *MMS*) break;

                $mms = max(MMS(m, r - 1), nodes (m + 1) * 2^{j-m})$

                if ($mms <$ min*MMS*) min*MMS* = *mms*;

            }

            $MMS(j, r) = $ min*MMS*;

        }

    return *MMS*(*W* – 1, *k*);

}

Figure 5.21 **DP algorithm to compute *MMS*(*W* – 1, *k*). (From Kim, K. and Sahni, S., *IEEE Transactions on Computers*, 56, 1, 2007. With permission.)**

UB. For real-world database and practical values for the number k of pipeline stages, Lemma 5.2.4 applies with $l = W - 1$ and $r = k$. So this Lemma provides a tighter upper bound for the binary search than the bound $2^W - 1$.

To successfully perform a binary search in the stated search range, we need a way to determine for any value p in this range whether or not there is an FST with at most k levels whose per-stage memory requirement is at most p. We call this a feasibility test of p. Note that from Lemma 5.2.3 it follows that such an FST exists iff such an FST with exactly k levels exists. We use Algorithm GreedyCover (Fig. 5.22) for the feasibility test. For a given memory constraint p, GreedyCover covers as many levels of the 1-bit trie as possible by level 0 of the FST. Then it covers as many of the remaining levels as possible by level 1 of the FST, and so on. The algorithm returns the value **true** if it is able to cover all W levels by at most k FST levels.

The correctness of Algorithm GreedyCover as a feasibility test may be established using standard proof methods for greedy algorithms [11]. The time complexity of GreedyCover is $O(W)$.

Let BinarySearchMMS be a binary search algorithm to determine $MMS(W - 1, k)$. Rather than using the upper bound from Lemma 5.2.4, this algorithm uses a doubling procedure to determine a suitable upper bound. Experimentally, we observed that this strategy worked better using the upper bound of Lemma 5.2.4 except for small k such as $k = 2$ and $k = 3$. The binary search algorithm invokes GreedyCover $O(W)$ times. So its complexity is $O(W^2)$.

Parametric Search. It was developed by Frederickson [12]. Kim and Sahni used the development of parametric search in [13] to compute $MMS(W - 1, k)$.

We start with an implicitly specified sorted matrix M of $O(W^2)$ candidate values for $MMS(W - 1, k)$. This implicit matrix of candidate values is obtained by observing that $MMS(W - 1, k) = nodes(i) * 2^j$ for some i and j, $0 \leq i \leq W$ and $1 \leq j \leq W$. Here i denotes a possible expansion level and j denotes a possible stride for this expansion level. To obtain the implicit matrix M, we sort the distinct $nodes(i)$ values into ascending order. Let the sorted values be $n_1 < n_2 < \cdots < n_t$, where t is the number of distinct $nodes(i)$ values. The matrix value $M_{i,j}$ is $n_i * 2^j$. Because we do not explicitly compute the

```
Algorithm GreedyCover(k, p)
// k is maximum number of expansion levels (i.e., pipeline stages) desired.
// p is maximum memory for each stage.
// W is length of longest prefix.
// Return true iff p is feasible.
{
    stages = eLevel = 0;
    while (stages < k){
        i = 1;
        while (nodes(eLevel) 2^i ≤ p && eLevel + i ≤ W) i + +;
        if (eLevel + i > W) return true;
        if (i == 1) return false;
        // start next stage
        eLevel + = i - 1;
        stages + +; }
    return false ; // stages≥k. So, infeasible
}
```

Figure 5.22 Feasibility test for *p*. (From Kim, K. and Sahni, S., *IEEE Transactions on Computers*, 56, 1, 2007. With permission.)

$M_{i,j}$, the implicit specification can be obtained in $O(W \log W)$ time (i.e., the time needed to sort the *nodes*(*i*) values and eliminate duplicates).

We see that M is a sorted matrix. That is, $M_{i,j} \leq M_{i,j+1}$, $1 \leq i \leq t$, $1 \leq j \leq W$ and $M_{i,j} \leq M_{i+1,j}$, $1 \leq i \leq t$, $1 \leq j \leq W$. Again, it is important to note that the matrix M is provided implicitly. That is, we are given a way to compute $M_{i,j}$ in constant time, for any value of i and j. We are required to find the least $M_{i,j}$ that satisfies the feasibility test of Algorithm *GreedyCover* (Fig. 5.22). Henceforth, we refer to this feasibility test as criterion GC.

We know that the criterion GC has the following property. If $GC(x)$ is false (i.e., x is infeasible), then $GC(y)$ is false (i.e., y is infeasible) for all $y \leq x$. Similarly, if $GC(x)$ is true (i.e., x is feasible), then $GC(y)$ is true for all $y \geq x$. In parametric search, the minimum $M_{i,j}$ for which GC is true is found by trying out some of the $M_{i,j}$'s. As different $M_{i,j}$'s are tried, we maintain two values λ_1 and λ_2, $\lambda_1 < \lambda_2$, with the properties: (a) $GC(\lambda_1)$ is false and (b) $GC(\lambda_2)$ is true.

Initially, $\lambda_1 = 0$ and $\lambda_2 = \infty$ (we may use the bounds $LB - 1$ and UB as tighter bounds). With M specified implicitly, a feasibility test in place, and the initial values of λ_1 and λ_2, defined, we may use a parametric search to determine the least candidate in M that is feasible. Details of the parametric search process to be employed appear in [13]. Frederickson [12] has shown that parametric search of a sorted $W \times W$ matrix M performs at most $O(\log W)$ feasibility tests. Because each feasibility test takes $O(W)$ time, $MMS(W - 1, k)$ is determined in $O(W \log W)$ time.

5.2.3 Computing T(W − 1, k)

Algorithm *FixedStrides*. Let $C((i, j), r)$ be minimum total memory required by an FST that uses exactly r expansion levels, covers levels i through j of the 1-bit trie, and each level of which requires at most $p = MMS(W - 1, k)$ memory. A simple DP recurrence for C is:

$$C((i,j),r) = \min_{m \in X(i,j,r)} \{C((i,m),r-1) + nodes(m+1) * 2^{j-m}\}, \quad r > 1, \ j-i+1 \geq 1 \quad (5.18)$$

$$C((i,j),1) = \begin{cases} nodes(i) * 2^{j-i+1} & \text{if } nodes(i) * 2^{j-i+1} \leq p \\ \infty & \text{otherwise} \end{cases} \quad (5.19)$$

where $X(i, j, r) = \{m \mid i + r - 2 \leq m < j \text{ and } nodes(m + 1) * 2^{j-m} \leq p\}$.

Let $M(j, r)$, $r > 1$, be the largest m that minimizes $C((i,m), r - 1) + nodes(m + 1) * 2^{j-m}$, in Equation 5.18.

Let FixedStrides be the resulting algorithm to compute $T(W - 1, k) = C((0, W - 1), k)$ for a given 1-bit trie, the desired number of expansion levels k, and given per-stage memory constraint $MMS(W - 1, k)$. The complexity of this algorithm is $O(kW^2)$.

Algorithm PFST: An alternative, and faster by a constant factor, algorithm to determine $T: (W - 1, k)$ results from the following observations.

O1: At least one level of the FST F that satisfies constraints C1–C3 requires exactly $MMS (W - 1, k)$ memory.

O2: If there is exactly one i for which $MMS(W - 1, k)/nodes(i) = 2^j$ for some j, then one of the levels of F must cover levels i through $i + j - 1$ of the 1-bit trie.

We may use Algorithm *CheckUnique* (Fig. 5.23) to determine whether there is exactly one i such that $MMS(W - 1, k)/nodes(i)$ is power of 2. This algorithm can check whether q is a power

```
Algorithm CheckUnique(p)
// p is MMS(W – 1, k).
// Return true iff the expansion level for p is unique.
{
    count = 0;
    for (i = 0; i < W; i + +){
        q = p/nodes(i);
        r = p%nodes(i); // % is mod operator
        if (r > 0) continue;
        if (q is a power of 2){
                if (count > 0) return false;
                    else count + +; }}
    return true;
}
```

Figure 5.23   Algorithm to check whether expansion level for *p* is unique. (From Kim, K. and Sahni, S., *IEEE Transactions on Computers*, 56, 1, 2007. With permission.)

of 2 by performing a binary search over the range 1 through $W - 1$. The complexity of *CheckUnique* is $O(W \log W)$.

Suppose that *CheckUnique* $(MMS(W - 1, k))$ is true and that *GreedyCover* $(k, MMS(W - 1, k))$ covers the W levels of the 1-bit trie using #*Stages* stages and that stage *sMax* is the unique stage that requires exactly $MMS(W - 1, k)$ memory. Let this stage cover levels *lStart* through *lFinish* of the 1-bit trie. Note that the stages are numbered 0 through #*Stages* $- 1$. It is easy to see that in a k-PFST (i.e., a k-level FST that satisfies C1–C3),

1. One level covers levels *lStart* through *lFinish*.
2. The k-PFST uses at least *sMax* levels to cover levels 0 through *lStart* - 1 of the 1-bit trie (this follows from the correctness of *GreedyCover*; it is not possible to cover these levels using fewer levels of the k-PFST without using more per-stage memory than $MMS(W - 1, k)$). An additional FST level if required to cover levels *lStart* through *lFinish* of the 1-bit trie. So at most $k - sMax - 1$ levels are available for levels *lFinish* $+ 1$ through $W - 1$ of the 1-bit trie.
3. The k-PFST uses at least #*Stages* $- sMax - 1$ levels to cover levels *lFinish* $+ 1$ through $W - 1$ of the 1-bit trie plus an additional stage for levels *lStart* through *lFinish*. So at most $k + sMax - $#*Stages* levels are available for levels 0 through *lStart* $- 1$ of the 1-bit trie.

Let PFST be the resulting algorithm to compute $T(W - 1, k) = C((0, W - 1), k)$. The complexity of this algorithm is $O(kW^2)$.

Accelerating Algorithm PFST

When *CheckUnique* returns true, Algorithm PFST invokes *FixedStrides* to compute $C((0, lStart -1), j)$, $j \leq k + sMax - $#*Stages* and $C((lFinish + 1, W - 1), j)$, $j \leq k - sMax - 1$. When $k + sMax - $#*Stages* ≤ 3 run time is reduced by invoking a customized algorithm *FixedStrides K123* [5]. Similarly, when $k - sMax - 1 \leq 3$, faster run time results using this customized algorithm rather than *FixedStrides*.

5.2.4 A Faster Two-Phase Algorithm for k = 2, 3

The two-phase algorithms described above determine the optimal pipelined FST in $O(kW^2)$ time. When $k = 2$, this optimal FST may be determined in $O(W)$ time by performing an exhaustive examination of the choices for the second expansion level $M(W-1, 2)$ (note that the first expansion level is 0).

From Theorem 5.2.1 and Equation 5.13, we see that the search for $M(W-1, 2)$ may be limited to the range $[\lceil (W-1)/2 \rceil, W-2]$. So for $k = 2$, Equation 5.13 becomes

$$MMS(W-1,2) = \min_{m\in\{\lceil (W-1)/2\rceil-1\cdots W-2\}} \{\max\{MMS(m,1), nodes(m+1)*2^{W-1-m}\}\}$$

$$= \min_{m\in\{\lceil (W-1)/2\rceil-1\cdots W-2\}} \{2^{m+1}, nodes(m+1)*2^{W-1-m}\}\} \qquad (5.20)$$

The run time for $k = 2$ may further be reduced to $O(\log W)$ (assuming that 2^i, $1 \le i < W$ are precomputed) using the observations (a) 2^{m+1} is an increasing function of m and (b) $nodes(m+1)* 2^{W-1-m}$ is a nonincreasing function of m (Lemma 5.2.1). Algorithm *PFSTK2* (Fig. 5.24) uses these observations to compute $MMS(W-1, 2)$, $T(W-1, 2)$, and $M(W-1, 2)$ in $O(\log W)$ time. The stride for the root level of the PFST is $M(W-1, 2) + 1$ and that for the only other level of the PFST is $W - M(W-1, 2) - 1$.

Although Algorithm *PFSTK2* employs a binary search to achieve the stated $O(\log W)$ asymptotic complexity, in practice, it may be more effective to use a serial search. This is so as the values being searched for are expected to be close to either the lower or the upper end of the range being searched.

We will find an $O(\log^k W)$ algorithm to compute $MMS(W-1, k)$, $k \ge 2$. First consider the case $k = 3$. Applying Theorem 5.2.1 to Equation 5.13 results in

$$MMS(W-1,3) = \min_{m\in\{\lceil (W-1)/2\rceil-1\cdots W-2\}} \{\max\{MMS(m,2), nodes(m+1)*2^{W-1-m}\}\} \qquad (5.21)$$

With respect to the two terms on the right-hand side of Equation 5.21, we make the two observations: (a) $MMS(m,2)$ is a nondecreasing function of m and (b) $nodes(m + 1) * 2^{W-1-m}$ is a nonincreasing function of m (Lemma 5.2.1). The algorithm shown in Figure 5.25 does $O(\log W)$ $MMS(m, 2)$ computations at a total cost of $O(\log^2 W)$. It is easy to see that this adaptation generalizes to arbitrary k (replace all occurrences of $MMS(*, 2)$ with $MMS(*, k-1)$ and of $MMS(*, 3)$ with $MMS(*, k)$). The resulting complexity is $O(\log^k W)$. Hence, this generalization provides a faster way to compute $MMS(W-1, k)$ than the parametric search scheme described in Section 5.2.2 for small k.

When $k = 3$, we may determine the best PFST in $O(W^2)$ time using the value of $MMS(W-1, 3)$ computed by the algorithm shown in Figure 5.25. Although this is the same asymptotic complexity as that achieved by the general algorithms described in Sections 5.2.2 and 5.2.3, we expect the new algorithm to be faster by a constant factor.

Let $T(j, 2, P)$ be the minimum total memory required by a 2-level FST for levels 0 through j of the 1-bit trie under the constraint that no level of the FST requires more than P memory. Let $M(j, 2, P) + 1$ be the stride of the root level of this FST.

$T(W-1, 3)$ and the expansion levels of the PFST may be computed using the algorithm shown in Figure 5.26. Its complexity is $O(W^2)$. Once again, in practice, better run-time performance may be observed using a serial rather than a binary search.

Algorithm *PFSTK 2()*
//Compute $MMS(W - 1, 2)$ and $T(W - 1, 2)$ values as well as optimal
//strides for two pipeline stages.
// $M(W - 1, 2) + 1$ is the stride for the first pipeline stage.
{

Phase 1: [Determine $MMS(W - 1, 2)$]

  Step1: [Determine *leastM*] Perform a binary search in the range $[\lceil (W - 1)/2 \rceil - 1, W - 2]$ to
    determine the least m, *leastM*, for which $2^{m+1} > nodes(m + 1) \star 2^{W-1-m}$.

  Step2: [Determine $MMS(W - 1, 2)$ and case] If there is no such m, set $MMS(W - 1, 2) = 2 \star nodes(W - 1)$, $bestM = W - 2$, and case = "noM". Go to Phase 2.
    Set $X = nodes(leastM) \star 2^{W-leastM}$.
    If $X < 2^{leastM+1}$, set $MMS(W - 1, 2) = X$, $bestM = leastM - 1$, and *case* = "$X <$".
    If $X = 2^{leastM+1}$, set $MMS(W - 1, 2) = X$, $bestM = leastM - 1$, and *case* = "$X =$".
    Otherwise, set $MMS(W - 1, 2) = 2^{leastM+1}$ and *case* = "$X >$".

Phase 2: [Determine $M(W - 1, 2)$ and $T(W - 1, 2)$]

  case = "$X >$"
    Set $M(W - 1, 2) = leastM$ and $T(W - 1, 2) = 2^{leastM+1} + nodes(leastM + 1) \star 2^{W-1-leastM}$.
  Remaining Cases
    Use binary search in the range $[\lceil (W - 1)/2 \rceil - 1, bestM]$ to find the least m for which $nodes(m + 1) \star 2^{W-1-m} = nodes(bestM + 1) \star 2^{W-1-bestM}$.
    Set $M(W - 1, 2)$ to be this value of m.
    Set $T(W - 1, 2) = 2^{m+1} + nodes(m + 1) \; 2^{W-1-m}$, where $m = M(W - 1, 2)$.
    Set $Y = 2^{leastM+1} + nodes(leastM + 1) \star 2^{W-1-leastM}$.
    If *case* = "$X =$" and $T(W - 1, 2) > Y$ then set $M(W - 1, 2) = leastM$ and $T(W - 1, 2) = Y$.

}

Figure 5.24 Algorithm to compute $MMS(W - 1, 2)$ and $T(W - 1, 2)$ using Equation 5.21. (From Kim, K. and Sahni, S., *IEEE Transactions on Computers*, 56, 1, 2007. With permission.)

Step 1: [Determine *leastM*] Perform a binary search in the range $[\star (W - 1)/2, \star W - 2]$ to
    determine the least m, *leastM*, for which $MMS(m, 2) > nodes(m + 1) \star 2^{W-1-m}$.

Step 2: [Determine $MMS(W - 1, 3)$ and case] If there is no such m, set $MMS(W - 1, 3) = 2 \star nodes(W - 1)$, $bestM = W - 2$, and case = "noM". Done.
    Set $X = nodes(leastM) \star 2^{W-leastM}$.
    If $X < MMS(leastM, 2)$, set $MMS(W - 1, 3) = X$, $bestM = leastM - 1$, and case = "$X <$".
    If $X = MMS(leastM, 2)$, set $MMS(W - 1, 3) = X$ and case = "$X =$".
    Otherwise, set $MMS(W - 1, 3) = MMS(leastM, 2)$ and case = "$X >$".

Figure 5.25 Compute $MMS(W - 1, 3)$. (From Kim, K. and Sahni, S., *IEEE Transactions on Computers*, 56, 1, 2007. With permission.)

Step 1 : [Determine *mSmall* and *mLarge*]

   case = "*X* >"

      Use binary search in the range [*leastM*, *W*– 2] to find the largest *m*, *mLarge*, for which
$MMS(m, 2) = MMS(leastM, 2)$.

      Set *mSmall* = *leastM*

   case = "*X* <" or "*noM*"

      Use binary search in the range [[(*W*–1)/2]–1, *bestM*] to find the least *m*, *mSmall*,

      for which $nodes(m + 1)*2^{W-1-m} = nodes(bestM + 1)*2^{W-1-bestM}$

     Set *mLarge* = *bestM*

   case = "*X* ="

      Determine *mLarge* as for the case "*X*>" and *mSmall* as for the case "*X*<" or "*noM*".

Step 2 : [Determine $T(W – 1, 3)$ and the expansion levels]

      Compute $T(j, 2, W – 1, 3)$. $mSmall \le j \le mLarge$ and determine the *j* value, *jMin*, that
minimizes $T(j, 2, MMS(W – 1, 3)) + nodes(j + 1)*2^{W-1-j}$

      Set $T(W – 1, 3) = T(jMin, 2, MMS(W – 1, 3)) + nodes(jMin+ 1)*2^{W-1-jMin}$.

      The expansion levels are 0, $M(jMin, 2, MMS(W – 1, 3) + 1$, and *jMin* + 1.

Figure 5.26 Compute $T(W – 1, 3)$ and expansion levels. (From Kim, K. and Sahni, S., *IEEE Transactions on Computers*, 56, 1, 2007. With permission.)

5.2.5 A Partitioning Scheme

Basu and Narlikar propose a node pullup scheme that results in improved multibit tries that are not FSTs [4]. Kim and Sahni propose a partitioning strategy to construct the pipelined multibit tries [5]. In the partitioning scheme, the prefixes in the forwarding tables are partitioned into two groups *AB* and *CD*. Group *AB* comprises all prefixes that begin with 0 or 10. These prefixes represent the traditional Classes *A* and *B* networks. The remaining prefixes form the group *CD*. The multibit trie is constructed using the following steps.

Step 1: Find the optimal pipelined FST for the prefixes in group *CD* using any of the algorithms discussed earlier.

Step 2: Let *mms*(*CD*) be the *MMS* for the FST of Step 1. Use a binary search scheme to determine $mms'(AB + CD)$. The binary search first tests $2*mms(CD)$, $4*mms(CD)$, $8*mms(CD)$, and so on for feasibility and stops at the first feasible value, *u*, found. Then it does a standard binary search in the range [*mms*(*CD*), *u*] to determine the smallest feasible value. The feasibility test is done using, as a heuristic, a modified version of GreedyCover, which does not change the FST for *CD* obtained in Step 1. The heuristic covers as many levels of the 1-bit trie for *AB* as possible using at most $mms'(AB + CD) – 2^{S_0}$ space. Then as many of the remaining levels of *AB* as possible are covered using at most $mms'(AB + CD) – nodes(CD,1)*2^{S_1}$ space, where *nodes*(*CD*, 1) is the number of nodes at level 1 of the FST for *CD*, and so on, where s_i is the stride for level *i* of the FST for *CD*.

Step 3: Merge the roots of the FSTs for *AB* and *CD* obtained as above into a single node. The stride of the merged root is max{*stride*(*AB*), *stride*(*CD*)}, where *stride*(*AB*) is the stride of the root of the FST for *AB*. If *stride*(*AB*) > *stride*(*CD*), then the stride of the level 1 nodes of the *CD* FST is reduced by *stride*(*AB*) – *stride*(*CD*). In practice, the new stride is ≥ 1 and so the stride reduction does not propagate further down the *CD* FST. When *stride*(*AB*) < *stride*(*CD*), the stride reduction takes place in the *AB* FST. When *stride*(*AB*) = *stride*(*CD*), there is no stride reduction in the lower levels of either FST.

Step 1 of the partitioning scheme takes $O(kW^2)$ time, Step 2 takes $O(W^2)$ time, and Step 3 takes $O(k)$ time (in practice, Step 3 takes $O(1)$ time). The overall complexity, therefore, is $O(kW^2)$.

5.2.6 Experimental Results

Kim and Sahni programmed the DP algorithms in C and compared their performance against that of the algorithm of Basu and Narlikar [4]. All codes were run on a 2.26 GHz Pentium 4 PC and were compiled using Microsoft Visual C++ 6.0 and optimization level O2. The six IPv4 prefix databases used as the test data are shown in Table 5.3. These databases correspond to backbone-router prefix sets.

Memory Comparison. Let FST denote the case when *R* is the FST that minimizes total memory [3], and PFST denote the case when *R* is the FST for which $MS(W-1, k, R) = MMS(W-1, k)$. Kim and Sahni experimented with seven values of k (2–8) for each of the six data sets. Note that the case $k = 1$ corresponds to the use of a single processor, as studied in [3]. The experiments show that, for six data sets, the reduction in maximum per-stage memory and total memory requirement is quite small when we go for $k = 7$ to $k = 8$ (Fig. 5.27). This suggests that using more than eight pipeline stages is of little value as far as the memory metric is concerned. Hence, limiting our experiments to k in the range 2 through 8 is justified. Of course, in practice, the router data structure will be implemented on a pipeline with a fixed number of stages. The experiments suggest that a good choice for this fixed number of stages is between 5 and 8.

In 25 of the 42 tests, $MMS(W-1, k) = MS(W-1, k, PFST) < MS(W-1, k, FST)$. In the remaining 17 tests, the FST that minimizes the total memory also minimizes the $MS(W-1, k, R)$. The maximum reduction in $MS(W-1, k, R)$ observed for test set is approximately 44 percent. Figure 5.27a shows the $MS(W-1, k, R)$ for the RRC01 data set.

Table 5.3   Prefix Database Obtained from RIS and IPMA

Database	No. of Prefixes	No. of 16-bit Prefixes	No. of 24-bit Prefixes	No. of Nodes
RRC04	109,600	7000	62,840	217,749
RRC03b	108,267	7069	62,065	216,462
RRC01	103,555	6938	59,854	206,974
MW02	87,618	6306	51,509	174,379
PA	85,987	6606	49,993	173,626
ME02	70,306	6496	39,620	145,868

Note: The last column shows the number of nodes in the 1-bit trie representation of the prefix database.

Source: RRC04, RRC03b, and RRC01 are from RIS (http://data.ris.ripe.net,2003). MW02, PA, and ME02 are from IPMA (http://nic.merit.edu/ipma, 2003). Kim, K. and Sahni, S., *IEEE Transactions on Computers*, 56, 1, 2007. With permission.

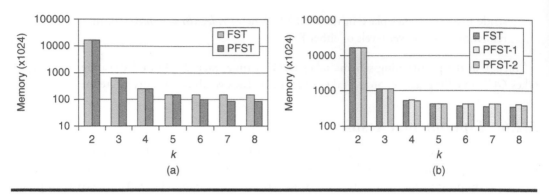

Figure 5.27 Maximum per-stage and total memory required (kBytes) for RRC01. (a) *MS(W* − 1, *k*, *R*) and (b) *T(W* − 1; *k*). (From Kim, K. and Sahni, S., *IEEE Transactions on Computers*, 56, 1, 2007. With permission.)

The three FSTs are labeled FST (the FST that minimizes total memory [3]), PFST-1 (the pipelined FST generated using the one-phase algorithm of [4]) and PFST-2 (the pipelined FST generated using any of the two-phase algorithms). Note that PFST-1 and PFST-2 minimize the $MS(W − 1, k, R)$ and that the total memory required by the PFST generated by the various two-phase algorithms is the same. Figure 5.27b shows $T(W − 1, k)$ for RRC01 data set. As expected, whenever $MS(W − 1, k, PFST) < MS(W − 1, k, FST)$, the $T(W − 1, k)$s for PFST-1 and PFST-2 are greater than that for FST. That is the reduction in maximum per-stage memory comes with a cost of increased total memory required. The maximum increase observed on the data set is about 35 percent for PFST-1 and about 30 percent for PFST-2. Although $MS(W − 1, k, PFST-1) = MS(W − 1, k, PFST-2) = MMS(W − 1, k)$, the total memory required by PFST-2 is less than that required by PFST-1 for 12 of the 42 test cases (shown in boldface). On several of these 12 cases, PFST-1 takes 7 percent more memory than that required by PFST-2.

Execution Time Comparison.   First, we compare the execution time of the Basu algorithm [4] and that of the reduced-range one-phase algorithm in Section 5.2.1. Note that both the algorithms generate identical FSTs. The reduced-range one-phase algorithm achieves a speedup, relative to the Basu algorithm [4], of about 2.5 when $k = 2$ and about 3 when $k = 8$. Figure 5.28 plots the run-times for the RRC04 data set.

Figure 5.29 shows the time taken by two—DP and binary search—algorithms to compute $MMS(W − 1, k)$ in Section 5.2.2. For $k ≤ 3$, digital programming algorithm is faster. When $k ≥ 4$, the binary search algorithm is faster. In fact, when $k ≥ 7$, the binary search algorithm took about 1/20th the time on some of the data sets. The run-time for a binary search algorithm is quite small for most of our data sets when $k = 7$ or 8. This is because, in the implementation of *BinarySearchMMS*, we begin by checking whether or not *GreedyCover*(*k*, *p*) is true. In case it is, we can omit the search. For the stated combinations of data set and *k*, *GreedyCover*(*k*, *p*) is true.

Figure 5.30 gives the execution times for the three—fixed strides (FS), PFST, and acclerating PFST (k123)—algorithms described in Section 5.2.3 to compute $T(W − 1, k)$. Algorithm K123 is the fastest of the three algorithms. On RRC04 with $k = 5$, FS took about 4.6 as much time as taken by K123 and PFST took 2.9 times as much time as taken by K123.

The run-times for the two-phase algorithms with $k = 2$ or 3 are given in Table 5.4. On RRC04 with $k = 2$, K123 takes 24 times as much time as taken by the two-phase algorithm with $k = 2$. For $k = 3$, K123 took 1.9 times the time taken by the two-phase algorithm with $k = 3$. Note that K123

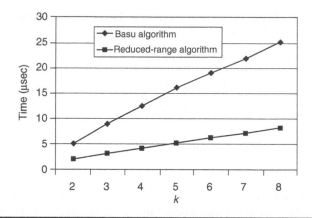

Figure 5.28 Execution time (in μs) for the Basu algorithm in [4] and the reduced-range one-phase algorithm in Sections 5.2.2 for RRC04 and 5.2.1 for RRC04. (From Kim, K. and Sahni, S., *IEEE Transactions on Computers*, 56, 1, 2007. With permission.)

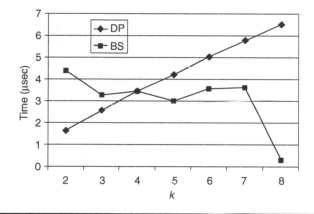

Figure 5.29 Execution time for computing $MMS(W - 1, k)$ for RRC04. DP: dynamic programming, BS: binary search. (From Kim, K. and Sahni, S., *IEEE Transactions on Computers*, 56, 1, 2007. With permission.)

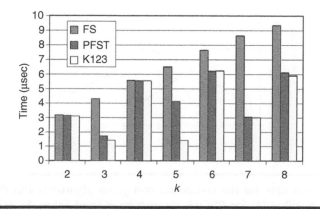

Figure 5.30 Execution time for computing $T(W - 1, k)$ for RRC04. (From Kim, K. and Sahni, S., *IEEE Transactions on Computers*, 56, 1, 2007. With permission.)

Table 5.4   Execution Time (in μs) for Two-Phase Algorithms with $k = 2$ and $k = 3$

k	RRC04	RRC03b	RRC01	MW02	PA	ME02
2	0.13	0.12	0.13	0.13	0.13	0.13
3	0.17	0.76	0.76	0.76	0.76	0.75

Source: Kim, K. and Sahni, S., *IEEE Transactions on Computers*, 56, 1, 2007. With permission.

determines $T(W-1, k)$ only and must be preceded by a computation of $MMS(W-1, k)$. However, the two-phase algorithms with $k = 2$ and 3 determine both $MMS(W-1, k)$ and $T(W-1, k)$.

To compare the run-time performance of the algorithm with that of [4], we use the times for implementation of faster $k = 2$ and $k = 3$ algorithms when $k = 2$ or 3 and the times for the faster implementations of the binary search algorithms when $k > 3$. That is, we compare our best times with the times for the algorithm of [4].

Figure 5.31 plots the run-time of the one-phase algorithm of [4] and that of our composite algorithm, which uses the $k = 2$ and $k = 3$ algorithms when $2 \leq k \leq 3$. When $k > 3$, the composite algorithm uses the binary search to compute $MMS(W-1; k)$ and *FixedStridesK*123 to compute $T(W-1; k)$. The plot shown in Figure 5.31 is for the largest database, RRC04. The new algorithm takes less than 1/38th the time taken by the algorithm of [4] when $k = 2$ and less than 1/4th the time when $k = 8$.

5.3   Pipelined Variable-Stride Multibit Tries

Algorithms to construct optimal pipelined FST (i.e., FST that satisfy constraints C1–C3 in Section 5.2) have been developed in [4] and [5] (see Sections 5.1 and 5.2). The memory required by the best VST for a given prefix set P and number of expansion levels k is less than or equal to

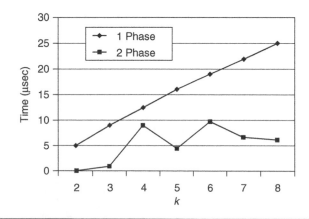

Figure 5.31   Execution time for the composite two-phase algorithm and the one-phase algorithm of Basu and Narlikar [4] for RRC04. (From Kim, K. and Sahni, S., *IEEE Transactions on Computers*, 56, 1, 2007. With permission.)

that required by the best FST for P and k. Lu and Sahni [6] consider the construction of PVSTs that satisfy constraints C1–C3 (although, in C1, "FST" is replaced with "variable-stride trie") and propose a heuristic to construct an "approximately optimal" PVST. This section is from [6] (© 2006 IEEE).

5.3.1   Construction of Optimal PVST

Let O be the 1-bit trie for the given prefix set, let N be a node of O, and let $ST(N)$ be the subtree of O that is rooted at N. Let $Opt'(N, r)$ denote the approximately optimal PVST for the subtree $ST(N)$; this PVST has at most r levels. Let $Opt'(N, r).E(l)$ be the (total) number of elements at level l of $Opt'(N, r)$, $0 \leq l < r$. We seek to construct $Opt'(root(O), k)$, the approximately optimal PVST for O that has at most k levels, where k is the number of available pipeline stages.

Let $D_i(N)$ denote the descendents of N that are at level i of $ST(N)$. So, for example, $D_0(N) = \{N\}$ and $D_1(N)$ denotes the children, in O, of N. The approximately optimal VST has the property that its subtrees are approximately optimal for those of N that they represent. So, for example, if the root of $Opt'(N, r)$ represents levels 0 through $i - 1$ of $ST(N)$, then the subtrees of (the root of) $Opt'(N, r)$ are $Opt'(M, r-1)$ for $M \in D_i(N)$.

When $r = 1$, $Opt'(N, 1)$ has only a root node; this root represents all the levels of $ST(N)$. So,

$$Opt'(N,r) \cdot E(l) = \begin{cases} 2^{height(N)} & l = 0 \\ 0 & l > 0 \end{cases} \tag{5.22}$$

where $height(N)$ is the height of $ST(N)$.

When $r > 1$, the number, q, of levels of $ST(N)$ represented by the root of $Opt'(N, r)$ is between 1 and $height(N)$. From the definition of $Opt'(N, r)$, it follows that

$$Opt'(N,r) \cdot E(l) = \begin{cases} 2^q & l = 0 \\ \displaystyle\sum_{M \in D_q(N)} Opt'(M,r-1) \cdot E(l-1) & 1 \leq l < r \end{cases} \tag{5.23}$$

where q is as defined below

$$q = \arg\min_{1 \leq i \leq height(N)} \left\{ \max \left\{ 2^i, \max_{0 \leq l \leq r-1} \left\{ \sum_{M \in D_i(N)} Opt'(M,r-1) \cdot E(l) \right\} \right\} \right\} \tag{5.24}$$

Although the DP recurrences of Equations 5.22 through 5.24 may be solved directly to determine $Opt'(root(O), k)$, the time complexity of the resulting algorithm is reduced by defining auxiliary equations. For this purpose, let $Opt'STs(N, i, r-1)$, $i > 0$, $r > 1$, denote the set of approximately optimal VSTs for $D_i(N)$ ($Opt'STs(N, i, r-1)$ has one VST for each member of $D_i(N)$); each VST has at most $r - 1$ levels. Let $Opt'STs(N, i, r-1).E(l)$ be the sum of the number of elements at level l of each VST of $Opt'STs(N, i, r-1)$. So,

$$Opt'STs(N,i,r-1) \cdot E(l) = \sum_{M \in D_i(N)} Opt'(M,r-1) \cdot E(l), \quad 0 \leq l < r-1 \tag{5.25}$$

For $Opt'STs(N, i, r-1) \cdot E(l)$, $i > 0$, $r > 1$, $0 \le l < r-1$, we obtain the following recurrence

$$Opt'STs(N, i, r-1) \cdot E(l) =$$
$$\begin{cases} Opt'(LC(N), r-1) \cdot E(l) + Opt'(RC(N), r-1) \cdot E(l) & i = 1 \\ Opt'STs(LC(N), i-1, r-1) \cdot E(l) + Opt'STs(RC(N), i-1, r-1) \cdot E(l) & i > 1 \end{cases} \quad (5.26)$$

where $LC(N)$ and $RC(N)$, respectively, are the left and right children, in $ST(N)$, of N.

Because the number of nodes in O is $O(nW)$, the total number of $Opt'(N, r)$ and $Opt'STs(N, i, r-1)$ values is $O(nW^2k)$. For each, $O(k)$ $E(l)$ values are computed. Hence, to compute all $Opt'(root(O), k) \cdot E(l)$ values, we must compute $O(nW^2k^2)$ $Opt'(N, r) \cdot E(l)$ and $Opt'STs(N, i, r-1) \cdot E(l)$ values. Using Equations 5.22 through 5.26, the total time for this is $O(nW^2k^2)$.

5.3.2 Mapping onto a Pipeline Architecture

The approximately optimal VST $Opt'(root(O), k)$ can be mapped onto a k stage pipeline in the most straightforward way (i.e., nodes at level 1 of the VST are packed into stage $l + 1$, $0 \le l < k$, of the pipeline), the maximum per-stage memory is

$$\max_{0 \le l < k} \{Opt'(root(O), k) \cdot E(l)\}$$

We can do quite a bit better than this by employing a more sophisticated mapping strategy. For correct pipeline operation, we require only that if a node N of the VST is assigned to stage q of the pipeline, then each descendent of N be assigned to a stage r such that $r > q$. Hence, we are motivated to solve the following tree packing problem.

Tree Packing (TP). Input: Two integers $k > 0$ and $M > 0$ and a tree T, each of whose nodes has a positive size. Output: "Yes" iff the nodes of T can be packed into k bins, each of capacity M. The bins are indexed 1 through k and the packing is constrained so that for every node packed into bin q, each of its descendent nodes is packed index more than q.

By performing a binary (or other) search over M, we may use an algorithm for TP to determine an optimal packing (i.e., one with least M) of $Opt'(root(O), k)$ into a k-stage pipeline. Unfortunately, problem TP is NP-complete. This may be shown by using a reduction from the partition problem [11]. In the partition problem, we are given n positive integers s_i, $1 \le i \le n$, whose sum is $2B$ and we are to determine whether any subset of the given s_is sums to B.

Theorem 5.3.1: TP is NP-complete.

Proof: It is easy to see that TP is in NP. So we simply show the reduction from the partition problem. Let n, s_i, $1 \le i \le n$, and B ($\sum s_i = 2B$) be an instance of the partition problem. We may transform, in polynomial time, this partition instance into a TP instance that has a k-bin tree packing with bin capacity M iff there is a partition of the s_is. The TP instance has $M = 2B + 1$ and $k = 3$. The tree T for this instance has three levels. The size of the root is M, the root has n children, the size of the ith child is $2s_i$, $1 \le i \le n$, and the root has one grandchild whose size is 1 (the grandchild may be made a child of any one of the n children of the root).

It is easy to see that T may be packed into three capacities M bins if the given s_is have a subset whose sum is B.

All VSTs have nodes whose size is a power of 2 (more precisely, some constant times a power of 2). The TP construction of Theorem 5.3.1 results in node sizes that are not necessarily a power of 2. Despite Theorem 5.3.1, it is possible that TP restricted to nodes whose size is a power of 2 is polynomially solvable. However, we have been unable to develop a polynomial-time algorithm for this restricted version of TP. Instead, we propose a heuristic, which is motivated by the optimality of the First Fit Decreasing (FFD) algorithm to pack bins when the size of each item is a power of a, where $a \geq 2$ is an integer. In FFD [11], items are packed in the decreasing order of size; when an item is considered for packing, it is packed into the first bin into which it fits; if the item fits in no existing bin, a new bin is started. Although this packing strategy does not guarantee to minimize the number of bins into which the items are packed when item sizes are arbitrary integers, the strategy works for the restricted case when the size of each item is of the form a^i, where $a \geq 2$ is an integer. Theorem 5.3.2 establishes this by considering a related problem–restricted max packing (RMP). Let $a \geq 2$ be an integer. Let s_i, a power of a, be the size of the ith item, $1 \leq i \leq n$. Let c_i be the capacity of the ith bin, $1 \leq i \leq k$. In the RMP problem, we are to maximize the sum of the sizes of the items packed into the k bins. We call this version of max packing restricted because the item sizes must be a power of a.

Theorem 5.3.2:   *FFD solves the RMP problem.*

Proof: Let a, n, c_i, $1 \leq i \leq k$ and s_i, $1 \leq i \leq n$ define an instance of RMP. Suppose that in the FFD packing the sum of the sizes of items packed into bin i is b_i. Clearly, $b_i \leq c_i$, $1 \leq i \leq k$. Let S be the subset of items not packed in any bin. If $S = \phi$, all the items have been packed and the packing is necessarily optimal. So, assume that $S \neq \phi$. Let A be the size of the smallest item in S. Let x_i and y_i be non-negative integers such that $b_i = x_i A + y_i$ and $0 \leq y_i < A$, $1 \leq i \leq k$. We make the following observations:

(a) $(x_i + 1) * A > c_i$, $1 \leq i \leq k$. This follows from the definition of FFD and the fact that S has an unpacked item whose size is A.

(b) Each item that contributes to a y_i has size less than A. This follows from the fact that all item sizes (and hence A) are a power of a. In particular, note that every item size $\geq A$ is a multiple of A.

(c) Each item that contributes to the $x_i A$ component of a b_i has size $\geq A$. Though at first glance, it may seem that many small items could collectively contribute to this component of b_i, this is not the case when FFD is used on items whose size is a power of a. We prove this by contradiction. Suppose that for some i, $x_i A = B + C$, where $B > 0$ is the contribution of items whose size is less than A and C is the contribution of items whose size is $\geq A$. As noted in (b), every size $\geq A$ is a multiple of A. So, C is a multiple of A. Hence B is a multiple of A formed by items whose size is smaller than A. However, S has an item whose size is A. FFD should have packed this item of size A into bin i before attempting to pack the smaller size items that constitute B.

The sum of the sizes of items packed by FFD is

$$FFDSize = \sum_{1 \leq i \leq k} x_i A + \sum_{1 \leq i \leq k} y_i \qquad (5.27)$$

For any other k-bin packing of the items, let $b_i' = x_i' A + y_i'$, where x_i' and y_i' are non-negative integers and $0 \leq y_i' < A$, $1 \leq i \leq k$, be the sum of the sizes of items packed into bin i. For this other packing, we have

$$OtherSize = \sum_{1 \leq i \leq k} x_i A + \sum_{1 \leq i \leq k} y_i \qquad (5.28)$$

From observation (a), it follows that $x_i' \leq x_i$, $1 \leq i \leq k$. So, the first sum of Equation 5.28 is less than the first sum of Equation 5.28.

From observations (b) and (c) and the fact that A is the smallest size in S, it follows that every item whose size is less than A is packed into a bin by FFD and contributes to the second sum in Equation 5.27. Because all item sizes are a power of a, no item whose size is more than A can contribute to a y_i'. Hence, the second sum of Equation 5.28 is the second sum of Equation 5.27. So, *OtherSize* $\leq$ *FFDSize* and FFD solve the RMP problem.

The optimality of FFD for RMP motivates our tree packing heuristic shown in Figure 5.32, which attempts to pack a tree into k bins each of size M. It is assumed that the tree height is $\leq k$. The heuristic uses the notions of a ready node and a critical node. A ready node is one whose ancestors have been packed into prior bins. Only a ready node may be packed into the current bin. A critical node is an, as yet, unpacked node whose height equals the number of bins remaining for packing. Clearly, all critical nodes must be ready nodes and must be packed into the current bin if we are to successfully pack all tree nodes into the given k bins. So, the heuristic ensures that critical nodes are ready nodes. Further, it first packs all critical nodes into the current bin and then packs the remaining ready nodes in the decreasing order of node size. We may use the binary search technique to determine the smallest M for which the heuristic is successful in packing the given tree.

5.3.3 Experimental Results

Because the algorithm of Kim and Sahni [5] is superior to that of Basu and Narlikar [4], Lu and Sahni compared the tree packing heuristic in Figure 5.32 with the algorithms of Kim and Sahni. Let PFST denote a 2-stage algorithm that results in pipelined FST that minimizes the total memory subject to minimizing the maximum per-stage memory. Let VST denote the algorithm of [14], which constructs variable-stride tries with minimum total memory, and let PVST be the heuristic described in Sections 5.3.1 and 5.3.2.

The algorithms are programmed in C++, and run on a 2.80 GHz Pentium 4 PC. The data sets are the same as that in Section 5.2.8—RRC04, RRC03b, RRC01, MW02, PA, and ME02. Table 5.5

Step 1: [Initialize]
 currentBin = 1; *readyNodes* = tree root;

Step 2: [Pack into current bin]
 Pack all critical ready nodes into *currentBin*;
 if bin capacity is exceeded **return failure**;
 Pack remaining ready nodes into *currentBin* in decreasing order of node size;

Step 3: [Update Lists]
 if all tree nodes have been packed return success;
 if *currentBin* == k **return failure**;
 Remove all nodes packed in Step 2 from *readyNodes*;
 Add to *readyNodes* the children of all nodes packed in Step 2;
 currentBin + +;
 Go to Step 2;

Figure 5.32 Tree packing heuristic. (From Wencheng Lu, Sartaj Sahni, *Proceedings of the 11th IEEE Symposium on Computers and Communication*, pp. 802–807, 2006. With permission.)

Table 5.5 Reduction in Maximum Per-Stage Memory Resulting from Tree Packing Heuristic

Algorithm	Min (%)	Max (%)	Mean (%)	Standard Deviation (%)
PFST	0	41	18	15
VST	0	41	17	15
PVST	0	31	11	10

Source: Kim, K. and Sahni, S., *IEEE Transactions on Computers*, 56, 1, 2007. With permission.

gives the reduction in maximum per-stage memory when we use the tree packing heuristic rather than the straightforward mapping. For example, on six data sets, the tree packing heuristic reduced the maximum per-stage memory required by the multibit trie generated by PVST by between 0 percent and 31 percent; the mean reduction was 11 percent and the standard deviation was 10 percent. The reduction obtained by the tree packing heuristic was as high as 41 percent when applied to the tries constructed by the algorithms of [5].

Tables 5.6 and 5.7 give the maximum per-stage and total memory required by the three algorithms normalized by the requirements for PVST. The maximum per-stage memory requirement for PFST are up to 32 times that of PVST, whereas the requirement for VST is up to 35 percent more than that of PVST. On average, the total memory required by the multibit tries produced by VST was 13 percent less than that required by the PVST tries; the tries generated by PFST required, on average, about 3.5 times the total memory required by the PVST tries.

Table 5.6 Maximum Per-Stage Memory Normalized by PVST's Maximum Per-Stage Memory

Algorithm	Min	Max	Mean	Standard Deviation
PFST	1.00	32.00	5.16	8.90
VST	0.91	1.35	1.18	0.11
PVST	1	1	1	0

Source: Kim, K. and Sahni, S., *IEEE Transactions on Computers*, 56, 1, 2007. With permission.

Table 5.7 Total Memory Normalized by PVST's Total Memory

Algorithm	Min	Max	Mean	Standard Deviation
PFST	1.09	20.90	3.57	5.22
VST	0.79	1.00	0.87	0.06
PVST	1	1	1	0

Source: Kim, K. and Sahni, S., *IEEE Transactions on Computers*, 56, 1, 2007. With permission.

Table 5.8 Execution Time (ms) for Computing 8-Level Multibit-Stride Tries

Algorithm	RRC04	RCC03b	RRC01	MW02	PA	ME02
PFST	303	302	372	244	274	1373
VST	450	441	428	350	334	296
PVST	1256	1253	1196	995	993	838

Source: Kim, K. and Sahni, S., *IEEE Transactions on Computers*, 56, 1, 2007. With permission.

Table 5.8 shows the time taken by the various algorithms to determine the pipelined multibit tries for the case $k = 8$. The shown time includes the time for the tree packing heuristic. VST has an execution time comparable to that of PFST but produces significantly superior pipelined tries. Although PVST takes about three times as much time as does VST, it usually generates tries that require significantly less maximum per-stage memory.

References

1. Gupta, P., Lin, S. and McKeown, N., Routing lookups in hardware at memory access speeds, *Proc. IEEE INFOCOM '98*, Apr. 1998, pp. 1240–1247.
2. Sikka, S. and Varghese, G. Memory-efficient state lookups with fast updates, *Proceedings of SIGCOMM '00*, Stockholm, Sweden, August 2000, pp. 335–347.
3. Srinivasan, V. and Varghese, G., Fast address lookups using controlled prefix expansion. *ACM Transaction on Computer Systems* 1999; 17(1):1–40.
4. Basu, A. and Narlikar, G., Fast incremental updates for pipelined forwarding engines. *IEEE or ACM Transactions on Networking* 2005; 13(3):609–704.
5. Kim, K. and Sahni, S., Efficient construction of pipelined multibit-trie router-tables. *IEEE Transactions on Computers* 2007; 56(1):32–43.
6. Wencheng Lu, Sartaj Sahni, Packet forwarding using pipelined multibit tries. *Proceedings of the 11th IEEE Symposium on Computers and Communications*, 2006, pp. 802–807.
7. Huston, G., Analyzing the internet's BGP routing table. *The Internet Protocol Journal* 2001; 4(1):2–15.
8. Labovitz, C., Ahuja, A., and Jahanian, F., Experimental study of internet stability and wide-area backbone failures, *Proceedings of the Twenty-Ninth Annual International Symposium on Fault-Tolerant Computing*, Madison, WI, June 1999, pp. 278–285.
9. Labovitz, C., Malan, G.R., and Jahanian, F., Origins of internet routing instability, *Proceedings of Infocom '99*, New York, NY, March 1999, pp. 218–226.
10. Nisson, S. and Karlsson, G., IP-address lookup using LC-trie, *IEEE JSAC* 1999; 17(6):1083–1092.
11. Horowitz, E., Sahni, S., and Rajasekaran, S., *Computer Algorithms*, W.H. Freeman, NY, 1997, 769 pages.
12. Frederickson, G., Optimal parametric search algorithms in trees I: Tree partitioning, *Technical Report*, CS-TR-1029, Purdue University, West Lafayette, IN, 1992.
13. Thanvantri, V. and Sahni, S., Optimal folding of standard and custom cells, *ACM Transactions on Design Automation of Electronic Systems* 1996; 1(1):123–143.
14. Sahni, S. and Kim, K., Efficient construction of multibit tries for IP lookup, *IEEE or ACM Transactions on Networking* 2003; 11(4):650–662.

Chapter 6

Efficient Data Structures for Bursty Access Patterns

In a tree data structure, if a recently accessed leaf node is moved up to the root, the next access time should be shorter than the last. Therefore, we can improve the throughput of a data structure by turning it according to lookup biases. Some researchers have proposed efficient schemes, such as the table-driven scheme [1], the near-optimal binary tree scheme [2], the biased skip list (BSL) [3], the collection of trees [4] etc. In this chapter, we will describe these schemes.

6.1 Table-Driven Schemes

When a routing table tree can be represented as a set of tables, a route lookup can be cast as a table lookup, where a single table is indexed by the destination address to yield the routing entry. Because a lookup operation takes constant time, this simple algorithm has complexity $O(1)$. Then the challenge is to devise an algorithm to produce a "good" set of tables for a routing table. Cheung and McCanne developed a model for table-driven route lookup [1]. This section is partly from [1]; portions have been reprinted with permission (© 1999 IEEE).

6.1.1 Table-Driven Models

If the IP address space precludes a single table, then it will produce a large table. Such large memories are impractical, expensive, and slow. We should therefore use a set of small tables to represent the forwarding table such that they can fit into small, fast memories, where a lookup operation may need to access more than one table. There is an optimal layout problem: how to choose small tables layouts (STL)? We call it the STL problem.

Cheung et al. [5] developed a model for the optimal layout problem: given are three types of hierarchical memories, with memory size and access speed (S_1, T_1), (S_2, T_2), and (S_3, T_3), respectively. Let f_j be the access probability of prefix p_j. The average search time is:

$$C = \sum_j f_j(a_j T_1 + b_j T_2 + c_j T_3) = w_1 T_1 + w_2 T_2 + w_3 T_3 \tag{6.1}$$

where a_j, b_j, c_j are the number of three type memory access needed to search prefix p_j. The weights $w_1 = \sum_j f_j a_j$, $w_2 = \sum_j f_j b_j$, and $w_3 = \sum_j f_j c_j$ are the access probability mass.

Now, the optimal layout problem can be presented as follows: how to structure a set of tables that minimize Equation 6.1 while satisfying the memory size constraints. It is proved that finding the optimal set of tables is an NP-complete problem [5]. Cheung et al. devised two algorithms, using dynamic programming and Lagrange multipliers, to solve the optimization problem optimally and approximately, respectively. We will next discuss two algorithms.

For simplicity, we only consider the STL problem with two memory types ($c_j = 0$ in Equation 6.1), type 1 fast, type 2 slow; that is, for every lookup there are no more than two access memories. Then we select a data structure for IP-address lookup. A binary tree is one of the simplest representations of a set of prefixes. Each prefix is represented by a node in the tree. Each node has at most two children. The implementation of a binary tree involves bit-by-bit manipulation, which is inefficient for long prefixes. A trie is a more general data structure than a binary tree, and each node in it can have any number of children. A level-compressed trie [6] is a trie in which each complete subtree of height h is collapsed into a subtree of height 1 with 2^h children, and all the intermediate nodes from level 1 to level $h - 1$ can be replaced by 2^h children. We can represent a set of prefixes with some small tables; they fit into different memory types. For the lookup tables, we make the following design decision: given a set of prefixes by a binary tree, stage 1 transforms it to a complete prefixes tree; stage 2 transforms it to a level-compressed trie.

For example, for a set of prefixes {00, 01, 1, 110}, the generated level-compressed trie is shown in Figure 6.1a, and there are two lookup tables. They are assigned to memory type 1 and type 2, respectively. The corresponding set of parameters, a's and b's, is shown in Figure 6.1b. The average search time is:

$$C = (f_1 + f_2 + f_3 + f_4 + f_5)T_1 + (f_4 + f_5)T_2 = w_1 T_1 + w_2 T_2 \tag{6.2}$$

Now, the STL problem can be described as follows: given a complete prefix binary tree and the size of two memory types—the size of type 1 memory s_1 is finite and the size of type 2 memory s_2 is infinite—what is the form of generalized level-compressed trie and the configuration of prefixes on hierarchical memories so that the average lookup time per prefix, defined in Equation 6.1, is minimal?

Here, we first find the form of the generalized level-compressed trie, that is, decide how many levels to compress from the root of the tree, to produce a number of small tables. Then we decide where each small table should go: type 1 or type 2 memory.

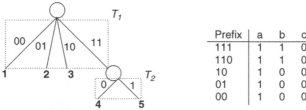

Prefix	a	b	c
111	1	1	0
110	1	1	0
10	1	0	0
01	1	0	0
00	1	0	0

(a) generalized level-compressed trie   (b) memory assignment parameters

Figure 6.1 Example of a configuration. (From Cheung, G. and McCanne, S., *Proceedings of Eighteenth Annual Joint Conference of the IEEE Computer and Communications Societies*, IEEE Press, 1999. With permission.)

6.1.2 Dynamic Programming Algorithm

In a generalized level-compressed trie, a subtree is rooted at node i, and with height h and weight w_i we can construct a table with the size 2^h. There are two choices: (i) the table can be put in type memory of size s_1, resulting in cost $w_i T_1$, and $s_1 - 2^h$ remains type 1 memory for the rest of the tree; (ii) the table can be put in type 2 memory, resulting in cost $w_i T_2$, and s_1 remains types memory. Let $TP_1(i, s_1)$ be a recursive "tree-packing" cost function that returns the minimum cost of packing a tree rooted at node i. The minimum of these two choices is:

$$TP_1(i, s_1) = \min_{1 \le h \le H_i} \{\min[w_i T_1 + TP_2(L_{h,i}, s_1 - 2^h), w_i T_2 + TP_2(L_{h,i}, s_1)]\} \tag{6.3}$$

where H_i is the maximum height of the tree rooted at node i, $L_{h,i}$ is the set of internal nodes at height h of the tree rooted at a node i, and $TP_2(L_{h,i}, s_1)$ is a sister "tree-packing" minimum cost function packing a set of trees of root nodes $\{i \in L\}$ into type 1 memory of s_1. To complete the analysis, we will also need the following base cases:

$$TP_1(., -) = \infty, \ TP_2(., -) = \infty, \ TP_2(\{\}, .) = \infty.$$

The first base case says that if the memory constraint is violated, that is, gives a negative value as the second argument, then this configuration is not valid and the cost is infinite. The second base case says that if the first argument is the empty set, then the cost is zero. In Equation 6.3, TP_1 calls TP_2, a function with possibly multiple node argument, $L_{h,i}$; we need to provide a mean to solve TP_2. If the set of nodes $L_{h,i}$ contain only one node i, then $TP_2(L_{h,i}, s_1) = TP_1(i, s_1)$. If $L_{h,i}$ has more than node, we first split up the set L into two disjoint subsets, L_1 and L_2. We then have to answer two questions: (i) given available type 1 memory space, how do we best divvy up the memory space between the two subsets, (ii) given the division of memory space, how do we perform generalized level-compression on the two subsets of nodes to yield the overall optimal solution. The optimality principle states that a globally optimal solution must also be a locally optimal solution. So if we give s memory space to set L_1, to be locally optimal, the optimal generalized level-compression must be performed on that set, returning a minimum cost of $TP_2(L_1, s)$. That leaves $s_1 - s$ memory space for L_2, with minimum cost of $TP_2(L_2, s_1 - s)$. The globally optimal solution is the minimum of all locally optimal solutions, that is, the minimum of all possible values of s. We can express this idea in the following equations:

$$TP_2(L_{h,i}, s_1) = TP_1(i, s_1) \quad \text{if } |L_{h,i}| = 1, \quad L_{h,i} = \{i\} \tag{6.4}$$

$$TP_2(L_{h,i}, s_1) = \min_{0 \le s \le s_1} \{TP_2(L_1, s) + TP_2(L_2, s_1 - s)\} \quad s.t. \ L_1 \cap L_2 = \varnothing, \ L_1 \cup L_2 = L_{h,i} \tag{6.5}$$

There are many ways to split up the set L into subsets L_1 and L_2. For efficiency reasons soon to be discussed, we will divides in such a way that all nodes in L_1 have a common ancestor, all nodes in L_2 have a common ancestor, and that the two ancestors of the two sets are distinct (if possible) and on the same level of the tree.

Because Equations 6.3 and 6.5 lead to overlapping subproblems, the subproblems should have been solved before. Cheung used a dynamic programming algorithm to compute TP_1 (TP_2). For example, a complete prefix is shown in Figure 6.2a. Assume that $(s_1, T_1) = (4, 1)$ and $(s_2, T_2) = (\infty, 1)$. The resulting dynamic programming table is shown in Figure 6.2b. Note the overlapping

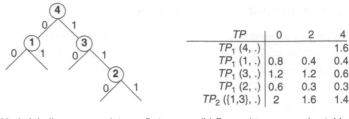

TP	0	2	4
$TP_1 (4, .)$			1.6
$TP_1 (1, .)$	0.8	0.4	0.4
$TP_1 (3, .)$	1.2	1.2	0.6
$TP_1 (2, .)$	0.6	0.3	0.3
$TP_2 (\{1,3\}, .)$	2	1.6	1.4

(a) Node labeling on complete prefix tree   (b) Dynamic programming table

Figure 6.2   Example of dynamic programming. (From Cheung, G. and McCanne, S., *Proceedings of Eighteenth Annual Joint Conference of the IEEE Computer and Communications Societies,* **IEEE Press, 1999. With permission.)**

subproblem in this case: for node 4, $TP_1(4, 4)$ calls $TP_1(2, 0)$ when $h = 2$. $TP_1(2, 0)$ has been solved from a previous call to $TP_1(3, 2)$ when $h = 1$ already. The optimal configuration is shown in Figure 6.1.

We now show that the running time of the dynamic programming algorithm is $O(HnS_1^2)$ where H is the height of the complete binary tree, n is the number of nodes, and S_1 is the size of the type 1 memory. The running time is the amount of time it takes to construct the dynamic programming table. Suppose we label the rows of the table using the first argument of the TP functions (see Fig. 6.2b). There are two types of row labels: single-node labels and multiply-node labels. The number of single-node labeled rows is simply the number of nodes, n. The number of columns is at most S_1, and therefore the size of half the table with single-node labeled rows is $O(nS_1)$. To solve each entry in this half of the table, we use Equation 6.3 and need at most H_i comparisons. H is the height of the complete prefix tree. Therefore, the running time for this half table with single-node labeled rows is $O(HnS_1)$.

For the half table with multiple-node labeled rows, let the number of these rows be fixed. For a fixed number of nodes, the tree that maximizes the number of multiple-node rows in a dynamic programming table is a full tree, because each call from Equation 6.3 to its children nodes at height h will result in a multiple-node argument. We will count the number of multiple-node rows for the full tree as follows. Let $f(h)$ be a function that returns the number of multiple-node labeled rows in a dynamic programming table for a full tree of height h. Suppose node i is the root of that full tree, with children nodes j and k, as shown in Figure 6.3. Because nodes j and k have height 1 less than i, they will each contribute $f(h - 1)$ to the multi-node labeled rows. In addition, as the call to

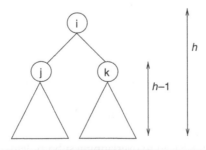

Figure 6.3   Counting rows of DP table for full tree case. (From Cheung, G. and McCanne, S., *Proceedings of Eighteenth Annual Joint Conference of the IEEE Computer and Communications Societies,* **IEEE Press, 1999. With permission.)**

$TP_1(i, S_1)$ swings h from 1 to $H_i - 1$, each new call to TP_2 with a root node argument $L_{h,i}$ creates a new row. Therefore, the total number of multiple-node labeled rows is

$$f(h) = 2 * f(h-1) + h - 1 \tag{6.6}$$
$$f(1) = 0$$

After rewriting Equation 6.6, we get:

$$f(h) = 2^{h-1} \sum_{i=1}^{h-1} \frac{i}{2^i}. \tag{6.7}$$

Notice that the summation in Equation 6.7 converges for any h and can be bounded above by 2. Therefore, the number of multiple-node labeled rows is $O(2^h)$. Because the tree is full, $h = \lceil \log_2 n \rceil$. The number of multiple-node labeled rows for a full tree is $O(2^{\log_2 n}) = O(n)$. The size of this half of the table is again $O(nS_1)$, and each entry takes at most S_1 comparisons using Equation 6.5. So the running time of a full tree, for the half table with multiple-node labeled rows, is $O(nS_1^2)$. Because this is the worst-case, the general tree with n nodes will also have a running time of $O(nS_1^2)$ for this half of the table. Combined with the part of the table with single-node labeled rows, the running time is $O(HnS_1^2)$.

For a nontrivial type 1 memory, the algorithm may be impractical [1]. Cheung et al. proposed a Lagrange approximation algorithm to improve time complexity.

6.1.3 Lagrange Approximation Algorithm

The original problem can be formulated as a constrained optimization problem which we can recast as follows:

$$\min_{b \in B} H(b) \quad \text{s.t. } R(b) \le S_1 \tag{6.8}$$

where B is the set of possible configurations, $H(b)$ is the average lookup time for a particular configuration b, $R(b)$ is the total size of type 1 table sizes, and S_1 is the size of type 1 memory. The corresponding dual problem is:

$$\min_{b \in B} H(b) + \lambda R(b). \tag{6.9}$$

Because the dual is unconstrained, solving it is significantly easier than solving the original problem. However, there is an additional step: we must adjust the multiplier value λ so that the constraint parameter, $R(b)$, satisfies Equation 6.8.

In Equation 6.9, the multiplier-rate constraint product, $\lambda R(b)$, represents the penalty for placing a lookup table is type 1 memory, and is linearly proportional to the size of the table. For each λ, we solve the dual problem, denoted as LP, as follows. For each table rooted at node i and height h, we have two choices: (*i*) place it in type 1 memory with cost plus penalty $w_i T_1 + 2^h \lambda$; or, (*ii*) place it in type 2 memory with cost $w_i T_2$. The minimum of these two costs plus the recursive cost of the children nodes will be the cost of the function at node i for a particular height. The cost of the function at node i will then be the minimum cost for all possible heights, which we can express as follows:

$$LP_1(i, \lambda) = \min_{1 \le h \le H_i} \{LP_2(L_{h,i}, \lambda) + \min[w_i T_1 + \lambda 2^h, w_i T_2]\} \tag{6.10}$$

where LP_1 and LP_2 are sister functions defined similarly to TP_1 and TP_2. Because we are optimizing with respect to λ, which does not change for each iteration, LP_2 with a multiple-node argument is simply the sum of calls to LP_1 of individual nodes:

$$LP_2(L,\lambda) = \sum_{j \in L} LP_1(j,\lambda) \qquad (6.11)$$

Similar to TP_2, we can also solve P_2 using recursion:

$$LP_2(L,\lambda) = LP_1(i,\lambda) \quad if\ |L| = 1, \quad L = i$$

$$LP_2(L,\lambda) = LP_2(L_1,\lambda) + LP_2(L_2,\lambda) \quad s.t.\ L_1 \cap L_2 = \varnothing, \quad L_1 \cup L_2 = L \qquad (6.12)$$

After solving the dual problem using Equation 6.10 and 6.12, we have an "optimal" configuration of tables from the prefix tree, denoted by b^*, that minimizes the dual problem for a particular multiplier value. The sum of sizes of tables assigned to type 1 memory will be denoted $R(b^*)$. If $R(b^*) = S_1$, then the optimal solution to the dual is the same as the optimal solution to the original problem. If $R(b^*) < S_1$, then the optimal solution to the dual becomes an approximate solution with a bounded error—in general, the closer $R(b^*)$ is to S_1, the smaller the error.

From the above subsection, we see that in the case of single node arguments, computing an entry i for each multiplier value λ, needs at most H_i operations using Equation 6.10, (H_i is the height of a tree rooted at node i). Let H be the height of the complete prefix tree; computing a single node needs $O(Hn)$ operations. In the case of multiple node arguments, computing each entry takes a constant number of operations using Equation 6.12. Because the constraint variable, $R(b^*)$, is inversely proportional to the multiplier λ, a simple strategy is to search for the appropriate multiplier value using a binary search on the real line to drive $R(b^*)$ towards S_1. Let A be the number of iterations of a binary search for multiplier values. Then the complexity of the Lagrange approximation algorithm is $O(HnA)$. In experiments, A is found to be around 10 [1].

To compare the performance, Chueng et al. used the forwarding table from sites PAIX and AADS and a Pentium II 266 MHz processor, with L1 cache 16 kB and L2 cache 512 kB. The results are shown in Table 6.1. There is an 8.9 percent increase in the speed of the Lagrange programming algorithm over the S&V algorithm [7].

Table 6.1   Performance Comparison between S&V and LP

Routing Table	No. of Prefixes	Million Lookups/Second	
		S&V (i.i.d/Scaled)	*Lagrange Programming (i.i.d/Scaled)*
PAIX	2638	1.626/1.695	1.770/2.198
AADS	23,068	1.252/1.681	1.570/1.869

Note: i.i.d: Each prefix is independent and identically distributed. Scaled: Each prefix is exponentially scaled according to its length.

Source: Cheung, G. and McCanne, S., *Proceedings of Eighteenth Annual Joint Conference of the IEEE Computer and Communications Societies*, IEEE Press, 1999. With permission.

6.2 Near-Optimal Scheme with Bounded Worst-Case Performance

In the data structure-trie, the accessed time of each node depends on the level in which it is, and the average time for all nodes is related to the distribution of the levels of all nodes and the frequency with which all nodes are accessed. To design a routing lookup scheme for minimizing the average lookup time, Gupta et al. proposed a binary search tree built on the intervals created by prefixes of the routing table [2]. Intuitively, in a binary search tree, if the node with the higher frequency has a shorter accessed time, the average time decreases. Thus, given the access frequency, it is important to constraint the maximum depth of the binary tree to some small prespecified number. To find the depth-constrained binary search trees, Gupta used information theory ideas and convex optimization techniques. We will describe them in the following section. This section is from [2]. Portions are reprinted with permission (© 2000 IEEE).

6.2.1 Definition

Each prefix in a forwarding table can be viewed as an interval on the number line [0, 2^{32}], called the IP number line. After the duplicates are eliminated, the intervals are disjointed. For example, a forwarding table with 4-bit prefixes is shown in Figure 6.4. The disjoint intervals on the IP number line are I1–I5. Any IP address belongs to only one interval. The longest-prefix matching problem becomes an exact matching problem. The binary tree is a very good data structure for the exact matching problem.

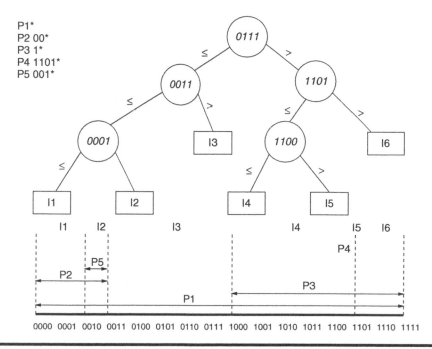

Figure 6.4   The binary search tree corresponding to an example routing table. (From Gupta, P., Prabhakar, B., and Boyd, S., *Proceedings of Nineteenth Annual Joint Conference of the IEEE Computer and Communications Societies*, IEEE Infocom, IEEE Press, 2000. With permission.)

But the use of a binary tree data structure brings up the worst-case depth. Today, there are more than 100,000 prefixes in the forwarding table of some backbone routers [8]; these prefixes generate a large number of intervals, and can lead to a binary search tree with a maximum depth [2], resulting in long lookup times for some intervals. Therefore, it is necessary to find the optimal binary search tree that minimizes the average lookup time while keeping the worst-case lookup time within a fixed bound.

Suppose the prefixes in a forwarding table generate n intervals, let f_i be the access probability of interval i. In the binary search tree, the lookup time of an IP address is proportional to the number of comparisons, which is equal to the depth (l_i) of the (i^{th}) interval in the binary search tree. The original problem can be formulated as follows:

$$\min_{\{l_i\}} C = \sum_{i=1}^{n} p_i l_i \quad s.t. \quad l_i \le D \quad \forall i \qquad (6.13)$$

where D is the fixed bound of the depth.

For example, the intervals $I1$ through $I6$ in Figure 6.4 are accessed with probabilities {1/2, 1/4, 1/8, 1/16, 1/32, 1/32}, respectively. We can construct two binary search trees, as shown in Figure 6.5. In Figure 6.5a, the maximum depth D is less than 5, the average depth is 1.9375; in Figure 6.5b, the maximum depth D is less than 4, the average depth is 2.

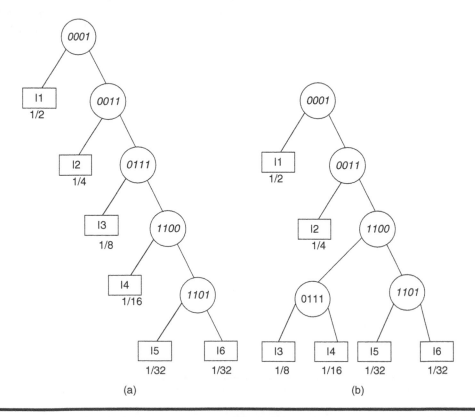

(a)   (b)

Figure 6.5 Optimal alphabetic tree corresponding to the tree in Figure 6.4 with leaf probabilities as shown: (a) unconstrainted depth, (b) depth-constrained to 4. (From Gupta, P., Prabhakar, B., and Boyd, S., *Proceedings of Nineteenth Annual Joint Conference of the IEEE Computer and Communications Societies*, IEEE Infocom, IEEE Press, 2000. With permission.)

The Huffman coding tree [9] is a well-known algorithm to minimize the average length of a binary tree for given access probabilities. But a Huffman solution changes the alphabetic order of the input data set. Because of the difficulty in finding the solution to the original problem, researchers proposed approximate solutions [10,11]. However, their algorithm is very complicated to implement. Gupta et al. proposed two near-optimal algorithms [2].

6.2.2 Algorithm MINDPQ

In the language of Information Theory, the original problem can be stated as: Find a minimum average length alphabetic code (or tree) for an m-letter alphabet, where each letter corresponds to an interval in Equation 6.13. Yeung [12] propose the existence of an alphabetic code with specified code word lengths and a method for constructing near-optimal trees that are not depth-constrained. They are described in Lemma 6.2.1 and 6.2.2, respectively.

Lemma 6.2.1: (The Characteristic Inequality): There exists an alphabetic code with code word lengths $\{l_k\}$ if and only if $s_n \leq 1$ where $s_n(L) = c(s_{k-1}(L), 2^{-l_k}) + 2^{-l_k}$ and c is defined by $c(a,b) = \lceil a/b \rceil b$.

Proof: The basic idea is to construct a canonical coding tree, a tree in which the codewords are chosen lexicographically using the length $\{l_i\}$. For instance, suppose that $l_i = 4$ for some i, and in drawing the canonical tree we find the codeword corresponding to the letter i to be 0010. If $l_{i+1} = 4$, then the codeword for the letter $i + 1$ will be chosen to be 0011; if $l_{i+1} = 3$, then the codeword for the letter $i + 1$ will be chosen to be 010; and if $l_{i+1} = 5$, then the codeword for the letter $i + 1$ will be chosen to be 00110. Clearly, the resulting tree will be alphabetic and Yeung's result verifies that this is possible if and only if the characteristic inequality defined above is satisfied by the lengths $\{l_i\}$. (A complete proof is given in [12].)

Lemma 6.2.2: The minimum average length, C_{min}, of an alphabetic code on n letters, where the ith letter occurs with probability f_i satisfies $H(f) \leq C_{min} \leq H(f) + 2 - f_1 - f_n$, where $H(f) = -\sum f_i \log_2 f_i$.

Proof: The lower bound, $H(f)$, is obvious. For the upper bound, the code length l_k of the kth letter occurring with probability f_k is chosen to be:

$$l_k = \begin{cases} \lceil -\log f_k \rceil & k = 1, n \\ \lceil -\log f_k \rceil + 1 & 2 \leq k \leq n - 1 \end{cases}$$

The proof in [12] verifies that these lengths satisfy the characteristic inequality, and shows that a canonical coding tree constructed with these lengths has an average depth satisfying the upper bound.

Because the given set of probabilities $\{f_k\}$ might be such that $f_{min} = \min_k f_k < 2^{-D}$, a direct application of Lemma 6.2.2 could yield a tree where the maximum depth is bigger than D. Gupta et al. transform the given probabilities $\{f_k\}$ into another set of probabilities $\{q_k\}$ such that $f_{min} = \min_k f_k \geq 2^{-D}$.

Given the set of probabilities $\{q_k\}$ such that $q_{\min} \geq 2^{-D}$, we construct a canonical alphabetic coding tree with the codeword length assignment to the kth letter given by:

$$l_k^* = \begin{cases} \min\left(\lceil -\log_2 q_k \rceil, D\right) & k = 1, n \\ \min\left(\lceil -\log_2 q_k \rceil + 1, D\right) & 2 \leq k \leq n-1 \end{cases} \tag{6.14}$$

The average length C can be formulated as:

$$C = \sum_{k=1}^{n} f_k l_k^* < \sum_{k=1}^{n} f_k \log_2 \frac{1}{q_k} + 2 = \sum_{k=1}^{n} f_k \log_2 \frac{f_k}{q_k}$$
$$- \sum_{k=1}^{n} f_k \log_2 f_k + 2 = D(f\|q) + H(f) + 2 \tag{6.15}$$

where $D(f\|q)$ is the "relative entropy" between the distributions f and q and $H(f)$ is the entropy of the distribution f. To minimize C, we must therefore choose $\{q_i^*\} = \{q_i\}$, so as to minimize $D(f\|q)$.

We are thus led to the following optimization problem:
Minimize

$$DPQ = D(f\|q) = \sum_{k=1}^{n} f_k \log_2 \frac{f_k}{q_k}, \quad \text{s.t.} \sum_{k=1}^{n} q_k = 1 \text{ and } q_k \geq Q = 2^{-D} \quad \forall k \tag{6.16}$$

Cover and Thomas have proved that the cost function $D(f\|q)$ is convex in (f, q) [13] and that the constraint set is defined by linear inequalities and is convex and compact. Minimizing convex cost functions with linear constraints is a standard problem in optimization theory and is easily solved by using Lagrange multiplier methods:

$$L(q, \lambda, \mu) = \sum_{k=1}^{n} f_k \log \frac{f_k}{q_k} + \sum_{k=1}^{n} \lambda(Q - q_k) + \mu\left(\sum_{k=1}^{n} q_i - 1\right)$$

Setting the partial derivatives with respect to q_k to zero at q_k^*, we get

$$\frac{\partial L}{\partial q_k} = 0 \Rightarrow q_k^* = \frac{f_k}{\mu - \lambda_k} \tag{6.17}$$

Putting q_k^* in $L(q, \lambda, \mu)$, we get the dual.
Minimize

$$G(\lambda, \mu) = \sum_{k=1}^{n} (f_k \log_2(\mu - \lambda_k) + \lambda_k Q) + (1 - \mu), \quad \text{s.t.} \lambda_i \geq 0 \text{ and } \mu > \lambda_i \; \forall_i$$

Setting the partial derivatives with μ and λ_i to zero, respectively, we get:

$$\frac{\partial G}{\partial \mu} = 0 \Rightarrow \sum_{k=1}^{n} \frac{f_k}{\mu - \lambda_k} = 1 \Rightarrow \sum_{k=1}^{n} q_k^* = 1$$

$$\frac{\partial G}{\partial \lambda_k} = 0 \Rightarrow Q = \frac{f_i}{\mu - \lambda_k} \Rightarrow \lambda_k^* = \max(0, \mu - f_k/Q) \quad \forall k$$

Substituting λ_k^* in Equation 6.17, we get

$$q_k^* = \max(f_i/\mu, Q) \tag{6.18}$$

To finish, we need to solve Equation 6.18 for $\mu = \mu^*$ under the constraint that $\sum_{i=1}^{n} q_i^* = 1$. $\{q_i^*\}$ will then be the desired transformed probability distribution. It turns out that we can find a closed form expression for μ^*, using which we can solve Equation 6.18 by an $O(n \log n)$ time and $O(n)$ space algorithm. The algorithm first sorts the original probabilities $\{f_i\}$ to get $\{\hat{f}_i\}$ such that $\hat{f}_1$ is the largest and $\hat{f}_n$ the smallest probability. Call the transformed (sorted) probability distribution $\{\hat{q}_i\}$. Then the algorithm solves for μ^* such that $F(\mu^*) = 0$ where

$$F(\mu) = \sum_{k=1}^{n} q_k^* - 1 = \sum_{k=1}^{k_\mu} \frac{\hat{f}_k}{\mu} + (n - k_\mu)Q - 1 \tag{6.19}$$

where the second equality follows from Equation 6.18, and k_μ is the number of letters with probability greater than μQ. Figure 6.6 shows the relationship between μ and k_μ. For all letters to the left of μ in Figure 6.6, $q_i^* = Q$ and for others, $q_i^* = \hat{f}_i/\mu$.

Lemma 6.2.3: $F(\mu)$ is a monotonically decreasing function of μ.

Proof: If μ increases in the interval $(\hat{f}_{r+1}/Q, \hat{f}_r/Q)$ such that k_μ does not change, $F(\mu)$ decreases monotonically. Similarly, if μ increases from $\hat{f}_r/Q - \varepsilon$ to $\hat{f}_r/Q + \varepsilon$ so that k_μ decreases by 1, it is easy to verify that $F(\mu)$ decreases monotonically.

From Lemma 6.2.3, we can do a binary search for finding a suitable value of r such that $\mu \in (f_r/Q, f_{r-1}/Q)$ and $F(f_r/Q) \geq 0$ and $F(f_{r-1}/Q) < 0$. This will take us only $O(\log n)$ time. Once we know that μ belongs to this half-closed interval, we know the exact value of $k_\mu = K$ and we can then directly solve for μ^* using Equation 6.19 to get $\mu^* = (\sum_{i=1}^{i=K} \hat{f}_i)(1 - (n - K)Q)$. Putting this value of μ^* in Equation 6.18 will then give us the transformed set of probabilities $\{\hat{q}_i^*\}$. Given such $\{\hat{q}_i\}$, the algorithm then constructs a canonical alphabetic coding tree as in [12] with the codeword length l_k^* as chosen in Equation 6.14. This tree then clearly has a maximum depth of no more than D, and its average weighted length is worse than that of the optimum algorithm by no more than

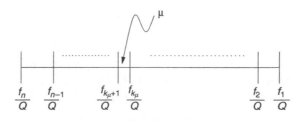

Figure 6.6 Showing the position of μ and k_μ. (From Gupta, P., Prabhakar, B., and Boyd, S., *Proceedings of Nineteenth Annual Joint Conference of the IEEE Computer and Communications Societies,* **IEEE Infocom, IEEE Press, 2000. With permission.)**

2 bits. To see this, let us refer to the code lengths in the optimum tree as $\{l_k^{opt}\}$. Then $C_{opt} = \sum_k f_k l_k = H(f) + D(f \| 2^{-l_k^{opt}})$. As we have chosen q^* to be such that $D(f \| q^*) \leq D(f \| q)$ for all probability distributions q, it follows from Equation 6.15 that $C_{\min dpq} \leq C_{opt} + 2$. We have thus proved the following main theorem:

Theorem 6.2.1:   Given a set of n probabilities $\{f_i\}$ in a specified order an alphabetic tree with a depth constraint D can be constructed in $O(n \log n)$ time and $O(n)$ space such that the average codeword length is at most 2 bits away from the optimum depth-constrained alphabetic tree. Further, if the probabilities are given in sorted order, such a tree can be constructed in linear time.

6.2.3   Depth-Constrained Weight Balanced Tree

Horibe proposed a heuristic algorithm to generate a weight balancing tree [14]. The heuristic algorithm is easy to implement and can generate trees that have lower average length than those generated by algorithm MINDPQ in Section 6.2.2, but have no maximum depth constraint. Gupta et al. proposed a heuristic algorithm to generate depth-constrained weight balanced trees (DCWBT) based on Horibe's algorithm.

In Horibe's algorithm, suppose the weight of each letter i is the access probabilities f_i in a tree, the leaves of a particular subtree correspond to letters numbered r through t, and the weight of the subtree is $\sum_{i=r}^{t} f_i$. The root node of this subtree represents the probabilities $\{f_i\}_{i=r}^{i=t}$. The trees can be constructed from root to leaf, called "top–down," and at each subtree the weight of the root node is split into two parts representing the weights of its two children in the most balanced manner possible. We can describe it as follows: given the representing probabilities of the root node of a subtree $\{f_i\}_{i=r}^{i=t}$, find s such that

$$\Delta(r,t) = \left| \sum_{i=r}^{s} f_i - \sum_{i=s+1}^{t} f_i \right| = \min_{\forall u : r \leq u \leq t} \left| \sum_{i=r}^{u} f_i - \sum_{i=u+1}^{t} f_i \right|, \quad r \leq s \leq t$$

The left and right children of the subtree represent the probabilities $\{f_i\}_{i=r}^{i=s}$ and $\{f_i\}_{i=s+1}^{i=t}$, respectively.

The average depth of the tree generated by Horibe's algorithm is greater than the entropy of the underlying probabilities distribution $\{f_i\}$ by no more than $2 - (n+2)f_{\min}$, where $f_{\min} = \min\{f_i\}$. Because there is no maximum depth constraint in Horibe's algorithm, sometimes it can generate a tree of which the average depth is lower, but the maximum depth is greater. An example is shown in Figure 6.5a. The average depth is 1.9375, the maximum depth is 5. The tree is highly skewed.

Gupta et al. follow Horibe's algorithm, constructing the tree in the normal top-down weight-balancing manner until we reach a node such that if we were to split the weight of the node further in the most balanced manner, the maximum depth constraint would be violated. Instead, we split the node maintaining as much balance as we can while respecting the depth constraint. If this happens at a node at depth d representing the probabilities $\{f_r \ldots f_t\}$, we take the left and right children as representing the probabilities $\{f_r \ldots f_s\}$ and $\{f_{s+1} \ldots f_t\}$, provided s is such that

$$\Delta(r,t) = \left| \sum_{i=r}^{s} f_i - \sum_{i=s+1}^{t} f_i \right| = \min_{\forall u : a \leq u \leq b} \left| \sum_{i=r}^{u} f_i - \sum_{i=u+1}^{t} f_i \right|, \quad a \leq s \leq b$$

where $a = t - 2^{D-d-1}$ and $b = r + 2^{D-d-1}$. Therefore, this idea is to use the weight-balancing heuristic as far down into the tree as possible. Intuitively, any node where we are unable to use the heuristic is expected to be very deep down in the tree. This would mean that the total weight of this node is small enough so that approximating the weight-balancing heuristic does not cause any substantial effect on the average path length. For instance, Figure 6.5b shows the depth-constrained weight balanced tree from a maximum depth constraint of 4 for the tree in Figure 6.4a.

There are $n - 1$ internal nodes in a binary tree with n leaves. A suitable split may be found by a binary search in $O(\log n)$ time at each internal node, the time complexity of the DCWBT algorithm is $O(n \log n)$, and the space complexity is the complexity of storing the binary tree $O(n)$.

6.2.4 Simulation

To measure the performance of MINPQ and DCWBT, Gupta et al. took three routing tables, MaeWest and MaeEast from IPMA (http://www.merit.edu/ipma 2000), VBNS from Fix-west (http://www.vbns.net/route/, 2000), and a traffic trace from NLANR (http://moat.nlanr.net, 2000). Table 6.2 shows the characteristics of the three routing tables, the entropy values of the probability distribution obtained from the trace, and the number of memory accesses required in an unoptimized binary search ($\lceil \log(\#Intervals) \rceil$).

Figure 6.7 shows the average number of memory accesses (tree depth) versus the maximum depth constraint value for the trace. As the maximum depth constraint is relaxed from $\lceil \log_2 n \rceil$ to higher values, the average tree depth quickly approaches the entropy of the corresponding distribution. The simple weight-balancing heuristic DCWBT almost always performs better than the MINDPQ algorithm, especially at higher values of maximum depth constraint.

Because the forwarding tables and the access patterns are not static, the of the tree data structure should be reconstructed in the MINDPQ and DCWBT algorithms periodically. Thus, the time to compute a new tree data structure is an important consideration. If the optimal tree is recomputed at a fixed interval regardless of the changes, the MINDPQ and DCWBT are of practical use.

6.3 Dynamic Biased Skip List

Skip List [15] is a simple data structure that uses probabilistic balancing rather than strictly enforced balancing, and can be used in place of balanced trees for most applications. Balanced trees can do everything that can be done with Skip List. But Skip List algorithms are very easy to

Table 6.2 Characteristics of Three Routing Tables and Two Traffic Traces

Routing Table	No. of Prefixes	No. of Intervals	No. of unopt_srch	Entropy
VBNS	1307	2243	12	6.63
Mae_west	24681	39277	16	7.89
Mae_east	43435	65330	16	8.02

Note: No. of unopt_srch is the number of memory accesses ($\lceil \log(\#Intervals) \rceil$) required in a naïve, unoptimized binary search tree.

Source: Gupta, P., Prabhakar, B., and Boyd, S., *Proceedings of Nineteenth Annual Joint Conference of the IEEE Computer and Communications Societies,* IEEE Infocom, IEEE Press, 2000. With permission.

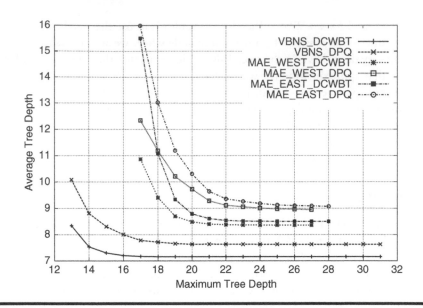

Figure 6.7 The average tree depth versus maximum tree depth. (From Gupta, P., Prabhakar, B., and Boyd, S., *Proceedings of Nineteenth Annual Joint Conference of the IEEE Computer and Communications Societies*, IEEE Infocom, IEEE Press, 2000. With permission.)

implement, extend and modify, and support $O(\log n)$ expected time search, insert, and delete operations for uniform access patterns (n is the number of items).

Skip Lists are balanced by consulting a random number generator; there is no distinction between keys based on the access frequency. Ergun et al. improve Skip List by exploiting the bias in the access pattern, called the Biased Skip List (BSL) [3], which can self adjust to access probabilities dynamically without resorting to complete reconstruction.

In the Internet, there are two fundamentally important access patterns: Independently skewed access patterns and bursty access patterns. For the first pattern, the access frequencies of certain items are high and stable over time. For the second pattern, a site gets "hot" for short periods of time, its access frequencies increase at high speed in a short time. Ergun et al. adapt BSL to the dynamic Internet environment to obtain efficient IP Lookup algorithms [3].

A prefix in a forwarding table is a set of continuous IP addresses. Two prefixes may overlap. To facilitate IP lookup with BSL, Ergun et al. transform the prefixes as nonoverlapping intervals in the IP address space, as in [2]. Each interval has two end points and is stored as a key in BSL. An IP address matches with a key if it is greater than the left end point and less than the right end point. All keys can be defined in increasing ordering as follows: keys $i <$ key j if any IP address in key i is less than that in key j. Because there is no overlapping key, any IP address has no more than one matching key, and IP lookup can perform exact matching. Based on the ordering key and the relation between IP address and key, Ergun et al. proposed a lookup scheme—Dynamic BSL. This section is from [3]. Portions are reprinted with permission (© 2001 IEEE).

6.3.1 Regular Skip List

A Skip List is a search data structure for n ordered keys. To construct a Skip List with a set of n ordered keys, we first make up level $\log n$, which is sorted linked list of all keys, then level $(\log n) - 1$, which is a sorted linked list consisting of a subset of the keys in level log n, and so on. Each key in level $i + 1$ is copied to level i independently with probability 1/2, and has vertical pointers to and from its copies on adjacent levels. Fifty percent of all keys are in level (log n) − 1, 25 percent in level (log n) − 2, 12.5 percent in level (log n) − 3, and so on. There is a key in level 1 at least.

To search for a key k, we start from the smallest (left most) key in the top level (level 1). On each level, we go right until we encounter a key, which is greater than k. We then take a step to left and go down one level (which leaves us in a key less than k). The search ends as soon as k is found, or when, at the lowest level, a key greater than k is reached. To delete a key, we perform a search to find its highest occurrence in the structure and delete it from every level it appears. To insert a new key k, we first search for it in the data structure to locate the correct place to insert it in the bottom level. Once the key is inserted into the bottom level, a fair coin is tossed to decide whether to copy it to the level above. If k is indeed copied, the procedure is repeated iteratively for the next level, otherwise the insertion is complete.

6.3.2 Biased Skip List

The regular Skip List has multiple levels, and the level of each key is determined at random. Because each search starts at the top level, it takes less time to search for keys appearing on higher levels than those appearing on lower levels. If the key in the lower level is frequently accessed, these average search time becomes long. To remedy these shortcomings, the BSL is proposed. In BSL, the level of each key is related to its access frequency. The keys with high access frequencies are selected to copy from the level below to the top level, and the more frequently the key is accessed, the nearer it is to the top level, and the less time it takes to search the key.

6.3.2.1 Data Structure

There is a set of n keys. Each key (k) is assigned a different rank $r(k)$ ($1 \leq r(k) \leq n$) depending on the access frequency. If key a is accessed more frequently than key b, then $r(a) < r(b)$. The keys are partitioned into classes $C_1, C_2, \ldots, C_{\log n}$ as follows: if key x and y are in classes C_k and C_{k+1} respectively, then $r(x) < r(y)$. The class sizes are geometric, $|C_i| = 2^{i-1}$.

BSL is constructed in a randomized way; it contains a sorted doubly linked list keys for each level. The levels are labeled from bottom up as: $L_{\log n}$, $L'_{\log n-1}$, $L_{\log n-1}$, ..., L_2, L'_1, L_1. The bottom level, $L_{\log n}$, includes all of the keys. We copy keys from L_{i+1} to L'_i as follows: (*i*) all $2^i - 1$ keys from classes $C_i, C_{i-1}, \ldots, C_1$, and (*ii*) a subset of remaining keys of L_{i+1} picked independently with probability 1/2 each. The keys in L_i consist of the following keys from L'_i: (*i*) all keys from classes $C_{i-1}, C_{i-2}, \ldots, C_1$, and (*ii*) a subset of the remaining keys in L'_i picked independently with probability 1/2 each. Each key in each level has a horizontal pointer that points to the key in the same level, and a vertical pointer that points to the key to and from its copies on the adjunct level. The data structure of BSL is a sorted doubly linked list of keys at each level. An example of a BSL is presented in Figure 6.8. The complexity of building a BSL is given as follows.

Lemma 6.3.1: Given a set of n keys in sorted order, a BSL can be constructed in $O(n)$ expected time.

Proof: To construct a BSL, we use two kinds of operations: the random copy and the automatic copy.

Random copy operation means that each key in each level is copied independently with probability 1/2. This process generates an expected single copy (not counting the original) of each key subjected to it; therefore, the expected total is $O(n)$.

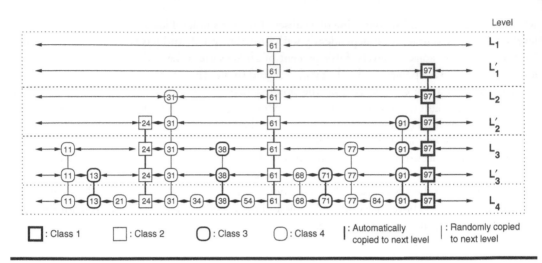

Figure 6.8 BSL and its levels. For simplicity, only the left endpoints of the keys are shown. (From Ergun, F., et al., *Proceedings of 20th Annual Joint Conference of the IEEE Computer and Communications Societies*, IEEE Infocom, IEEE Press, 2001. With permission.)

In the process of the automatic copy, none of the $(k + 1)/2$ keys in $C_{\log n}$ are automatically copied. The $(n + 1)/4$ keys in $C_{\log n-1}$ are automatically copied to two levels. The keys in $C_{\log n-2}$ are copied to four levels. In general, keys in $C_{\log n-p}$ are copied to $2p$ levels. Summing all automatic copies for each key in each class, we find a total of $O(n)$ copies. Thus the construction time is $O(n)$.

6.3.2.2 Search Algorithm

The search algorithm in a BSL is similar to that in a regular Skip List. For each IP address A, we start from the top level. On each level i, we follow the linked list from left to right until we encounter a key which is greater than A. Then we take a step back (left) and go to a key in level $i - 1$ following vertical pointers. The search time is given in the following theorem.

Theorem 6.3.1:   A search for an address matched by a key k in a BSL of n keys takes $O(\log r(k))$ expected time.

Proof: For an IP address, the total search time can be measured as the number of horizontal and vertical links to be traversed. Consider searching for an address that matches a key k, where k belongs to class C_c, with $c \leq \log r(k) + 1$. By construction, all of the keys in C_c are present on level L'_c, which is $2c - 1$ levels down from the top. This binds the vertical distance that we travel. Let us now consider the horizontal links that we traverse on each level. The top level has an expected single key, thus we traverse at most two links. Now let our current level be L_m (resp. L'_m). We must have come down from level L'_{m-1} (resp. L_m) following a vertical link on some key s. Let t be the key immediately following s on the level above, that is, L'_{m-1} (resp. L_m). Then, it must be that $s < k < t$ (recall the ordering between intervals). Thus, on the current level, we need to go right at most until we hit a copy of t. The number of links that we traverse on the current level is at most 1 plus the number of keys between s and t. Note that these keys do not exist on the level above, even though s and t do. The problem then is to determine the expected number of keys between s and t on this level. In a scheme with no automatic copying (all copying is random with probability 1/2), given two adjacent keys on some level l, the expected number of keys between them on the level below (not copied to l) is 1. This is because, due to the 1/2 probability, one expects on the average to "skip" (not copy upwards) one key for each that one does copy. In our scheme, some keys get copied automatically; therefore, we are less likely to see less than 1 uncopied key between two copied keys. Therefore, the expected number of horizontal links that we travel on the current level between s and t is at most 2. Because we visit at most $2c - 1$ levels, traversing the expected two links on each level, the total (expected) running time is $O(c) = O(\log r(k))$.

Note that, for an unsuccessful search, the horizontal distance analysis is the same. However, one has to go all the way down to the bottom level. The vertical distance traveled is thus $O(\log n)$ levels, which gives a total expected running time of $O(\log n)$.

6.3.3 Dynamic BSL

In BSL, the keys of small rank are copied from bottom to top, for example, the key "61" in Figure 6.8 is copied in each level. Because we start our search at the top level each time, if there is no result, then we go down level. The keys appearing at the top level will not be found as the search result at the lower level. Therefore, it is wasteful to copy the keys with small rank all the way to the bottom. On the other hand, if we insert a key with rank 1, it must be copied from bottom to top level. This makes more operations for inserting a new key. To improve BSL, the keys in class C_i need not be copied in all the levels below L'_i, but are randomly copied down for only a few levels. We call this improved BSL Dynamic BSL.

6.3.3.1 Constructing Data Structure

To construct dynamic BSL, we start with the lowest level $L_{\log n}$ that is made up of all the keys in class $C_{\log n}$. For $i < \log n$, an upper level L'_i includes the keys chosen from L_{i+1} independently with probability 1/2, and the keys in class C_i (called the default keys of L'_i). To facilitate an efficient search, some of the default keys of L'_i are chosen and copied to the lower level L_{i+1} independently with probability 1/2. Each key copied to level L_{i+1} may further be copied to lower levels $L'_{i+1}, L_{i+2}, L'_{i+2}, \ldots$ using the same randomized process. Once we have level L'_i, we construct level L_i by simply choosing and copying keys from level L'_i independently with probability 1/2 for each key. For a dynamic BSL, see Figure 6.9. Lemma 6.3.2 shows that the construction time for the dynamic BSL is the same as that for BSL.

Lemma 6.3.2:   A dynamic BSL with n keys can be constructed in $O(n)$ expected time, given the keys in sorted order.

Proof: The expected number of keys on each level is bounded by that in the static BSL. In static BSL, level L'_i contains all keys in Class C_1 through C_i, which number $O(|C_i|)$. L_i contains (an expected) half of those. Thus, L_i and L'_i contain $O(2^i)$ keys.

Let us look at the cost per level. L'_i is formed from L_{i+1} and Class C_i in time $O(|C_i|)$. L_i is formed from L'_i in time $O(|C_i|)$. The only other cost is for copying down the default keys of L'_i. The expected number of times that a key will be copied down is just under 1. Thus, the expected number of copies down from C_i below L'_i is $O(|C_i|)$. To copy a key k down, we go to its left neighbor on the same level, and keep going left until we find a down link, which we take. Then, we go right until we come to a key greater than k and then insert k before it. The expected number of left steps is less than 2, because the probability that a key that exists on this level will not exist on the

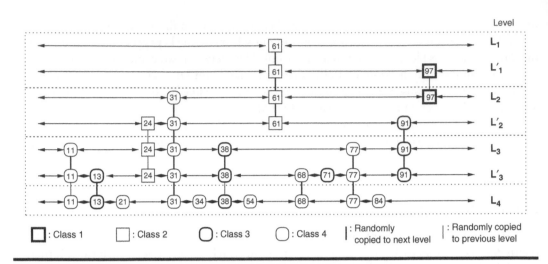

Figure 6.9 Dynamic BSL. Only the left endpoints of the keys are shown. (From Ergun, F., et al., *Proceedings of 20th Annual Joint Conference of the IEEE Computer and Communications Societies*, IEEE Infocom, IEEE Press, 2001. With permission.)

previous level is less than 1/2. Likewise, the expected number of right steps is less than 2. Thus, we spend $O(1)$ time per copy, and $O(|C_i|)$ time for the entire C_i. The costs associated with L'_i and L_i add up to $O(|C_i|)$; summing over i we get $O(n)$.

In theory, the construction time for dynamic BSL is the same as that for BSL. In practice, the number of copy operations in dynamic BSL is less than that in BSL because there is no automatic copying of keys to an upper level in dynamic BSL. The search algorithm in dynamic BSL is the same as that in BSL.

6.3.3.2 Dynamic Self-Adjustment

When a key is accessed (through search/insertion/deletion), its rank changes. In this subsection, the main problem is how to adjust the data structure to the rank changes for independently skewed access patterns. For simplicity, the search operation and insert/delete operation are discussed respectively.

Search: Let C_i be the class of a key k before it was accessed. After the key k is accessed due to a search, we check whether the rank of k is changed. If so, k should be moved to Class C_{i-1}. We need to assign L'_{i-1} as the default level of k and shift the bottommost and topmost levels of k up by 2 by manipulating the pointers to and from the copies of k on the level involved. Once k is moved up to class C_{i-1}, another key l from C_{i-1} is moved down to C_i to preserve class sizes. We need to change the default level of l and push down the topmost and bottommost levels of l by two as well.

Insertion/deletion: When inserting a key k to BSL, it is assigned rank $n + 1$, and inserted at the bottom. If $n = 2^l - 1$ for some integer l, then it means class $C_{\log n}$ is full, and we need to establish a new class $C_{\log n+1}$ with k as its sole member. To do this, we create two levels $L'_{\log n}$ and $L_{\log n+1}$ and simply insert k in level $L_{\log n+1}$. For promoting k to upper levels, we again use the independent random process (fair coin) with probability 1/2.

To delete key k, we first identify its location in BSL, then delete all of its copies from the BSL as well as from the master-list. This changes the ranks of keys with rank greater than $r(k)$.

We update the default levels of all keys whose classes change—there can be at most log n of them—by pushing their top- and bottommost levels up by two.

Theorem 6.3.5: For the BSL with n keys, insertion and deletion can be done in $O(\log n)$ expected time.

6.3.3.3 Lazy Updating Scheme

So far, the BSL handles insertions and deletions in $O(\log n)$ time. This means that the time of insertion/deletion depends on the number of keys. For a large number of keys, the time may too long, resulting in blocking. In many applications, the majority of the keys have a very short life span. Once inserted, these keys are rapidly accessed a few times, then deleted after a short period; its rank always remain close to 1. In such applications, insertion/deletions are the main bottleneck.

To facilitate a more efficient implementation of insertions and deletions, Ergun et al. employ a lazy updating scheme of levels and allow flexibility in class sizes [3] for bursty access patterns. The size of a class C_i is allowed to be in the range $(2^{i-2}, \ldots, 2^i - 1)$; recall that the default size of C_i is 2^{i-2}. The lazy updating allows us to postpone updates until the size of a level becomes too large or too small.

For example, let the number of elements in classes $C_1, C_2, \ldots, C_5$ be 1, 2, 4, 8, 16, respectively (the default sizes). A sequence of eight insertions would respectively yield the following class sizes.

1	1	1	1	1	1	1	1	1
2	3	2	3	2	3	2	3	2
4	4	6	6	4	4	6	6	4
8	8	8	8	12	12	12	12	8
16	16	16	16	16	16	16	16	24

It is observed that the effects of most insertions and deletions are confined to the upper levels. As a result, the insertion or deletion of a key k takes $O(\log r_{\max}(k))$ time, where $r_{\max}(k)$ is the maximum rank of key k in its lifespan.

For bursty access patterns, when a key k is searched for, its rank is set to be 1 and the ranks of all keys whose ranks are smaller are incremented by one. The rank changes are reflected in the BSL. Although the class sizes are not rigid, to make the analysis simple, we maintain the class size after a search without affecting the overall cost. This is possible by shifting the largest ranked key down by one class for all the classes numbered lower than the initial class of k; k is moved to the top class.

When a key k is inserted into the BSL, it is assigned a rank of 1 and the ranks of all keys are incremented by one. After an insertion, if the size of a class C_i reaches its upper limit of 2^i, then half of the keys in C_i with the largest rank change their default level from L'_i to L'_{i+1}. This is done by moving the topmost and bottommost levels of each such key by two. One can observe that such an operation can be very costly; for instance, if all classes $C_1, C_2, \ldots, C_l$ are full, an insertion will change the default levels of 1 key in class C_1, 2 keys in class C_2, and in general 2^{i-1} keys from class C_i. However, a tighter amortized analysis is possible by charging more costly insertions to less costly ones. The amortized analysis gives the average insertion time for a key k as $O(\log r_{max}(k))$, where $r_{max}(k)$ is the maximum rank of key k in its lifespan.

To delete a key k, we first search for it in the BSL, which changes its rank to 1 and then we delete it from all the levels where it has copies. If the number of keys in class C_i after the deletion is above its lower limit of 2^{i-2} then we stop. Otherwise, we go to the next class C_{i+1} by moving their default levels upwards by two. We continue this process until all levels have a legal number of keys. As with insertions, it is possible to do amortizations to show that the average deletion time for a key k is $O(\log r_{max}(k))$. Thus, the average insertion/deletion time depends on the rank rather than the total number of keys.

6.3.3.4 Experimental Results

To evaluate the performance of dynamic BSL, Ergun et al. used the simulated data with geometric access distribution. In the experiments on the geometric access distribution, at each step, a key k with rank $r(k)$ was accessed with probability $P(k) = p^{r(k)}(1 - p)$, where p is the bias parameter.

The performance of several variants of dynamic BSL is compared with the regular Skip Lists and a simple move-to-front (MFT) linked list. There are a number of parameters that affect the practical performance of a BSL implementation, such as the size of the topmost class C_1, the rank of keys etc. Here, two schemes are built: BSL with $|C_1| = 8$ and BSL with $|C_1| = 512$. The change of the search time with the average rank is shown in Figure 6.10.

When the size of the topmost class C_1 is increased from 8 to 512, the difference of the search time is enlarged with the growth of the average rank. A BSL with $|C_1| = 512$ consistently outperforms the regular Skip List. When the bias is low, BSL implementations outperform the MTF list by at least one order of magnitude. However, MTF lists perform better when there is high bias. This is because the MTF list allows a large number of keys to be stored in the fast cache memory. The characteristics of MTF lists allow it to be combined with BSL in a hybrid scheme: before a search is forwarded to BSL, a short MTF list can be used as an initial filter. The BSL with initial MTF outperforms the BSL with $|C_1| = 8$.

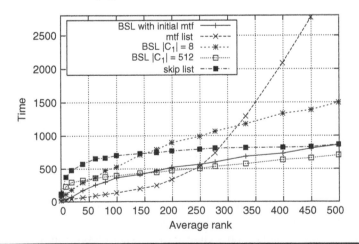

Figure 6.10 Search times on geometrically distributed data with varying biased; a total of 106 searches were performed on 198K keys. (From Ergun, F., et al., *Proceedings of 20th Annual Joint Conference of the IEEE Computer and Communications Societies*, IEEE Infocom, IEEE Press, 2001. With permission.)

6.4 Collection of Trees for Bursty Access Patterns

A BSL has a self-update mechanism that reflects the changes in the access patterns efficiently and immediately, without any need for rebuilding. To facilitate the longest-prefix matching prefix with BSL, the prefixes in the forwarding table are represented as nonoverlapping intervals in the IP-address space. Given an ordering between intervals, each of the nonoverlapping intervals is stored as key in the BSL, and we can perform an exact matching search in $O(\log n)$ expected time. But the addition and removal of prefixes requires us to reconstruct these intervals. To overcome this problem, Sahni and Kim propose two data structures: a collection of splay trees (CSTs) and BSLs with prefix trees (BSLPT) [4]. They are based on the definition of ranges in [16] and a collection of red-black trees in [17]. This section is from [4] (© 2006 WorldSci).

6.4.1 Prefix and Range

Lampson et al. [16] have proposed a binary search scheme for IP-address lookup, in which prefixes are encoded as ranges. Any prefix P is a set of the continuous IP addresses, and is represented as a range $[s, f]$ in the IP number line, where s is the start, f is the finish, and each is called an end point. The starts and finishes of all prefixes in a forwarding table can partition the IP number line into intervals. For example, Table 6.3 shows a set of five prefixes with the start and the finish. They partition the IP number line into 8 intervals: [0, 10], [10, 11], [11, 16], [16, 18], [18, 19], [19, 26], [26, 27], [27, 31]. Let r_i be an end point, $1 \leq i \leq q \leq 2n$, and $r_i < r_{i+1}$; each of the intervals $I_i = [r_i, r_{i+1}]$ is called a basic interval, where n is the number of prefixes, q is the number of the end points. Each end point and interval correspond to a unique prefix. The corresponding relation between the prefixes and the end point and intervals are shown in Figure 6.11. Any destination address belongs to a unique interval or is equal to an end point. Lampson et al. construct a binary search table using $\{r_i\}$. For the destination address d, we can perform a binary search to find a unique i such that $r_i \leq d < r_{i+1}$. If $r_i = d$, the longest matching prefix for the destination address d, $LMP(d)$, is given by the "=" entry; otherwise, it is given by the ">" entry. For example, if $d = 20$ satisfies $19 < d < 23$, then $LMP(d)$ is P1, which is given by the ">" entry of the end point 19.

Here, the basic intervals are given by the start and finish of prefixes directly and do not overlap, and the addition or deletion of a prefix requires us to add or delete at most two end points. In [3],

Table 6.3   Prefixes with Start and Finish

No.	Prefix	Start	Finish
P1	*	0	31
P2	0101*	10	11
P3	100*	16	19
P4	1001*	18	19
P5	10111	23	23

Note: The length of address is 5-bits.

Source: Sahni, S. and Kim, K.S., *International Journal of Foundation Computer Science*, 15, 4, 2006. With permission.

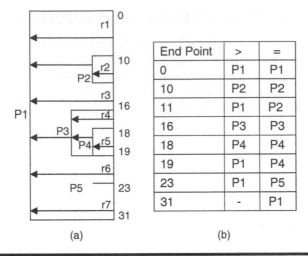

End Point	>	=
0	P1	P1
10	P2	P2
11	P1	P2
16	P3	P3
18	P4	P4
19	P1	P4
23	P1	P5
31	-	P1

(a)   (b)

Figure 6.11 (a) Pictorial representation of prefixes and ranges; (b) table for binary search. (From Sahni, S. and Kim, K.S., *International Journal of Foundation Computer Science*, 15, 4, 2006. With permission.)

the intervals are given by the translating prefixes; the addition and removal of prefixes requires us to reconstruct these intervals.

6.4.2 Collection of Red-Black Trees (CRBT)

The collection of red-black trees is proposed by Sahni and Kim [17]. It consists of a basic interval tree (BIT) and n prefix trees that are called a collection of prefix trees (CPT). For any destination IP-address d, define the matching basic interval to be a basic interval with the property $r_i \leq d \leq r_{i+1}$.

BIT is a binary search tree that is used to search for a matching basic internal for any IP address d and comprises internal and external nodes, with one internal node for each interval r_i. It has q internal nodes and $q + 1$ external nodes. The first and last of these, in order, have no significance. The remaining $q - 1$ external nodes, in order, represent the $q - 1$ basic intervals of the given prefix set. Figure 6.12a gives a possible BIT for the five-prefix example of Figure 6.11a. Internal nodes are shown as rectangles while circles denote external nodes. Every external node has three pointers: *startPointer*, *finishPointer*, and *basicInternalPointer*. For an external node that represents the basic interval $[r_i, r_{i+1}]$, *startPointer* (*finishPointer*) points to the header node of the prefix tree (in the back-end structure) for the prefix (if any) whose range start and finish points are $r_i(r_{i+1})$. Note that only prefixes whose length is W can have this property. *basicInternalPointer* points to a prefix node in a prefix tree of the back-end structure. In Figure 6.12a, the labels in the external (circular) nodes identify the represented basic interval. The external node with r1 in it, for example, has a *basicIntervalPointer* to the rectangular node labeled r1 in the prefix tree of Figure 6.12b.

For each prefix and basic interval, x, define $next(x)$ to be the smallest range prefix (i.e., the longest-prefix) whose range includes the range of x. For the example in Figure 6.11a, the $next()$ values for the basic intervals r1 through r7 are, respectively, P1, P2, P1, P3, P4, P1, and P1. Notice that the next value for the range $[r_i, r_{i+1}]$ is the same as the ">" value for r_i in Figure 6.11b, $1 \leq i < q$. The $next()$ values for

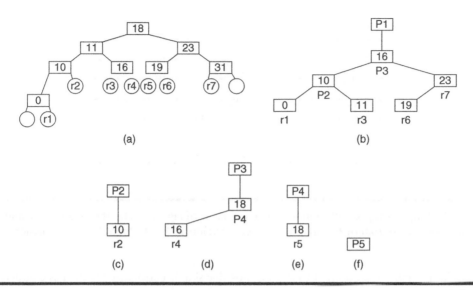

Figure 6.12 **CBST (a) base interval tree, (b) prefix tree for P1, (c) prefix tree for P2, (d) prefix tree for P3, (e) prefix tree for P4, (f) prefix tree for P5. (From Sahni, S. and Kim, K.S.,** *International Journal of Foundation Computer Science*, **15, 4, 2006. With permission.)**

the nontrivial prefixes P1 through P4 of Figure 6.11a are, respectively, "-," P1, P1, and P3. The *next()* values for the basic interval and the nontrivial prefixes are shown in Figure 6.11a as left arrows.

The back-end structure, which is a CPT, has one prefix tree for each of the prefixes in the routing table. Each prefix tree is a red-black tree. The prefix tree for prefix *P* comprises a header node plus one node, called a prefix node, for every nontrivial prefix or basic interval *x* such that *next(x) = P*. The header node identifies the prefix *P* for which this is the prefix tree. The prefix trees for each of the five prefixes of Figure 6.11a are shown in Figure 6.12b through f. Notice that prefix trees do not have external nodes and that the prefix nodes of a prefix tree store the start points of the basic intervals and prefixes are shown inside the prefix nodes while the basic interval or prefix name is shown outside the node.

The search for *LMP(d)* begins with a search of the BIT for the matching basic interval for *d*. Suppose that the external node *Q* of the BIT represents this matching basic interval. When the destination address equals the left (right) end-point of the matching basic interval and *start-Pointer (finishPointer)* is not null, *LMP(d)* is pointed to by *startPointer (finishPointer)*. Otherwise, the back-end CPT is searched for *LMP(d)*. The search of the back-end structure begins at the node *Q.basicIntervalPointer*. By following parent pointers from *Q.basicIntervalPointer*, we reach the header node of the prefix tree that corresponds to *LMP(d)*.

6.4.3 Biased Skip Lists with Prefix Trees (BSLPT)

BSL is a dynamic lookup data structure that adapts to changes in the access pattern without the need for an explicit and costly reconstruction, described in Section 6.3. To construct the initial BSL, the *n* prefixes in the forwarding table are translated to $2n - 1$ nonoverlapping intervals in the address space. The longest matching prefix for IP addresses that falls within each basic interval as well as for IP addresses that equal an end point is determined. A master-list of basic intervals along with the determined longest matching prefix information is constructed. This list is indexed into

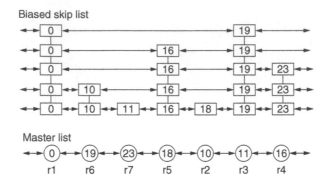

Figure 6.13   Skip list representation for basic intervals of Figure 6.11a. (From Sahni, S. and Kim, K.S., *International Journal of Foundation Computer Science*, 15, 4, 2006. With permission.)

using a skip list structure. Figure 6.13 shows a possible skip list structure for the basic intervals of Figure 6.11a.

In a BSL, ranks are assigned to the basic intervals. $rank(a) < rank(b)$, whenever interval a is accessed more recently than interval b. Basic interval ranks are used to bias the selection of intervals that are represented in skip list chains that are close to the top. Hence, searching for a destination address that is matched by a basic interval of smaller rank takes less time than when the matching interval has a higher rank. If the destination of the current packet is d and the matching basic interval for d as determined by the skip list search algorithm is a, then $rank(a)$ becomes 1 and all previous ranks between 1 and $oldRank(a)$ are increased by 1. The skip list structure is updated following each search to reflect the current ranks. Consequently, searching for basic intervals that were recently accessed is faster than searching for those that were last accessed a long time ago. This property makes the BSL structure [3] perform better on bursty access patterns than on random access patterns. Regardless of the nature of the access, the expected time for a search in a BSL is $O(\log n)$.

When a router table has to support the insertion and deletion of prefixes, the BSL structure becomes prohibitively expensive. This is because an insert or delete can require us to change the longest matching prefix information stored in $O(n)$ basic intervals. To improve the performance of the insertion/deletion in BSL [3], Sahni and Kim used the CPT as the back-end data structure, and configured the master-list node exactly the same as the external nodes of a BIT. Each master-list node represents a basic interval $[r_i, r_{i+1}]$ and has three pointers: *startPointer*, *finishPointer*, and *basicIntervalPointer*. *startPointer* (*finishPointer*) points to the header node of the prefix tree (in the back-end structure) for the prefix (if any) whose range start and finish points are $r_i(r_{i+1})$; *basicIntervalPointer* points to the prefix node for the basic interval $[r_i, r_{i+1}]$. This prefix node is in a prefix tree of the back-end structure.

To find the longest-matching prefix $LMP(d)$ for IP address d in BSLPT, at first we search the front-end BSL using the search scheme in [3], then get a matching basic interval that is the master-list node Q. If d equals the left (right) end point of the matching basic interval and *startPointer* (*finishPointer*) is not null, $LMP(d)$ is pointed to by *startPointer* (*finishPointer*). Otherwise, we follow the *basicIntervalPointer* in Q to get into a prefix tree, and $LMP(d)$ is determined by the parent pointers to the header of the prefix tree.

For a new prefix $P = [s, f]$, at least one of start point s and finish point f is a new end point. For example, the end point s is new. Because the default prefix * is always present, there must be a matching basic interval [*a*, *b*] for the new end point *s*. The basic interval *i* = [*a*, *b*] is replaced by *i*1 = [*a*, *s*] and *i*2 = [*s*, *b*]. The end point *s* is inserted in BSL and the master-list using the algorithm in [15]. The pointers of the corresponding node are determined as follows:

1. *i*1.*startPointer* = *i*.*startPointer* and *i*2.*finishPointer* = *i*.*finishPointer*.
2. When |*P*| = *W* (|*P*| is the length of prefix *P*, *W* is the length of IP address), *i*1.*finishPointer* = *i*2.*startPointer* = pointer to header node of prefix tree for tree for *P*; *i*1.*basicIntervalPointer* and *i*2.*basicIntervalPointer* point to prefix nodes for *i*1 and *i*2, respectively. Both of these prefix nodes are in the same prefix tree as was the prefix node for *i*. This prefix tree can be found by the following parent pointers from *i*.*basicIntervalPointer*.
3. When |*P*| < *W*, *i*1.*finishPointer* = *i*2.*startPointer* = null. *i*1.*basicIntervalPointer* = *i*.*basicIntervalPointer*, *i*2.*basicIntervalPointer* points to a new prefix node for interval *i*2. This new prefix node will be the prefix tree for *P*.

The deletion of a prefix is the inverse of what happens during an insert. Because the prefix trees are red-black trees, the expected complexity of a search, insert, and delete is *O*(log *n*).

6.4.4 Collection of Splay Trees

Splay trees are the binary search trees that self adjust so that the deepest node can become the root. That is to say, the node with high access frequency can be moved up to the root in a splay tree; each node in a splay tree can be the internal node or the leaf (external) node. The purpose of splaying is to minimize the number of access operations. This self-adjusting property makes the splay tree a promising data structure for bursty applications.

Unfortunately, we cannot simply replace the use of red-black trees in the CRBT structure with splay trees, because there are two types of nodes: internal node and external node in BIT. Following the splay operation, the reached external node should become the root. This is not permissible in BIT. So, Sahni and Kim propose an alternative BIT (ABIT) structure in which we have only internal nodes.

Each (internal) node of the ABIT represents a single basic interval, and has a left and right child pointers to *LMP*(*d*) when IP address *d* equals either the left or right end point of the basic interval, and a pointer to the corresponding basic-internal node in a prefix tree for the case when *d* lies between the left and right end points of the basic interval. All nodes can be an internal node or an external node. This is the same as the *basicIntervalPointers* used in CRBT and BSLPT. Figure 6.14 shows the ABIT for the basic interval of Figure 6.11. In this figure, only the left and

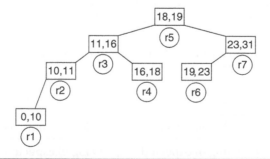

Figure 6.14 **Alternative base interval tree corresponding to Figure 6.11. (From Sahni, S. and Kim, K.S.,** *International Journal of Foundation Computer Science,* **15, 4, 2006. With permission.)**

right end points for each basic interval are shown, the *LMP*(d) values and the basicIntervalPointers are not shown.

The CST structure uses the ABIT structure as the front end; the back end is the same prefix tree structure as in [17] except that each prefix tree is implemented as a splay tree rather than as a red-black tree. The algorithms to search, insert, and delete in the CST structure are simple adaptations of those used in the BSLPT structure. The amortized complexity of algorithms is $O(\log n)$. In the following, we will compare of BITs and ABITs.

Space Complexity. For an n-prefix forwarding table, the BIT has $I(<2n)$ internal nodes and $I-1$ external nodes. Each internal node has a field (d bytes) for an interval end point, and two children and one parent field (each p byte), each external node has three pointers (each p byte): *startPointer*, *finishPointer*, and *basicIntervalPointer*. Additionally, every node has a type field (t bytes) to distinguish between internal and external nodes and a color field (c bytes) to distinguish between red and black nodes. The storage requirement of a BIT is $(d + t + c + 3p)I + (t + 3p)$ $(I-1)$ bytes. Each node in an ABIT has two end-points one color field, and six pointer fields (one parent, two children, and three back-end nodes). The storage requirement of an ABIT with $I-1$ nodes is $(2d + c + 6p)(I-1)$ bytes. For IPv4, $d = 4$ bytes, $t = c = 1/8$ bytes (i.e., 1-bit), $p = 4$ bytes, then *Memory* (*BIT*) = 28.375I − 12.125 bytes, and *Memory*(*ABIT*) = 32.125I − 32.125 bytes. The BIT structure requires about 12 percent less space than the ABIT structure. In practice, when coding in a language such as C++, we find it convenient to implement the type and color field using the data type byte ($t = c = 1$). For IPv4, *memory*(*BIT*) = 32I bytes and *memory*(*ABIT*) = 33I Bytes. BIT takes 6% less memory than is taken by ABIT.

Time Complexity. For an n-prefix forwarding table, BIT and ABIT have almost the same number of internal nodes. A BIT has external nodes, whereas an ABIT does not, then *height*(*ABIT*) ≈ *height*(*BIT*) − 1. For the longest matching prefix, the worst-case number of key comparisons in BIT is *height*(*BIT*), whereas the number in ABIT is 2*height*(*ABIT*), because it needs two comparisons in each internal node. In theory, the amortized complexity of the longest-prefix matching, insertion, and deletion is $O(\log n)$ in BIT and ABIT. The performance characteristics of CRBT, CST, and BSLPT are shown in Table 6.4. In practice, the average number in an ABIT should be less that in a BIT, because every search in a BIT necessarily goes all the way to an external node, whereas in an ABIT, a search may terminate at any internal node.

6.4.5 Experiments

To measure the performance of the three data structures CRBT, CST, and BSLPT, Sahni and Kim did experiments on a SUN Ultra Enterprise 4000/5000 computer, used the five IPv4 prefix

Table 6.4 Performance of Data Structures for Longest Matching-Prefix

Data Structure	Search	Update	Memory
CRBT	$O(\log n)$	$O(\log n)$	$O(n)$
CST	$O(\log n)$ amortized	$O(\log n)$ amortized	$O(n)$
BSLPT	$O(\log n)$ expected	$O(\log n)$ expected	$O(n)$

Source: Sahni, S. and Kim, K.S., *International Journal of Foundation Computer Science*, 15, 4, 2006. With permission.

Table 6.5 Routing Tables from IPMA Project

Routing Tables	Paix	Pb	MaeWest	Aads	MaeEast
No. of prefixes	85,988	35,303	30,719	27,069	22,713
No. of 32-bit prefixes	1	0	1	0	1
No. of end points	167,434	69,280	60,105	53,064	44,463

Source: Sahni, S. and Kim, K.S., *International Journal of Foundation Computer Science,* 15, 4, 2006. With permission.

databases obtained from the IPMA project (http://nic.merit.edu/ipma, snapshot on September 13, 2000), shown in Table 6.5.

In CRBT, the BIT structure is implemented using conventional red-black trees. In CST, the ABIT is implemented using a top-down splay tree, and the back-end prefix trees are implemented using bottom-up splay trees. In BSLPT, the back-end prefix trees are also implemented using bottom-up splay trees. The total memory requirement is shown in Table 6.6. The memory required by the CST structure is about 9 percent less than that required by the CRBT structure. This is because the front-end splay tree requires no parent, no type field, and no color field, and the back-end splay tree requires no color field. BSLPT requires about twice the memory required by each of the CRBT structure. Figure 6.15 histograms show the total memory required by each data structure. The ratio of total memory to the number of prefixes is about 0.134, 0.123, and 0.287 for CRBT, CST, BSLPT, respectively. This is consistent with the theory analysis in Section 6.4.4. The CST is most efficient on memory.

Search Time.   Sahni and Kim used six sets of test data to measure the average search time. The first data set, NODUP, comprised the end points of the basic intervals corresponding to the database being searched. These end points were randomly permuted. The second (third) data set, DUP10 (DUP20), was constructed from NODUP by making 10 (20) consecutive copies of each destination address. The fourth (fifth) data set, RAN 10(RUN20), was permuted for every block of 50 (100) destination addresses from DUP10 (DUP20) [each such block of 50 (100) destination addresses]. The trace sequences (DEC-PKT) were obtained from Lawrence Berkeley Laboratory (http://ita.ee.lbl.gov/html/comtrib/DEC-PKT.html, 1995). These data sets represented different degrees of burstiness. In NODUP, all search addresses are different. So, this access pattern represented the lowest possible degree of burstiness. In DUP20, every block of 20 consecutive IP

Table 6.6 Memory Requirement (kB)

Schemes	Paix	Pb	MaeWest	Aads	MaeEast
CRBT	11,534	4761	4134	3648	3056
CST	10,548	4354	3781	3336	2796
BSLPT	23,508	10,158	9152	8381	6159

Source: Sahni, S. and Kim, K.S., *International Journal of Foundation Computer Science*, 15, 4, 2006. With permission.

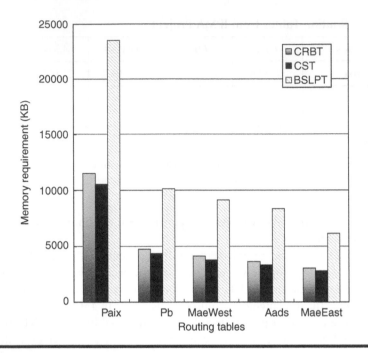

Figure 6.15 Total memory requirement.

packets has the same destination address; this access pattern represented a high degree of burstiness. In DEC-PKT, each destination address occurs approximately 379 times, this trace represented a much higher degree of burstiness then the DUP and RAN data sets.

The total search time for each data set was measured 10 times. The average time is shown in Table 6.7. The first thing to notice is that even though the CRBT is not designed for the access pattern, it actually performs better on bursty access patterns than on nonbursty ones. For example, the average search time on NODUP is about four times that on DUP10. This is because the computer caches automatically result in improved performance for bursty access patterns. The performance improvement between RAN10 and NODUP is not as much, because of the increased cache conflicts caused by intermediate searches for different destinations.

The data structures, CST and BSLPT, are designed for bursty access pattern. On the forwarding table Paix, CST and BSLPT took about eight to nine times as much time on the NODUP data set as they did on the DUP10 data set. There is a further 26–46 percent reduction in average search-time when going from DUP10 to DUP20 and from RAN10 to RAN20. For the trace data set, CST took about 7 times as much time on the NODUP data set as on the DEC-PKT data set, BSLPT took 17 times. Although CST and BSLPT perform better on the access pattern, on the data set with a high degree of burstiness, BSLPT takes five times as much time as does CST.

Update Time.   To measure the average insert time, at first the prefixes in each of the databases were randomly permuted. Next, the first 75 percent of the prefixes were inserted into an initially empty data structure. The time to insert the remaining 25 percent of the prefixes was measured. Then the time to delete the last 25 percent of the prefixes was measured. This timing experiment

Table 6.7 Average Search Time (in μ sec)

Data Structure	Data Set	Paix	Pb	MaeWest	Aads	MaeEast
CRBT	NODUP	6.51	5.77	5.50	5.29	5.15
	DUP10	1.55	1.41	1.36	1.36	1.31
	DUP20	1.29	1.21	1.17	1.17	1.13
	RAN10	2.08	1.54	1.58	1.39	1.49
	RAN20	1.89	1.44	1.37	1.28	1.22
	DEC	3.91	3.44	3.03	2.91	2.95
CST	NODUP	9.91	8.68	8.45	8.23	7.78
	DUP10	1.23	1.13	1.11	1.09	1.04
	DUP20	0.76	0.71	0.69	0.68	0.66
	RAN10	1.69	1.59	1.56	1.53	1.50
	RAN20	1.24	1.18	1.18	1.16	1.12
	DEC	1.35	1.41	1.46	1.43	1.42
BSLPT	NODUP	88.20	77.70	75.16	73.10	66.14
	DUP10	9.36	8.28	8.13	7.90	7.13
	DUP20	5.01	4.38	4.26	4.15	3.81
	RAN10	9.73	8.50	8.22	8.21	7.43
	RAN20	5.39	4.79	4.72	4.60	4.34
	DEC	5.27	5.03	4.95	4.74	4.98

Source: Sahni, S. and Kim, K.S., *International Journal of Foundation Computer Science*, 15, 4, 2006. With permission.

was repeated ten times. The average update times (AVG) and the standard deviations (SD) are shown in Tables 6.8 and 6.9.

In Tables 6.8 and 6.9, BSLPT takes about twice as much time as do other data structures when we insert (delete) a prefix. The update operation is faster in CST than in CRBT. For example, in the Paix database, the insert time for CST is about 30 percent less than for CRBT and the delete time for CST is about 15 percent less than for CRBT.

Although BSLPT is designed for bursty access patterns, and takes about eight to nine times as much time on the no-bursty data set as it does on the bursty data set, experiments indicate that BSLPT is highly noncompetitive with the CRBT, which is not designed for bursty access patterns, and is always significantly inferior to CST on all operations. Therefore, of the three data structures tested CRBT is recommended for nonbursty to moderately bursty applications and CST is recommended for highly bursty applications.

Table 6.8 Average Time to Insert a Prefix (in μ sec)

Data Structure		Paix	Pb	MaeWest	Aads	MaeEast
CRBT	AVG	35.12	32.86	31.51	31.62	29.94
	SD	1.35	0.00	0.54	1.03	0.00
CST	AVG	24.60	23.68	22.39	22.60	21.66
	SD	0.14	0.35	0.54	1.40	0.85
BSLPT	AVG	64.10	58.47	67.84	67.82	51.60
	SD	0.65	1.52	0.73	0.46	0.85

Source: Sahni, S. and Kim, K.S., *International Journal of Foundation Computer Science*, 15, 4, 2006. With permission.

Table 6.9 Average Time to Delete a Prefix (in μ sec)

Data Structure		Paix	Pb	Mae-West	Aads	Mae-East
CRBT	AVG	38.89	36.03	34.90	35.31	33.81
	SD	1.30	0.71	0.54	0.83	0.74
CST	AVG	32.88	31.72	30.86	31.18	29.41
	SD	0.31	0.00	0.62	1.62	0.85
BSLPT	AVG	74.01	65.49	63.68	63.10	57.76
	SD	0.77	0.89	0.96	0.99	1.11

Source: Sahni, S. and Kim, K.S., *International Journal of Foundation Computer Science*, 15, 4, 2006. With permission.

References

1. Cheung, G. and McCanne, S., Optimal routing table design for IP address lookups under memory constraints, *Proceedings of Eighteenth Annual Joint Conference of the IEEE Computer and Communications Societies*, March 1999, IEEE Press, pp. 1437–1444.
2. Gupta, P., Prabhakar, B., and Boyd, S., Near-optimal routing lookups with bounded worst case performance, *Proceedings of Nineteenth Annual Joint Conference of the IEEE Computer and Communications Societies*, IEEE Infocom, 2000, IEEE Press, pp. 1184–1192.
3. Ergun, F., Mitra, S., Sahinalp, S., Sharp, J., and Sinha, A., A dynamic lookup scheme for bursty access patterns, *Proceedings of 20th Annual Joint Conference of the IEEE Computer and Communications Societies*, IEEE Infocom, 2000, IEEE Press, 2001, pp. 1441–1453.
4. Sahni, S. and Kim, K.S., Efficient dynamic lookup for bursty access patterns. *International Journal of Foundation Computer Science* 2006; 15(4):567–591.
5. Cheung, G., McCanne, S., Papadimitriou, C., Software synthesis of variable-length code decoder using a mixture of programmed logic and table lookups, *Proceedings of Data Compression Conference*, 1999 (DCC '99), 29–31 Mar 1999, pp. 121–130.
6. Nisson, S. and Karlsson, G., IP-address lookup using LC-trie. *IEEE Journal on Selected Areas on Communication* 1999; 17(6):1083–1092.
7. Srinivasan, V. and Varghese, G., Faster IP lookups using controlled prefix expansion, *Proceedings of ACM Sigmetrics 98*, June 1998, pp. 1–11.
8. Huston, G., Analyzing the internet's BGP routing table. *The Internet Protocol Journal* 2001; 4(1):2–15.

9. Huffman, D.A., A method for the construction of minimum redundancy codes, *Proc. Inst. Radio Engineers*, 1952; 40(10):1098–1101.

10. Milidiu, R.L. and Laber, E.S., Warm-up algorithm: a Lagrangean construction of length restricted Huffman codes. *Siam Journal on Computing* 1999; 30(5):1405–1426.

11. Larmore, L.L. and Przytycka, T.M., A fast algorithm for optimum height limited alphabetic binary trees. *SIAM Journal on Computing* 1994; 23(6):1283–1312.

12. Yeung, R.W., Alphabetic codes revisited. *IEEE Transactions on Information Theory* 1991; 37(3): 564–572.

13. Cover, T.M. and Thomas J.A., Elements of Information Theory. *Wiley Series in Telecommunications*, 1995.

14. Horibe, Y., an improved bound for weight-balanced tree. *Information and Control* 1977; 34:148–151.

15. Pugh, W., Skip lists: a probabilistic alternative to balanced trees. *Communication of the ACM* 1990; 33(6):668–676.

16. Lampson, B., Srinivasan, V., and Varghese, G., IP lookups using multiway and multicolumn search. *Proc. IEEE INFOCOM '98*, April 1998, pp. 1248–1256.

17. Sahni, S. and Kim, K., An O(log n) dynamic router-table design. *IEEE Transactions on Computers* 2004; 53(3):351–363.

Chapter 7

Caching Technologies

Route caching techniques can speed up packet forwarding. But their performance is influenced by traffic locality, the size of the cache, the replacement algorithm etc. Researchers have proposed many efficient schemes to improve a route caching techniques. The basic caching schemes have been discussed in Section 3.2. In this chapter, we describe the Suez lookup algorithm [1–3], prefix caching [4,5], and multi-zone caches [6–9] etc.

7.1 Suez Lookup Algorithm

Suez is a cluster-based parallel Internet protocol (IP) router built on a hardware platform consisting of a cluster of commodity PCs connected by a gigabit/second system area network. One of the major research focuses of *Suez* is to exploit CPU caching as a hardware assistant to speed up routing table lookup significantly. Chiueh et al. proposed three schemes: *host address cache* (HAC), *host address range cache* (HARC), *intelligent host address range cache* (IHARC).

7.1.1 Host Address Cache

Typically, multiple packets are transferred during a network connection's lifetime, and the destination IP address stream seen by a router exhibits temporal locality. IP address caching can speed up routing table lookup significantly. Chiueh et al. treated IP addresses as virtual memory addresses and proposed a lookup algorithm based on two data structures: a destination HAC and a destination network address routing table (NART) [1]. This section is from [1]. Portions are reprinted with permission (© 1999 IEEE).

The algorithm first looks up the HAC; if the lookup succeeds, the corresponding output port is returned. Otherwise, the algorithm further consults the NART to complete the lookup. Therefore, minimizing the HAC hit access time is crucial to the overall routing table lookup performance.

7.1.1.1 HAC Architecture

Rather than using a software data structure such as a hash table, HAC is designed to be resident in the Level-1 (L1) cache of the CPU at all times, and to be able to exploit the cache hardware's lookup capability directly. As a first cut, the 32-bit IP addresses can be considered as 32-bit virtual memory addresses and simply looked up in the L1 cache. If the lookup succeeds in the L1 cache, it is completed in one CPU cycle; otherwise an NART lookup is required. However, this approach exhibits several disadvantages. First, there is much less spatial locality in the destination host address stream compared to the memory reference stream in a typical program execution. As a result, the address space consumption or access pattern is going to be sparse, which may lead to a very large page table size and page fault rate. Second, unless special measures are in place, there is no guarantee that HAC is L1 cache-resident all the time. In particular, uncoordinated virtual-to-physical address mapping may result in unnecessary conflict cache misses, because of interference between HAC and other data structures used in the lookup algorithm. Third, the cache block size of the modern L1 cache is too large for the HAC. That is, due to the lack of spatial locality in the network packet reference stream, individual cache blocks tend to be under-utilized, leading to zero or one HAC entry in the cache block most of the time. Consequently, the overall caching efficiency is less than what it should be for a fixed cache size.

To address these problems, the HAC lookup algorithm takes a combined software or hardware approach. To reduce virtual address space consumption, we only use a certain portion of each IP address to form a virtual address, and leave the remaining bits of the IP address as tags to be compared by software. This approach makes it possible to restrict the number of virtual pages reserved for the HAC to a small number.

To ensure that the HAC is always L1 cache-resident, a portion of the L1 cache is *reserved*. For the purpose of exposition, let us assume that the page size is 4 kBytes, the L1 cache is 16 kBytes direct mapped with a block size of 32-bytes, and the first 4 kBytes of the cache are reserved for the HAC. This means that given a physical address, its block number is given by bits 5 through 13, whereas its physical page number is given by 12-bits and higher. Thus, the physical page number and the block number for the address overlap in two bits, and hence the physical addresses that get mapped to the first 4 kBytes of the cache would be those that lie in physical pages whose page numbers are a multiple of four.

Now, to prevent other data structures from polluting the portion of the L1 cache reserved for HAC, all the other pages whose physical page number is an integral multiple of four are marked *uncacheable*, so that they would never be brought into the L1/L2 cache at run time. This would ensure that each HAC access is always a cache hit and hence completes in one cycle. Thus, by reserving one page in the cache for the HAC and remapping IP addresses to virtual addresses lying within the HAC page, we ensure that all IP addresses are checked against the HAC and that these checks are guaranteed to involve only L1 cache accesses. There is thus a performance tradeoff between the HAC hit ratio and the utilization efficiency of the L2 cache. The larger the percentage of the L1 cache reserved for HAC, the lesser is the percentage of L2 cache usable for other purposes, for example, a NART search. However, for a given HAC size, say one page, utilization efficiency increases as L1 cache size increases.

Finally, to improve cache utilization efficiency, a software-driven set-associative caching mechanism is adopted. The fact that the HAC address space is mapped to a single physical page means that IP addresses with different "virtual page numbers" can now co-reside in a cache block. Therefore, each cache block may contain multiple HAC entries. The cache hardware first identifies the block of interest, and the software examines each HAC entry of that block in turn until a hit or exhaustion. To further exploit temporal locality, the software starts the examination of each cache block from the last HAC entry it accesses in the previous visit. Because associativity is implemented in software, HAC hit time is variable, whereas miss time increases with the degree of associativity. The performance tradeoff in choosing the degree of associativity then lies in the gain in higher HAC hit ratios versus the loss in longer miss handling time.

Assume the L1 data cache is 16 kBytes direct-mapped and physically addressed, with 32-byte blocks, and the page size is 4 kBytes. Each HAC entry is 4 bytes wide, containing a 23-bit tag, one unused bit, and an 8-bit output port identifier. Therefore each cache block can contain at most eight HAC entries. Finally, assume that one quarter of the L1 cache, that is, 128 out of 512 cache sets, is reserved for HAC.

Assume that the HAC page is allocated with a virtual page number V_{HAC}. Given a destination IP address, DA, the $DA_{5,11}$ is the substring of DA ranging from the 5th-bit to the 11th-bit, used as an offset into the HAC page to form a 32-bit virtual address $VA = V_{HAC} + DA_{5,11}$. The virtual address thus formed fetches the first word, CW, of the cache set specified by $DA_{5,11}$. Meanwhile, a 32-bit tag, DAT, is constructed from a concatenation of $DA_{12,31}$ and $DA_{2,4}$. Because, realistically, none of the routing tables contain network addresses with prefix lengths longer than 30, the last two bits of the destination host address are ignored while forming the tag.

The first word of each HAC cache set is a duplicated HAC entry that corresponds to the HAC entry last accessed in the previous visit to the cache set. If DAT matches $CW_{9,31}$, the lookup is completed and $CW_{0,7}$ is the output port identifier. Otherwise, the software search continues with the second word of the cache set and so on, until a hit or the end of the set. The way VA is formed from DA means that only one virtual page is allocated to hold the entire HAC. Figure 7.1 illustrates the flow of the HAC lookup process (we assume that L1 cache is direct mapped with 512 32-byte cache sets, among which 128 cache sets are reserved for the HAC. Each HAC entry is 4 bytes wide. The search through the HAC entries within a cache set is performed by software).

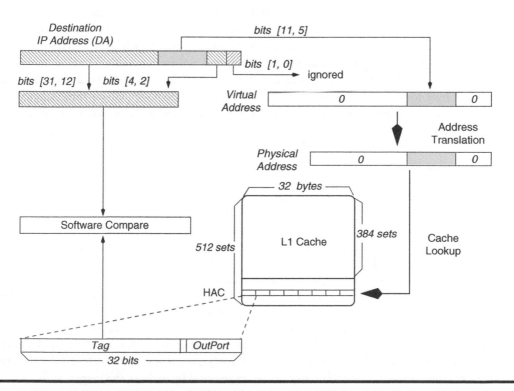

Figure 7.1 Data flow of HAC lookup. (From Tzi-Cker Chiueh and Prashant Pradhan, *IEEE INFOCOM*, 3, 1999. With permission.)

We assume that the CPU is three-way superscalar, and the load delay stall is one cycle. The algorithm gives a 5-cycle best-case latency for an IP address lookup, and takes three cycles for each additional HAC entry access in the cache set. So it takes $5 + 3 * 7 + 1 = 27$ cycles to detect a HAC miss. By pipelining the lookups of consecutive IP packets, that is, overlapping the 5-cycle work of each lookup, the best-case throughput can be improved to three cycles per lookup. Because the HAC is guaranteed to be in the L1 cache, the 3-cycle per lookup estimate is exact, because the HAC lookup itself is a hit.

7.1.1.2 Network Address Routing Table

If the HAC access results in a miss, a full-scale NART lookup is required. The guiding principle of NART design is to trade the table size for lookup performance. This principle is rooted in the observation that the L2 cache size of modern microprocessors is comparatively large for storing routing tables and the associated search data structures, and it is expected to continue increasing over time.

Note that the generality of Classless Inter-Domain Routing complicates full-scale routing lookups because the network address part of a destination IP address must be determined by the longest matching prefix rule. We present the construction and lookup procedure for the NART data structure below.

NART Construction.   Let us classify the network addresses in an IP routing table into three types according to their length: smaller than or equal to 16-bits (called Class AB), between 17 and 24-bits (called Class BC), and higher than 24-bits (called Class CD). To represent a given routing table, NART includes three levels of tables: one *Level-1 table*, one *Level-3 table* and a variable number of *Level-2 tables*. The Level-1 table has 64 K entries, each of which is 2 bytes wide, and contains either an 8-bit output port identifier or an offset pointer to a Level-2 table. These two cases are distinguished by the entry's most significant bit. Each Level-2 table has 256 entries, each of which is 1 byte wide, and contains either a special indicator ($0xff$) or an 8-bit output port identifier. The Level-3 table has a variable number of entries, each of which contains a 4-byte network mask field, a 4-byte network address, and an 8-bit output port identifier. The number depends on the given IP routing table and is typically small.

An IP routing table is converted to the NART representation as follows. Although insertion in any order can be handled, for the purpose of exposition, assume that Class AB addresses are processed first, then Class BC addresses, followed by Class CD addresses. Let NA denote a network address, L its length and OPT its output port identifier in the routing table. Each entry in the Level-1 table is initialized to hold the *default* output port identifier. Each Class AB network address takes 2^{16-L} entries of the Level-1 table starting from Level-1 table $[NA_{0,L-1} * 2^{16-L}]$, with each of these entries filled with OPT. A Level-1 table entry that corresponds to more than one class AB address holds the output port corresponding to the longest address.

For each Class BC address, if a Level-1 table $[NA_{L-16,L-1}]$ contains an 8-bit output port identifier, a Level-2 table is created and populated with this identifier and the Level-1 table $[NA_{L-16,L-1}]$ is changed to be an offset to the base of the newly created Level-2 table. If the Level-1 table $[NA_{L-16,L-1}]$ already contains an offset to a Level-2 table, no table is created. In both cases, 2^{24-L} entries of the Level-2 table, starting from Level-2 table $[NA_{0,L-17} * 2^{24-L}]$, are filled with OPT. Multiple Class BC addresses mapping to a Level-2 table entry are again resolved by picking the longest address.

For each Class CD address, if the Level-1 table $[NA_{L-16,L-1}]$ contains an 8-bit output port identifier, a Level-2 table is created and populated with this identifier, and the Level-1 table $[NA_{L-16,L-1}]$ is changed to be an offset to the base of the newly created Level-2 table. If the Level-1 table [$NA_{L-16,L-1}$] already contains an offset to a Level-2 table, this Level-2 table is used in the next step. If the Level-2 table[$NA_{L-24,L-17}$] is not 0x*ff*, the output port value therein is used to create a new Level-3 table entry with a network mask of 0x*ffffff00* and a network address of $NA_{L-24,L-1}$, and the Level-2 table [$NA_{L-24,L-17}$] is changed to 0x*ff*. Regardless of the contents of the Level-2 table [$NA_{L-24,L-17}$], a Level-3 table entry with a network mask the same as that of the, *NA* a network address of *NA* and an output port identifier of *OPT* is added to the global Level-3 table.

NART Lookup. Given a destination IP address, *DA*, the most significant 16-bits, $DA_{16,31}$, are used to index into the Level-1 table. If *DA* has a Class AB network address, one lookup into the Level-1 table is sufficient to complete the routing tables' search. If *DA* has a Class BC network address, Level-1 table [$DA_{16,31}$] contains an offset from which the base of the Level-2 table is computed. If Level-2 table [$DA_{8,15}$] is not 0x*ff*, the lookup is completed. Otherwise the network address of *DA* is a Class CD address, and a search through the Level-3 table is required.

We chose to use a single global Level-3 table, rather than multiple Level-3 tables like Level-2 tables, because it reduces the storage space requirement and because it allows the Level-2 table entries to specify up to 255 output port identifiers by not requiring them to have offsets to Level-3 tables. Because the Level-3 table entries are sorted according to the length of their network addresses, the linear search can stop at the first match. Figure 7.2 illustrates the data flow of the NART lookup process.

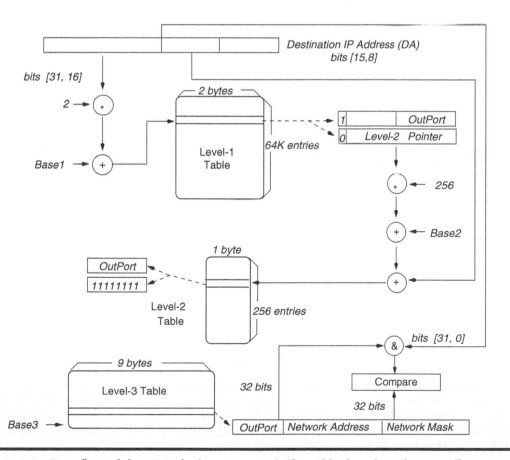

Figure 7.2   Data flow of the NART lookup. (From Tzi-Cker Chiueh and Prashant Pradhan, *IEEE INFOCOM,* **3, 1999. With permission.)**

We assume that the CPU is 3-way superscalar and the load delay stall is one cycle. For Class AB, BC, and CD addresses, the lookup takes 7, 12, and $20 + 4 * K$ cycles to complete respectively, where K is the number of Level-3 table entries that have to be examined before the first match. Unlike the HAC lookup, these cycle estimates may not be precise because there is no guarantee that the memory accesses in the NART lookup always result in cache hits.

7.1.1.3 Simulations

A HAC architecture is shown in Figure 7.3a. Chiueh et al. use a trace-driven simulation methodology to evaluate the performance of HAC. The packet trace is collected from the periphery link that connects the Brookhaven National Laboratory (BNL) to the Internet. From 9 AM on March 3, 1998 to 5 PM on March 6, 1998, the total number of packets in the trace was 184,400,259. The forwarding table is from the IPMA project (http://nic.merit.edu/ipma). Table 7.1 shows the miss ratio under varying cache size, cache block sizes and degrees of associativity. The performance difference between cache configurations that are identical, except for the block size, could be dramatic. For example, the miss ratios of a 4-way set associative 8K-entry cache with a 32-entry block size and one with a 1-entry block size is nearly an order of magnitude apart, 38.05% vs. 3.29%. As cache size increases, the performance impact of the block size decreases, (although still significant) because the space utilization efficiency is less of an issue with larger caches.

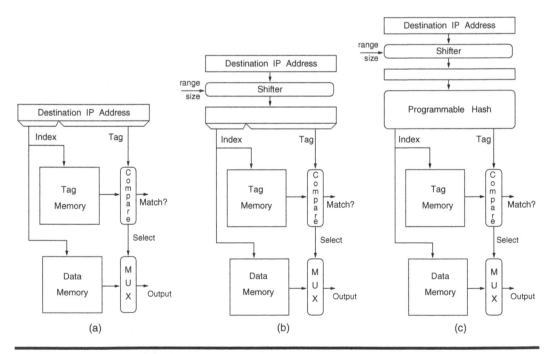

Figure 7.3 Network processor cache architectures. (a) Host address cache (HAC), (b) host address range cache (HARC), (c) intelligent host address range cache (IHARC). (From Kartik Gopalan and Tzi-Cker Chiueh, *Proceedings of the 2002 ACM/IEEE Conference on Supercomputing,* **Baltimore, Maryland, 2002. With permission.)**

Table 7.1 Miss Ratio for HAC under Varying Cache Size, Cache Block Sizes, and Degrees of Associativity

Cache Size	Block Size	Associativity	Miss Ratio (Percent)
4K	32	1	57.09
		2	53.25
		4	50.92
	8	1	36.51
		2	31.29
		4	29.00
	1	1	12.71
		2	8.42
		4	6.86
8K	32	1	43.70
		2	40.78
		4	38.05
	8	1	26.35
		2	21.72
		4	19.33
	1	1	7.57
		2	4.59
		4	3.29
32K	32	1	18.65
		2	16.52
		4	15.58
	8	1	9.59
		2	6.66
		4	5.49
	1	1	2.39
		2	1.07
		4	0.75

Note: Cache sizes are reported in numbers of entries rather than numbers of bytes.

Source: Tzi-Cker Chiueh and Prashant Pradhan, *Proc. IEEE Sixth International Symposium on High-Performance Computer Architecture*, Los Alamitos, California, 2000. With permission.

7.1.2 Host Address Range Cache

In a forwarding table, the prefixes represent a set of a contiguous ranges of the IP address space. Chiueh et al. proposed an algorithm to increase the effective coverage of a HAC by caching host address ranges, called HARC [2]. But there are two additional processing steps before HARC can be put to practical use. This subsection is from [2] (© 2000 IEEE).

First, with CIDR, it is possible that a prefix in a forwarding table is a parent of another prefix. If a prefix and its parent are cached, there is more than one result for a lookup. With the longest prefix match requirement, we need to mask the parents of any prefix as non-cacheable. This step ensures that any IP address is covered by at most one prefix in a cache.

Second, to improve the efficiency of the HARC, adjacent prefixes with the same next-hop should be merged into shorter prefixes as much as possible. Once this merging is done, the minimum number of all prefixes is calculated, called the *minimum_range_granularity* parameter of the HARC. Range size, which is defined as log(*minimum_range_granularity*), thus represents the number of least significant bits of an IP address that can be ignored during routing-table lookup, because destination addresses falling within a minimum address range size are guaranteed to have the same lookup result. Figure 7.3b shows the hardware architecture of the HARC, which has a logical shifter. The destination address of an incoming packet is logically right-shifted by *range size* before being fed to the cache.

Compared with HAC, HARC needs a logical shift to take additional processing steps. Table 7.2 shows the comparison between HARC and HAC, assuming that block size is one entry wide. HAC's miss ratio is between 1.68 and 2.10 times higher than that of HARC. The average lookup time in HARC is 58–78 percent faster than that in HAC, assuming that the hit access time in one cycle and the miss penalty is 120 cycles.

7.1.3 Intelligent HARC

A traditional CPU cache of size 2^K and block size one directly takes the least significant K-bits of a given address to index into the data and to tag arrays. It is possible to further increase every cache entry's coverage of the IP address space by choosing a more appropriate hash function for cache lookup. Chiueh et al. proposed an intelligent host address range cache (IHARC) [2]. This subsection is taken from [3]. Portions are reprinted with permission (© 2002 IEEE).

7.1.3.1 Index Bit Selection

Consider the routing table in Figure 7.4, where there are sixteen 4-bit host addresses with three distinct output interfaces, 1, 2, and 3. The merging algorithm used in calculating the *range size* of

Table 7.2 Miss Ratio Comparison

Cache Size	Associativity	Miss Ratio (Percent)		Ratio of Miss Ratio		Ratio of Average Lookup Time		
		HARC	IHARC	HAC/ HARC	HARC/ IHARC	HAC/ HARC	HARC/ IHARC	HARC/ IHARC
4K	1	7.54	2.30	1.69	3.28	1.62	2.67	4.31
	2	4.58	1.12	1.84	4.09	1.71	2.77	4.72
	4	3.64	0.57	1.88	6.39	1.72	3.18	5.46
8K	1	4.48	1.54	1.69	2.91	1.58	2.24	3.53
	2	2.20	0.48	2.09	4.58	1.78	2.30	4.11
	4	1.56	0.22	2.11	7.09	1.72	2.26	3.90

Source: Tzi-Cker Chiueh and Prashant Pradhan, *Proc. IEEE Sixth International Symposium on High-Performance Computer Architecture*, Los Alamitos, California, 2000. With permission.

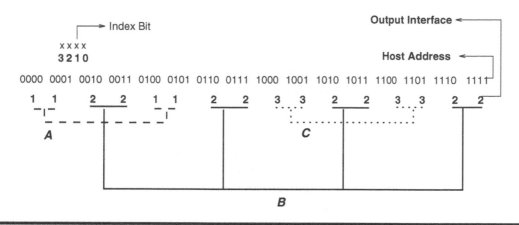

Figure 7.4   A routing table that illustrates the usefulness of carefully choosing index bits. (From Tzi-Cker Chiueh and Prashant Pradhan, *Proc. IEEE Sixth International Symposium on High-Performance Computer Architecture,* **Los Alamitos, California, 2000. With permission.)**

the HARC will stop after all adjacent address ranges with identical output interfaces are combined. In this case, the total number of address ranges is 8, because the minimum range granularity is 2. To further grow the address range that a cache entry can cover, one could choose the index bits carefully such that when the index bits are ignored, some of the identically labeled address ranges are now "adjacent" and can thus be combined. For example, if bit 1 (with bit 0 being least significant) is chosen as the index bit into the data/tag array, then the host addresses 0000, 0001, 0100, and 0101 can be merged into an address range because they have the same output port, 1, and when bit 1 it ignored, they form a continuous sequence, 000, 001, 010, and 011. Similarly, 1000, 1001, 1100, and 1101 also can be merged into an address range, as well as all the host addresses whose output port is 2. With this choice of the index bit, the total number of address ranges to be distinguished during cache lookup is reduced from 8 to 3. Note that index bit 1 induced a partitioning of the address space such that in each partition, some address ranges that were not adjacent in the original address space become adjacent.

Intuitively, IHARC provides more opportunities of merging identically labeled address ranges by decomposing the address space into partitions, based upon a set of index bits, and merging identically labeled ranges that are adjacent within a partition. Note that HARC insists on merging ranges that are adjacent in the *original* IP address space, and thus is a special case of IHARC.

IHARC selects a set of K index bits in the destination address that corresponds to 2^K cache sets. Each cache set corresponds to a partition of the IP address space. In a partition, some address ranges that were not originally adjacent in the IP address space will become adjacent. Any adjacent ranges that are identically labeled are then merged into larger ranges. Thus, we get a set of distinct address ranges for every partition (or cache set). Because distinct address ranges in a cache set need unique tags, the number of distinct address ranges in a cache set represents the degree of contention in the cache set. Thus, the index bits are selected in such a way that after the merging operation, the total number of address ranges and the difference between the numbers of address ranges across cache sets is minimized.

We first describe the index bit selection algorithm. Assume N and K are the number of bits in the input address and the index key, respectively. In general, any subset of K bits in the input addresses could be used as the index bits, except the least significant range size bits as determined

```
S = Null;
for ( 1≤ i ≤ K ) {
    score = ∞;
    candidate = 0;
    for ( range_size + 1≤ j ≤ N ) {
        if (!( j∉ S )) {
            currentscore = Score (S, j);
                if (currentscore < score) {
                score = Score (S, j);
                candidate = j;
            }}}
    S = S∪{candidate};

}
```

Figure 7.5 A greedy index bit selection algorithm used to pick bits in the input addresses for cache lookup. (From Tzi-Cker Chiueh and Prashant Pradhan, *Proc. IEEE Sixth International Symposium on High-Performance Computer Architecture*, Los Alamitos, California, 2000. With permission.)

by the basic merging step in constructing the HARC. We use a greedy algorithm to select the K index bits, as shown in Figure 7.5. S represents the set of index bits chosen by the algorithm so far. $Score(S, j)$ is a heuristic function that calculates the desirability of including the j-th bit, given that the bits in S have already been chosen to be included in the index set. For each partition of the IP address space included by the bits in $S \cup \{j\}$, the algorithm first merges adjacent identically labeled ranges in that partition. This step gives us, for every partition, the number of distinct address ranges that need to be uniquely tagged.

For a candidate bit set $S \cup \{j\}$, we define the i-th partition's metric $M_{i,(S,j)}$, $\forall i$, as the number of distinct address ranges in partition i. Then, the algorithm minimizes $Score(S, j)$, given by

$$Score(S, j) = \sum_i M_{i,(S,j)} + W \sum_i \left| \overline{M_{S,j}} - M_{i,(S,j)} \right|$$

where $\overline{M_{S, j}}$ is the mean of $M_{i,(S,j)}$ over all partitions i, and W is a parameter that determines the relative weight of the two terms in the minimization. Note that the second term of the weighted sum minimizes the standard deviation and is included to prevent the occurrence of the hot-spot partitions, and thus excessive conflict misses, in the IHARC caches sets. Figure 7.3c shows the hardware architecture of IHARC. The programmable hash function engine allows tailoring the choice of the index bit set to individual routing tables.

7.1.3.2 Comparisons between IHARC and HARC

While HAC and HARC use a fixed three-level-table NART structure (16, 8, and 8-bits) that is independent of the hardware cache configurations, the NART associated with IHARC depends on the hardware configuration. In particular, the number of entries in the Level-1 table is equal to 2^K, where K is the number of index bits in IHARC, and the set of K selected index bits is used to index into the Level-1 table. As a result, the cache miss penalty for IHARC may be different for different cache configurations. However, because the miss penalty is dominated by the number of memory accesses made in software NART lookup (three lookups in the worst case), which is comparable in all configurations, measurements from the prototype implementation show that the average miss penalty is almost the same for all IHARC cache configurations we experimented with, and moreover is close to that of HAC and HARC, that is, 120 cycles.

Another important difference is that in addition to the output interface, a leaf NART entry must contain the address range it corresponds to, so that after an NART lookup following a cache miss, the cache set can be populated with the appropriate address range as the cache tag. Given an N-bit address, the K index bits select a particular cache set, say C. The remaining $N - K$ bits of the address form a value, say T, which lies in one of the address ranges in this partition. Initially, when a cache set is not populated, T is looked up in software using the NART, and the address range A in which T falls becomes the tag of the cache set C. If the cache entry was already populated with an address range A, a range check is required to figure out whether the lookup is a hit (which corresponds to checking that T lies in the range A). However, a general range check is still too expensive to be incorporated into caching hardware. By guaranteeing that each address range size is a power of two and the starting address of each range is aligned with a multiple of its size during the merge step, one can perform the range check simply by a mask-and-compare operation. Therefore, each tag memory entry in the IHARC includes a tag field as well as a mask field, which specifies the bits in the address to be used in the "tag match." The price of simplifying cache lookup hardware is an increase in the number of resulting address ranges, as compared to the case when no such alignment requirement is imposed. If the range check results in a miss, the NART data structure is looked up and the cache is populated with the appropriate address range.

Compared to HARC, IHARC reduces the number of distinct address ranges that need to be distinguished by a careful choice of the index bits. Table 7.3 shows the number of distinct address ranges that result after applying the index bit set selection algorithm to the IPMA routing table, for different numbers of index bits. The number of the address ranges is less than that from HARC, 2^{27} (= 134,217,728).

Table 7.3   Number of Address Ranges and Index Bits

Number of Index Bits	Bits Chosen	No. of Ranges (Without Constraint)	No. of Ranges (With Constraint)
9	13 14 16 17 18 19 21 22 25	76,867	116,789
10	13 14 16 17 18 19 21 22 25 30	77,379	116,789
11	13 14 16 17 18 19 20 21 22 25 30	88,465	134,093
12	13 14 16 17 18 19 20 21 22 24 25 30	92,579	138,756
13	13 14 16 17 18 19 20 21 22 24 25 26 30	98,563	145,538

Note: The constraint is that each address range's size has to be a power of 2.

Source: Tzi-Cker Chiueh and Prashant Pradhan, *Proc. IEEE Sixth International Symposium on High-Performance Computer Architecture*, Los Alamitos, California, 2000. With permission.

Chiueh and Pradhan have implemented IHARC on a Pentium-II 233MHz machine. For the packet traces from BNL and the forwarding table from IPMA, the miss ratios for IHARC is shown in Table 7.2, assuming that the block size is one entry wide. HARC's miss ratios are 2.91 to 7.09 times larger than IHARC. The average lookup time of HARC is between 2.24 and 3.18 times slower than that of IHARC. Compared to HAC, IHARC reduces the average lookup time by up to a factor of 5.

7.1.3.3 Selective Cache Invalidation

IHARC has focused on designing a compact forwarding table, and the cache performance can be improved over HAC. Because the forwarding tables in real routers are dynamic, there is a problem of how to update the forwarding table in IHARC.

When a prefix is updated, the entire cache is invalidated. This scheme can effectively wipe out any performance gains from the caching technique. Gopalan and Chiueh proposed a selective-cache invalidation technique to minimize the performance overhead due to frequent route updates [3]. This approach invalidates only those cache sets that are actually affected by the route update. If a P-bit prefix that could cross 2^m partitions is updated, the selective invalidation approach invalidates the cache sets corresponding to these 2^m partitions, where m is the number of index bits that are not in the most significant P bits of the IP address. More concretely, if the number of index bits is K, those cache sets that have the same remaining K-m index bit values as in the updated P-bit address prefix have to be invalidated.

To examine the effect of route update on cache miss ratio using the two invalidation techniques, a forwarding table is used in simulations, which is a combination of the routing table from the Route Views project router and the Taiwan router, which has 1,53,000 entries. The route entry chosen as a target of each delete or update operation is picked randomly from the set of most recently accessed forwarding table entries. In the simulations, the average lookup latency is 2.95 ns (or 339 MPLS) with no update. At the frequency of one update every 10,000 packet lookups, the average lookup latency using whole-cache invalidation technique degrades to 8.5 ns. On the other hand, the average lookup latency using selective-cache invalidation degrades to only 2.98 ns.

In the selective-cache invalidation scheme, a greedy index bit selection algorithm is used to determine the set of index bits into cache. After a route update, the current set of index bits no longer remains the ideal choice. Because the index bit selection algorithm is expensive, the set of index bits cannot be recomputed at the frequency of route updates. But it can be done once every few hours. In the simulation, using old index bits as the number of updates from 0 to 25,000, the average lookup latency increases from 2.95 ns to 2.96 ns. If the routes are updated even at the rate of five updates per second, then 25,000 updates correspond to 1.4 hours, which is long enough to allow one to recompute the index bits without adversely affecting the hit ratio.

7.2 Prefix Caching Schemes

Caching techniques can improve lookup speed by storing the most recently used destination IP address in a fast local cache. The cache hit rate relies on the locality in traffic and the cache size. IP-address caching techniques worked well in the past; however, rapid growth of the Internet will soon reduce its effectiveness. Newman et al. found that the faster the router operates, the less likely that there will be two packets with an identical destination IP address in any given time interval [10]. This means that there will be little or no locality at all in a backbone router. That is, more IP addresses will be swapped in and out of the cache resulting in a higher miss rate.

Thus, it would seem useless to cache IP address in a high-speed backbone router. But it will still be very beneficial to cache routing prefixes. There are several reasons we believe this to be true [4].

1. The possible number of IP routing prefixes is much smaller than the number of possible individual IP addresses. A check of routing prefixes from the University of Oregon Route View project (http://www.antc.uoregon.edu/route-views/, 2000) showed 61,832 entries as compared to up to 2^{32} possible distinct IP addresses. The smaller range means that there is a greater probability that any one routing prefix will be accessed in short succession.
2. The number of routing prefixes seen at a router is even smaller. The reason is that CIDR is designed with geographical reach in mind. A border router for an AS is supposed to aggregate all routing prefixes within its domain. Thus a router would only see a small number of routing prefixes from each AS. Meanwhile, the IP address does not benefit from the aggregation. It actually increases quickly as more computers are connected to the Internet.
3. Web content servers are increasingly being hosted in a central data center. They offer a cheaper solution by amortizing the cost of network maintenance and administration over a large number of clients. With the World Wide Web dominating Internet traffic, it is not surprising that a large portion of the traffic will be moving toward data centers. These centers host thousands of web sites, but only share one single routing prefix. One would argue that routing prefixes for data centers should be permanently stored in the cache of any Internet router.
4. Routing prefixes were assigned to Internet Service Providers (ISPs) in blocks. For routing efficiency, ISPs further allocate prefixes to different geographical locations. Thus a routing prefix typically represents a local area. Because traffic from New York will certainly peak at a different hour than those from Los Angeles, routing prefixes for those two areas will not be frequently used at the same time.
5. Caching routing prefixes can benefit from routing table compaction techniques [11,12]. A compacted table is much smaller, resulting in higher probability that any cached routing prefix would be accessed again.

Given the above observations, it is natural that a routing prefix cache would produce better results than an IP address cache; a routing prefix cache can more effectively deal with explosive Internet growth, because the routing table size grows slower than the number of IP addresses in use as the number of Internet hosts grow. Some researchers proposed lookup schemes based on prefix caching, such as Liu's scheme [4], aligned prefix cache [13], reverse routing cache [5] etc.

7.2.1 Liu's Scheme

7.2.1.1 Prefix Cache

In the forwarding engine (FE) under which the cache system is employed, the network processor uses the cache to store the most recent lookup results. If an IP address matches an entry in the cache, the next-hop information could be quickly returned, otherwise we lookup the IP address in the complete routing table. There are two types of memories to store the prefixes: the cache that stores the most recent matched prefixes is called *Prefix Cache*; the main memory that stores the all prefixes in a forwarding table is called *Prefix Memory*. This subsection is from [4]. Portions are reprinted with permission (© 2001 IEEE).

A prefix cache is similar to TCAM; each entry contains the tag and the content. The tag has two fields: routing prefix and its corresponding mask. The content is the next-hop information,

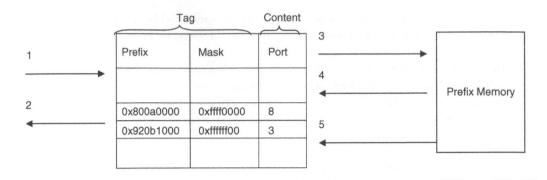

Figure 7.6 Prefix cache lookup flow. (From Huan Liu, *Proc. International Conference on Computer Communications and Networks (ICCCN)*, **Phoenix, Arizona, 2001. With permission.)**

showing which interface of the router the packet should be forwarded to. An example prefix cache is shown in Figure 7.6. A prefix cache works as follows.

1. The prefix cache receives a destination IP address which is first masked and then compared with all the prefixes in the prefix cache.
2. If the comparison succeeds (cache hit), the corresponding next hop is returned. The search stops. For example, IP address 128.10.15.3 matches the first entry, and port 8 is the lookup result.
3. If there is no matching prefix in the prefix cache, the IP address will be forwarded onto the prefix memory.
4. The prefix memory looks up the IP address in its full forwarding table. If there is a match prefix, its next hop is returned as the lookup result, and the whole entry including the prefix, mask, and next hop is inserted into the first available free entry in the prefix cache using a replacement policy. If there is no match in the prefix memory, the packet with the IP address cannot be forwarded.
5. A prefix cache also supports invalidation commands, because it is guaranteed that there is no more than one match in the prefix cache for any IP address. If there is more than one match in the prefix cache, the lookup becomes complex.
6. If a prefix is inserted into the prefix cache, the cache must be notified by the prefix memory as to which entry in it is no longer valid.
7. Avoid unnecessary invalidation, the prefix memory will keep track of which entry is cached and which invalidated.

The prefix cache is easy to implement using regular cache or TCAM. But prefix memory become more complex, because it stores the whole forwarding table, performs the longest-prefix match when there is no match in the prefix cache, and maintains the entries in the prefix cache.

7.2.1.2 *Prefix Memory*

One of the main functions of the prefix memory is to serve the IP address lookup that was missed in the prefix cache. The prefix memory stores the whole forwarding table and performs the longest-prefix match. There are many schemes for the longest-prefix match, and any can be used. The second main function of the prefix memory is to manage the prefix cache. To guarantee that there is no more than one matching prefix in the prefix cache for an IP address, a regular prefix in the forwarding table cannot be directly cached because an IP address could match multiple prefixes. Liu proposed three designs to transform the forwarding table so that a correct lookup result is always guaranteed [4].

Complete Prefix Tree Expansion (CPTE): Transform the routing table into a complete prefix tree such that all prefixes can be cached. The newly created prefix gets the next hop from its nearest ancestor. Such a transformation guarantees that there is no more than one lookup result for any IP address, but increases the forwarding table size.

No Prefix Expansion (NPE): Because the subroot prefix is cached, an IP address could match multiple prefixes in a cache. If the subroot prefixes in the forwarding table are always not cached, marked as noncacheable, any IP address can match no more than one prefix in a cache. All prefixes are not transformed, and the forwarding table size will remain the same.

Partial Prefix Tree Expansion (PPTE): As a tradeoff between CPTE and NPE, we can expand partial non-cacheable prefixes. In CPTE, most inflation is caused by prefixes being expanded into several levels. We can stop the expansion after the first level; that is, it is only expanded once. For example, a routing table with two prefixes, 0* and 0101*—the prefix 0* is a subroot (non-cacheable) prefix—is expanded once. The prefix 00* is the newly created prefix; it is cacheable. The subroot prefix 0* cannot be moved. In PPTE, the forwarding table inflation is reduced, and caching effectiveness is much better compared with the IP address cache.

7.2.1.3 Experiments

To evaluate the performance of a prefix cache, Liu used these forwarding tables: MacEast, MaeWest, Paix, Aads, and PacBell from IPMA (www.merit.edu/ipma), and Oix from the University of Oregon (www.antc.uoregon.edu/route-views). The expansion percentage is shown in Table 7.4. CPTE greatly increases the table size. In most cases, it grows about 50 percent. In contrast, PPTE results in about 20 percent growth. In general, the expansion percentage of CPTE and PPTE is related with the distribution of the subroot prefixes in the forwarding table.

Liu, implementing a cache simulator with a least recently used (LRU) replacement policy, used the MaeEast forwarding table and three separate traces from NLANR MOAT (http://moat.nlanr.net/ipma). Liu compared the CPTE, PPTE, and NPE with IP address caching (called "IP only"). The cache miss ratio is shown in Table 7.5. As expected, CPTE outperforms PPTE, and PPTE in turn outperforms NPE; all consistently outperform IP only.

Table 7.4   Number of Prefixes after CPTE and PPTE

Routing Table	MaeEast	MaeWest	Paix	Aads	PacBell	Oix
NPE	23,554	32,139	15,906	29,195	38,791	61,832
CPTE (percentage)	34,139	48,875	23,215	41,846	58,026	134,930
	145	152	146	143	150	218
PPTE (percentage)	26,613	36,662	18,000	32,541	44,173	74,363
	113	114	113	111	114	120

Source: Huan Liu, *Proc. International Conference on Computer Communications and Networks (ICCCN)*, Phoenix, Arizona, 2001. With permission.

Table 7.5   Cache Miss Ratio

Scheme	Cache Trace	512 (%)	1024 (%)	2048 (%)	4096 (%)	8192 (%)
CPTE	SDC964	1.9	0.9	0.38	0.18	0.1
	SDC958	1.8	0.77	0.36	0.27	0.26
	970222	4	2	0.88	0.57	0.57
PPTE	SDC964	2.3	1.2	0.5	0.29	0.17
	SDC958	2.3	1.0	0.5	0.34	0.32
	970222	4.4	2.3	1.1	0.66	0.66
NPE	SDC964	2.8	1.6	0.8	0.48	0.29
	SDC958	2.8	1.4	0.7	0.5	0.4
	970222	4.9	2.7	1.5	0.8	0.8
IP only	SDC964	4.9	3	2	1.2	0.8
	SDC958	5.5	2.5	1.6	0.95	0.75
	970222	9	5.8	3.7	2.4	1.6

Source: Huan Liu, *Proc. International Conference on Computer Communications and Networks (ICCCN)*, Phoenix, Arizona, 2001. With permission.

7.2.2   Reverse Routing Cache (RRC)

Because any prefix in the prefix cache must be disjoint, the subroot (parent) prefixes cannot be directly placed in the cache. Akhbarizadeh and Nourani proposed an approach—reverse routing cache (RRC) [5]. With RRC, mostly original prefixes are paced into the cache directly, and any specific organization or transformation of the routing table is not needed. This subsection is taken from [5]. Portions are reprinted with permission (© 2004 IEEE).

7.2.2.1   RRC Structure

Because the RRC stores prefixes, it is best to implement it as a Ternary CAM (TCAM). Conventional TCAM sorts the prefixes on the basis of their lengths and uses a priority encoder to choose the longest-matching prefix among multiple matches. Because parent prefixes are not allowed in cache, each RRC lookup generates no more than one match per search. This means that prefixes in RRC need not be sorted, the priority encoder can be eliminated, and a new prefix can be added to any available location in the RRC.

7.2.2.2   Handling Parent Prefixes

Parent prefixes cannot be directly placed in RRC. To handle parent prefixes, Akhbarizadeh et al. proposed two different approaches: RRC with parent restriction (RRC-PR) and RRC with minimal expansion (RRC-ME) [5].

RRC-PR.   Akhbarizadeh et al. studied a few samples from AADS, MaeWest, and PacBell, and found that only less than 8 percent of prefixes in the forwarding table are parent prefixes.

All the parent prefixes are non-cacheable; only the disjoint prefixes are allowed to be mirrored in the RRC-PR. So, the RRC-PR is the same as the NPE in Liu's scheme.

RRC-ME. For many Internet service providers, it might not be acceptable to let a traffic flow that corresponds to a parent prefix always go through the slower path because the parent prefix cannot be mirrored in the RRC. Also despite being scarce, the parent prefixes might be more popular than others, because they cover a relatively wider portion of the network. For these reasons, it is necessary to handle the parent prefixes in RRC more intelligently. A parent prefix is partially represented with the shortest expanded disjoint child that matches the given search key. Such a prefix is called the *Minimal Expansion Prefix* (MEP). For example, there is a forwarding table with a parent prefix 10* and a child prefix 1010* (5-bit address space). The given IP address $D = 10110$ is looked up in the main forwarding table. The longest-matching prefix is the parent prefix 10*. The minimal expansion technique generates a new prefix $P' = 1011*$, which is the disjoint prefix with the prefix in the RRC. It can be added to the cache. Over time, minimal expansion may occur to the same parent prefix multiple times. For a parent prefix with n-bit length, if the length of its child prefix is m-bits, then the RRC-ME will generate at most 2^{m-n} prefixes. Compared with Liu's PPTE scheme, the RRC-ME advocates gradual, adaptive, and dynamic expansion of only parent prefixes.

7.2.2.3 Updating RRC

A semi-LRU method that conceptually uses a descending age counter for each RRC entry was implemented. The counter range is equal to the size of the cache and is set to maximum at the placement time and at any subsequent matches. If the RRC is full and every virtual counter is at a value bigger than zero, then the replacement has failed.

Insertion. When a new prefix is being added to the routing table, it might convert one of the existing disjoint prefixes to a parent prefix. For example, if 129.110/16 exists in the routing table as a disjoint prefix, then adding 129.110.128/17 will make 129.110/16 a parent prefix. Now, if the prefix 129.110/16 has been previously mirrored in the RRC, then it must be wiped off because only one parent prefix is allowed to reside in RRC.

To confidently prove that a parent of the new prefix resides in the RRC, both zero padded and one padded IP numbers of the prefix should be looked up in it. For prefix 129.110.128/17, the zero padded IP number and the one padded IP number would be 129.110.128.0 and 129.110.255.255, respectively. Three distinct cases may occur.

1. None of the IP numbers match. The new prefix can be added to the routing table without modifying the RRC.
2. Only one of the IP numbers match or they match different RRC entries. This happens only if the new prefix is itself a parent. No further action is needed. The new prefix can be added to the routing table without modifying the RRC.
3. Both IP numbers match the same RRC entry. In this case the RRC entry is definitely the mirror of a parent prefix. It must be removed before the new prefix can be added to the routing table. Using the above example, if 129.110/16 was mirrored in the RRC, then both 129.110.128.0 and 129.110.255.255 would match it. Then, 129.110/16 would have to be removed from the RRC. If 129.110/16 stays popular after this point then future address lookups that match this prefix in the routing table will initiate minimal expansion that brings its right-grafted child to the RRC. It is easy to see that each prefix cannot have more than one parent in the RRC.

Three memory accesses are needed for RRC coherence in insertion procedures: two accesses to find the undesirable prefix and one access to remove it. This makes the complexity of this procedure $O(1)$.

Deletion. When a disjoint prefix is being removed from the routing table, its mirror in the RRC (if any) should also be eliminated. A single search with the zero padded IP number of the prefix is enough to reveal if it is mirrored in the RRC. Because, if such a search matches a prefix, then that prefix is either the mirror of this same prefix or one of its parents. We know that no parent prefix can be directly mirrored in the RRC. Then the match has to be the mirror of the given prefix and nothing else.

However, if the prefix being removed from the routing table is itself a parent, then one or more of its minimal expansions might exist in the RRC-ME and should be found and eliminated (the problem does not exist for RRC-PR). Finding the minimal expansions of a parent prefix P can be done by finding any children of P in the RRC-ME that have the same data field (egress number, etc.). This requires an associative search in the RRC-ME for both the prefix and its data field. If the hardware implementation of such a combined search is not affordable, then the method can be further simplified by removing all the children of P from the RRC-ME. The price of this simplification is the extra number of entries to delete from the RRC-ME upon deleting parent prefixes from the routing table.

To find all the children of a given prefix P_1 in the RRC-ME, one can take advantage of the capability of TCAM structures to search for a prefix (masked IP address). Specifically, many TCAM circuits can accept an input mask in addition to the searchable key. The input mask is applied to the key and all the masked bits of the key will be considered *do not care* during the search, the same way that the masked bits of the TCAM words are considered *do not care*. Consequently, any match with such a prefix key will be either its parent or its exact equal or its children. In the case of RRC-ME, the first two cases are not possible for a parent prefix. Therefore, any RRC-ME matches with prefix P would be its children.

If the TCAM module that implements RRC-ME can support a combined *search and destroy* operation in which all the search results can be marked *invalid* in the same cycle, then the RRC-ME coherence procedure for table deletion can be done in one cycle. Otherwise, each search result has to be removed from the RRC-ME in a separate cycle. For a disjoint prefix, it will still take only one cycle, but a parent prefix needs as many as the number of child prefixes mirrored in the RRC-ME. Because a parent prefix can have $N - 1$ children, where N is the number of entries in the routing table, the procedure requires $O(N_{RRC})$ cycles in the worst case. N_{RRC} is the size of the RRC-ME, which is significantly smaller than N. Such worst case situations are extremely rare as the following analysis and experiments show.

In the absolute majority of cases, deletion of a parent prefix causes removal of less than a handful of entries from the RRC-ME. Moreover, because parent prefixes constitute a tiny minority in the routing table, the average number of RRC-ME removal operations needed as a consequence of a table deletion is always very close to one. To assess this empirically, Akhbarizadeh and Nourani counted the number of RRC-ME prefixes with a parent in the routing table at the end of a simulation run. The maximum number of child prefixes was 17, which means that the maximum number of cycles that it takes to synchronize the RRC-ME's contents when a prefix is deleted from the routing table will be 17. The average number of child prefixes in the RRC-ME per parent prefix in the same simulation was 2.8 (including minimal expansion prefixes and others), and 31 percent of the RRC-ME was occupied by these child prefixes. That means the average number of cycles per coherence task when all routing table entries with a mirror in RRC-ME are deleted sequentially is $1/[(0.31/2.8) + (1-0.31)] = 1.25$ [5].

7.2.2.4   Performance Evaluation

Akhbarizadeh et al. developed models RRC-PR, RRC-ME, and IP cache. The packet trace was collected from the periphery link that connects the Brookhaven National Laboratory to the Internet from 9 AM on March 3, 1998 to 5 PM on March 6, 1998. The total number of packets in the trace is 180,400,259 same as [2]. The routing tables were from the IPMA project website (http://www.merit.edu, 2002) and contained 26,786 entries.

Figure 7.7 demonstrates the outcome of the simulations. When the cache size exceeds 2048, the hit ratio of RRC-ME and RRC-PR reaches 0.998 and 0.956, respectively. For an IP cache, the hit ratio is only 0.917 when its size is 2048, and it reaches the hit ratio of 0.999 only when its size reaches 26,000, which is almost the size as of the routing table itself. To get an average hit ratio of 0.928, the required RRC-ME size is 128 while the size of the IP cache has to be 3500. This means that RRC-ME needs 27.3 times less memory. To get the highlighted hit ratio of 0.962, the required RRC-ME size is 256, but the IP cache size is almost 8600. This translates to 33.6 timesless memory.

Table 7.6 shows the miss ratio ($m = 1 - h$), the relative parent miss ratio, and the average search times for the same two RRC-ME size. Average search times are reported for two different systems, one with the minimum K_p value of 1, the other with the maximum K_p value of 2. The values of parent miss ratio (α_p) are obtained through simulation and show the fraction of the missed lookups that match a parent prefix in the routing table. The table shows the effects of cache size as well as K_p for two RRC-ME size designs providing typical hit ratio. Doubling the RRC-ME size reduces the average search time by 42 percent and 41 percent for $K_p = 1$ and $K_p = 2$, respectively. The average search times are reported for a miss penalty of 120 time units and a cache access time of 1.

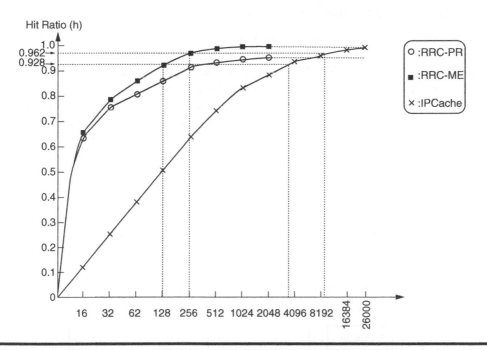

Figure 7.7   Hit ratio for various sizes of RRC-ME, PRC-PR, and IP cache. (From Akhbarizadeh, M.J. and Nourani M., *12th Annual IEEE Symposium on High Performance Interconnects*, 2004. With permission.)

Table 7.6 Average Search Time for Two Different RRC-ME Sizes

RCC-ME Size	Miss Ratio (m)	Parent Miss Ratio (α_p)	T_{srch}-ME [Time unit]	
			$K_p = 1$	$K_p = 2$
128	0.072	0.167	9.56	11.00
256	0.038	0.201	5.52	6.44

Source: Akhbarizadeh, M.J. and Nourani, M., *12th Annual IEEE Symposium on High Performance Interconnects*, 2004. With permission.

7.3 Multi-Zone Caches

In 1993, a three-level system of hierarchical caching was used in Cisco routers: AGS+ router and 7000 router. The main idea was to distribute small cache tables into every interface so to deal with incoming route requests locally. In 1998, Besson et al. [6] proposed a three-level system with a major table, a central cache table, and a local cache table, shown in Figure 7.8. The experiments showed that the system better performances when subject to unbalanced and bursty traffic [6]. After 2000, some researchers paid more attention to multi-cache techniques such as two-zone full address cache [7], multi-zone pipelined cache [8], and design algorithm of multi-zone cache based on spatial locality [9] etc.

7.3.1 Two-Zone Full Address Cache

Chvets and MacGregor analyzed temporal and spatial locality in IP traffic, and observed the almost 95 percent of the references come from less than half of the IP address space. On the basis of spatial locality, Chvets and MacGregor divided the cache into two zones where IP addresses are stored in each zone according to their lookup result prefix lengths [7]. For example, the prefixes in zone 1 are 16-bits long. If packets for 192.168.3.58 should be routed via 192.168/14, this address will be cached in zone 1 because the mask for its route is less than 16-bits long. If the packet the length of the prefix is less than or equal to the maximum prefix length (m) of an entry in zone 1 the IP address will be stored in zone 1, otherwise it will be stored in zone 2. The configuration of a two-zone cache is known as Scheme [m: 32].

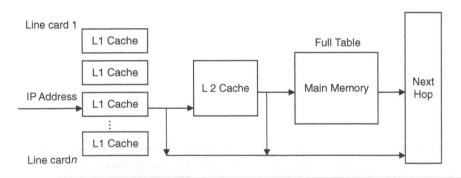

Figure 7.8 Three-level routing tables.

A simple experiment was performed to evaluate the improvement of cache hit rates using multi-zones. The experiment data were collected at the University of Auckland and at the San Diego Supercomputing Center. Three different configurations of two-zone caches are Scheme [20: 32], Scheme [18: 32], and Scheme [16: 32]. The results are summarized in Figure 7.9, in which the miss ratio is plotted as a function of the number of entries in zone 1. The horizontal line shows the miss ratio for the single zone cache, which was 0.2568. The multi-zone cache schemes significantly outperform the regular single zone cache. The optimal number of entries in zone 1 for the three multi-zone schemes range from about 350 to 370, and in this region the miss ratio for the multi-zone schemes is half that of the single zone cache. This is a significant improvement in miss ratio, so that the simple strategy of just increasing the capacity of the cache will not yield an appreciable decrease in miss ratio. An architectural change—such as the use of a multi-zone cache—is required to obtain performance improvements in this region.

7.3.2   Multi-Zone Pipelined Cache

Based on the two-zone full address cache, Kasnavi et al. proposed a multi-zone pipelined cache (MPC), where one zone stores IP prefixes and the second zone stores full addresses [8]. The MPC adopts a nonblocking buffer to reduce the effective cache miss penalty. A pipelined design implements a novel search and reduces power consumption. This subsection is from [8], and has been included with permission from Springer Science and Business Media.

7.3.2.1   Architecture of MPC

Figure 7.10 presents a structural description of the MPC, which has two zones: *Prefix Zone* and *Full Address Zone*. The prefix zone stores short IP prefixes, which are 16 or fewer bits long.

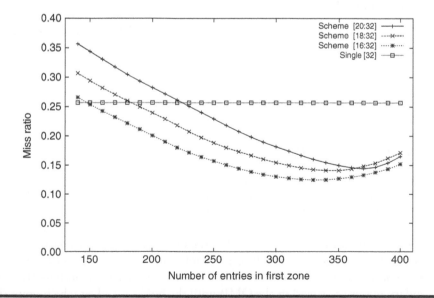

Figure 7.9   Miss ratios for varying cache configurations. (From Chvets, I.L. and MacGregor, M., *Proc. IEEE High Performance Switching and Routing,* **Kobe, Japan, 2002. With permission.)**

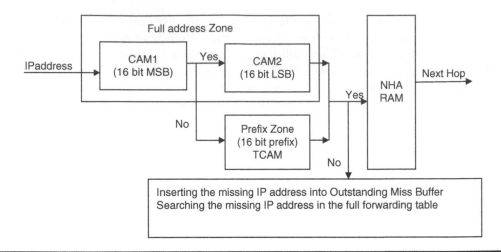

Figure 7.10 Architecture of MPC. (From Springer Science and Business Media. With permission.)

A TCAM is used to implement the prefix zone of the cache because a TCAM can store *do not care* states in addition to 0s and 1s. The full address zone stores full 32-bit IPv4 destination addresses. It is further divided vertically into CAM1 and CAM2, which have each entry split in half. CAM1 stores the 16 most significant bits of the IP addresses, while CAM2 stores the 16 least significant bits. A CAM is used to implement the full address zone because it is the fully associative binary memory capable of matching a specific pattern of data against all its entries in parallel. Next-hop information is stored in the next hop array (NHA).

7.3.2.2 Search in MPC

If a packet arrives, the 16 most significant bits of its destination IP address are applied to CAM1 (Stage 1). If there any matches in CAM1, the 16 least significant bits are searched for in CAM2. Otherwise, the 16 most significant bits are searched for in prefix zone (Stage 2). If there is a match in either CAM2 or the prefix zone, the NHA returns the next hop. Otherwise, the IP address is stored in the *outstanding miss buffer* (OMB) and the full forwarding table is searched (Stage 3). There are three stages. The pipeline flow diagram for a cache search is shown in Figure 7.11a.

7.3.2.3 Outstanding Miss Buffer

The MPC uses the OMB to store recent misses until the processor returns their lookup results. Without the OMB, the MPC would need to stall while each cache miss is serviced. Blocking hinders cache throughput because further cache searches cannot proceed until the lookup is performed and the cache updated, even if a pending request would hit the cache. When a miss occurs in the nonblocking MPC, the address is stored in the OMB and the cache continues performing lookups. If subsequent IP addresses hit the cache while a miss is being serviced, a *hit under miss* occurs. A *miss under miss* (secondary miss) occurs when a subsequent IP address also misses the cache. Secondary misses are stored in the OMB until the buffer is full, at which point the MPC blocks and the processor stalls until misses are serviced and removed from the OMB. An example

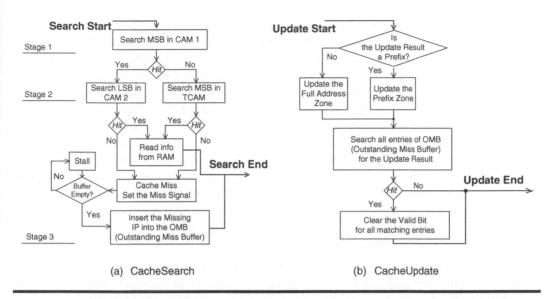

Figure 7.11 Flow diagram of the cache performance. (From Soraya Kasnavi, Paul Berube, Vincent C. Gaudet, and José Nelson Amaral, *A Multi-Zone Pipelined Cache for IP Routing*, Springer Berlin/Heidelberg, 2005. With permission.)

of MPC functionality with a two-entry OMB is given in Figure 7.12. In this example, IP2, IP4, and IP6 are cache misses. The MPC is able to search for IP3 and IP5 and forward their corresponding information to the processor while the main table lookup for IP2 is in progress. The MPC stalls after searching for IP6 because the OMB is full. No new IP can be located until IP2 is serviced and removed from the OMB to make room for IP6.

When the lookup result of a pending IP address in the OMB comes back from the main memory lookup, the MPC updates either the prefix zone or the full address zone with the corresponding information according to the MPC replacement policy. An expansion of the lookup table ensures that the result is either a short prefix that updates the prefix zone, or a full address with 32-bits that updates the full address zone.

	1	2	3	4	5	6	Time (clock cycles)	
IP$_1$	C1	C2/T	NHA					s : Stall
IP$_2$		C1	C2/T	B	*Latency Clock Cycles*	B	UC1/s UC2/T UNHA	B : Buffer
IP$_3$			C1	C2/T	NHA			C1 : CAM1
IP$_4$				C1	C2/T	B · · ·		IP : IP Address
IP$_5$					C1	C2/T	NHA	NHA : Next Hop Array
IP$_6$					C1	C2/T s s	B · · ·	C2 / T : CAM 2 or TCAM
IP$_7$						s	s C1 C2/T	UC1 / s : Update CAM 1 or Stall

UC2 / s : Update CAM 2 or Stall
UNHA : Update Next Hop Array

Figure 7.12 Pipeline diagram of the cache. (From Soraya Kasnavi, Paul Berube, Vincent C. Gaudet, and José Nelson Amaral, *A Multi-Zone Pipelined Cache for IP Routing*, Springer Berlin/Heidelberg, 2005. With permission.)

The MPC requires a pipeline stall to update the data stage by stage. After an update is complete, the missing IP address is removed from the OMB. However, other pending IP addresses in the OMB might be identical to the recently updated IP address. Additionally, an update result might be a prefix covering multiple pending IP addresses in the OMB. To ensure that the same lookup result is not written into the cache multiple times, we implement the OMB as a 33-bit CAM. Each OMB entry stores a 32-bit address and a *valid bit*. Only valid entries require a software lookup and cache update. After each update, an associative search of the OMB identifies all matching entries. If the lookup result is a prefix, its *don't care* bits are externally masked to ensure that they match with the data in the OMB. The valid bits of all matching entries are cleared. A second search for those matching OMB entries will now hit the cache and provide the processor with the next-hop information. The flow diagram for an MPC update is shown in Figure 7.11b.

7.3.2.4 Lookup Table Transformation

The MPC fully expands the routing table for short prefixes, pushing them down the tree until they become either leaf nodes or 17-bits long. All short prefixes are thus cacheable in the prefix zone of the MPC. Any destination address matching a long prefix in the lookup table is stored in full in the full address zone of the MPC. This table transformation has the following advantages.

1. Routing table expansion is limited to those prefixes that provide the greatest coverage of the IP address space.
2. At most one prefix stored in the prefix zone can match a destination address, because all short prefixes are cacheable.
3. Because the short prefixes in the prefix zone are cacheable, no address can hit both the prefix zone and the full address zone. Therefore, any address that could hit the prefix zone is guaranteed to miss the full address zone.
4. A length check is sufficient to determine if a prefix is cacheable.

Note that points 2 and 3 guarantee that there cannot be multiple hits for a search in the cache, which simplifies cache design. Also, point 3 enables our power-saving pipelined search. Table 7.7 shows the total number of prefixes after the expansion for the prefix cache in [4] and the MPC. There are 10,219 entries in the forwarding table from ISP1, and after the expansion for prefix

Table 7.7 The Number of Prefixes after Table Expansion

ISP	ISP1 Entries (Percent Larger)	ISP2 Entries (Percent Larger)
Original table	10,219	6355
Prefix cache	30,620 (199)	7313 (15)
MPC	17,485 (71)	6469 (2)

Source: Soraya Kasnavi, Paul Berube, Vincent C. Gaudet, and José Nelson Amaral, *A Multi-Zone Pipelined Cache for IP Routing*, Springer Berlin/Heidelberg, 2005. With permission.

cache, the number of entries is 30,620. After the expansion for MPC, the number is 17,485. Because the prefix cache requires full expansion, and the MPC only a partial expansion, the MPC has half as many entries as compared to a prefix cache.

7.3.2.5 Performance Evaluation

To evaluate the performance improvements achieved by the MPC, Kasnavi et al. compared its miss rates with a full address IP cache and a full prefix cache. In the experiment, the IP cache is simulated as a two-zone two-stage pipelined cache with equal sized zones; the prefix cache is a 32-bit Ternary-CAM that stores prefixes.

The IP traces data for simulations are collected from the distribution routers of Internet service providers: ISP1 and ISP2. The number of packets from ISP1 and ISP2 is 99117 and 98142, respectively. The average miss rates are given in Table 7.8.

Clearly, prefix cache outperforms MPC. Because the prefix cache requires full expansion, the number of entries in the prefix cache is almost twice that in the MPC for the same forwarding table. A prefix cache must be implemented in a 32-bit TCAM. The area required for a TCAM cell is almost twice the area of a CAM cell; the MPC and IP cache use half the area of a prefix cache with the same number of entries. Therefore, an MPC requires less memory than a prefix cache.

The MPC uses a small buffer, the OMB, to hide the miss penalty. To evaluate the performance of the OMB, Kasnavi et al. use the average number of clock cycles taken to provide the next hop as a metric, called clock per output (CPO). Figure 7.13 shows CPO versus latency. Without the OMB, CPO increases latency linearly. CPO is less with OMB than without OMB.

7.3.3 Design Method of Multi-Zone Cache

The design variables for a conventional single-zone IP address cache are degree of associativity, number of cache lines, and the size of each line. IP address caches are small enough that the cache can be fully associative, so that removes one design variable. The number of cache lines is the

Table 7.8 Miss Ratio (Percent) versus Cache Size

Cache Size		IP Cache	MPC	Prefix Cache
512	ISP1	22.7	15.5	7.4
	ISP2	3.6	3.0	0.5
1024	ISP1	15.4	7.9	2.5
	ISP2	2.2	2.0	0.5
2048	ISP1	10.5	3.7	1.4
	ISP2	1.9	1.6	0.5

Source: Soraya Kasnavi, Paul Berube, Vincent C. Gaudet, and José Nelson Amaral, *A Multi-Zone Pipelined Cache for IP Routing,* Springer Berlin/Heidelberg, 2005. With permission.

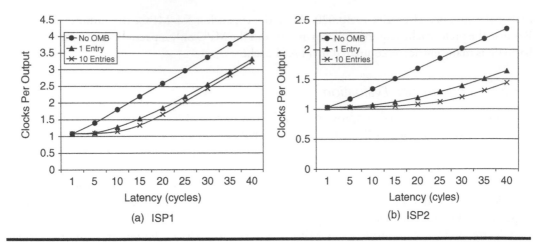

Figure 7.13 CPO versus latency for 1K-entry (equally sized zones) MPC. (From Soraya Kasnavi, Paul Berube, Vincent C. Gaudet, and José Nelson Amaral, *A Multi-Zone Pipelined Cache for IP Routing,* Springer Berlin/Heidelberg, 2005. With permission.)

prime design variable because line size is fixed by the format of an IP address plus the format of the internal specifier for the selected output port. As a first approximation, if we allow a router to have up to 1024 interfaces, then cache lines of 42-bits will be sufficient—32-bits for an IPv4 address and 10-bits to denote the outbound interface. Thus, for a conventional single-zone IP-address cache, the design problem devolves into choosing the number of cache lines.

Selecting a multi-zone design means that choices must be made for the number of zones, the prefix lengths of the addresses kept in each zone, and the number of cache lines in each zone. As with the single-zone cache, full associativity will be used to maximize hit rates, and the address and interface identifier formats will set the size of each line. The design space can be simplified by assuming that contiguous prefix lengths will be kept in each zone, and that all possible prefix lengths will be covered by the cache. If we assume that the designer has a fixed amount of fast memory available to be used by the cache, then the problem can also be simplified by assuming that we have a known, fixed cache size. In [9], MacGregor presents analytical models that can be used to make decisions about these parameters. This subsection is from [9]. Portions have been reprinted with permission (© 2003 IEEE).

7.3.3.1 Design Model

The central feature of the design problem is that of predicting the hit rate of a cache zone of known size, containing addresses with a given range of prefix lengths. Once the hit rates of the individual zones are known, the global hit rate can be calculated.

The hit rate of an individual zone must be evaluated with reference to the traffic that will be seen by the cache because the locality characteristics of IP destination address streams change tremendously depending on the location at which the traffic is observed. We concentrate on a general method for a multi-zone cache design that is independent of the statistical character of any individual trace.

One method for characterizing reference stream locality is the footprint function proposed in [14]. If n is the sequence number of a reference in a stream of references (e.g., the nth packet in a stream of IP packets) then the footprint function $u(n)$ gives the number of unique references in the stream up to that point. For example, if $u(120) = 14$ then at the time of the arrival of packet 120, we have 14 unique destination addresses.

Thiebaut et al. [14] propose that the footprint function be fitted by regression using

$$u = An^{1/\theta}. \tag{7.1}$$

They also note that the miss rate for a cache of size C is the value of the first derivative of the footprint function at the point that $u(n) = C$. Clearly, a cache of size C will contain at most C addresses, and each address will be unique because the cache will maintain only a single entry per address. The rate at which new unique addresses arrive in the reference stream, giving rise to cache misses at that same rate, is the first derivative of $u(n)$ at a cache size of C. The inverse of Equation 7.1 is:

$$n = (u/A)^{\theta}. \tag{7.1a}$$

The derivative of Equation 7.1 is:

$$\frac{du}{dn} = \left[\frac{A}{\theta}\right] n^{\left(\frac{1}{\theta}-1\right)} = \left[\frac{A}{\theta}\right] n^{\frac{1-\theta}{\theta}} \tag{7.1b}$$

Substituting Equation 7.1a into Equation 7.1b, we have:

$$\frac{du}{dn} = \frac{A}{\theta} \left[\left(\frac{u}{A}\right)^{\theta}\right]^{\frac{1-\theta}{\theta}} = \frac{A}{\theta} \left(\frac{u}{A}\right)^{(1-\theta)} = \left[\frac{A^{\theta}}{\theta}\right] u^{(1-\theta)}.$$

If we denote the value of the first derivative—the miss rate as $m(C)$, and evaluate it at the point $u = C$, where C denotes the cache capacity, then we have:

$$m(C) = \frac{A^{\theta}}{\theta} C^{(1-\theta)}. \tag{7.2}$$

A serious shortcoming of the original footprint function now becomes apparent: in reality, the cache miss rate must be identically 1 when $C = 0$ because if the cache capacity is zero then all references must be misses! However Equation 7.2 cannot supply this behavior because it is obliged to predict a miss rate of zero for a cache size of zero, which is nonsensical. This problem is easily corrected: if we modify the function slightly, substituting $(C + 1)$ for C, then the function can produce a miss rate of 1 when $C = 0$ as long as $A^{\theta}/\theta = 1$. With this change, Equation 7.2 becomes:

$$m(C) = (C + 1)^{(1-\theta)} \tag{7.3}$$

Integrating and substituting we find that we should use:

$$u(n) = (\theta n)^{(1/\theta)} - 1 \tag{7.4}$$

to fit the footprint function. With this single-parameter model we can be sure that the miss rate will behave sensibly as the cache size decreases.

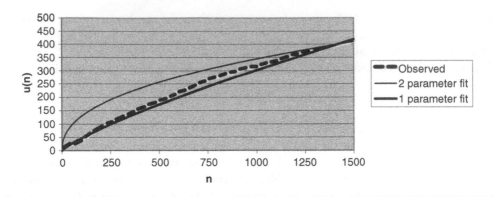

Figure 7.14 Models for the footprint function. (From MacGregor M.H., *Proc. High Perf. Switching and Routing (HPSR 2003)*, Torino, Italy, 2003. With permission.)

To evaluate the design model, MacGregor used an IP address trace obtained from a campus gateway at the University of Alberta. There are one million packets in the trace, and MacGregor used the footprint function from the first 10,000 packets to fit the parameters for the models of Equations 7.1 and 7.4. The data are shown in Figure 7.14. At cache sizes below about 500 lines, the two-parameter model fits the observations very poorly, and in fact has a slope much greater than 1.0 (implying miss rates in excess of 100 percent!). In contrast, the one-parameter model fits the observations very well in this range, and exhibits the correct slope as it approaches $C = 0$.

7.3.3.2 Two-Zone Design

Almost 25 percent of the references in the U of A trace are to destinations whose prefixes are 23-bits in length. This suggests using of a cache with two zones, where addresses whose prefixes are 22-bits or shorter are kept in zone 1 and addresses whose prefixes are 23-bits or longer are kept in zone 2. We split the 10,000 packet subtrace by prefix length, and fit Equation 7.4 to these two portions. For the portion with addresses whose prefixes are 22-bits and less we find $\theta = 1.3538$ and for the portion with addresses whose prefixes are 23-bits and longer, we find $\theta = 1.6915$. The design problem is to find the best allocation of cache lines between the two zones while keeping the total number of cache lines fixed. Based on other considerations, we selected a total cache size of 400 lines.

This particular problem is simple enough that an exhaustive search can be used, requiring only 400 evaluations of the global hit rate. To validate the model's predictions we also used a simulation to measure the hit rate of 39 cache configurations, beginning with 10 lines in zone 1 and 390 lines in zone 2, and then increasing the allocation to zone 1 by 10 lines (and decreasing the allocation to zone 2) until zone 1 reached 390 lines in size. The predictions and measurements are shown in Figure 7.15.

Firstly, it should be noted that the predictions in Figure 7.15 are based on the parameter values from a 10,000 packet subsample while the measured values from the simulation are based on the full one million packet trace. This was done to test whether the parameter values for the footprint function yield reasonable predictions for the entire trace. The model holds reasonably well, based on the match between the predicted and measured values. The predicted optimum is at 320 lines in zone 1, with a predicted global hit rate of 0.8922, while the measured optimum is at 330 lines in zone 1 with a global hit rate of 0.9211. The absolute values of the hit rate are quite close, and the location of the optimum is within the step size (10 lines) of the simulated cache configurations.

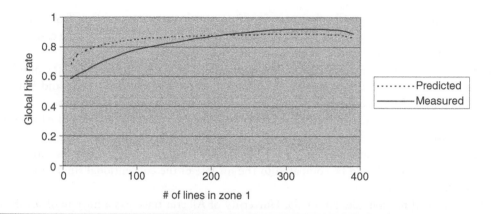

Figure 7.15 **Global hit rate—predicted versus measured. (From MacGregor M.H.,** *Proc. High Perf. Switching and Routing (HPSR 2003)***, Torino, Italy, 2003. With permission.)**

7.3.3.3 *Optimization Tableau*

The multi-zone cache design problem can be formulated as a constrained integer nonlinear optimization problem. Considering the case of a two-zone design first, we have:

$$\min(F_1 m_1(C_1) + F_2 m_2(C_2))$$

subject to:

$$F_1 = \sum_{i=1}^{B} f_i, \quad F_2 = \sum_{i=B+1}^{32} f_i,$$

$$m_i(C_i) = (C_i + 1)^{(1-\Phi_i)}, \quad i = 1, 2, \ C_1 + C_2 = C;$$

$$p_j \in \{0, 1\}, \quad j = 1, 2, 3, \ldots, 31, \quad \sum p_j = 1, \quad B = \sum j p_j.$$

$$\Phi_1 = \sum p_j \theta_j, \quad \Phi_2 = \sum p_j \theta'_{j+1}.$$

F_1 is the fraction of references to zone 1 and F_2 is the fraction of references to zone 2. f_i is the fraction of references with prefixes of length i. p_j selects the prefix lengths for zone 1 and B is the maximum prefix length for zone 1. θ_j is the parameter for the footprint function of references with prefixes up to length j, and θ'_j is the parameter for the footprint function of references with prefixes of length j and greater. We assume that a given reference falls into only one zone.

The tableau for caches with more than two zones is only slightly more complex, but there are some added constraints due to the additional breakpoints in prefix length. In the case of a cache with four zones, we let $B_1 = \sum j p_{1,j}$, $B_2 = \sum j p_{2,j}$, $B_3 = \sum j p_{3,j}$ denote the prefix lengths of zones 1, 2, and 3, respectively. Zone 4 caches addresses whose prefixes are from $(B_3 + 1)$ to 32-bits long. We also add $B_1 < B_2 < B_3$ to establish the ordering of the zones. Finally, we need parameters for the footprint functions of the two middle zones:

$$\Phi_2 = \sum\sum p_{1,j} p_{2,j} \hat{\theta}_{j,k-1}, \quad \Phi_3 = \sum\sum p_{2,j} p_{3,j} \hat{\theta}_{j,k-1}$$

where $\hat{\theta}_{j,k}$ is the parameter for the footprint function of references with prefixes from length j to length k, inclusive. The constraint for Φ_1 is similar to that for Φ_1 in the two-zone case, and the constraint for Φ_4 is similar to that for Φ_2 from the two-zone case.

It is possible to write the optimization tableau for the general case and optimize the number of zones as part of the tableau, but it is much less complex to solve the two, three, and four-zone designs separately. This also quantifies the incremental benefits of using extra zones.

The optimization surface for the design of a two-zone cache for the University of Alberta trace is presented in Figure 7.16. The vertical axis denotes the global hit rate. The best design has a global hit rate of 0.933, and allocates addresses 382 lines to zone 1, where it keeps addresses with prefixes of 23-bits or less. This should be compared to the hit rate of the conventional single-zone cache, which is 0.851.

The optimal three-zone design for the University of Alberta trace has a hit rate of 0.958, and keeps addresses whose routes have prefix lengths of at most 11, 22, and 32-bits in zones 1, 2, and 3, respectively. The cache allocation is 126 lines in zone 1, 190 lines in zone 2, and 84 lines in zone 3. The optimal four-zone design for the University of Alberta trace has a hit rate of 0.979, and keeps addresses whose routes have prefix lengths of at most 11, 13, 22, and 32-bits in zones 1, 2, 3, and 4, respectively. The cache allocation is 134 lines in zone 1, 78 lines in zone 2, 89 lines in zone 3, and 99 lines in zone 4. For a hypothetical million packet-per-second interface, the conventional cache reduces the software lookup rate to under 150 thousand packets per second, and the four-zone cache reduces the software lookup rate to about 20 thousand packets per second. The cost of this improvement is relatively trivial, being of the order of 12,000-bits of fast memory plus associated control.

7.4 Cache-Oriented Multistage Structure

In a parallel-router architecture, the FEs are often separated from the linecards (LC). IP-address lookups are done by multiple forwarding engines independently and concurrently. LCs share an array of FEs. For example, the Juniper M160 backbone router contains four FEs, which are shared

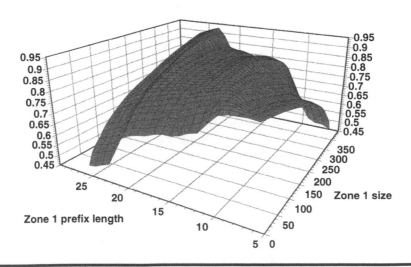

Figure 7.16 Two-zone optimization surface. (From MacGregor M.H., *Proc. High Perf. Switching and Routing (HPSR 2003)*, Torino, Italy, 2003. With permission.)

by up to eight LCs. Tzang proposed a cache-oriented multistage structure (COMS) to provide connections dynamically between LCs and FEs in a parallel router [15]. COMS is based on a bi-directional multistage interconnect (BMI), and comprises multiple stages of 2×2 switching elements (SEs) equipped with caches. In the following, we will describe COMS in detail. This section is from [15]. Portions are reprinted with permission (© 2003 IEEE).

7.4.1 Bi-Directional Multistage Interconnection

Figure 7.17 shows a BMI of size 8, where the basic SE is of size 2×2, and there are $\log_2 8$ stages of SEs interconnected by bi-directional links. The routing decision in each SE is based on one bit of the routing tag. In the upward direction of the BMI depicted in Figure 7.17 (i.e., from bottom to top), a routing tag is one top port number. For example, the routing tag from BP_2 to TP_6 is 110_2 (= 6), with the first bit for setting an SE in the bottom stage, such that a "0" (or "1") makes a connection to the left (or right) upstream port. Similarly, the second (or third) bit is to control an SE in the next (or top) stage. The path from BP_2 to TP_6 is marked in Figure 7.17, so is the path from BP_4 to TP_3 (with its tag being 011_2 = 3). Routing in the downward direction takes the same path as in the opposite direction; it can easily be done by keeping the arrival port information in each visited SE when the package was routed upward, as will be elaborated in Section 7.4.4. This enables downward routing without using a tag and is applicable to any interconnect topology.

7.4.2 COMS Operations

Figure 7.18 shows the proposed cache-oriented multistage structure. It comprises multiple stages of SEs, each of which is equipped with (on-chip) SRAM for caching lookup results returning from FEs.

In COMS, each FE contains a forwarding table to enable lookups. The forwarding tables of all FEs reflect changes caused by updates to the core routing table. To ensure appropriate lookups, we assume that all cache contents in COMS are flushed entirely after each table update. In the next

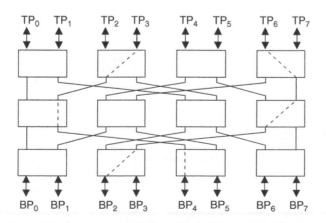

Figure 7.17 Bi-directional multistage interconnection. (From Tzeng, N.-F., *Proc. 2003 International Conference on Parallel Processing*, 2003. With permission.)

section, COMS will be demonstrated by simulation to arrive at very high forwarding performance, even under this simple cache invalidation (by flushing) method after each table update.

Consider a packet arrival immediately after a table update. The packet terminates at one LC (referred to as the incoming LC), where the packet header is extracted and delivered through the COMS to one FE for the longest-prefix matching lookup. Because the cache contents in COMS are all flushed at that time, the packet header cannot have a hit in any SE and will finally reach one FE (which is dictated by selected bits of the IP address), where the table lookup is conducted according to the longest-prefix matching algorithm implemented therein. After the lookup result is obtained, it is sent along the same path in the reverse direction back to the incoming LC. On its way back, the result is written to the cache of every visited SE. This cached result will satisfy later packet lookups for an identical destination address much faster. In general, a packet header delivered along COMS is checked against the cached entries in each visited SE; if there is a hit, a reply is produced by the SE and sent along the same path back to the incoming LC. Again, the reply is cached at each SE on its way back.

COMS is expected to yield high hit rates, as packets arriving at an LC closely together (in time) will have a good chance to head for the same destination. Additionally, COMS can serve multiple packets (from different LCs) with the same destination concurrently at SEs in different stages of COMS, enjoying parallelism in packet lookups. In Figure 7.18, for example, packet p_2 has a hit at SE_b after the lookup result of packet p_1 has flown back to its originating LC (i.e., LC_0).

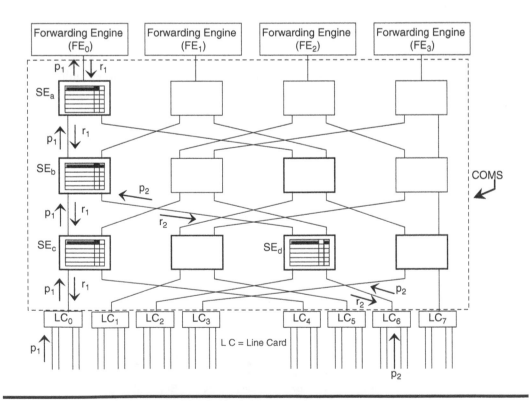

Figure 7.18 Proposed cache-oriented multistage structure (COMS), derived from the BMI shown in Figure 7.17. A tree of SEs rooted at FE$_0$ is highlighted. (From Tzeng, N.-F., *Proc. 2003 International Conference on Parallel Processing*, 2003. With permission.)

At the same time, another packet destined for the same address (not shown in Fig. 7.18) from LC_1 can be served by SE_a. A lookup result obtained at an FE (say, FE_0 in Fig. 7.18) may be cached in COMS along a tree of SEs, rooted at FE_0 and spanned across all LCs. Thus, subsequent lookups of the same address are enabled for packets arriving at any LC. Because every SE has potential to serve one packet (with any destination address) in a cycle, the degree of parallelism offered by COMS could be massive, in particular for a large COMS. COMS keeps arrival port information of every packet which is recorded and waiting in the cache for completion, so as to simplify routing the lookup reply along the reverse direction. As a result, no routing tags are needed for delivering lookup replies. In addition, a simple but effective cache replacement mechanism is devised for COMS to enhance its hit rates, as elaborated next.

7.4.3 Cache Management

Cache management affects the hit rate of COMS, and thus the effective lookup times of packet addresses. To enhance the performance of COMS, appropriate status bits are required in each cache entry for implementing an efficient cache replacement mechanism. As a cache entry is in either an *invalid state* or a *shared state*, its status is denoted by one bit. In addition, *two duplication status bits*, called the *L-bit* and the *R-bit*, are introduced to each cache entry in an SE (say, SE_t) that indicates whether the left-child and the right-child of SE_t contains a copy of the same entry in their respective caches. These two bits reflect the redundancy degree of a cache entry and are helpful to achieve higher hit rates under a given cache organization, resulting from better cache replacement.

In COMS, the cache block size is chosen to hold only *one address lookup result*, because devices with contiguous IP addresses usually have little direct temporal correlation of network activities. The cache size and the degree of set associativity are left as design parameters for investigation. When a lookup result is sent back to the incoming LC, it is cached at each visited SE. If the SE has no free entry in the set of interest (decided by the destination IP address), one entry in the set has to be chosen for replacement, provided the degree of set associativity is larger than 1 (which is common). To minimize the adverse impact of replacement, it is obvious that an entry with *both* *L-bit* and *R-bit* set, if any, will be chosen as the one to be evicted because replacing the entry will have little impact on future hits. These history status bits are examined first to decide which entry is to be replaced. If multiple entries have both their duplication status bits set, a conventional replacement strategy (such as LRU, FIFO, or random) is applied to break the tie. On the other hand, when no entry has both its bits set, a conventional replacement strategy is followed to choose the entry for eviction. After a cache entry is created completely in an SE, its associated L-bit (or R-bit) is set if the return path is along the left (or right) downstream link of the SE.

The 2×2 SEs are assumed to operate at 200 MHz. Each visit to an SE thus takes 5 ns (which is possible because on-chip SRAM is employed in SEs and the access time of such SRAM can be as low as 1 ns, if its size is small, as is the case of SEs). Each SE cache also incorporates a *victim cache* to keep blocks which are evicted from a cache due to conflict misses. A victim cache is a small fully associative cache [16], aiming to hold those blocks that get replaced so that they are not lost. Entry replacement in the victim cache follows a conventional replacement mechanism. When a packet is checked against an SE cache, its corresponding victim cache is also examined simultaneously. A hit, if any, will be in either the cache itself or its corresponding victim cache, but not in both. The victim cache normally contains four to eight entries, and it can effectively improve the hit rates by avoiding most conflict misses.

Early Cache Block Recording. To lower traffic toward FEs and the loads of FEs, a packet is recorded in the cache of an SE where a miss occurs. This early cache block recording prevents subsequent packets with the same destination from proceeding beyond the SE, and also makes it possible to deliver a packet reply back along the same path without resorting to a routing tag. COMS performance is thus enhanced and the SE design simplified. As this recorded cache entry is not complete until its corresponding reply is back, a status bit (waiting bit, i.e., W-bit) is added to the cache entry, with the bit set until a reply is received and the entry filled. In addition, an indicator is employed to record the incoming (downstream) port from which the packet arrives at the SE, referred to as the A-bit. When a subsequent packet hits a cache entry whose W-bit is set in an SE, the packet is stopped from proceeding forward and is put in the waiting list of the arrival port. The packet is allowed to advance after the W-bit of the hit cache entry is cleared (by a reply). Once a reply comes back and hits the cache entry recorded earlier, the entry is completed with the lookup result and its W-bit is cleared. The reply is then moved to the (downstream) outgoing port indicated by the A-bit, ready for delivery backward.

7.4.4 Details of SEs

The constituent SEs of COMS are bi-directional, as depicted in Figure 7.19. Each incoming buffer from the downstream (denoted by I_d–buffer) can hold one packet, so can an incoming buffer from the upstream (I_u–buffer). On the other hand, each outgoing buffer to the downstream (O_d–buffer) has the capacity to hold two packets, so does an outgoing buffer to the upstream (O_u–buffer). When two requests (or replies) compete for the same upstream (or downstream) link in a cycle, the two requests (or replies) are moved to the corresponding O_u–buffer (or O_d–buffer).

The cache in an SE is of on-chip SRAM and organized as a set-associative cache, with a block to hold one lookup result (i.e., <IP address, Next_hop_LC#>). When a packet arrives at

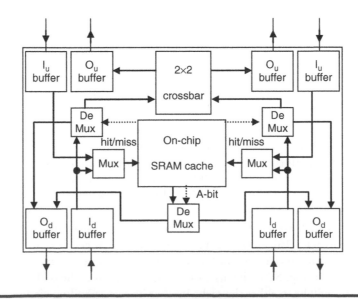

Figure 7.19   Block diagram of bi-directional 2×2 SEs. (From Tzeng, N.-F., *Proc. 2003 International Conference on Parallel Processing*, 2003. With permission.)

the SE from the downstream, it is checked against the cache and the associated victim cache simultaneously. This check results in different possible outcomes, as follows.

■ The packet hits a cache entry with its W-bit not set. The reply is produced directly from the packet plus Next_hop_LC# found in the hit entry. The reply is then moved to the corresponding O_d–buffer (see Fig. 7.19). If O_d–buffer is full, the packet is left in its incoming Id–buffer, waiting for the next cycle.

■ The packet hits a cache entry with its W-bit set. The packet cannot proceed and is put in the waiting list associated with the hit entry, no matter whether the hit happens to the cache or its victim cache. After the reply for the hit cache entry comes back, its W-bit will be reset and the packet may advance (according to the above scenario).

■ The packet misses in a cache (and also in its corresponding victim cache). The packet takes up a cache entry, with its W-bit set. In addition, the A-bit of the entry records the downstream incoming port from which the packet arrived. This A-bit will be used for directing the reply of the packet to its appropriate downstream outgoing buffer. If this cache entry creation is to be done in a set which has no available block, one block in the set has to be chosen for replacement. The availability of a cache block is reflected by its associated I/S bit, with "I" (or "S") indicating the invalid (or shared) state. A block is available if it is in the invalid state. Our replacement examines the L-bit and the R-bit of each block in the set to choose the block to be replaced, before resorting to a conventional replacement strategy.

When a reply (i.e., lookup result) arrives at an SE (from the upstream), it is checked against the cache (and the victim cache). This check must produce a hit, and the hit cache entry is completed by this lookup result by filling its Next_hop_LC# field and setting its L-bit or R-bit accordingly (i.e., based on the A-bit). The W-bit of the hit entry is cleared, and then the reply is forwarded to the O_d–buffer specified by the entry's A-bit. Also, replies for elements in the associated waiting list are produced and forwarded to their corresponding O_d–buffers.

7.4.5 Routing Table Partitioning

Routing tables in many backbone routers currently contain 120K+ prefixes, and their size is expected to grow rapidly. Here, we have obtained one routing table with 140,838 prefixes recently from AS1221 (http://bgp.potaroo.net, 2003) for use. COMS allows each FE to hold only a subset of all prefixes, increasing the address space coverage of each FE and alleviating the memory requirement of FEs. This is realized by partitioning the set of prefixes in a routing table into subsets (of roughly equal size) so that each subset involves a fraction of all prefixes and constitutes a forwarding table held in one FE. The number of bits chosen for partitioning the set of prefixes depends on the number of FEs, and the bits chosen are determined by the prefixes themselves, namely, those bits that result in the size difference between the largest partition and the smallest one being minimum. Those same bits of the destination addresses of lookup packets are employed to control the packets routed through COMS. They separate packets across COMS into groups, one at a top output of COMS (connecting to an FE). As a larger (higher-end) router involves more FEs, each FE may carry out faster lookups and contain fewer prefixes, enjoying better address space coverage and thus increase cache hit rates and overall COMS performance.

Partitioning is done by searching for those bits that result in the size difference between the largest partition and the smallest one being minimum. It is intended to balance and minimize the number of prefixes held in each FE, according to the routing table of a router; the partitioning is done irrespective of traffic over the router and is not meant to balance the load on FEs. For a given routing table and the COMS size, however, a desirable partition often gives rise to reasonably balanced load at FEs for all the traces we examined [15].

To evaluate the performance of COMS, Tzeng implemented a scalable router equipped with COMS [15]. The packet streams are derived from the Lawrence Berkeley National Lab (http://ita.ee.lbl.gov, Apr. 2002) and the National Laboratory for Applied Network Research (http://pma.nlanr.net, Sept. 2002). Tzeng investigated the impact of the cache size (β) on COMS behavior. The hit rates exceed 90 percent for all traces when β is of no less than 1024 blocks, and a larger β consistently yields a shorter lookup time. With $\beta = 4K$, the mean lookup times drop below 2.8 cycles, translating to a lookup speed of more than 70 millions packets per second for each LC. A COMS-based router sized 16 can thus forward more than 1120 millions packets per second, provided that $\beta \geq 4K$. When compared with an existing router, the COMS-based parallel router achieves faster mean forwarding performance by a factor more than 10, despite its reduced total SRAM amount and a possibly shorter worst-case lookup time [15].

References

1. Tzi-Cker Chiueh and Prashant Pradhan, High performance IP routing table lookup using CPU caching. *IEEE INFOCOM* 1999; 3:1421–1428.
2. Tzi-Cker Chiueh and Prashant Pradhan, Cache Memory Design for Network Processors. *Proc. IEEE Sixth International Symposium on High-Performance Computer Architecture*, January 2000, pp. 409–419, Los Alamitos, California.
3. Kartik Gopalan and Tzi-Cker Chiueh, Improving route lookup performance using network processor cache. *Proceedings of the 2002 ACM/IEEE Conference on Supercomputing*, 2002, Baltimore, Maryland, pp. 1–10.
4. Huan Liu, Routing prefix caching in network processor design. *Proceedings International Conference on Computer Communications and Networks (ICCCN)*, 2001, Phoenix, Arizona, pp. 18–23.
5. Akhbarizadeh, M.J. and Nourani, M., Efficient prefix cache for network processors. *12th Annual IEEE Symposium on High Performance Interconnects*, Aug 2004, pp. 41–46.
6. Besson, E. and Brown, P., Performance evaluation of hierarchical caching in high-speed routers, *Proceedings Globecom*, 1998, pp. 2640–2545.
7. Chvets, I.L. and MacGregor, M., Multi-zone caches for accelerating IP routing table lookups. *Merging Optical and IP Technologies Workshop on High Performance Switching and Routing*, May 2002, pp. 121–126.
8. Soraya Kasnavi, Paul Berube, Vincent C., Gaudet, and José Nelson Amaral, *A Multi-Zone Pipelined Cache for IP Routing*, Springer Berlin/Heidelberg, Volume 3462/2005, 2005, pp. 574–585.
9. MacGregor, M.H., Design algorithms for multi-zone IP address caches, *Proc. High Perf. Switching and Routing (HPSR 2003)*, Torino, Italy, June, 2003, pages 281–285.
10. Newman, P., Minshall, G., Lyon, T., and Huston, L., IP switching and gigabit routers, *IEEE Commun. Mag.*, 1997; 35:64–69.
11. Liu, H., Routing table compaction in ternary CAM, *IEEE Micro*, 2002; 22(1):58–64.
12. Richard P. Draves, et al., Constructing optimal IP routing tables, *Proceedings IEEE Infocom'99*, New York, U.S.A., 21–25 March 1999, pp. 88–97.
13. Shyu, W.L., Wu, C.S., and Hou, T.C., Multilevel aligned IP prefix caching based on singleton information. *GLOBECOM 02*, November 2002; 3:2345–2349.

14. Thiebaut, D., Wolf J.L., Stone, H.S., Synthetic traces for trace-driven simulation of cache memories, *IEEE Transactions. Computer* 1992; 41(4):388–410.

15. Tzeng, N.-F., Hardware-assisted design for fast packet forwarding in parallel routers, *Proc. 2003 International Conference on Parallel Processing*, October 2003, pp. 11–18.

16. Jouppi, N., Improving direct-mapped cache performance by the addition of a small fully-associative Cache and Prefetch Buffers, *Proceedings 17th Annual Int'l Symposium on Computer Architecture*, May 1990, pp. 364–373.

Chapter 8

Hashing Schemes

Hash table is one of the standard techniques for exact matching, which has a good performance. It cannot directly be used for the Internet address lookups because Classless Inter-Domain Routing was deployed in 1993, and the Internet routers have to perform the longest prefix matching (LPM). In this chapter, we introduce the up-to-date hashing algorithms for the LPM.

8.1 Binary Search on Hash Tables

Because any two prefixes that an IP address matches have different lengths, Waldvogel et al. proposed an algorithm to the LPM, using a binary search over hash tables organized by the length of the prefix [1]. It requires a worst-case search time of $\log_2(\textit{addressbits})$ and scales very well as IP address and routing table increase.

8.1.1 Linear Search of Hash Tables

The prefixes in a forwarding table are divided into small tables according to the lengths of the prefixes. A hashing table is constructed based on each small table. If the maximum length of the prefix is L, there are L hashing tables. For an IP address D, at first, we extract the first L bits of D and find the exact matching prefix in the Lth hashing table. If there is a match, it is the longest-matching prefix. Otherwise, we go to the next hashing table.

Let T be an array of records with two fields: length and hash. $T[i].\textit{length}$ is the length of the prefixes found at position i, and $T[i].\textit{hash}$ is a pointer to a hash table containing all prefixes of length $L[i].\textit{length}$. Array $T[i]$ is in the increasing order of prefix length: $T[i].\textit{length}<T[i+1].\textit{length}$. The following is the search algorithm. The search complexity is $O(W)$, where W is the length of the IP address.

```
Function Linear Search(D)  (*D is an IP address*) {
    LMP := Null; (LMP is the longest-matching prefix)
    i:= Highest index in array T;
```

```
        While (LMP = Null & i >= 0) {
            D' = The first T[i].length bits of D;
            LMP := Search(D', T[i].hash);
            i := i-1;
        }
    Return LMP;
}
```

8.1.2 Binary Search of Hash Tables

To reduce the complexity of Linear Search, $O(W)$, Waldvogel et al. used the binary search on the array T to cut down the complexity to $O(\log W)$. However, for a binary search to make its branching decision, it requires the result of an ordered comparison, returning whether the probed entry is less than, equal, or greater than our search key. If the probed entry is equal to the search key, the match is found. If the probed entry is less than or greater than our search key, we cannot be sure that the search should proceed in the direction of shorter lengths, because the longest-matching prefix could be in the direction of longer length as well. Waldvogel et al. insert extra prefixes of adequate length, called *marker*, to be sure that, when no match is found, the search must proceed necessarily in the direction of shorter prefixes.

For example, there are three prefixes: P0 = 0, P1 = 00, and P2 = 111. We can construct three hash tables corresponding to their length (Fig. 8.1b). To search for 111, we start at the middle hash table corresponding to the length 2, there is no match in the hash table. If the search proceeds to the hash table corresponding to the length 1, we cannot find the longest-matching prefix. After the marker P* = 11 is inserted into the middle hash table, the search can proceed to the hash table corresponding to the length 3, then the longest-matching prefix is found. The binary search on hash tables is shown in Figure 8.1a.

But where do we need markers, and how many are they? In general, for every prefix, there would be a marker at all other prefix lengths. Because a binary search only checks a maximum of $\log_2 W$ levels, each entry will generate a maximum of $\log_2 W$ markers. In fact, the number of markers will be much smaller for two reasons: the prefix corresponding to a marker already exists as an entry in hash tables, and some prefixes have the same marker. For a real forwarding table from MaeEast (http://www.merit.edu/ipma/, 12/19/1996), Waldvogel et al. reported that the storage required would increase by 25 percent [1].

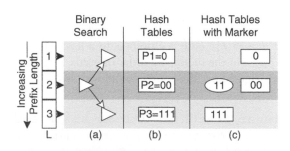

Figure 8.1 Binary search on hash tables.

The marker can guide the search to potentially better prefixes lower in the table, but they can also give an incorrect result. For example, there are three prefixes: P0 = 1, P1 = 00, and P2 = 111. The marker 11* is inserted (Fig. 8.1c).

When we search for 110, at first, we search in hash table 2, it matches the marker 11*, we will go into hash table 3. In fact, the longest-matching prefix is not in hash table 3, but in hash table 1. The marker 11* guide the incorrect direction. We have to go back and search the upper half of the range (called backtracking).

Each hash table (markers plus real prefixes) can be thought of as a horizontal layer of a trie corresponding to some length L (except that the hash table contains the complete path to that layer of each entry in that layer). The basic scheme is essentially doing a binary search on the levels of a trie (Fig. 8.2). We start by doing a hash on prefixes corresponding to the median length of the trie. If we match, we search the upper half of the trie; if we fail, we search the lower half of the trie.

Figure 8.2 and other figures describing the search order contain several elements: (*i*) horizontal stripes grouping all the elements of a specified prefix length, (*ii*) a trie containing the prefixes, shown on the right of the figure and rooted on the top of the figure, and (*iii*) a binary tree, shown on the left of the figure and rooted at the left, which depicts all possible paths that a binary search can follow. We will use the *upper half* to mean the half of the trie with prefix lengths strictly less than the median length. We also use the *lower half* for the portion of the trie with prefix lengths strictly greater than the median length. It is important to understand the conventions in Figure 8.2 to understand the later figures and text.

Although each prefix may bring at most $\log_2 W$ markers, the worst-case number of backtrackings is $O(W)$. As an example, we have a prefix P_i of length i ($1 \leq i \leq W$) that contains all 0s, and the prefix Q whose first $W - 1$ bits are all 0s, but whose last bit is 1. If we search for the W bit address containing all 0s, then we can show that a binary search with backtracking will take $O(W)$ time and visit every level in the table.

8.1.3   *Precomputation to Avoid Backtracking*

To avoid backtracking, when a marker (M) is inserted into its hash table, the value of the best-matching prefix of the markers ($M.bmp$) is recomputed. When we find M at the midpoint of R (the current range), we remember $M.bmp$ as the current best-matching prefix and search the lower half

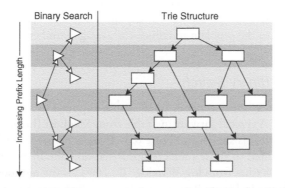

Figure 8.2   Binary search on trie levels.

```
Function Binary Search (D) (*search for address D*)
Initialize search range R to cover the whole array L;
Initialize BMP found so far to null string;
While R is not empty do
    Let i correspond to the middle level in range R;
    Extract the first L[i].length bits of D into D';
    M:= Search(D', L[i].length); (*search hash for D*)
    If M is nil then set R := upper half of R; (*not found*)
        Elseif M is a prefix and not a marker
        Then BMP := M.bmp; break; (*exit loop*)
        Else (*M is a pure marker, or marker and prefix*)
            BMP := M.bmp; (*update best matching prefix so far*)
            R := lower half of R;
    Endif
Endwhile
```

Figure 8.3   Binary Search with recomputation. (From Waldvogel, M., et al., *Proc. of ACM SIGCOMM'97*, September 1997. With permission.)

of R. If the searching fails, we need not backtrack, because the results of backtracking are already summarized in the value of $M.bmp$. The search algorithm is shown in Figure 8.3.

The standard invariant for binary search when searching for key K is: "K is in range R." We then shrink R while preserving this invariant. The invariant for this algorithm, when searching for key K, is "EITHER (the best matching prefix of K is BMP) OR (there is a longer matching prefix in R)." It is easy to see that initialization preserves this invariant, and each of the search cases preserves this invariant. Finally, the invariant implies the correct result when the range shrinks to 1. Thus, the algorithm works correctly and has no backtracking; it takes $O(\log_2 W_{dist})$ time.

8.1.4   Refinements to Basic Scheme

The basic scheme of the binary search on a hash table has been described previously. It takes just seven hash computations, in the worst case, for 128-bit IPv6 addresses. Each hash computation takes at least one access to memory. The average number of hash computations is dependent on the distribution of the prefix lengths. Thus, we will explore a series of optimizations that exploit the deeper structure inherent in the problem to reduce the average number of hash computations.

8.1.4.1   Asymmetric Binary Search

The basic scheme is a fast, yet very general, BMP search engine. For a particular data set, can we improve its performance? A typical forwarding table from MaeEast (http://www.merit.edu/ipma/routing table/, January 1996) is shown in Figure 8.4. There is no prefix with lengths 1–7, 25, and 31. The search can be limited to those prefix lengths that do contain at least one entry, reducing the worst-case number of hashes from $\log_2 W$ to $\log_2 W_{dist}$ (W_{dist} is the number of nonempty buckets in the histogram), as shown in Figure 8.5.

In Figure 8.4, there is another phenomenon: the prefixes are not equally distributed over the different prefix lengths. We can improve the average performance based on the access frequency.

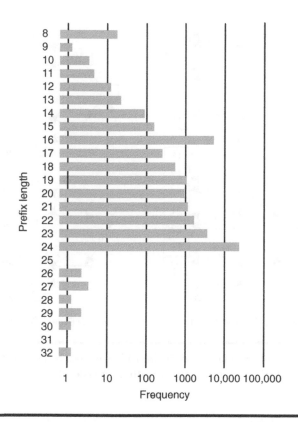

Figure 8.4 Histogram of the prefix length distribution. (From Waldvogel, M., et al., *Proc. of ACM SIGCOMM'97*, September 1997. With permission.)

Waldvogel et al. introduced the asymmetric binary tree. Some searches will make a few more steps, which have a negative impact on the worst case. The search for a BMP can only be terminated early if we have a "terminal" condition stored in the node. This condition is signaled by a node being a prefix but not a marker.

On the other hand, with a large number of frequently accessed entries, building an optimal tree is a complex optimization problem, especially because restructuring the tree also removes the terminal condition on many markers and adds it to others.

To build a useful asymmetrical tree, we can recursively split both the upper and lower parts of the binary search tree's current node's search space, at a point selected by a heuristic weighting function. Two different weighting functions with different goals (one strictly picking the level covering most addresses, the other maximizing the entries while keeping the worst case bound) are shown in Figure 8.6, with coverage and average/worst-case analysis for both weighting functions in Table 8.1. As can be seen, balancing gives faster increases after the second step, resulting in generally better performance than "narrow-minded" algorithms.

Now we can see why our first attempt, while improving the worst case, makes the average case worse: the prefixes with lengths 8, 16, and 24 are very common and also cover a big part of the address space, so they should be reached in early stages of the binary tree. In the original binary search, they were reached in steps 2, 1, and 2, respectively. In the new "optimized" approach, they were moved to steps 4, 3, and 5, respectively (Fig. 8.5, to the bottom of the tree). Besides slowing

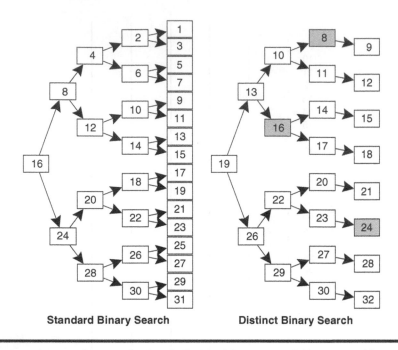

Standard Binary Search | Distinct Binary Search

Figure 8.5 **Search trees for standard and distinct binary search. (From Waldvogel, M., et al., *Proc. of ACM SIGCOMM'97*, September 1997. With permission.)**

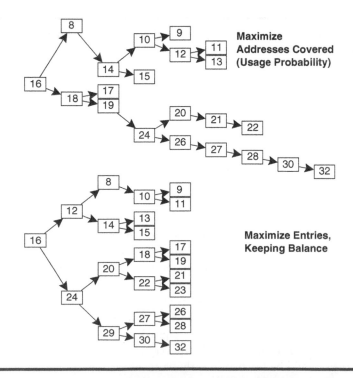

Figure 8.6 **Asymmetric trees produced by two weighing functions. (From Waldvogel, M., et al., *Proc. of ACM SIGCOMM'97*, September 1997. With permission.)**

Table 8.1   Address (*A*) and Entry (*E*) Coverage for Asymmetric

Steps	Usage		Balance	
	A	*E*	*A*	*E*
1	43%	14%	43%	14%
2	83%	16%	46%	77%
3	88%	19%	88%	80%
4	93%	83%	95%	87%
5	97%	86%	100%	100%
Average	2.1	3.1	2.3	2.5
Worst case	9	9	5	5

Source: Waldvogel, M., et al., *Proc. of ACM SIGCOMM'97*, September 1997. With permission.

down the search, this increased the number of pure markers required to exceed the real prefixes, resulting in a large growth in memory requirements and insertion time.

8.1.4.2   Mutating Binary Search

The histogram of prefix distributions led us to propose asymmetric binary search, which can improve the average speed. Further information about prefix distributions can be extracted by dissecting the histogram: For each possible n bits prefix, we could draw 2^n individual histograms with possibly fewer nonempty buckets, thus reducing the depth of the search tree.

When partitioning according to 16-bit prefixes and counting the number of distinct prefix length in the partitions, we discover a nice property of the routing data (Table 8.2). Though the whole histogram (Fig. 8.4) shows 23 distinct prefix lengths with many buckets containing a significant number of entries, none of the "sliced" histograms contain more than 12 distinct prefixes; in fact, the vast majority only contain one prefix, which often happens to be in the 16 bits in the binary search, and to get a match, we need only do binary search on a set of lengths that is much smaller than the 16 possible lengths we would have to search in naïve binary search.

In general, every match in the binary search with some marker X means that we need only search among the set of prefixes for which X is a prefix. This is illustrated in Figure 8.7. On a match, we need only search the subtrie rooted at X (rather than search the entire lower half of the trie, which is what naïve binary search would do). Thus the whole idea in mutating binary search is as follows: "whenever we get a match and move to a new subtrie, we only need to do binary search on the levels of new subtrie." In other words, the binary search mutates or changes the levels on which it searches dynamically (in a way that always reduces the levels to be searched), as it gets more and more match information.

Thus, each entry E in the search table could contain a description of a search tree specialized for all prefixes that start with E. This simple optimization cuts the average search time to below two steps (Table 8.3), assuming probability proportional to the covered address space. Also with other probability distributions (i.e., according to actual measurements), we expect the average number of lookups to be around two.

Table 8.2 Number of Distinct Prefix Lengths in the 16-Bit Partitions (Histogram)

Distinct Lengths	Frequency
1	4977
2	608
3	365
4	249
5	165
6	118
7	78
8	46
9	35
10	15
11	9
12	3

Source: Waldvogel, M., et al., *Proc. of ACM SIG-COMM'97*, September 1997. With permission.

As an example, consider binary search to be operating on a tree of levels starting with a root level, say 16. If we get a match which is a marker, we go "down" to the level pointed by the down child of the current node; if we get a match which is a prefix and not a marker, we are done; finally, if we get no match, we go "up." In the basic scheme without mutation, we start with root level 16; if we get a marker match we go down to level 24, and go up to level 8 if we get no match.

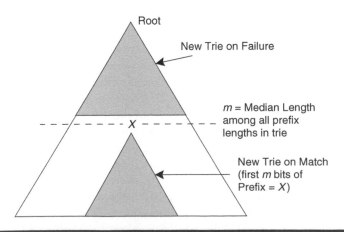

Figure 8.7 Showing how mutating binary search for prefix P dynamically changes the trie on which it will do binary search of hash tables. (From Waldvogel, M., et al., *Proc. of ACM SIGCOMM'97*, September 1997. With permission.)

Table 8.3   Address (*A*) and Entry (*E*) Coverage for Mutating Binary

	Usage		Balance	
Steps	*A*	*E*	*A*	*E*
1	43.9%	14.2%	43.9%	14.2%
2	98.4%	65.5%	97.4%	73.5%
3	99.5%	84.9%	99.1%	93.5%
4	99.8%	93.6%	99.9%	99.5%
5	99.9%	97.8%	100.0%	100.0%
Average	1.6	2.4	1.6	2.2
Worst case	6	6	5	5

Source: Waldvogel, M., et al., *Proc. of ACM SIGCOMM'97*, September 1997. With permission.

Doing basic binary search for an IPv4 address whose BMP has length 21 requires checking the prefix lengths 16 (hit), 24 (miss), 20 (hit), 22 (miss), and finally 21. On each hit, we go down, and on misses up.

Using mutating binary search, looking for an address (Fig. 8.8) is different. First, we explain some new conventions for reading Figure 8.8. As in Figure 8.2, we continue to draw a trie on the right. However, in this figure, we now have multiple binary trees drawn on the left of the figure, labeled as Tree 1, Tree 2, and so on. This is because the search process will move from tree to tree. Each binary tree has the root level (i.e., the first length to be searched) at the left; the upper child of each binary tree node is the length to be searched on failure, and whenever there is a match, the search switches to the more specific tree.

Finally, Figure 8.8 has a number of prefixes and markers that are labeled as *E*, *F*, *G*, *H*, and *J* for convenience. Every such entry in the example has *E* as a prefix. Thus rather than describe all

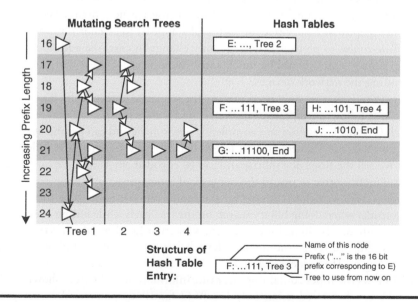

Figure 8.8   Mutating Binary Search example. (From Waldvogel, M., et al., *Proc. of ACM SIGCOMM'97*, September 1997. With permission.)

the bits in E, we denote the bits E as ...; the bits in say F are denoted as ...111, which denotes the concatenation of the bits in E with the suffix 111. Finally, each hash table entry consists of the name of the node, followed by the bits representing the entry, followed by the label of the binary tree to follow if we get a match on this entry. The *bmp* values are not shown for brevity.

Consider now a search for an address whose BMP is G in the database of Figure 8.8. The search starts with a generic tree, Tree 1, so length 16 is checked, finding E. Among the prefixes starting with E, there are known to be only six distinct lengths (say 17–22). So E contains a description of the new tree, Tree 2, limiting the search appropriately. Using Tree 2, we find F, giving a new tree with only a single length, leading to G. The binary tree has mutated from the original tree of 32 lengths, to a secondary tree of 5 lengths, to a tertiary "tree" containing just a single length.

Looking for J is similar. Using Tree 1, we find E. Switching to Tree 2, we find H, but after switching to Tree 4, we miss at length 21. Because a miss (no entry found) cannot update a tree, we follow the current tree upwards to length 20, where we find J.

In general, whenever we go down in the current tree, we can potentially move to a specialized binary tree, because each match in the binary search is longer than any previous matches, and hence may contain more specialized information. Mutating binary trees arise naturally in our application (unlink classical binary search), because each level in the binary search has multiple entries stored in a hash table. As opposed to a single entry in classical binary search, each of the multiple entries can point to a more specialized binary tree.

In other words, the search is no longer walking through a single binary search tree, but through a whole network of interconnected trees. Branching decisions are based not only on the current prefix length and whether or not a match is found, but also on what the best match so far is (which in turn is based on the address we are looking for). Thus at each branching point, you not only select which way to branch, but also change to the most optimal tree. This additional information about optimal tree branches is derived by precomputation based on the distribution of prefixes in the current data set. This gives us a faster search pattern than just searching on either prefix length or address alone.

Two possible disadvantages of mutating binary search immediately present themselves. First, precomputing optimal trees can increase the time to insert a new prefix. Second, the storage required to store an optimal binary tree for each prefix appears to be enormous. For now, we only observe that while routes to prefixes may frequently change in cost, the addition of a new prefix (which is the expensive case) should be much rarer. We proceed to deal with the space issue by compactly encoding the network of trees.

A key observation is that we only need to store the sequence of levels which a binary search on a given subtrie will follow on repeated failures to find a match. This is because when we get a successful match (Fig. 8.7), we move to a completely new subtrie and can get the new binary search path from the new subtrie. The sequence of levels, that a binary search would follow on repeated failures is what we call the Rope of a subtrie and can be encoded efficiently. We call it rope, because it allows us to swing from tree to tree in the network of interconnected binary trees.

If we consider a trie, we define the *rope* for the root of the trie node to be the sequence of trie levels we will consider when doing binary search on the trie levels while failing at every point. This is illustrated in Figure 8.9. In doing binary search, we start at Level m, which is the median length of the trie. If we fail, we try at the quartile length (say n), and if we fail at n, we try at the one-eighth-level (say o). The sequence $m, n, o, ...$ is the rope for the trie.

Figure 8.10 shows the ropes containing the same information as the trees shown in Figure 8.8. Note that a rope can be stored using only $\log_2 W$ (7 for IPv6) pointers. Because each pointer needs to discriminate only among at most W possible levels, each pointer requires only $\log_2 W$ bits.

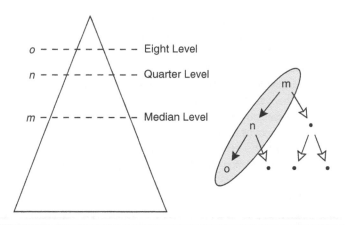

Figure 8.9 In terms of a trie, a rope for the trie node is the sequence of lengths starting from the median length, the quartile length, and so on, which is the same as the series of left children (see dotted oval in binary tree on right) of a perfectly balanced binary tree on tries levels. (From Waldvogel, M., et al., *Proc. of ACM SIGCOMM'97,* **September 1997. With permission.)**

For IPv6, 64 bits of rope is more than sufficient, though it seems possible to get away with 32 bits of rope in most practical cases. Thus a rope is usually not longer than the storage required to store a pointer. To minimize storage in the forwarding database, a single bit can be used to decide whether the rope or only a pointer to a rope is stored in a node.

Using the rope as the data structure has a second advantage: it simplifies the algorithm. A rope can be easily followed, by just picking pointer after pointer in the rope, until the next hit. Each strand in the rope is followed in turn, until there is a hit (which starts a new rope), or the end of the rope is reached.

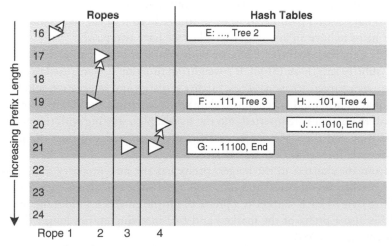

Note: Rope 1 contains the partially invisible layers 16, 8, 4, 2, and 1.

Figure 8.10   Sample ropes. (From Waldvogel, M., et al., *Proc. of ACM SIGCOMM'97,* **September 1997. With permission.)**

Function Rope Search(*D*) (*search for address *D**)
Initialize Rope *R* containing the default search sequence;
Initialize *BMP* so far to null string;
While R is not empty do
    Pull the first strand (pointer) off *R* and store it in *i*;
    Extract the first *L[i].length* bits of *D* into*D'*;
    M:= Search (*D'*, *L[i].hash*); (*Search hash table for *D**);
    If M is not nil then
        BMP: = *M.bmp*; (*update best matching prefix so far*)
        R: = *M.rope*; (*get the new Rope, possibly empty*)
    Endif
Endwhile

Figure 8.11   Rope search. (From Waldvogel, M., et al., *Proc. of ACM SIGCOMM'97*, September 1997. With permission.)

Pseudo-code for the rope variation of a mutating binary search is shown in Figure 8.11. An element that is a prefix but not a marker (i.e., the "terminal" condition) specifies an empty rope, which leads to the search termination. The algorithm is initialized with a starting rope. The starting rope corresponds to the default binary search tree. For example, using 32-bit IPv4 addresses, the starting rope contains the starting level 16 followed by levels 8, 4, 2, and 1. The levels 8, 4, 2, and 1 correspond to the "up" pointers to follow when no matches are found in the default tree. The resulting pseudo-code (Fig. 8.11) is elegant and simple to implement. It appears to be simpler than the basic algorithm.

8.1.5   Performance Evaluation

Building a rope search data structure balanced for optimal search speed is more complex, because every possible binary search path needs to be optimized. To find the *bmp* values associated with the markers, it helps to have an auxiliary trie. Thus we have two passes.

Pass 1 builds a conventional trie. Each trie node contains a list of all prefix lengths used by its "child" nodes [subtree length set (SLS)]. If a weighting function is being used to optimize accesses based on known or assumed access patterns, further statistics and forecasts should be summarized. All this additional information is kept up-to-date while inserting, in $O(NW)$ time.

In Pass 2, all the prefixes are inserted into the hash tables, starting with the shortest prefix: for each prefix, its rope and the BMP for its markers are calculated and then the markers and the prefixes are inserted. This takes $O(N \log^2 W)$.

Inserting from the shortest to the longest prefix has the nice property that all BMPs for the newly inserted markers are identical and thus only need to be calculated once. This can be easily seen by recalling that (*i*) each marker is a prefix of all the entries it guides the search to, (*ii*) the marker's BMP is also a prefix of the marker, and (*iii*) inserting entries longer than the marker's length cannot change its BMP.

There are at most $O(\log W)$ markers to insert for each real prefix, and each prefix and marker need a rope, which can be calculated from the SLS in $O(\log W)$. Thus, the overall work is $O(N \max(W, \log^2 W))$. We are working on a faster and more elegant algorithm to build the rope search data structure in time $O(N \log^2 W)$ (that also does not require building a trie).

One problem for the insertion is that the number of markers for each length is not known in advance, which makes it difficult to allocate memory for the hash table in advance. This problem can be avoided by putting all the entries in a *single* hash table and including the prefix length in the hash calculation. Because there is an upper limit of log W markers per real prefix, we can size the single hash table. For typical IPv4 forwarding tables, about half of this maximum number is being used.

Adding and removing single entry from the tree can also be done, but since no rebalancing occurs, the performance of the lookups might slowly degrade over time. However, addition and deletion are not trivial. Adding or deleting a single prefix can change the *bmp* values of a large number of markers, and thus insertion is potentially expensive in the worst case. Similarly, adding or deleting a new prefix that causes a new prefix length to be added or deleted can cause the ropes of a number of entries to change. The simplest solution is to batch a number of changes and do a complete build of the search structure. Such solutions will have adequate throughput (because whenever the build process falls behind, we will batch more efficiently), but have poor latency. We are working on fast incremental insertion and deletion algorithms, but we do not describe them here. The incremental insertion and deletion algorithms still require the tree to be rebuilt after a large number of different inserts and deletes.

In the worst case, log W markers are necessary per prefix; hence, memory complexity is $O(N \log W)$. In practice, many prefixes will share markers, reducing the memory.

To measure the performance of the rope search, Waldvogel et al. did experiments on a 200 MHz Pentium Pro from C code using the complier's maximum optimization. The forwarding table with 33,000 entries is from IPMA project (http://www.merit.edu/ipma/routing table/, January 1996). Memory usage is close to 1.4 MBytes, the average lookup time is about 100 ns, and the worst-case lookup time is about 600 ns [1].

8.2   Parallel Hashing in Prefix Length

The binary search of hash tables takes the complexity of the lookup operation $O(\log_2 W)$. The scheme assumes to use a perfect hashing hardware and does not consider the occasion of collision in hashing. The most intuitive method to implement an address lookup is to use IP address itself as a memory pointer. In other words, set up a memory with 2^{32} entries knowing that the IPv4 destination address consists of 32 bits. Unfortunately, this method is impractical because it requires huge memory space. Hashing function provides the capability of taking the longer field of addresses and producing a shorter field that can be used as an index to a subset of the table in memory. Hence, the required size of memory can be significantly reduced when the hashing output is used as an index. Hashing has been effectively used for a search with exact matching. However, there is an issue to apply hashing functions to search the longest prefix. In the LPM scheme, it is required to locate the entry matched with the input address in maximum length, and hence it is not known how many bits of address should be used as an input of hash function in advance.

Lim et al. proposed a scheme to apply parallel hashing to each address table of a single prefix and apply a binary search for collided entries [2]. Figure 8.12 shows the proposed scheme.

8.2.1   Parallel Architecture

As shown in Figure 8.12, each plane represents a separate hardware employing hashing search in each prefix length. Hash tables are composed of the main table and the subtable. Because the main tables include finite entries, two or more hash values may locate to the same entry, and the subtables

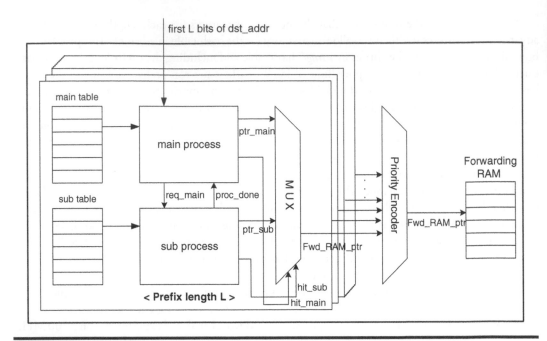

Figure 8.12 A parallel hashing architecture. (From Lim, H., Seo, J., and Jung, Y., *IEEE Communication Letter*, 7, 10, 2003. With permission.)

are provided to solve the collision. Each main entry includes a prefix, a pointer to the forwarding RAM, a pointer to the subtable, and the number of collisions corresponding to this entry. The hashing result locates one entry in the main table, and if the prefix used as a hashing input is matched with the prefix stored in the pointed entry, then the search for this prefix is done, and the forwarding RAM pointer is directed to the priority encoder. If the input prefix is not matched with the prefix stored in this entry, then the pointer to the subtable is used to direct the search to it. Each entry in the subtable includes a prefix and a pointer to the forwarding RAM. A binary search is performed in the subtable for the collided entries. Once address lookups in each prefix return matching entries, the entry which has the longest prefix among them is selected in the priority encoder. The forwarding RAM has the output interface information, and the information retrieved from it and pointed to by the selected entry is used to forward the packet. The proposed algorithm can be expressed as shown in Figure 8.13.

The proposed scheme has the advantage in updating the address table. Because main tables are separated by prefix lengths and only a limited number of entries of a subtable is involved in a binary search, by distributing empty entries between colliding prefixes of subtables, inserting or deleting subentries can be more efficient.

8.2.2 Simulation

A simulation is performed to evaluate the performance of the proposed scheme in terms of required memory size, average memory accesses, and minimum and maximum number of memory accesses using data gathered from MaeWest router (http://www.merit.edu, March 15, 2002). Hashing function is selected based on its hardware simplicity because the proposed scheme requires separate hashings in each prefix.

```
Function Search_with_Hashing /* search for address D */
Do parallel ( L = 8 ~ 32 )
      Extracts the first L bits of D into D ;
      Hash_value = Hash(D ) ;
      Compare D    and the prefix value pointed by Hash_value;
      if   ( not same )   begin                    // a collision occurred
                 use pointer to sub-table and # of entries;
                 perform binary search for the collided entries in sub-table;
      end
      Send the forwarding RAM pointer and ACK to the Priority Encoder;
End Do parallel
Select the entry with longest prefix on the Priority Encoder;
Retrieve the forwarding information from forwarding RAM;
```

Figure 8.13 Search algorithm in the parallel hashing architecture. (From Lim, H., Seo, J., and Jung, Y., *IEEE Communication Letter*, 7, 10, 2003. With permission.)

The routing table has a prefix length ranging between 8 and 32 bits. Because address lookups in the main table and the subtable are performed successively, they can be constructed using a single SRAM. The proposed architecture is implemented using 24 SRAMs for address lookup tables in prefix lengths of 8 to 32 (except the length of 31).

The size of memories in each prefix is adjusted depending on the number of different routes, and the number of bits of hashing output is calculated accordingly. For instance, a small memory of main table with four entries is allocated for prefix length 10, and hence hashing hardware receives 10 bits as an input and generates 2 bits as an output. Hash function is selected based on its hardware simplicity. Hashing hardware performs Exclusive-OR logic after grouping the bits of prefix into the bits required in hashing result. If the prefix length is not divided into the bits of the required result, arbitrary bits (0101 ... in the proposed scheme) are padded to the end.

Each main entry has four fields: 8 through 32 bits of prefix, 5 bits of pointer to the forwarding RAM, 15 bits of pointer to the subtable, and 4 bits of the number of colliding entries. The subtable entries only include a prefix and a pointer to the forwarding RAM. A forwarding RAM is composed of 24 entries and each entry has 16 bits of output interface information. The number of total entries included in the simulation is about 37,000 and the required memory size is 189 kBytes including the forwarding RAM.

Figure 8.14 shows the simulation results of the number of memory accesses versus the number of prefixes. From the simulation results, address lookups of more than 78 percent of prefixes are achieved within two memory accesses, and address lookups of more than 95 percent of prefixes are achieved within three memory accesses. Table 8.4 describes the performance comparison of the proposed scheme with other schemes in terms of the number of memory accesses and the required memory size. Table 8.4 shows the comparison with Huang's scheme [3], DIR-24-8 [4], DIR 21-3-8 [4], and SFT [5]; the proposed scheme requires the smallest memory space except SFT which is based on software scheme. Even though the proposed scheme has the overhead caused by requiring multiple SRAMs and multiple hashing logics, multiple small size SRAMs are simply included on a chip in the current technology and hashing using Exclusive-OR logic is easily implemented and does not require much area. The simulation shows that the number of average memory accesses per packet of the proposed scheme is 1.93 times, which performance is comparable with other hardware-based schemes that require huge memory.

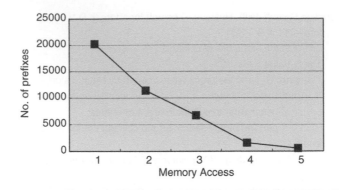

Figure 8.14 Distributions of the number of memory accesses. (From Lim, H., Seo, J., and Jung, Y., *IEEE Communication Letter*, 7, 10, 2003. With permission.)

The maximum number of memory accesses of the proposed scheme is 5. It is the case where a lot of collisions occurred due to small memory allocation to the main table. It is possible to adjust the maximum number of memory accesses by allocating more entries to the main table. Hardware pipelining also can be considered to reduce the number of memory accesses. In other words, if main tables and subtables are implemented using separate SRAMs, then the address lookups in main table and the subtable can be performed in parallel for the incoming stream of datagrams.

8.3 Multiple Hashing Schemes

For some time, it has been known that using multiple hash functions can lead to different performance behaviors than using a single hash function. In this section, we will introduce the concept of multiple hash functions and multiple hashing schemes.

8.3.1 Multiple Hash Function

A single hash function takes a data element (or key) as the input. It provides an index into an array of buckets, usually implemented as linked lists, as the output. The key is stored in the bucket associated with its hash index.

Table 8.4 Performance Comparison with Other Schemes

Address Lookup Scheme	Number of Memory Accesses (Minimum/Maximum)	Forwarding Table Size
Huang's scheme	1/3	450 kB–470 kB
DIR-24-8	1/2	33 MB
DIR-21-3-8	1/3	9 MB
SFT	2/9	150 kB–160 kB
Proposed Architecture	1/5	189 kB

Source: Lim, H., Seo, J., and Jung, Y., *IEEE Communication Letter*, 7, 10, 2003. With permission.

Linked lists have an advantage in that you do not need to set aside memory ahead of time for the buckets. Instead, you assign memory to buckets as keys are placed. They have the following disadvantages:

a) Searching for a key requires several slow accesses to memory, because the linked list may not exhibit good locality for a cache. Because the time for memory accesses generally dominates the computation time for large hash tables, this lack of locality can create a bottleneck.

b) Additional memory for the pointers for the linked list is required.

c) Finding a suitable semi-perfect hash function can be a slow process. Srinivasan and Varghese reported that constructing such a hash function took almost 13 minutes for the forwarding table from MaeEast [6].

d) There is a case where buckets overflow. For example, a cache line holds 32 bytes, and each key requires 8 bytes to store. It only works if at most four items end up in every bucket.

e) There is the potential to waste a lot of memory. A bucket, which can store four keys, contains only one key, three-fourths of the bucket are wasted.

The problems d) and e) are related to the distribution of keys among the buckets. To avoid wasting memory, we need an even spread of keys among the buckets. To avoid overflow, we need a strong guarantee about how the distribution of keys behaves. In fact, the number of keys per bucket is very dramatically, in a hash table, we need a lot of wasted memory to avoid overflow.

To a more even spread of keys among buckets, Broder and Mitzenmacher proposed two hash functions instead of one [7]. Suppose there are n buckets in a hash table (n is even), we split the n buckets into two disjoint equal parts and use two different hash functions to obtain two indices. When a key is inserted, we call both hash functions to find a bucket, the key is placed in the bucket with small number of existing keys; in case of a tie, the key is placed in the bucket on the first part. To search for a key now requires looking in two buckets corresponding to the two hashes.

Assuming that the hash function distributes keys independently and uniformly at random into buckets, for n keys and m buckets, the probability of a bucket ending up with k keys is simply

$$\binom{n}{k}\left(1-\frac{1}{m}\right)^{n-k}\left(\frac{1}{m}\right)^k \approx \frac{e^{-n/m}(n/m)^k}{k!}$$

where $\binom{n}{k}$ is the number of combinations that we choose k out of the n keys to fall in the bucket, $\left(1-\frac{1}{m}\right)^{n-k}$ the probability that $n-k$ keys do not fall in the bucket, and $\left(\frac{1}{m}\right)^k$ the probability that k keys fall in the bucket.

Table 8.5 shows the probability of a bucket having k keys for various n/m for a single hash function [7]. If we have the same number of keys as buckets ($n = m$), then on average there is one key per bucket, but over 0.3 percent of all the buckets will have five or more keys. If $n = m/2$, about 1.6 out of every 10,000 buckets will have five or more keys. If hash tables have tens of thousands or more, there is a lot of wasted memory to avoid overflow.

Table 8.6 shows the approximate probability a bucket has k keys when each key chooses the best of two buckets [7]. Blank entries contain quantities with a value less than $1e-100$, which seemed small enough to discount. The tails of the distribution shown in Table 8.6 decrease much more quickly than that shown in Table 8.5. The power of using two hash functions is rather surprising. For example, when n keys are hashed into n buckets, Table 8.5 shows that $1e-06$ of buckets will have load at least 9 keys if a single hash function is used; in Table 8.6, only about $5.2e-08$ ($\approx 5.2e-08 + 1.2e-21 + 5.3e-58$) of the buckets will have load four or more with two hash functions.

Table 8.5   Probability of a Bucket Having *k* Keys for Various *n/m* for a Single Hash Function

m= k	n/2	n	2n	3n	4n
0	6.1e − 01	3.7e − 01	1.4e − 01	5.0e − 02	1.8e − 02
1	3.0e − 01	3.7e − 01	2.7e − 01	1.5e − 01	7.3e − 02
2	7.6e − 02	1.8e − 01	2.7e − 01	2.2e − 01	1.5e − 01
3	1.3e − 02	6.1e − 02	1.8e − 01	2.2e − 01	2.0e − 01
4	1.6e − 03	1.5e − 02	9.0e − 02	1.7e − 01	2.0e − 01
5	1.6e − 04	3.1e − 03	3.6e − 02	1.0e − 01	1.6e − 01
6	1.3e − 05	5.1e − 04	1.2e − 02	5.0e − 02	1.0e − 01
7	9.4e − 07	7.3e − 05	3.4e − 03	2.2e − 02	6.0e − 02
8	5.9e − 08	9.1e − 06	8.6e − 04	8.1e − 03	3.0e − 02
9	3.3e − 09	1.0e − 06	1.9e − 04	2.7e − 03	1.3e − 02
10	1.6e − 10	1.0e − 07	3.8e − 05	8.1e − 04	5.3e − 03
11	7.4e − 12	9.2e − 06	6.9e − 06	2.2e − 04	1.9e − 03
12	3.1e − 13	7.7e − 06	5.5e − 05	6.4e − 05	6.4e − 04
13	1.2e − 14	5.9e − 11	1.8e − 07	1.3e − 05	2.0e − 04
14	4.2e − 16	4.2e − 12	2.5e − 08	2.7e − 06	5.6e − 05
15	1.4e − 17	2.8e − 13	3.4e − 09	5.5e − 07	1.5e − 05

Source: Broder, A. and Mitzenmacher, M., *IEEE INFOCOM*, 2001. With permission.

An important issue in using hashing for an IP address lookup is how to minimize the collisions. Border and Mitzenmacher proposed multiple hashing to solve the collision problem and discussed the applicability of multiple hashing to IP routing [7]. But they did not build a complete architecture for testing the multiple hashing and comparing its performance against other approaches. Lim and Jung proposed a parallel multiple-hashing architecture (PMHA) to improve the IP lookup performance [8]. Compared with other approaches, it has excellent characteristics in routing table update and in scalability to IPv6. We will describe it in detail.

8.3.2   Multiple Hashing Using Cyclic Redundancy Code

In PMHA, the forwarding table is partitioned into multiple tables by the prefix length, each table is stored into a separated memory, and the address lookup in each table is performed in parallel using hashing. The longest-matching prefix is finally determined among matches gathered from all processes. The multiple-hashing architecture is shown in Figure 8.15. This subsection is from [8]. Portions are reprinted with permission (© 2004 IEEE).

Table 8.6 Approximate Probability a Bucket has *k* Keys When Each Key Chooses the Best of Two Buckets

m= k	n/2	n	2n	3n	4n
0	5.3e − 01	2.3e − 01	3.4e − 02	4.6e − 03	6.2e − 04
1	4.4e − 01	5.5e − 01	2.1e − 01	4.0e − 02	6.9e − 03
2	3.0e − 02	2.2e − 01	5.0e − 01	2.0e − 01	4.3e − 02
3	8.6e − 06	4.4e − 03	2.6e − 01	4.8e − 01	1.9e − 01
4	9.2e − 16	5.2e − 08	9.1e − 03	2.7e − 01	4.7e − 01
5	1.4e − 42	1.2e − 21	5.0e − 07	1.2e − 02	2.8e − 01
6		5.3e − 58	7.2e − 19	1.1e − 06	1.3e − 02
7			1.5e − 50	6.6e − 18	1.6e − 06
8				5.7e − 48	1.8e − 17
9					8.4e − 47

Source: Broder, A. and Mitzenmacher, M., *IEEE INFOCOM*, 2001. With permission.

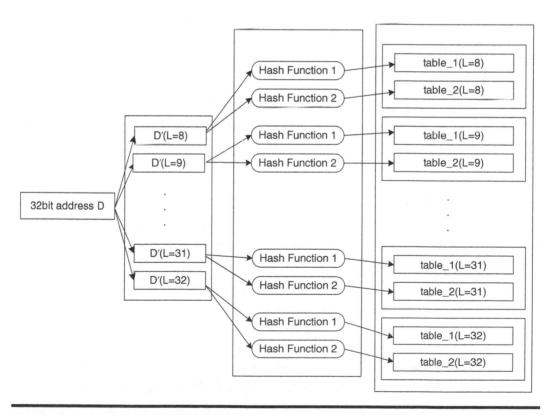

Figure 8.15 A parallel multiple-hashing architecture. (From Lim, H. and Jung, Y., *Proc. IEEE HPSR2004*, April 2004. With permission.)

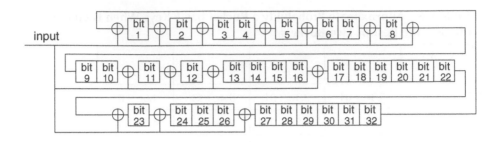

Figure 8.16 CRC-32. (From Lim, H. and Jung, Y., *Proc. IEEE HPSR2004*, April 2004. With permission.)

Hashing is implemented with cyclic redundancy code (CRC) checker, which is known as a semi-perfect hash function. Multiple hash indices for each prefix length are extracted from a single CRC hardware, shown in Figure 8.16.

Figure 8.16 depicts the CRC-32 hashing hardware structure [8]. Because the forwarding table in the proposed scheme is organized independently by the prefix length, separate hash indices are required as indices of routing tables in each prefix length. Extracting hash indices from CRC hashing hardware is explained as follows. First, each bit of destination IP address is serially entered into CRC hashing hardware. After L (for $L = 8, 9, ..., 32$) cycles, two fixed hash indices for prefix L are extracted from the register 0 to the register $L - 1$. By repeating the same procedure, hash indices for different prefix lengths are taken at different timings from CRC registers in the proposed scheme. For example, hash index for prefix length 8 is taken from CRC registers after eight cycles, and one clock cycle later, the hash index for prefix length 9 is chosen. All the hash indices for each routing table are available after 32 cycles.

If the forwarding engine operates at 100 MHz clock, the required time to obtain all the hash indices is 320 ns. Suppose that the minimum size packet length is 72 bytes, a router which has the aggregated bandwidth of up to 2.1 Gbps works at line rate. At 200 MHz clock, a router with the aggregated bandwidth of up to 4.2 Gbps works at line rate. Beyond the rate, multiple-hashing hardware is required. Therefore, the 32 cycles to obtain all the hash indices do not prevent routers from working at line rate.

8.3.3 Data Structure

When N prefixes are hashed into $N/2$ buckets using two hash indices, the analysis of [7] shows that the probability for a bucket to have two or more loads is 5.0e − 7. Figure 8.17 shows the bucket structure, which stores maximum two loads. Each bucket of the routing table consists of a field to indicate the number of items and multiple fields to store loads, and each load consists of a field for the prefix and a field for the forwarding RAM pointer.

The length of hash bit is determined according to the number of buckets. To store N prefixes into tables with total capacity of $2N$ loads, we need two tables composed of $N/2$ buckets, each bucket

Number of Items (2bits)	Prefix (8–32bits)	Forwarding RAM pointer (5bits)	Prefix (8–32bits)	Forwarding RAM pointer (5bits)

Figure 8.17 Data structure in PHMA. (From Lim, H. and Jung, Y., *Proc. IEEE HPSR2004*, April 2004. With permission.)

```
For prefix length L, P[L-1:0]
P[L-1:0] serially entered to CRC hash function
Extract H1(L), H2(L) from CRC registers after L cycles
Do
    table1_ptr=H1(L)
    table2_ptr=H2(L)
    If (( # of load(table1_ptr)) == (# of load(table2_ptr)) == 2)
        Then put P[L-1:0] to overflow table
    Else if ( # of load(table1_ptr) > # of load(table2_ptr))
        Then put P[L-1:0] to table2
    Else put P[L-1:0] to table1
End Do
```

Figure 8.18   Algorithm to build a forwarding table for PHMA. (From Lim, H. and Jung, Y., *Proc. IEEE HPSR2004*, April 2004. With permission.)

having two loads. Because hash index is used as an index of routing table, the required length of hash index is the nearest integer of $\log_2(N/2)$. We use the minimum length of hash index as 2.

Data structure is built using the algorithm shown in Figure 8.18. First, a prefix enters into CRC hashing hardware bit by bit, and after L (L is the prefix length) cycles, two hash indices are extracted from CRC registers between bit 0 and bit $L - 1$. Each hash index indicates a bucket of each table, and the new prefix is stored into the bucket, which has smaller number of loads. If two buckets have equal loads, the prefix is placed in the bucket of the first table in default. In case of overflows, the prefix is stored into the overflow table. In the proposed scheme, two hash tables per each length of prefixes are used, and hence total 48 hash tables (prefixes 8–32, except the prefix length 31) and an overflow table are used to insert a new prefix when both buckets are full and have no space to store more prefix.

8.3.4   Searching Algorithms

Searches in each table are executed in parallel using the hash indices obtained from CRC hash function. As mentioned earlier, 48 hash indices are obtained from a single CRC hardware. As shown in Figure 8.19, entries in each table are concurrently searched for the buckets indicated by hash indices. Additionally, overflow ternary content addressable memory (TCAM) is also searched in parallel. The LPM is selected by the priority encoder among matching entries resulting from the tables and the overflow TCAM. The packet is finally forwarded to the output port pointed by the forwarding RAM pointer indicated by the selected entry. Figure 8.20 is the block diagram of searching procedure. As shown in Figure 8.20, only a single memory access time is required because lookup in each prefix length is performed.

8.3.5   Update and Expansion to IPv6

Routing table update for the proposed scheme is incremental. Update process is the same as building process. The newly added prefix is located into the bucket with fewer loads. If both buckets are full, the newly added prefix is stored into the overflow TCAM. Prefix deletion is also incremental. After deleting the prefix from the bucket indicated by hash index, loads of the bucket are rearranged to make no invalid entry between the valid entries, and the number of items is reduced by one. If the

At cycle L(for L=8~32), let D[31:31-L+1] is L bits of destination address D.

D[31:31-L+1] serially entered to CRC hash function

Extract H1(L), H2(L) from CRC registers

Do Parallel (L=8~32)

    table1_ptr=H_1(L)

    table2_ptr=H_2(L)

    If(D[31:31-L+1]=prefix(table1_ptr))

       Then fwd_ptr=fwd_ptr(table1_ptr)

    Else if(D[31:31-L+1]=prefix(table2_ptr))

       Then fwd_ptr=fwd_ptr(table2_ptr)

End Do Parallel

Search from overflow CAM

Determine LPM among matching entries

Figure 8.19   Searching algorithm for PHMA. (From Lim, H. and Jung, Y., *Proc. IEEE HPSR2004*, April 2004. With permission.)

bucket has no matching prefix, the prefix is searched on an overflow table and deleted. It does not require long computation to build a new table, and hence fast update is achieved. Too many memories may be required toward IPv6 because the proposed scheme uses separate memory in each prefix length, and the problem is expected to be solved by prefix grouping.

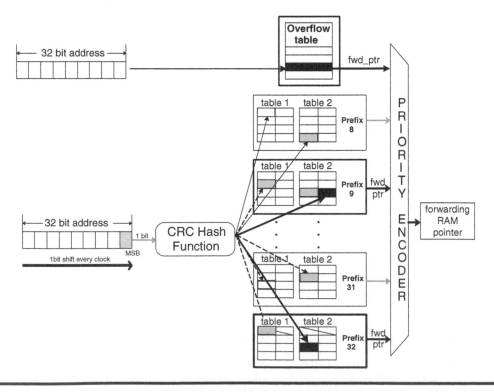

Figure 8.20   Search procedure for PHMA. (From Lim, H. and Jung, Y., *Proc. IEEE HPSR2004*, April 2004. With permission.)

8.3.6   Performance Comparison

Lim and Jung performed IP lookup simulation for PMHA using the forwarding table for a snapshot of the MaeWest (2002/03/15), which has 29,584 prefixes. Assuming that N items are hashed into $N/2$ buckets, Lim and Jung performed several test cases in which the number of entries per bucket and the number of hash functions are different. The results are shown in Table 8.7.

Case 1 is to use a single hash function and four entries per bucket. For each prefix length (except the prefix length 31), a single table is used, and hence a total of 24 tables and an overflow TCAM are used. About 203 kBytes of memory is required and the overflow rate is 3.4 percent. Case 2 uses two hash functions and two entries per bucket, and the overflow rate is significantly reduced to 0.52 percent (154 entries). The reason is that the prefixes are more evenly distributed using multiple hash functions. In Case 3, three entries per bucket is used, and the overflow rate is completely removed. Case 4 uses $N/4$ buckets for N items, three hash functions, and two entries per bucket. The case consumes 152 kBytes of memory and 136 overflows occurred, which can be chosen for optimum memory usage. The test cases show that multiple hash functions improve the hashing performance, and the memory efficiency is traded off with memory overflow rate.

The searching procedure of PMHA is shown in Figure 8.20; only single memory access time is required because lookup in each prefix length is performed in parallel. PMHA outperforms the algorithms based on trie structure. The memory access time is equal to that of TCAM. For a forwarding table with 29,584 prefixes, PMHA only needs 203 kBytes of memory and a few-hundred-entry TCAM. If we use TCAM to implement the forwarding engine, a large TCAM is needed and it has higher power consumption. For example, a typical 18M TCAM can store up to 512 K 32-bit prefixes and can consume up to 15 W of power when all the entries are searched. Therefore, PHMA has advantages in required memory size and the number of memory accesses.

8.4   Using Bloom Filter

8.4.1   Standard Bloom Filter

The Bloom filter, conceived by Bloom in 1970 [9], is a space-efficient probabilistic data structure that is used to test whether or not an element is a member of a set. A Bloom filter consists of two components: a set of k hash functions and a bit vector of m bits. Initially, all bits of the vector are set to 0. Given a set X with n members, for each member $x_i \in X$, k hash functions produce k values: $h_1^i, h_2^i, h_3^i, \ldots, h_k^i$ ($1 \le h_j^i \le m$, $1 \le j \le k$). Each of these values addresses a single bit in the

Table 8.7   Memory Size and Entry Efficiency Analysis

Case	No. of Bucket	No. of Hash	Entry/ Bucket	Memory Size (KB)	Memory Efficiency (%)	Overflow Rate (%)
1	$N/2$	1	4	203	49.85	3.4
2	$N/2$	2	2	203	49.85	0.52
3	$N/2$	2	3	303	33.41	0
4	$N/4$	3	2	152	66.3	0.46

N is the number of items.

Source: Lim, H. and Jung, Y., *Proc. IEEE HPSR2004*, April 2004. With permission.

m-bit vector. These bits in the *m*-bit vector are set to 1. A bit in the *m*-bit vector can be set to 1 multiple times, but only the first change has an effect.

For example, consider a Bloom filter with three hash functions and a bit vector of length 12. Initially, all bits are set to 0, as shown in Figure 8.21a. Let us now add the string "apple" into the filter. Three hash functions give the values: Hash_1(apple) = 3; Hash_2(apple) = 11; Hash_3(apple) = 12. The bits in positions 3, 10, and 11 of the bit vector are set to 1, as shown in Figure 8.21b. To add another key "banana," three functions give the values: Hash_1(banana) = 8; Hash_2(banana) = 1; Hash_3(banana) = 11. The bits in positions 3 and 8 of the bit vector are set to 1, the bit at position 11 was already set to 1 in the previous step, now is not set again, shown in Figure 8.21c. This means that the bit 11 may store information for both "apple" and "banana." That is to say, any one bit in the vector may be encoding multiple keys simultaneously.

Querying the Bloom filter for set membership of a given key is similar to adding a key. *k* hash functions give the *k* values. The bits at positions corresponding to the *k* hash values are checked. If at least one of the bits is 0, then the key is a nonmember of the set. If all bits are 1, then the key belongs to the set with a certain probability. For example, three hash functions give the values of the key "mango": Hash_1(mango) = 8; Hash_2(mango) = 3; Hash_3(mango) = 12. The bits at positions 3, 8, and 12 are 1. In fact, the key "mango" is not a member of the set: {apple, banana}.

Given a set X and a key x, the Bloom filter implies that the key x certainly does not belong to the set X and that the key x belongs to the set X with a certain probability, called the false-positive probability. The number of sets is n, and the Bloom filter is the *m*-bit vector with k hash functions. The probability that a random bit of the *m*-bit vector is set to 1 by a hash function is simply $\frac{1}{m}$. The probability that it is not set is $1 - \frac{1}{m}$. The probability that it is not set by any of the n members of X is $\left(1 - \frac{1}{m}\right)^n$. Because each key set k bits in the vector, it becomes $\left(1 - \frac{1}{m}\right)^{nk}$. Hence, the probability that this bit is found to be 1 is $1 - \left(1 - \frac{1}{m}\right)^{nk}$. For a message to be detected as a possible member of the set, all k bit positions generated by the hash functions need to be 1. The probability that this happens, f, is given by

$$f = \left(1 - \left(1 - \frac{1}{m}\right)^{nk}\right)^k \approx \left(1 - e^{-\frac{nk}{m}}\right)^k \qquad \text{(for large values of } m\text{)} \qquad (8.1)$$

1	2	3	4	5	6	7	8	9	10	11	12
0	0	0	0	0	0	0	0	0	0	0	0

(a) An empty bit vector

1	2	3	4	5	6	7	8	9	10	11	12
0	0	1	0	0	0	0	0	0	0	1	1

(b) An bit vector after adding "*apple*"

1	2	3	4	5	6	7	8	9	10	11	12
1	0	1	0	0	0	0	1	0	0	1	1

(c) An bit vector after adding "*apple*" and "*banana*"

Figure 8.21 An example of a standard Bloom filter.

Because this probability is independent of the input, it is termed the false-positive probability. It can be reduced by choosing appropriate values for m and k for a given size of the set, n. It is clear that the size of the bit vector, m, needs to be quite large compared with the size of the set, n. For a given ratio of $\frac{m}{n}$, the false-positive probability can be reduced by increasing the number of hash functions, k. In the optimal case, when the false-positive probability is minimized with respect to k, we get the following relationship

$$k = \frac{m}{n} \ln 2. \tag{8.3}$$

The false-positive probability at the optimal point is given by

$$f = \left(\frac{1}{2}\right)^k. \tag{8.4}$$

It should be noted that if the false-positive probability is to be fixed, then the size of the filter, m, needs to scale linearly with the size of the set, n.

8.4.2 Counting Bloom Filter

When a set is changing over time, inserting element into a Bloom filter is easy: hash the element k times and set the bits to 1. Unfortunately, one cannot perform a deletion in the Bloom filter. Deleting a particular entry requires that corresponding k hashed bits in the bit vector be set to 0. If the bit stores multiple keys, this could disturb the other key. To solve this problem, Fan et al. proposed the *Counting Bloom filters* [10]. In a counter Bloom filter, each entry in the vector is not a single bit but instead a small counter. When a key is inserted, the corresponding counters are incremented; when a key is deleted, the corresponding counters are decremented. To avoid counter overflow, we choose sufficiently large counters.

8.4.3 Basic Configuration of LPM Using Bloom Filter

A basic configuration of the longest prefix matching using Bloom filters is shown in Figure 8.22 [11]. The prefixes in a forwarding table are grouped into sets according to the prefix length. The system employs a set of W counting bloom filters, where W is the length of the input addresses, and associates one filter with each unique prefix length. While the bit-vector-associated set of prefixes with each Bloom filter are stored in embedded memory, the counters associated with each filter are maintained by a separate control processor responsible for managing route updates. Separate control processors with ample memory are common features of high-performance routers. This subsection is from [11]. Portions are reprinted with permission (© 2006 IEEE).

A hash table is also constructed for each distinct prefix length. Each hash table is initialized with the set of corresponding prefixes, where each hash entry is a [prefix, next hop] pair. While we assume that the result of a match is the next hop for the packet, more elaborate information may be associated with each prefix if so desired. The set of hash tables is stored off-chip in a separate memory device (assuming that it is a large, high-speed SRAM). The problem of constructing hash tables to minimize collisions with reasonable amounts of memory is well studied. We will discuss the number of memory access for an IP lookup. For the purpose of our discussion, we assume that probing a hash table stored in off-chip memory requires one memory access. A search proceeds as follows. The input IP address is used to probe the set of W Bloom filters in parallel. One-bit prefix

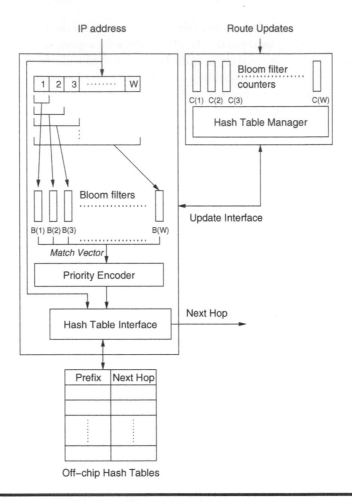

Figure 8.22 A basic configuration of longest prefix matching using Bloom filters. (From Dharmapurikar, S., Krishnamurthy, P., and Taylor, D., *IEEE/ACM Transactions on Networking,* 14, 2, 2006. With permission.)

of the address is used to probe the filter associated with length 1 prefixes, two-bit prefix of the address is used to probe the filter associated with length 2 prefixes, etc. Each filter simply indicates match or no match. By examining the outputs of all filters, we compose a vector of potentially matching prefix lengths for the given address, which we will refer to as match vector. Consider an IPv4 example in which the input address produces matches in the Bloom filters associated with prefix lengths 8, 17, 23, and 30; the resulting match vector would be {8, 17, 23, 30}. Remember that Bloom filters may produce false positives, but never produce false negatives; therefore, if a matching prefix exists in the database, it will be represented in the match vector. Note that the number of unique prefix lengths in the prefix database, W_{dist}, may be less than W. In this case, the Bloom filters representing empty sets will never contribute a match to the match vector, valid or false positive.

The search proceeds by probing the hash tables associated with the prefix lengths represented in the match vector in the order of the longest prefix to the shortest. The search continues until a match is found or the vector is exhausted. We now establish the relationship between the false-positive probability of each Bloom filter and the expected number of hash probes per lookup.

The number of hash probes required to determine the correct prefix length for an IP address is determined by the number of matching Bloom filters. For an IP address, which matches a prefix of length l, we first examine any prefix lengths greater than l represented in the *match vector*. Let B_l represent the number of bloom filters for the prefixes of length greater than l. Assuming that all Bloom filters share the same false-positive probability, f, the probability that exactly i filters associated with prefix lengths greater than l will generate false positives is given by

$$P_l = \binom{B_l}{i} f^i (1-f)^{B_l - i} \tag{8.5}$$

For each value of i, we would require i additional hash probes. The expected number of additional hash probes required when matching a length l prefix is

$$E_l = \sum_{i=1}^{B_l} i \binom{B_l}{i} f^i (1-f)^{B_l - i} \tag{8.6}$$

which is the mean for a binomial distribution with B_l elements and a probability of success f. Hence,

$$E_l = B_l f \tag{8.7}$$

Equation 8.7 shows that the expected number of additional hash probes for the prefixes of a particular length is equal to the number of Bloom filters for the longer prefixes times the false-positive probability (which is the same for all the filters). Let B be the total number of Bloom filters, the worst-case value of E_l, which we denote as E_{add}, can be expressed as

$$E_{add} = Bf \tag{8.8}$$

which is independent of the input address. Because these are the expected additional probes due to the false positives, the total number of expected hash probes per lookup for any input address is

$$E_{exp} = E_{add} + 1 = Bf + 1 \tag{8.9}$$

where the additional one probe accounts for the probe at the matching prefix length. Note that there is the possibility that an IP address creates false-positive matches in all the filters. In this case, the number of required hash probes is

$$E_{worst} = B + 1 \tag{8.10}$$

Because the expected number (E_{exp}) and the maximum number (E_{worst}) of hash probes for a LPM depend on the number of filters B, reducing B is important for limiting the worst case.

8.4.4   Optimization

In this section, we seek to develop a system configuration that provides a high performance independent of prefix database characteristics and input address patterns. The design goal is to architect a search engine that achieves an average of one hash probe per lookup, bounds the worst-case search, and utilizes a small amount of embedded memory. Several variables affect system performance and resource utilization:

- N, the target amount of prefixes supported by the system;
- M, the total amount of embedded memory available for Bloom filters;
- W_{dist}, the number of unique prefix lengths supported by the system;
- m_i, the size of each Bloom filter;
- k_i, the number of hash functions computed in each Bloom filter;
- n_i, the number of prefixes stored in each Bloom filter;
- f_i, the false-positive probability of the ith Bloom filter.

Assuming that each individual Bloom filter has the same false positive f, the system performance is independent of the prefix distribution. Given that a filter is allocated m_i bits of memory, stores n_i prefixes, and performs $k_i = \frac{m_i}{n_i} \ln 2$ hash functions, the expression for f_i becomes,

$$f_i = f = \left(\frac{1}{2}\right)^{\frac{m_i}{n_i} \ln 2} \qquad \forall i \in [1, 32].\tag{8.11}$$

This implies that

$$\frac{m_i}{n_i} = \frac{m_{i+1}}{n_{i+1}} = \frac{\sum m_i}{\sum n_i} = \frac{M}{N} \qquad \forall i \in [1, 32].\tag{8.12}$$

Therefore, the false-positive probability f_i for a given filter i may be expressed as

$$f_i = f = \left(\frac{1}{2}\right)^{\frac{M}{N} \ln 2}.\tag{8.13}$$

Based on the preceding analysis, the expected number of hash probes per lookup depends on the total amount of memory resources, M, and the total number of supported prefixes, N. It is important to note that this is independent of the number of unique prefix lengths and the distribution of prefixes among the prefix lengths.

8.4.4.1 Asymmetric Bloom Filters

Based on the examination of real IP forwarding tables from http://bgp.potaroo.net/ (February 2003), the distribution of prefixes is not uniform over the set of prefix lengths and demonstrated common trends such as large numbers of 24-bit prefixes and few prefixes of length less than 8 bits. An average prefix distribution for all the tables is shown in Figure 8.23.

If we use a static system configured uniformly distribution prefix lengths to search a database with nonuniform prefix length distribution, some filters have "over-allocated" memory, whereas others have "under-allocated"; thus the false-positive probabilities for the Bloom filters are no longer equal. Clearly, we need to proportionally allocate the amount of embedded memory per filter based on its current share of the total prefixes while adjusting the number of hash functions to maintain minimal false-positive probability. Dharmapurikar and Paxson refer to this configuration

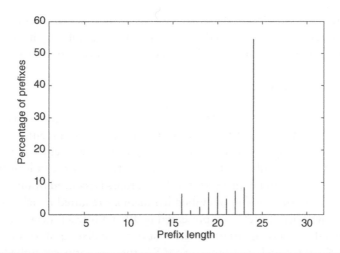

Figure 8.23 Average prefix length distribution for IPv4 BGP table snapshots. (From Dharmapurikar, S., Krishnamurthy, P., and Taylor, D., *IEEE/ACM Transactions on Networking,* 14, 2, 2006. With permission.)

as "asymmetric Bloom filters" [11]. For IPv4, $B = 32$, the expected number of hash probes per lookup, E_{exp} may be expressed as

$$E_{exp} = 32 \times \left(\frac{1}{2}\right)^{\frac{M \ln 2}{N}} + 1 \tag{8.14}$$

The worst-case number of dependent memory accesses is simply 33. Given the feasibility of asymmetric Bloom filters, the expected number of hash probes per lookup, E_{exp}, is shown in Figure 8.24. With a modest 2 MBytes embedded memory, the expected number of hash probes

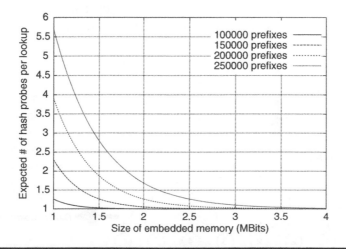

Figure 8.24 Expected number of hash probes per lookup (E_{exp}) versus total embedded memory size (M) for various values of total prefixes (N) using a basic configuration for IPv4 with asymmetric Bloom filters. (From Dharmapurikar, S., Krishnamurthy, P., and Taylor, D., *IEEE/ACM Transactions on Networking,* 14, 2, 2006. With permission.)

per lookup is less than two for 250,000 prefixes. We assert that such a system is also memory efficient, as it only requires 8 bits of embedded memory per prefix. Doubling the size of the embedded memory to 4 MBytes provides near-optimal average performance of one hash probe per lookup.

8.4.4.2 Direct Lookup Array

Based on the statistics that sets associated with the first few prefix lengths are typically empty and the first nonempty sets hold few prefixes shown in Figure 8.23, Dharmapurikar et al. suggest that utilizing a direct lookup array for the first *a* prefix lengths is an efficient way to represent shorter prefixes while reducing the number of Bloom filters [11]. For every prefix length, we represent in the direct lookup array, the number of worst-case hash probes is reduced by one. The use of a direct lookup array also reduces the amount of embedded memory required to achieve optimal average performance, as the number of prefixes represented by Bloom filters is decreased.

An example of a direct lookup array for the first $a = 3$ prefixes is shown in Figure 8.25. The prefixes of length $\leq a$ are stored in a binary trie. We then perform controlled prefix expansion (CPE) [6] for a stride length equal to *a*. The next hop associated with each leaf at level *a* is written to the array slot addressed by the bits labeling the path from the root to the leaf. The structure is searched by using the array. For example, an address with initial bits 101 would result in a next hop of 4. Note that this data structure required $2^a \times NH_{len}$ bits of memory, where NH_{len} is the number of bits required to represent the next hop.

For the purpose of a realistic analysis, we select $a = 20$ resulting in a direct lookup array with 1M slots. For a 256 port router where the next hop corresponds to the output port, 8 bits are required to represent the next hop value and the direct lookup array requires 1 MBytes of memory. The use of a lookup array for the first 20 prefix lengths leaves prefix length 21:32 to Bloom filters; hence, the expression for the expected number of hash probes per lookup becomes

$$E_{\exp} = 12 \times \left(\frac{1}{2}\right)^{\frac{M \ln 2}{N - N_{[0:20]}}} + 1 \tag{8.15}$$

where $N_{[0:20]}$ is the sum of the prefixes with lengths [0:20]. On average, the $N_{[0:20]}$ prefixes constituted 24.6 percent of the total prefixes in the sample IPv4 BGP tables; therefore, 75.4 percent

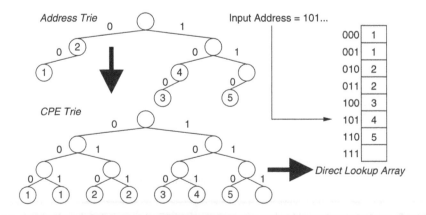

Figure 8.25 Example of a direct lookup array for the first three prefix lengths. (From Dharmapurikar, S., Krishnamurthy, P., and Taylor, D., *IEEE/ACM Transactions on Networking,* 14, 2, 2006. With permission.)

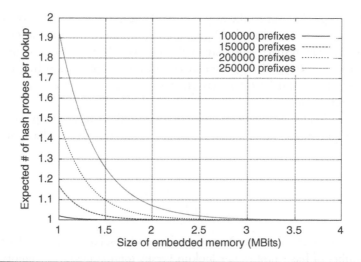

Figure 8.26 Expected number of hash probes per lookup (E_{exp}) versus total embedded memory size (M) for various values of total prefixes (N) using a direct lookup array for prefix lengths 1, ..., 20 and 12 Bloom filters for prefix lengths 21, ..., 32. (From Dharmapurikar, S., Krishnamurthy, P., and Taylor, D., *IEEE/ACM Transactions on Networking,* 14, 2, 2006. With permission.)

of the total prefixes N are represented in the Bloom filters. Given this distribution of prefixes, the expected number of hash probes per lookup versus total embedded memory size for various values of N is shown in Figure 8.26.

The expected number of hash probes per lookup for database containing 250,000 prefixes is less than two when using a small 1 MByte embedded memory. Doubling the size of the memory to 2 MBytes reduces the expected number of hash probes per lookup to less than 1.1 for 250,000 prefix database. While the amount of memory required to achieve good average performance has decreased to only 4 bits per prefix, the worst-case hash probes per lookup is still large. Using Equation 8.10, the worst-case number of dependent memory accesses becomes $E_{worst} = (32 - 20) + 1 = 13$. For an IPv4 database containing the maximum of 32 unique prefix lengths, the worst case is 13 dependent memory accesses per lookup. A high-performance implementation option is to make the direct lookup array the final stage in a pipelined search architecture. IP addresses, which reach this stage with a null next hop value, would use the next hop retrieved from the direct lookup array. A pipelined architecture does require a dedicated memory bank or port for the direct lookup array. The following section describes the additional steps to further improve the worst case.

8.4.4.3 Reducing the Number of Filters

We can reduce the number of remaining Bloom filters by limiting the number of distinct prefix lengths via further use of CPE. We would like to limit the worst-case hash probes as few as possible without prohibitively large embedded memory requirements. Clearly, the appropriate choice of CPE strides depends on the prefix distribution. In Figure 8.23, there is a significant concentration of prefixes from 21 to 24. On average, 24.6 percent of the prefixes fall in the range of 1–20, which are stored in the direct lookup array; 75.2 percent of the N prefixes fall in the range of 21–24; 0.2 percent of the N prefixes fall in the range 25–32.

Based on these observations, Dharmapurikar et al. divide the prefixes not covered by the direct lookup into two groups: G_1 and G_2 corresponding to prefix lengths 21–24 and 25–32, respectively

[11]. Each group is expanded out to the upper limit of the group so that G_1 contains only length 24 prefixes and G_2 contains only length 32 prefixes. Let N_{24} and N_{32} be the number of prefixes in G_1 and G_2 after expansion, respectively. Using CPE increases the number of prefixes in each group by an "expansion factor" α_{24} and α_{32}, respectively. In the sample forwarding table, the average value of α_{24} and α_{32} is 1.8 and 49.9, respectively. Such a large value of α_{32} is tolerable due to the small number of prefixes in G_2.

The use of this technique results in two Bloom filters and a direct lookup array, the worst-case lookup is two hash probes and an array lookup. The expected number of hash probes per lookup becomes

$$E_{exp} = 2 \times \left(\frac{1}{2}\right)^{\frac{M \ln 2}{\alpha_{24} N_{24} + \alpha_{32} N_{32}}} + 1 \tag{8.16}$$

Using the observed average distribution of prefixes and observed average values of α_{24} and α_{32}, the expected number of hash probes per lookup versus total embedded memory M for various values of N is shown in Figure 8.27.

The expected number of hash probes per lookup for database containing 250,000 prefixes is less than 1.6 when using a small 1 MByte embedded memory. Doubling the size of the memory to 2 MBytes reduces the expected number of hash probes per lookup to less than 1.2 for 250,000 prefix database. The use of CPE to reduce the number of Bloom filters provides for a maximum of two hash probes and one array access per lookup while maintaining near-optimal average performance with modest use of embedded memory resources.

If the scheme is implemented in current semiconductor technology and coupled with a commodity SRAM device operating at 333 MHz, it could achieve average performance of over 300 million lookups per second and worst-case performance of over 100 million lookups per second.

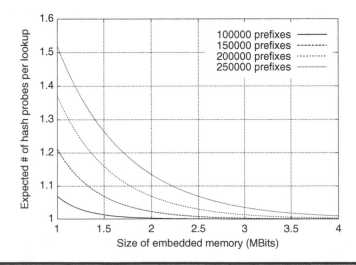

Figure 8.27 **Expected number of hash probes per lookup (E_{exp}) versus total embedded memory size, M, for various values of total prefixes (N) using a direct lookup array for prefix lengths 1, ..., 20 and two Bloom filters for prefix lengths 21, ..., 24 and 25, ..., 32. (From Dharmapurikar, S., Krishnamurthy, P., and Taylor, D., *IEEE/ACM Transactions on Networking*, 14, 2, 2006. With permission.)**

In comparison, state-of-the-art TCAM-based forwarding engines provide 100 million lookups per second, consume 150 times more power per bit of storage than SRAM, and cost approximately 30 times as much per bit of storage than SRAM.

8.4.5 Fast Hash Table Using Extended Bloom Filter

In counting Bloom filter, each bit of filter is replaced by a counter. A counter in the filter essentially gives us the number of items hashed in it, Song et al. used this information efficiently to minimize the search time [12]. The variant of counting Bloom filter is called the fast hash table. We will describe the basic algorithms and its optimization and compare them with the baseline—a naïve hash table (NHT) that consists of an array of m buckets pointing to the list of items hashed into it.

8.4.5.1 Basic Fast Hash Table

In a counter Bloom filter with m counters, each counter C_i is associated with bucket i of the hash table. We compute k hash functions $h_1(), ..., h_k()$ over an input item and increment the corresponding k counters indexed by these hash values. Then we store the item in the lists associated with each of the k buckets. Thus, a single item is stored k times in the off-chip memory. The following algorithm describes the insertion of an item in the table:

InsertItem$_{RFHT}$(x)

    1   for ($i = 1$ to k)
    2   if ($h_i(x) \neq h_j(x)\ \forall j < i$)
    3            $C_{hi}(x) + +$
    4            $X^{hi(x)} = X^{hi(x)} \cup x$

If more than one hash functions map to the same address, then we increment the counter only once and store just one copy of the item in that bucket. To check if the hash values conflict, we keep all the previously computed hash values for that item in registers and compare the new hash value against all of them (line 2).

Figure 8.28 shows a counting filter after four different items, x, y, z, and w have been inserted, sequentially. Each of the items is replicated in $k = 3$ different buckets and counter value associated with the bucket reflects the number of items in it.

Search procedure is similar to the insertion procedure: given an item x to be searched, we compute k hash values and read the corresponding counters. When all the counters are non-zero, the filter indicates the presence of input item in the table. Only after this step, we proceed to verify it in the off-chip table by comparing it with each item in the list of items associated with one of the buckets. Indeed, if the counters are kept in the fast on-chip memory such that all of the k random counters associated with the item can be checked in parallel, then in almost all cases we avoid an off-chip access if the table does not contain the given item. Given the recent advances in the embedded memory technologies, it is conceivable to implement these counters in a high-speed multi-port on-chip memory.

Secondly, the choice of the list to be inspected is critical because the list traversal time depends on the length of the list. Hence, we choose the list associated with the counter with smallest value to reduce off-chip memory accesses. The speedup comes from the fact that it can choose the smallest list to search, whereas an NHT does not have any choice but trace only one list, which can potentially have several items in it.

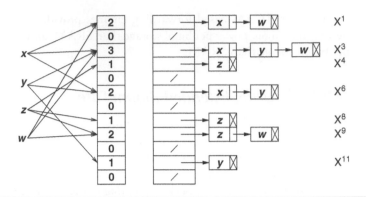

Figure 8.28 Basic fast hash table. The data structure after inserting x, y, z, and w. (From Song, H., et al., *ACM SIGCOMM Computer Communication Review*, 35, 4, 2005. With permission.)

In most of the case, for a carefully chosen value of the number of buckets, the minimum value counter has a value of 1 requiring just a single memory access to the off-chip memory. In Figure 8.28, if the item y is queried, we need to access only the list X^{11}, rather than X^3 or X^6, which are longer than X^{11}.

When multiple counters indexed by the input item have the same minimum value, then somehow the tie must be broken. We break this tie by simply picking the minimum value counter with the smallest index. For example, in Figure 8.28, item x has two bucket counters set to 2, which is also the smallest value; in this case, we always access the bucket X^1. The following pseudo-code summarizes the search algorithm on BFHT.

SearchItem$_{BFHT}$(x)

1    $C_{min} = \min\{C_{h1(x)}, \ldots, C_{hk(x)}\}$

2   if $(C_{min} == 0)$

3   return false

4   else

5        $i = \text{SmallestIndexOf}(C_{min})$

6   if $(x \in X^i)$ return true

7   else return false

Finally, if the input item is not present in the item list, then clearly it is a false-positive match indicated by the counting Bloom filter.

With the data structure above, deletion of an item is easy. We simply decrement the counters associated with the item and delete all the copies from the corresponding lists. The following is the pseudo-code of the deletion algorithm.

DeleteItem$_{BFHT}$(x)

1   for $(i = 1$ to $k)$

2   if $(h_i(x) \neq h_j(x) \; \forall j < i)$

3            $C_{h_i(x)} --$

4            $X^{h_i(x)} = X^{h_i(x)} - x$

8.4.5.2 Pruned Fast Hash Table

In BFHT, we need to maintain up to k copies of each item, which requires k times more memory compared with NHT. However, it can be observed that in a BFHT only one copy of each item—the copy associated with the minimum counter value—is accessed when the table is probed. The remaining $(k-1)$ copies of the item are never accessed. This observation offers us the first opportunity for significant memory optimization: all the other copies of an item except the one that is accessed during the search can now be deleted. Thus, after this pruning procedure, we have exactly one copy of the item, which reduces the memory requirement to the same as that of the NHT. We call the resulting hash table a pruned fast hash table (PFHT). The following pseudo-code summarizes the pruning algorithm.

PruneSet(x)

1   for (each $x \in X$)
2       $C_{\min} = \min\{C_{h_1 (x)}, \ldots, C_{h_k(x)}\}$
3       $i = \text{SmallestIndexOf}(C_{min})$
4   for ($l = 1$ to k)
5   if ($h_l (x) \neq i$) $X^{h_l (x)} = X^{h_l (x)} - x$

The pruning procedure is illustrated in Figure 8.29a. It is important to note that during the pruning procedure, the *counter values are not changed*. Hence, after pruning is completed, the counter value no longer reflects the number of items actually presenting in the list and is usually greater than that. However, for a given item, the bucket with the smallest counter value always contains this item. This property ensures the correctness of the search results. Another property of pruning is that it is independent of the sequence in which the items are pruned because it depends just on the counter values, which are not altered. Hence, pruning in sequence $x–y–z–w$ will yield the same result as pruning it in $z–y–x–w$.

A limitation of the pruning procedure is that now the incremental updates to the table are hard to perform. Because counter values no longer reflect the number of items in the list, if counters are incremented or decremented for any new insertion or deletion, respectively, then it can disturb the counter values corresponding to the existing items in the bucket which in turn will result in an incorrect search. For example, in Figure 8.29a, the item y maps to the lists $\{X^3, X^6, X^{11}\}$ with counter values $\{3, 2, 1\}$, respectively. If a new item, say v, is inserted which also happens to share the bucket 11 then the counter will be incremented to 2. Hence, the minimum counter value bucket with smallest index associated with y is no longer 11 but now it is 6, which does not contain y at all. Therefore, a search on y will result in an incorrect result. With this limitation, for any new insertion and deletion, the table must be reconstructed from scratch, which can make this algorithm impractical for variable item sets.

We now describe a version of InsertItem and DeleteItem algorithms, which can be performed incrementally. The basic idea used in these functions is to maintain the invariant that out of the k buckets indexed by an item, it should always be placed in a bucket with smallest counter value. In case of a tie, it should be placed in the one with smallest index. If this invariant is maintained at every point, then the resulting hash table configuration will always be the same irrespective of the order in which items are inserted.

To insert an item, we first increment the corresponding k counters. If there are any items already present in those buckets, then their corresponding smallest counter might be altered. However, the counter increments do not affect all other items. Hence, each of these items must be

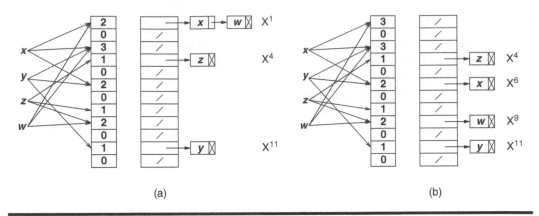

Figure 8.29   Illustration of PFHT. (a) The data structure after execution of pruning. (b) List balancing. Items *x* and *w* are readjusted into different buckets by incrementing counter 1. (From Song, H., et al., *ACM SIGCOMM Computer Communication Review*, 35, 4, 2005. With permission.)

reinserted in the table. In other words, for inserting one item, we need to reconsider all and only the items in those k buckets. The following pseudo-code describes the insertion algorithm.

InsertItem$_{PFHT}$(x)

1    $Y = x$

2   for ($i = 1$ to k)

3   if ($h_i(x) \neq h_j(x) \; \forall j < i$)

4                    $Y = Y \cup X^{h_i(x)}$

5                    $X^{h_i(x)} = \phi$

6                    $C_{h_i(x)}$ ++

7   for (each $y \in Y$)

8            $C_{min} = min\{C_{h_1(x)}, ..., C_{h_k(x)}\}$

9            $i = SmallestIndexOf(C_{min})$

10    $X^i = X^i \cup y$

In the pseudo-code, Y denotes the list of items to be considered or inserted. It is first initialized to x because that is definitely the item we want to insert (line 1). Then for each bucket x maps to, if the bucket was not already considered (line 3), we increment the counter (line 6), collect the list of items associated with it (line 4) because now all of them must be reconsidered, and also delete the lists from the bucket (line 5). Finally, all the collected items are reinserted (lines 8–10). It is important to note that we do not need to increment the counters while reinserting them because the items were already inserted earlier. Here we just change the bucket in which they go.

Because the data structure has n items stored in m buckets, the average number of items per bucket is n/m. Hence the total number of items read from buckets is nk/m requiring as many member accesses. Finally, $1 + nk/m$ items are inserted in the table which again requires as many memory accesses. Hence the isertion procedure has a complexity of the order $O(1 + 2nk/m)$ operations totally. Moreover, for an optimal Bloom filter configuration, $k = m(\ln2)/n$. Hence, the overall memory accesses required for insertion are $1 + 2\ln 2 \approx 2.44$.

Unfortunately, incremental deletion is not as trivial as insertion. When we delete an item, we need to decrement the corresponding counters. This might cause these counters to be eligible as the smallest counter for some items, which hashed to them. However, now that we keep just one copy of each item we cannot tell which items hash to a given bucket if the item is not in that bucket. This can be told with the help of only prepruning data structure, that is, BFHT in which an item is inserted in all the k buckets and hence we know which items hash to a given bucket. Hence to perform an incremental deletion, we must maintain an *off-line* BFHT like the one shown in Figure 8.28. Such a data structure can be maintained in router software, which is responsible for updates to the table.

To differentiate between the off-line BFHT and on-line PFHT, we denote the off-online lists by χ and the corresponding counter by ζ. Thus, χ^i denotes the list of items associated with bucket i, χ^i_j, the jth item in χ^i, and ζ_j the corresponding counter. The following pseudo-code describes the deletion algorithm.

$DeleteItem_{PFHT}(x)$

1 $\quad Y = \phi$

2 $\quad$ for ($i = 1$ to k)

3 $\quad\quad$ if ($h_i(x) \neq h_j(x) \; \forall j < i$)

4 $\quad\quad\quad \zeta_{h_i(x)} --$

5 $\quad\quad\quad \chi^{h_i(x)} = \chi^{h_i(x)} - x$

6 $\quad\quad\quad Y = Y \cup \chi^{h_i(x)}$

7 $\quad\quad\quad C_{h_i(x)} --$

8 $\quad\quad\quad X^{h_i(x)} = \phi$

9 $\quad$ for (each $y \in Y$)

10 $\quad\quad C_{min} = \min\{C_{h_1(x)}, \ldots, C_{h_k(x)}\}$

11 $\quad\quad i = \text{SmallestIndexOf}(C_{min})$

12 $\quad\quad X^i = X^i \cup y$

When we want to delete an item, we first perform deletion operation on off-line data structure using $DeleteItem_{BFHT}$ algorithm (lines 2–5). Then we collect all the items in all the affected buckets (buckets whose counters are decremented) of BFHT for reinsertion. At the same time, we delete the list of items associated with each bucket from the PFHT because each of them now must be reinserted (lines 7 and 8). Finally, for each item in the list of collected items, we reinsert it (lines 9–12) just as we did in $InsertItem_{PFHT}$. Notice the resemblance between the lines 6–12 of $DeleteItem_{PFHT}$ with lines 4–10 of $InsertItem_{PFHT}$. The only difference is that in $DeleteItem_{PFHT}$, we collect the items to be reinserted from the BFHT and we decrement the counter instead of incrementing it.

Before we derive the expressions for the complexity of $DeleteItem$ algorithm, we notice that we have two types of operations involved: on the BFHT and on the PFHT. We derive the complexity for only the PFHT operations because the BFHT operations can be performed in the background without impeding the normal operations on PFHT. With this consideration, we note that the number of items per nonempty bucket in BFHT is $2nk/m$ because only half the buckets in the following optimal configuration are nonempty. Because we collect the items from k buckets, we have totally $2nk^2/m$ items to be readjusted in the loop of line 9. For readjustment, we need to read as well as write each item. Hence the overall complexity of the deletion operation is $O(4nk^2/m)$. With optimal configuration of the table, it boils down to $4k \ln 2 \approx 2.8k$.

Optimizations

After the pruning procedure, more than one item can still reside in one bucket. We show a heuristic balancing scheme to further balance the bucket load by manipulating the counters and a few items. The reason that a bucket contains more than one item is because this bucket is the first least loaded bucket indicated by the counter values for the involved items that are also stored in this bucket. Based on this observation, if we artificially increment this counter, all the involved items will be forced to reconsider their destination buckets to maintain the correctness of the algorithm. There is hope that by rearranging these items, each of them can be put into an actually empty bucket. The feasibility is based on two facts: first, analysis and simulations show that for an optimal configuration of Bloom filter, there are very few collisions and even fewer collisions involving more than two items. Each item has k possible destination buckets, and in most case, the collided bucket is the only one they share. The sparsity of the table provides a good opportunity to resolve the collision by simply giving them second choice. Secondly, this process does not affect any other items, we need to only pay attention to the involved items in the collided bucket.

However, incrementing the counter and rearranging the items may potentially create other collisions. So we need to be careful to use this heuristics. Before we increment a counter, we first test the consequence. We perform this scheme only if this action does not result in any other collision. The algorithm scans the collided buckets for several rounds and terminates if no more progress can be made or the involved counters are saturated. We will show that this heuristics is quite effective and in the simulations all collisions are resolved and each nonempty bucket contains exactly one item. Figure 8.29b illustrates this list balancing optimization. By simply incrementing the counter in bucket X^1 and reinserting the involved items x and w, we resolve the collision and now each nonempty bucket contains exactly one item.

8.4.5.3 Shared-Node Fast Hash Table

In the previous section, we saw that to perform incremental updates, we need an off-line BFHT. However, with the assumption that the updates are relatively infrequent compared to the query procedure, we can afford to maintain such a data structure in control software, which will perform updates on the internal data structure (which is slow) and later update the pruned data structure accordingly. However, some applications involve time critical updates, which must be performed as quickly as possible. An example is the TCP/IP connection context table where connections get set up and broken frequently and the time for table query per packet is comparable to time for addition/deletion of connection records [13].

Song et al. present an alternative scheme [12], which allows easy incremental updates at the cost of a little more memory than that required for PFHT but significantly less than that of BFHT. The basic idea is to allow the multiple instances of the items to share the same item node using pointers. We call the resulting hash table as shared-node fast hash table (SFHT). The lookup performance of the resulting scheme is the same as that of the BFHT, but slightly worse than the PFHT. Moreover, with the reduced memory requirement, this data structure can be kept on-line.

The new algorithm can be illustrated with the help of Figure 8.30. We start with an empty table and insert items one by one. When the first item, x, is inserted, we just create a node for the item and instead of inserting it separately in each of the lists corresponding to the hash buckets, we simply make the buckets point to the item. This clearly results in a great deal of memory savings. When we insert the next item y, we create the node and make the empty buckets point to the item directly. However, two of the three buckets already have a pointer pointing to the earlier item, x.

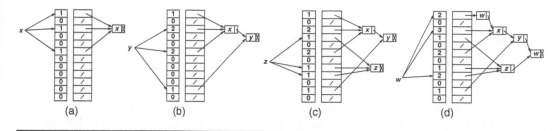

Figure 8.30 Illustration of SFHT. (From Song, H., et al., *ACM SIGCOMM Computer Communication Review***, 35, 4, 2005. With permission.)**

Hence, we make the item x point to y using the next pointer. Note that the counters are incremented at each insertion. More importantly, the counter values may not reflect the length of the linked list associated with a bucket. For instance, the first bucket has a value of 1 but there are two items, x and y in the linked list associated with this bucket. Nevertheless, it is guaranteed that we will find a given item in a bucket associated with that item by inspecting the number of items equal to the associated counter value. For instance, when we wish to locate x in the first bucket, it is guaranteed that we need to inspect only one item in the list although there are two items in it. The reason that we have more items in the linked list than indicated by the counter value is because multiple linked lists can get merged as in the case of x and y.

The insertion of item z is straightforward. However, an interesting situation occurs when we insert w. Notice that w is inserted in the first, third, and ninth bucket. We create a node for w, append it to the linked lists corresponding to the buckets, and increment the counters. For third and ninth bucket, w can be located exactly within the number of items indicated by the corresponding counter value. However, for the first bucket, this is not true: while the counter indicates two items, we need to inspect three to locate w. This inconsistency will go away if instead of appending the item we prepend it to the list having the property of counter value smaller than the number of items in the list. Thus, if we want to insert w in the first bucket and we find that the number of items in the list is two, but the counter value is one, we prepend w to the list. This will need replication of the item node. Once prepended, the consistency is maintained. Both the items in the first list can be located by inspecting at the most two items as indicated by the counter value.

The item node replication causes the memory requirement to be slightly more than what we need in NHT or PFHT where each item is stored just once. However, the overall memory requirement is significantly lower than the BFHT. The following pseudo-code describes the insertion algorithm.

InsertItem$_{SFHT}(x)$

1 for $(i = 1$ to $k)$
2   if $(h_i(x) \neq h_j(x) \; \forall j < i)$
3   if $(C_{h_i(x)} == 0)$
4   Append $(x, X^{h_i(x)})$
5   else
6              $l \leftarrow 0$
7   while $(l \neq C_{h_i(x)})$
8                  $l++$

9   read $X_l^{h_i(x)}$

10   if $(X_{l+1}^{h_i(x)} \neq NULL)$ Prepend $(x, X^{h_i(x)})$

11   else Append $(x, X^{h_i(x)})$

12   $C_{h_i(x)} ++$

In this pseudo-code, l is used as a counter to track the number of items searched in the list. We search up to $C_{h_i(x)}$ items in the list. If the list does not end after $C_{h_i(x)}$ items (line 10) then we prepend the new item to the list otherwise we append it. Note that prepending and appending simply involves scanning the list for at the most $C_{h_i(x)}$ items. Hence, the cost of insertion depends on the counter value and not on the actual linked list length. In SFHT, we have nk items stored in m buckets giving us an average counter value nk/m. We walk through nk/m items of each of the k lists and finally append or prepend the new item. Hence the complexity of the insertion is of the order $O(nk^2/m + k)$ [12]. Moreover, for an optimal counting Bloom filter, $k = m(\ln 2)/n$. Hence, the memory accesses for deletion are proportional to k.

The extra memory requirement due to node replication is hard to compute, but typically small. The simulations in [12] show that the memory consumption is typically one to three times that of NHT (or PFHT). This, however, is significantly smaller than that of BFHT. The pseudo-code for deletion on SFHT is as shown subsequently. We delete an item from all the lists by tracing each list. However, because the same item node is shared among multiple lists, after deleting a copy, we might not find that item again by tracing another list which was sharing it. In this case, we do not trace the list till the end. We just need to consider the number of items equal to the counter value. If the list ends before that then we simply start with the next list (line 4).

DeleteItem$_{SFHT}(x)$

1 for $(i = 1$ to $k)$

2 if $(h_i(x) \neq h_j(x) \; \forall j < i)$

3   $l \leftarrow 1$

4   while $(l \neq C_{h_i(x)}$ and $X_l^{h_i(x)} \neq NULL)$

5   if $(X_l^{h_i(x)} = x)$

6     $X^{h_i(x)} = X^{h_i(x)} - x$

7   break

8   $l++$

9 $C_{h_i(x)} --$

The analysis and simulations in [12] show that the FHTs are significantly faster than an NHT using the same amount of memory; hence, they can support better throughput for router applications that use hash tables. We can find the theoretical analysis and simulations in [12].

References

1. Waldvogel, M., Varghese, G., Turner, J., and Plattner, B., Scalable high speed IP routing lookups. *Proc. of ACM SIGCOMM'97*, September 1997, pp. 25–36, DOI: 10.1145/263105.26.
2. Lim, H., Seo, J., and Jung, Y., High speed IP address lookup architecture using hashing. *IEEE Communication Letter* 2003; 7(10):502–504.

3. Huang, N. and Ming, S., A noble IP-routing lookup scheme and hardware architecture for multigigabit switching routers. *IEEE Journal on Selected Areas in Communications* 1999; 17(6):1093–1104.
4. Gupta, P., Lin, S., and McKeown, N., Routing lookups in hardware at memory access speeds. *Proc. of IEEE INFOCOM*, March/April 1998, vol. 3, pp. 1240–1247.
5. Degermark, M., et al., Small forwarding tables for fast routing lookups. *Proc. ACM SIGCOMM'97*, September 1997, pp. 3–14.
6. Srinivasan, V. and Varghese, G., Fast address lookup using controlled prefix expansion. *ACM Transactions on Computer Systems* 1999; 17(1):1–40.
7. Broder, A. and Mitzenmacher, M., Using multiple hash functions to improve IP lookups, *IEEE INFOCOM* April 2001; 3:1454–1463.
8. Lim, H. and Jung, Y., A parallel multiple hashing architecture for IP address lookup. *Proc. IEEE HPSR2004*, April 2004, pp. 91–95.
9. Bloom, B., Space/time trade-offs in hash coding with allowable errors. *Communications of the ACM* 1970; 13(7):422–426.
10. Fan, L. et al., Summary cache: A scalable wide-area web cache sharing protocol. *IEEE/ACM Transactions on Networking* 2000; 8(3):281–293.
11. Dharmapurikar, S., Krishnamurthy, P., and Taylor, D., Longest prefix matching using Bloom filters. *IEEE/ACM Transactions on Networking* 2006; 14(2):397–409.
12. Song, H., Dharmapurikar, S., Turner, J., and Lockwood, J., Fast hash table lookup using extended bloom filter: an aid to network processing. *ACM SIGCOMM Computer Communication Review* 2005; 35(4):181–192. DOI: 10.1145/1090191.1080114.
13. Dharmapurikar, S. and Paxson, V., Robust TCP stream reassembly in the presence of adversaries. *Proc. of USENIX Security Symposium*, August 2005, pp. 65–80.

Chapter 9

TCAM-Based Forwarding Engine

9.1 Content-Address Memory

Random access memory (RAM) is an integrated circuit that stores data temporarily. Data are stored in RAM at a particular location which is called an address. Data are got by the address. Content-address memory (CAM) is a type of memory device that operates in an inverse manner to RAM. It uses data as the search key, which is provided to every memory cell in the CAM array. The CAM searches through the whole memory with parallel in one-clock cycle and returns the "highest priority" matching address, where the data are found.

Because of the inherent parallelism, CAM can accelerate any application requiring fast searches of databases, list, or patterns in image or voice recognition, computer and communication designs, where searching time is critical and must be very short. In network, applications, CAM is ideally suited for Ethernet address lookup, address filtering in firewalls, bridges, switches, and routers.

9.1.1 Basic Architectural Elements

The structure of CAM is similar to that of the normal RAM but has an additional circuit for its operation of comparison. Vendors aim to develop their CAM devices with a variety of features. Nevertheless, there are a number of basic architectural elements that all CAMs typically have, as shown in Figure 9.1.

Main Memory Array

Main memory array contains a large number of CAM cells, which are usually partitioned into multiple databases. The memory array is typically 72-bits wide and can support "word widths" of 36-, 72-, 144-, 288-, or 576-bits; it is used for a variety of applications. It is developed using the embedded SRAM (DRAM). An example of the typical cell is shown in Figure 9.2. The cell structure

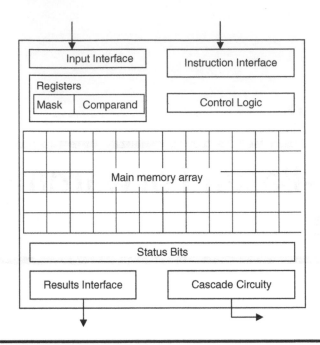

Figure 9.1   CAM architecture blocks.

is similar to that of a conventional SRAM cell but has three additional transistors (N1, N2, and N3) for the compare circuitry.

Control Logic

Control logic is used to implement various operations including search, learn, write, purge, etc.

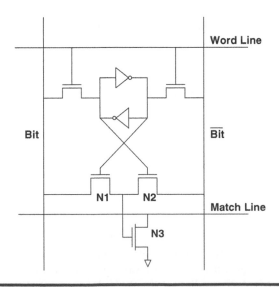

Figure 9.2   The structure of a basic CAM cell.

Comparand Register

Comparand register stores the search key that is provided to the CAM via an input interface. This interface can be proprietary, or follow industry standards for increased interoperability with a number of network processors or ASICs. The search key is typically compared against all entries in the CAM simultaneously in one search.

Mask Register

Mask register can be applied with a comparand to each CAM cell when conducting a search operation.

Results/Address

The result of a CAM search is the address in which the comparand was found. This is provided as output typically called on a "results interface." In some cases, a search will result in multiple matches. You can add the multi-match/priority resolution circuitry to select the most appropriate match.

Status Bits/User Defined Bits

Status bits/user defined bits can be considered as part of the memory array, but for each row, there are typically a number of bits that may be individually set to describe the nature of the contents of that row. Examples of this are Empty, Skip, Permanent, and Age bits.

Cascade Circuitry

Cascade circuitry is performed by cascading CAMs in either a daisy chain or by using a multi-drop bus approach. If the amount of memory stored in an individual CAM is insufficient for those applications, we can aggregate multiple CAMs using cascade circuitry.

Vendors may have other elements in their CAM architecture that are required to support their feature set.

9.1.2 Binary versus Ternary CAMs

The conventional CAMs are binary, in which each cell can take two logic states: 0 and 1. Binary CAM performs exact-match searches, whereas a ternary CAM (TCAM) allows pattern matching with the use of "do not care" (X). Each cell in TCAM can take three logic states: 0, 1, and X. X is 1 or 0. X acts a wildcard during a search. This means that TCAM can store a range of data as an entry. For example, the decimal range 0 to 255 can be represented in a TCAM by an entry 0XXXXXXXX, while binary CAM would require 256 distinct entries to represent the same range of values.

Because of this feature, TCAMs are well suited for network operations, where the action performed on a packet can be identical for an entire range of destination addresses, such as packet classification and IP address lookup. Because TCAM searches a key against all entries in parallel and returns the address of the matching entries, the search time is O(1). Thus, TCAM search function operates much faster than its counterpart in software, and TCAMs are used for search-intensive applications.

9.1.3 Longest-Prefix Match Using TCAM

The prefix in the routing table is a bit string specifying the initial substring of a network address and trailing bits as wildcards. For example, a prefix 100110**** represents the IP address ranges from 1001100000 to 1001101111. Thus, TCAM is very much suited for storing prefixes of different lengths.

In TCAM, each bit of the incoming data is compared with the same position bit of the stored data, and the result is the address of the memory location where the match passed. In some cases, a search will result in multiple matches. To perform the longest-prefix matching operation, all prefixes are stored in a TCAM in decreasing order of lengths. The TCAM searches the destination IP address of an incoming packet with all the prefixes in parallel. Several prefixes may match the destination IP address. A priority encoder logic then selects the first matching entry, that is, the entry with the matching prefix at the lowest physical memory address, that is the longest-matching prefix. The physical memory address is used to extract the corresponding forwarding information such as next-hop form an SRAM module. Figure 9.3 shows the basic architecture of longest-prefix match using TCAM.

Because TCAM can directly store the prefixes and find the longest-matching prefix in a single cycle (using a single TCAM access), it has become an attractive technology. Nonetheless, TCAM traditionally has disadvantages.

High Cost to Density Ratio

During the past few years, the cost of TCAMs has been dramatically reduced and density has been vastly increased. TCAM devices with high capacity (up to 18 Mbits) came to market with costs that are competitive with alternative technologies such as pipelined ASIC-based forwarding engine. For instance, the IDT's 2-Mbit TCAM device is priced at only $30 (http://www.idt.com, 2005).

Slow Update Operation

Because all TCAM entries must be in order of decreasing prefix length, adding or deleting a prefix in TCAM involves multiple TCAM entry moves. When a TCAM is updating, search operations must freeze until the update operations complete. The router in Internet core can reach 100–1000

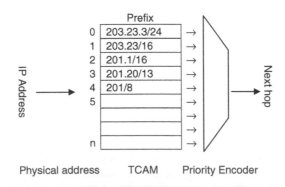

Figure 9.3   Basic architecture of longest-prefix match using TCAM.

updates per second [1]. Consequently, high frequent updating seriously limits a router's lookup performance.

High Power Consumption

A typical 18-Mbit TCAM device that can store up to 512K 32-bit prefixes consumes up to 15 W when all the entries are enabled for search. More than two devices need to store the routing tables in Internet core. High power consumption refers to the incremental overall power supply and cooling expense, and further results in fewer linecards to be packaged in a router because the TCAM power consumption on a linecard is a major cost overhead.

We will discuss the efficient updating schemes and the power-efficient TCAMs.

9.2 Efficient Updating on the Ordered TCAM

TCAM uses data as the search key and returns the matching physical address(es), and the priority encoder can only uses the physical address(es) to find the "highest priority" entry in TCAM. Storing prefix lengths is a simple and fast solution for LPM, called the prefix-length-ordering constraint. Two prefixes that are of the same length do not need to be in any specific order, because they cannot match an IP address. The prefix-length-ordering constraint can be described as follow: for any two prefixes P_1 and P_2, if $|P_1| < |P_2|$, then $\text{Loc}(P_1) > \text{Loc}(P_2)$, where $|P|$ is the length of P and $\text{Loc}(P)$ is the physical address. To insert a new prefix, it is necessary to find a free space. Under the prefix-length-ordering constraint, some prefixes should be moved. When a TCAM is updating, search operations must freeze until the update operations complete. Therefore, we use the number of the moved prefix for inserting a new prefix to measure the performance of the updating schemes on the ordered TCAM. Because the number of the moved prefixes is related to the constraint of prefixes and the configuration of free spaces in TCAM, our task is to loose the constraint of prefixes and find the optimal configuration of free space to minimize the number of the moved prefixes. In this section, we describe the algorithms for the prefix-length-ordering constraint, the algorithms for the chain-ordering constraint [2], and the level-partitioning techniques [3].

9.2.1 Algorithm for the Prefix-Length-Ordering Constraint

There are N prefixes in TCAM with the prefix-length-ordering constraint. If the free spaces are between the prefixes of different lengths, as shown in Figure 9.4a, the new prefix can be inserted into the free space which the new prefix belongs to, it is not necessary to move any prefixes. But there are more free spaces to waste.

If the free space is at the end of TCAM, as shown in Figure 9.4b, in the worst case, it is necessary to move N prefixes to add a new prefix. If the free space is in the middle of TCAM, as shown in Figure 9.4b, in the worst case, it is necessary to move $N/2$ prefixes to add a new prefix.

When we move one of the prefixes of the same length, if the free space is in the bottom, the maximum number of the moved prefixes is the number of different lengths (L). If the set of prefixes of length L, $L - 1, ..., L/2$ are always above the free space and the set of prefixes of length $L/2 - 1$, $L/2 - 2, ..., 1$ are always below the free space, as shown in Figure 9.4c, the maximum number of the moved prefixes is $L/2$.

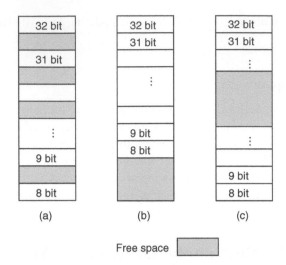

Figure 9.4   Distribution of free spaces in TCAM.

9.2.2   Algorithm for the Chain-Ancestor-Ordering Constraint (CAO_OPT)

In fact, the prefix-length-ordering constraint is more restrictive than what is required for correct longest prefix matching operation using a TCAM. From Proposition 2.2.2, if two prefixes P_1 and P_2 match a destination IP address, then $P_1 \subset P_2$ ($P_2 \subset P_1$), they are called overlapping prefixes. Hence, the constraint on ordering of prefixes in a TCAM can now be relaxed to only overlapping prefixes, that is, if $P_1 \subset P_2$, then P_2 must be at the lower physical address than P_1, called the chain-ancestor-ordering constraint. Figure 9.5 shows the distribution of prefixes under the chain-ancestor-ordering constraint. The free space is at the bottom.

Insertion of a new prefix q proceeds as follows: first, the prefix q is padded with "0," and then become an IP address q'; second, q' is compared with all prefixes on a TCAM, if there is no matched prefix, q has no parent prefix, it can be inserted into the free space directly, otherwise there are the matched prefixes: $P_m \subset \cdots \subset P_{i+1} \subset P_i \subset \cdots \subset P_2 \subset P_1$. If $P_{i+1} \subset q \subset P_i$, then q is inserted into the physical address of P_i, P_i is moved to the physical address of P_{i-1}, P_{i-1} is moved to the physical address of P_{i-2}, and so on; at last, P_1 is inserted the free space. At worst case, the number of the moved prefixes is the number of parent prefixes for a new prefix, called the chain length. For IPv4, the maximal chain length is 31. Figure 9.6 shows the statistics of the chain length on the real routing tables (20011101, 20021101) from RouteViews project (ftp://ftp. routeviews.org/bgpdat, 2003). The maximal chain length is 6 and the average length is less than 2.

9.2.3   Level-Partitioning Technology

If there are two prefixes P_i and P_k, and a new prefix P_0 such that $P_i \subset P_0 \subset P_k$, the prefix P_0 is inserted between P_i and P_k. If there is a free space between P_i and P_k, it is not necessary to move any prefix to insert a new prefix P_0.

Based on the structure of routing table in Section 2.5.1, we can partition the routing tables by level into small tables: Level 0 table (including *standalone* prefixes and *subroot* prefixes), Level 1

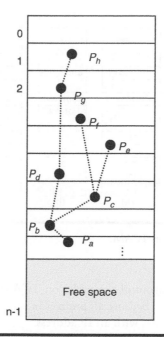

Figure 9.5   Distribution of prefixes under the chain-ancestor-ordering constraint on TCAM $(P_h \subset P_g \subset P_d \subset P_b \subset P_a, P_f \subset P_c \subset P_b \subset P_a, P_e \subset P_c \subset P_b \subset P_a)$.

table (including prefixes in Level 1), Level 2 table (including prefixes in Level 2), Level 3 table (including prefixes in Level 3), Level 4 table (including prefixes in Level 4), and Level 5 table (including prefixes in Level 5), called Level partitioning. A TCAM is partitioned all of them into six parts. From low physical address to high physical address, they are Level 5 table, Level 4 table, Level 3 table, Level 2 table, Level 1 table, and Level 0 table. The free spaces are in the lowest

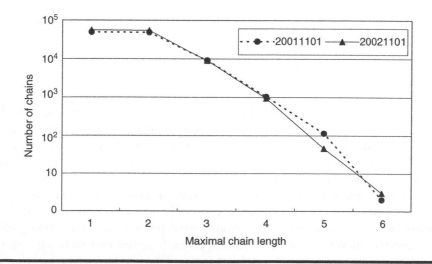

Figure 9.6   Statistics of the chain length.

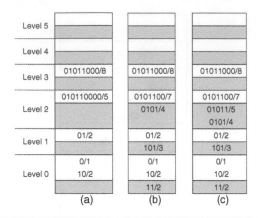

Figure 9.7   Layout of the tables on TCAM.

location of each level table. The layout of the tables on TCAM is shown in Figure 9.7a. In each part, the prefixes can be in disorder, because all prefixes in each level table are disjoin, and there is no more than one prefix that matches with an IP address.

If a new prefix *P* is inserted, we find its parent prefix and one of the shortest child prefixes. There are the following cases:

Case I: If there is no parent prefix, the new prefix will be inserted into the free space of Level 0 table. In Figure 9.7b, the new prefix 11/2 is an example.

Case II: If there is a parent prefix, no child prefix, the new prefix will be inserted into the free space of the next level table of the parent prefix. In Figure 9.7b, the new prefix 101/3 is an example.

Case III: If the parent prefix is the same as the child prefix, the next hop may be updated.

Case IV: If there are the parent prefix and the child prefix, and they are in different level tables, the new prefix will be inserted into the free space of the next level table of the parent prefix. In Figure 9.7b, the new prefix 0101/4 is an example.

Case V: If there are the parent prefix and the child prefix, and they are in the same level table, then the parent prefix will be moved down in the level table, the new prefix will be inserted into the location of the parent prefix. In Figure 9.7b, the prefix 0101/4 is moved to insert the new prefix 01011/5.

After all initial prefixes are partitioned by level, every initial prefix and its initial child prefixes are in different level tables. In Case V, only the inserted prefixes may be moved, and all initial prefixes are not moved. Suppose a new prefix *P* has the parent prefix P_1 and the shortest child prefix P_2, $P_2 \subset P \subset P_1$, they are in the same level table. P_1 should be moved down to insert the prefix *P*. From Level Partition algorithm, at least, one of the two prefixes is the inserted prefix. If P_1 is the initial prefix, P_2 is the inserted prefix, from Case II and IV, it is impossible that P_1 and P_2 are in the same level table. That is to say, the initial prefix is not in the same level table with its child prefixes. Thus P_1 must be the inserted prefix, that is, all moved prefixes are the inserted prefixes. The maximum number of movements is the number of parent (inserted) prefixes in a free space.

The number of movements depends on the sequence of the new inserted prefixes. For example, there are two initial prefixes 1/1 and 111111/6 in the routing table, and four new prefixes 11/2,

Table 9.1 Statistics of Routing Table

Routing table	200,306	200,308
Prefixes	134,223	117,886
Insert	8288	15,127
Total number of the moved prefixes	92	65

111/3, 1111/4, and 11111/5. If the inserting sequence is 11111/5, 1111/4, 111/3, 11/2, there is no movement (Case IV). If the inserting sequence is 11/2, 111/3, 1111/4, 11111/5, there is no movement for inserting 11/2 (Case IV), there is a movement for inserting 111/3 (Case V), two movements for inserting 1111/4 (Case V), and three movements for inserting 1111/5 (Case V).

Deletion is the same as Case III. At first, we find the prefix P that will be deleted, and then remove it from TCAM. If a prefix P is removed, the physical addresses of other prefixes are not changed, and hence no prefix is moved.

Two routing tables and updates are selected at random from RouteViews Project (ftp://ftp. routeviews.org/bgpdata, 2003). The statistics of the routing tables and route updates are shown in Table 9.1. For the routing table in June 2003, the number of the initial prefixes is 134,223, and there are 8288 new prefixes that are inserted. The total number of the moved prefixes is 92. There is a moved prefix per 100 new prefixes, the worst number of movements is 2. For the routing table in August 2003, the number of the initial prefixes is 117,886, and there are 15,127 new prefixes that are inserted. The total number of the moved prefix is 65. There are four moved prefixes per 1000 new prefixes, and the worst number of movements is 2.

9.3 VLMP Technique to Eliminate Sorting

9.3.1 VLMP Forwarding Engine Architecture

Kobayashi et al. [4] proposed a scheme—Vertical Logical Operation with Mask-encoded Prefix-length (VLMP) to remove the restriction that prefixes have to be stored in the order of their lengths in customary TCAM. Figure 9.8 depicts the forwarding engine architecture with parallel comparison and VLMP. Parallel comparison uses the existing TCAM to store prefixes. VLMP is used to determine the longest prefix among the matched prefixes which are stored in arbitrary order. This section is from [4]. Portions are reprinted with permission (© 2000 IEEE).

In parallel comparison, each prefix is represented a pair of bit strings: a data string and a mask string. For a prefix P with L-bit length, the data string contains a prefix from its leftmost L-bit; the rest of the bits are padded by 0. The match string contains an L-bit string of contiguous 1's; the rest of the bits are padded by 0. For example, assuming an IP address prefix is $P = 1010/4$, its data string is P_DS = 101000, its mask string is P_MS = 111100 (the length of IP address is 6 bit), as shown in Figure 9.8.

A mask bit string is used to indicate the portion of a stored data string in parallel comparison and the length of prefix (called horizontal AND operation). VLMP is logical OR operation applying to corresponding bits from different mask strings to obtain the maximum length of prefixes matching a given key (called vertical OR operation). Once the result of VLMP is obtained, the longest prefix match entry is found.

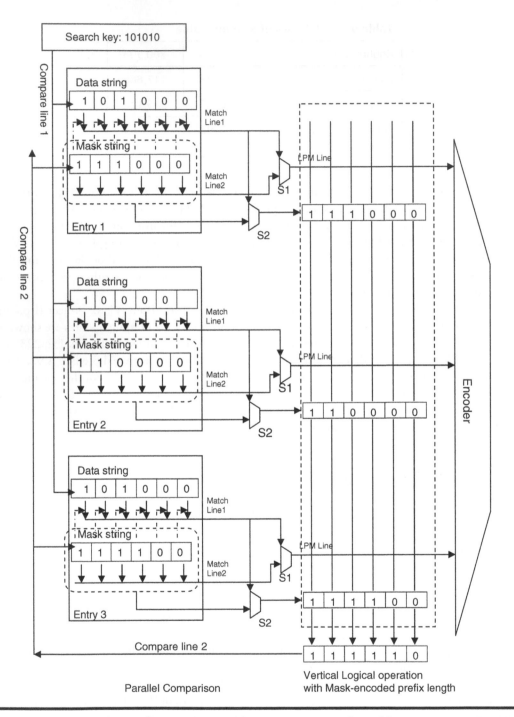

Figure 9.8 VLMP forwarding engine architecture. (From Kobayashi, M., Murase, T., and Kuriyama, A. *Proceedings of the International Conference on Communications (ICC 2000)*, New York: IEEE Press, New Orleans, LA, 2000. With permission.)

9.3.2   Search Algorithm

For a given search, at first, parallel comparison is performed, and is compared with all entries in TCAM. In each entry, each bit of a DS is compared with a corresponding bit of the given search key. The result is a bit string in which matched bit positions are expressed by 1's and others are expressed by 0's. If the result is identical to the mask string, then the match string is outputted to VLMP. VLMP is a logical OR operation applying to corresponding bits from different mask bit strings. The result of VLMP is the longest mask bit string. It is compared with all mask strings to find the physical address of the longest-matching prefix in TCAM. The physical address is used to find the forwarding information by encoder. The search algorithm is shown as follows.

9.3.2.1   First Stage

1-1 A search key, **K**, stored in the **Key Register**, is provided to each entry by means of **Comparand Line1**.

1-2 In each entry, a masked comparison,

$$R1 := (K \,\&\, MS) \text{ XOR } (DS \,\&\, MS)$$

is performed, where "**&**" and "**XOR**" are a bitwise **AND** operation and a bitwise exclusive **OR** operation, respectively. **DS** is the data string and **MS** is the mask string.

1-3 An **AND** operation in terms of all the bits in the **R1** is performed, and the result is provided to **Match Line1**.

1-4 If **Match Line1** is set to 1, a selector **S2** outputs **MS** onto the **VLMP Lines**. Otherwise, **S2** output all 0's onto the **VLMP Lines**.

1-5 On each bit position of the **VLMP Lines**, a vertical bitwise logical **OR** operation (**VLMP**) is performed. The result is referred to as **RV** in the following.

9.3.2.2   Second Stage

2-1 RV is provided to each entry by means of the Comparand Line2.

2-2 In each entry, two bit strings, **RV** and **MS**, are exactly compared, that is,

$$R2 := RV \text{ XOR } MS$$

is performed.

2-3 An **AND** operation in terms of all the bits in the **R2** is performed, and the result is provided to **Match Line2**.

2-4 If an entry's **Match Line1** and **Match Line2** are both 1, a selector **S1** will output 1 onto the **LPM Line**. If not, **S1** will output 0 onto the **LPM Line**.

2-5 The address of the entry whose **LPM Line** is 1 can then be obtained from the **Encoder**.

9.3.3   Performance of VLMP Architecture

VLMP frees the TCAM-based forwarding engine from the restriction that prefixes must be stored in order of their lengths, and the updating of the forwarding table is easy to be completed. Kobayashi et al. analyzed the performance of the VLMP architecture through delay simulation results obtained in a currently available 0.25-μm CMOS process.

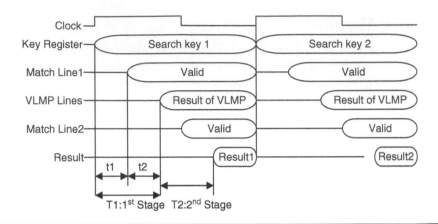

Figure 9.9 Timing chart. (From Kobayashi, M., Murase, T., and Kuriyama, A. *Proceedings of the International Conference on Communications (ICC 2000)*, New York: IEEE Press, New Orleans, LA, 2000. With permission.)

Figure 9.9 is a timing chart illustrating the search steps. Let $T1$ and $T2$ be the delay for the first and the second stages, respectively. $T1$ consists of $t1$ and $t2$, the delay of the horizontal **AND** operation on **Match Line1** of an entry and that of the **vertical OR** operation on **VLMP Lines**, respectively. $T2$ consists of the time consumed in the horizontal **AND** operation on the **Match Line2**.

For 4K entries with 64-bit length, simulation results show the following typical values: $t1 = 7.5$ ns, $t2 = 8.5$ ns, and $T2 = 15.0$ ns. The total delay is $T_{total} = T1 + T2 = t1 + t2 + T2 = 31.0$ ns. Taking into account other wiring delays that might be expected to occur in an LSI layout, we conservatively estimate the search latency to be less than 40 ns, which suggests that LSI architecture can be operated at 25 MHz. In other words, the throughput would be no less than 25 million search/second with a fixed latency of 40 ns.

When we apply a pipelining technique, the operational frequency of an LSI is determined by which delay in each of the two stages is larger. The estimated delay of the first stage is $T1 = t1 + t2 = 7.5 + 8.5 = 16.0$ ns. The estimated delay of the second stage is $T2 = 15.0$ ns. We estimate 20 ns to be long enough for either of the stages to be completed taking into account other wiring delays. That is to say, a pipelined architecture can be operated at 50 MHz. The performance of the VLMP architecture is adequate for wire-speed forwarding of OC-192 (9.6 Gb/s). Because there are the data string and match string and MS of each prefix in the VLMP architecture, the memory is more than the customary TCAM-based engine.

9.4 Power-Efficient TCAM

The inherent parallelism of TCAM can provide the search rate of over 100 million lookups per second, but it consumes higher power than SRAM and DRAM. A system using four TCAMs could consume upto 60 W. To support large number of prefixes in routing table, four to eight TCAM ships are often used. In fact, a longest prefix match involves issuing lookups to each of the TCAM chips. Given that the power consumption of a TCAM is linearly proportional to the number of searched entries, we use this number as a measure of the power consumed.

9.4.1 Pruned Search and Paged-TCAM

For a forwarding engine with multiple TCAM chips, if fewer chips need to be searched for each lookup, the lower power consumes. Panigrahy and Sharma proposed the Pruned Search that only one TCAM chip needs to be searched for each lookup, and Paged-TCAM that achieves significantly lower power consumption within each TCAM chip [5]. This subsection is from [5]. Portions are reprinted with permission (© 2002 IEEE).

9.4.1.1 Pruned Search

The basic idea is to partition the set of prefixes into eight groups so that the prefixes will match in only one of the groups for a lookup. We are thus pruning the search into one of the groups.

For IPv4, the most significant 3-bits of IP address is used as a group id, and the prefixes are divided equally into eight different groups. Each group resides a TCAM chip. For each lookup, a set of eight comparator pairs in TCAM interface is used to determine the group from the most significant 3-bits in IP address. Only one chip corresponding to this group is searched, and others can be inactive. The power consumption is reduced to a factor of 8. The architecture is shown in Figure 9.10.

If each TCAM uses a chip with 256 entries of 72-bits each, it consumes 14.3 W in the worst case. For a lookup, eight chips that are searched would consume about 14.3 W * 8 = 114 W. In the

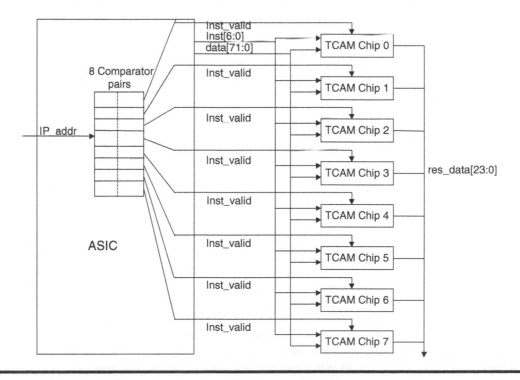

Figure 9.10   Pruned search. (From Panigrahy, R. and Sharma, S. *Proceedings of 10th Symp. High-Performance Interconnects (HOTI 02)*, New York: IEEE CS Press, Stanford, California, 2002. With permission.)

architecture of pruned search, the power consumption would be 14.3 W for effectively one TCAM chip and an idle power of 2.5 W for the remaining seven chips. So total worst-case power consumption would be 31.8 W (= 14.3 W + 7 * 2.5 W).

In practice, it is difficult to divide the prefixes equally into eight groups and search each TCAM chip evenly. In particular, given the distribution of traffic to destination IP addresses, it is possible to partition the prefixes into about 4 * 8 ranges, and distribute these ranges into eight TCAMs so that each TCAM gets close to one-eighth of the number of prefixes and close to one-eighth of the traffic. That is to say, if each TCAM is placed on separate bus, eight TCAMs can deliver close to 8 * 125 million packets per second (MPPS), where each TCAM can perform 125 MPPS (http://www.music-ic.com, 2002).

Assuming that no IP address gets more than 1/16 of total bandwidth (8 * 125 = 10,000 MPPS), and bandwidth of each IP address is given. To improve the throughput, we can partition the prefixes in routing table into eight TCAMs as follows:

1. The IP addresses are partitioned into 2 * 8 ranges based on traffic, where each range has at most one-eighth of the traffic.
2. Each range is divided into ranges based on the number of prefixes evenly. Total number of ranges is 32 (= 2 * 2 * 8).
3. These 32 ranges are divided between TCAMs so that each TCAM does not get too many prefixes and too much traffic. For example, four ranges of these 32 ranges are selected randomly to be inserted in first TCAM, another four ranges in second TCAM, and so on.
4. These ranges are put into 32 comparators.

Because the distribution is random, one can expect with reasonable probability that about one-eighth of the traffic goes to each TCAM. The partition algorithm can handle 250 MPPS in the worst case [5].

9.4.1.2 Paged TCAM

To further achieve significantly lower power consumption, Panigrahy and Sharma applied the idea of the pruned search inside a single TCAM chip. The modifications to the TCAM hardware are shown in Figure 9.11.

Assuming that the TCAM in Figure 9.11 has 256K entries, we organize a TCAM into pages where each page contains 256 consecutive entries. This will give rise to 1024 pages. For IP prefix lookup, supply 6-bits in addition to the 32-bit key. With each page in the TCAM, associate a 6-bit group ID. During a lookup, turn on a page only if its group id matches the first 6-bits of the input key. By generating the 6-bits appropriately, one can ensure that roughly 1/64th pages in the TCAM are active. The 6-bit id is generated by dividing the range of prefixes into 64 roughly equal sized chunks. This can be accomplished by using a set of comparator pairs external to the TCAM containing 64 ranges. We might also want to provide a null (could be thought of as a masked group match) group id with a page so that it is always active. This would be useful for prefixes that match multiple groups (e.g., 1******). That is, a group id could have "do not care" bits.

Put the array of comparator inside the paged TCAM (see Fig. 9.11). Each comparator array takes a different portion of the key and each array produces 6-bits. Each page can decide which arrays 6-bits it will use for comparing with its group id. This could be useful for organizing access control list (ACL) or quality of service (QoS) entries into pages. One of the comparator arrays

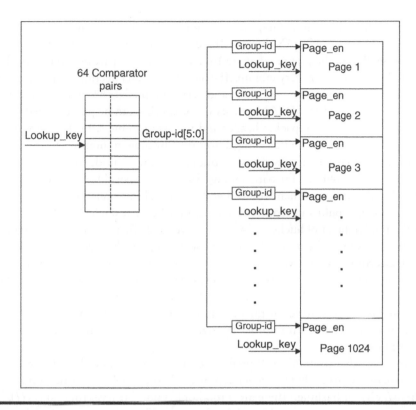

Figure 9.11   Paged TCAM. (From Panigrahy, R. and Sharma, S. *Proceedings of 10th Symp. High-Performance Interconnects (HOTI 02)*, New York: IEEE CS Press, Stanford, California, 2002. With permission.)

would classify based on the source IP address and the other based on the destination address. Each page could be configured to choose the group id produced by either of the comparator arrays.

9.4.2   Heuristic Partition Techniques

Several TCAM vendors (e.g., IDT) now provide mechanisms for searching only a part of the TCAM device during a lookup operation. Zane et al. [6] take advantage of this feature to provide two power-efficient TCAM-based architectures for IP address lookup: the bit-selection architecture and the trie-based architecture. The basic idea of the two architectures is to divide the TCAM device into partitions depending on the power budget. This raises two important problems: How to partition the TCAM into sub-tables for minimizing the size of the largest partition, and How to select the right partition and search it for each IP address lookup. This subsection is from [6]. Portions are reprinted with permission (© 2003 IEEE).

9.4.2.1   Bit-Selection Architecture

The forwarding engine design for the bit-selection architecture is based on a key observation: a very small percentage (<2%) of the prefixes in the core routing tables are either very short (<16-bits) or very long (>24-bits). In the architecture, the very short and very long prefixes are grouped into the minimum possible number of TCAM blocks. These blocks are searched for every lookup.

The remaining 98% of the prefixes that are 16–24-bits long are partitioned into buckets, one of which is selected by hashing for every lookup. The bit-selection architecture is shown in Figure 9.12. The TCAM blocks containing the very short and very long prefixes are not shown explicitly. The bit-selection logic in front of the TCAM is a set of muxes that can be programmed to extract the hashing bits from the incoming packet header and use them to index to the appropriate TCAM bucket. The set of hashing bits can be changed over time by reprogramming the muxes.

For simplicity, we make the following assumptions. First, we consider only the set of 16–24-bit long prefixes (called the split set) for partitioning. Second, it is possible that the routing table will span multiple TCAM devices, which would then be attached in parallel to the bit-selection logic. However, each lookup would still require searching a bucket in a single TCAM device. Third, we assume that the total number of buckets $K = 2^k$ is a power of 2. Then, the bit-selection logic extracts a set of k hashing bits from the packet header and selects a prefix bucket. This bucket, along with the TCAM blocks containing the very short and very long prefixes, is then searched.

The two main issues now are how to select these k hashing bits and how to allocate the different buckets among the various TCAM blocks. From the above assumption, the hashing bits must be chosen from the first 16-bits, which is the minimum length of a prefix in the split set. The "best" set of hashing bits is the one that minimizes the size of the biggest resulting bucket (the worst-case power consumption).

Given a routing table containing N prefixes, each of length more than L, we would like to calculate the size of the largest bucket generated by the best possible hash function that $k \leq L$ bits out of the first L bits for hashing. There are a total of $\binom{L}{k}$ possible hash functions (H). The set of N prefixes is represented as a set of L-bit weighted vectors. The weight wt(y) of an L-bit vector y is defined as the number of prefixes in the prefix set that has the first L-bits the same as y. $\hat{w} = \max_y(\text{wt}(y))$. Let supp($y$) denote the support (number of non-zero bits) of the vector y. The following

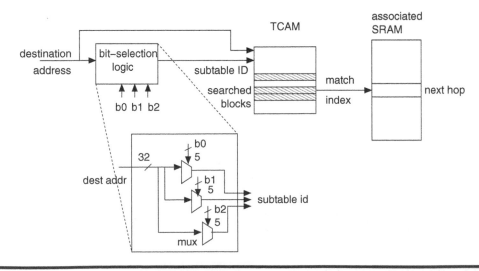

Figure 9.12 Forwarding engine architecture for using bit selection. The three hashing bits here are selected from the 32-bit destination address by setting the appropriate 5-bit values for b0, b1, and b2. (From Zane, F., Narlikar, G., and Basu, A., *Proceedings of IEEE INFOCOM*, New York: IEEE Press, San Francisco, CA, 2003. With permission.)

theorem states an upper bound on the size of the largest bucket for any input prefix set. The proof can be found in [6].

Theorem 9.4.1: For all $Y \subseteq \{0,1\}^L$, $\Sigma_{y \in Y} \text{wt}(y) = N$, there exists some hash function $h \in H$ that splits the set Y into buckets such that the size of the largest bucket is at most $\frac{[F(N,L,k)]}{\binom{l}{k}}$, where $F(N,L,K) = \hat{w} \Sigma_{a \in A} \binom{L-\text{supp}(a)}{k}$, $A \subset \{0,1\}^L$ is a set consisting of the first $N/\hat{w}$ vectors in order of increasing support, each with weight $\hat{w}$

In practice, it is difficult to find a real routing table that matches the worst-case input. But the bound on the worst-case input helps designers to determine the power budget.

A simple scheme is to check all possible subsets of k bits from the first 16-bits until the first subset that satisfies the power budget appears, called a brute force search. Because the scheme compares $\binom{16}{k}$ possible sets of k-bits, it requires a great amount of computation.

Zane et al. proposed a greedy algorithm to reduce the computation. To select k hashing bits, the greedy algorithm performs k iterations, selecting one hashing bit per iteration, doubling the number of buckets (partitions of the routing table) per iteration. To minimize the size of the biggest bucket, a bit is selected by the 2-way split in that iteration. The greedy algorithm is shown in Figure 9.13.

Zane et al. did experiments to evaluate the schemes with respect to two metrics—the running time and the quality of the splits. Two real core routing tables rrc04 and Oregon are used with the size of 109,600 and 121,883 prefixes, respectively. rrc04 is from Genva (11/01/2001), and Oregon is from University of Oregon (05/01/2002). All experiments were run on a 800 MHz PC and required less than 1 MB of memory. The running time for the brute force algorithm was less than 16 seconds for selecting up to 10 hashing bits, whereas the time for the greedy heuristic was as low as 0.05 seconds for selecting up to 10 hashing bits.

```
B = {};
bins = {P}; // P is the set of all prefixes in the routing table
for i = 1 to k
        minmax = ∞;
        foreach bit b∈ {1,...,16} − B
                binsb = { sb=0, sb=1 | s∈ bins};
                maxb = max (binsb);
                if (minmax > maxb) then
                        min_bit = b;
                        minmax = maxb;
                endif
        endforeach

        B = B ∪ min_ bit ;
        bins = binsmin_bit;
endfor
```

Figure 9.13 Greedy algorithm for selecting k hashing bits for a satisfying split. B is the set of bits selected. Here $s_b = j$ denotes the subset of prefixes in set s that have a value of j ($j = 0$ or 1) in bit position b. (From Zane, F., Narlikar, G., and Basu, A., *Proceedings of IEEE INFOCOM*, New York: IEEE Press, San Francisco, CA, 2003. With permission.)

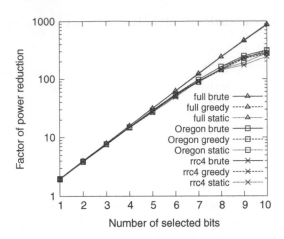

Figure 9.14 Power reduction factor plotted on a log scale, using the different algorithms. "brute" uses the brute force method, "greedy" uses the greedy algorithm, whereas "static" uses the last few consecutive bits out of the first 16-bits of a prefix. "Full" is the synthetic table that was generated by selecting the number of prefixes for each combination of the first 16-bits uniformly at random. (From Zane, F., Narlikar, G., and Basu, A., *Proceedings of IEEE INFOCOM*, New York: IEEE Press, San Francisco, CA, 2003. With permission.)

To explore the nature of the splits produced by the hashing bits, we use the ratio N/C_{max} as a measure of the quality (evenness) of the split, called power reduction factor, where N is the number of 16–24-bit prefixes in the routing table, and C_{max} is the maximum bucket size. Figure 9.14 shows a plot of N/C_{max} versus the number of hashing bits k.

From Figure 9.14, at $k = 6$, the power reduction factor for the brute force and greedy schemes is nearly 53. As the number of hashing bits (k) is increased, the differences of the power reduction factors between the three bit-selection schemes widen.

The bit-selection architecture provides a straightforward technique to reduce the power consumption of data TCAMs. But it requires the additional hardware—muxes and the assumption that the bulk of the prefixes lie in the 16–24-bit range. To overcome these drawbacks, Zane et al. proposed the trie-based algorithms that can provide tighter bounds on worst-case power consumption at the cost of some additional hardware.

9.4.2.2 Trie-Based Table Partitioning

The trie-based architecture uses a prefix trie in the first-stage lookup process instead of hashing on a set of input bits in the bit-selection architecture. Each IP address is first matched with respect to a small-sized TCAM (called index TCAM) that indexes into an associated SRAM (called index SRAM). We can find the ID of the TCAM bucket that will be searched. The second step is the same as that in the bit-selection architecture. The trie-based architecture is shown in Figure 9.15.

Now the main problem is how to build the trie in the index TCAM to minimize the power budget. Zane et al. proposed the two trie-based partition algorithms: subtree-split and postorder-split [6]. We will explain the algorithms using the example routing table shown in Figure 9.16.

Subtree-Split. This algorithm (see Fig. 9.17) takes as input a parameter b that denotes the maximum size of a TCAM bucket (in terms of number of prefixes). The output is a set of $K \in \left[\left\lceil \frac{N}{b} \right\rceil, \left\lceil \frac{2N}{b} \right\rceil\right]$ TCAM buckets, each with a size in the range $\left[\lceil b/2 \rceil, b\right]$, and an index TCAM of

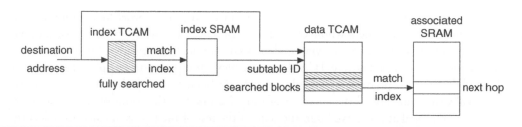

Figure 9.15 Forwarding engine architecture for the trie-based power reduction schemes. (From Zane, F., Narlikar, G., and Basu, A., *Proceedings of IEEE INFOCOM,* **New York: IEEE Press, San Francisco, CA, 2003. With permission.)**

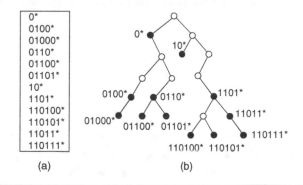

(a)   (b)

Figure 9.16 (a) An example routing table and (b) the corresponding 1-bit trie built from it. The prefixes of only the black nodes are in the routing table. A "*" in the prefixes denotes the start of do not care bits. (From Zane, F., Narlikar, G., and Basu, A., *Proceedings of IEEE INFOCOM,* **New York: IEEE Press, San Francisco, CA, 2003. With permission.)**

```
Subtree-split (b):
    While (there is a next node in post order)
        p = next node in post order;
        if (count(p) ≥ ⌈b/2⌉ and (count(parent(p)) > b))
            carve out subtree rooted at p
            put subtree in new TCAM bucket bu
            put prefix(p) in index TCAM
            bu = bu ∪ {cp(p)}
            foreach node u along path from root to p
                count(u) = count(u) − count(p)
                if (count(u) == 0) remove u endif
            endforeach
        endif
    endwhile
```

Figure 9.17 Algorithm subtree-split for carving the 1-bit trie into buckets of size in the range $[\lceil b/2 \rceil, b]$**. Here count(p) is the number of prefixes remaining under node p, prefix(p) is the prefix of node p, parent(p) is the parent node of p in the 1-bit trie, and cp(p) is the covering prefix of node p. (From Zane, F., Narlikar, G., and Basu, A.,** *Proceedings of IEEE INFOCOM,* **New York: IEEE Press, San Francisco, CA, 2003. With permission.)**

size *K*. During the partitioning step, the entire trie is traversed in post-order looking for a carving node. A carving node is a node *v* whose count is at least $\lceil b/2 \rceil$ and whose parent exists and has a count greater than *b*. Every time a carving node *v* is encountered, the entire subtree rooted at *v* is carved out and placed in a separate TCAM bucket. Next, the prefix of *v* is placed in the index TCAM, and the covering prefix of *v* is added to the TCAM bucket (we explain why in the next paragraph). Finally, the counts of all the ancestors of *v* are decreased by the count of *v*. In other words, once the subtree rooted at *v* is carved out, the state of the rest of the tree is updated to reflect that. When there are no more carving nodes left in the trie, the remaining prefixes (if any) are put in a new TCAM bucket with an index entry of * in the index TCAM. Note that the size of this last TCAM bucket is in the range [1, *b*].

Figure 9.18 shows how subtrees are carved out of the 1-bit trie from Figure 9.16. Note that the index (root) for a carved subtree need not hold a prefix from the routing table. Hence the index

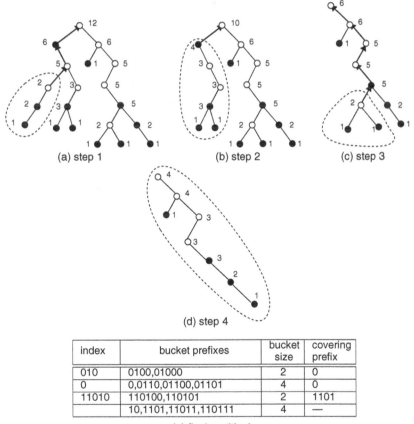

(a) step 1   (b) step 2   (c) step 3

(d) step 4

index	bucket prefixes	bucket size	covering prefix
010	0100,01000	2	0
0	0,0110,01100,01101	4	0
11010	110100,110101	2	1101
	10,1101,11011,110111	4	—

(e) final partitioning

Figure 9.18 **(a–d) Four iterations of the subtree-split algorithm (with parameter *b* set to 4) applied to the 1-bit trie from Figure 9.16. The number at each node *u* denotes the current value of count(*u*). The arrows show the path along which count(*u*) is updated in each iteration, whereas the dashed outline denotes the subtree that is carved. The table in (e) shows the five resulting buckets. Bucket sizes vary between *b*/2 and *b* prefixes. The covering prefix of each bucket, if not already in the bucket, is finally added to it. (From Zane, F., Narlikar, G., and Basu, A., *Proceedings of IEEE INFOCOM*, New York: IEEE Press, San Francisco, CA, 2003. With permission.)**

TCAM may include prefixes not in the original routing table. They simply serve as pointers to the buckets in the data TCAM that contains the corresponding routing table prefixes. Therefore an input address that matches an entry in the index TCAM may have no matching prefix in the corresponding subtree. The addition of the covering prefix to a bucket ensures that a correct result is returned in this case. For example, for the partitioning in Figure 9.18, the input address 01011111 matches 010* in the index TCAM, but has no matching prefix in the corresponding subtree. The covering prefix 0* is the correct longest-matching prefix for this input.

Because we perform a post-order traversal of the trie, the subtree indices must be added to the index TCAM in the order that the corresponding subtrees were carved out. In other words, the first subtree index must have the highest priority (lowest address) in the index TCAM, whereas the last subtree index must have the lowest priority.

The following properties can be proved for algorithm subtree-split when applied with parameter b to a table with N prefixes, the proofs are omitted.

Theorem 9.4.2: The size of each bucket created lies in the range $\left\lceil \lceil b/2 \rceil, b \right\rceil$, except for the last bucket, whose size is in the range $[1, b]$. In addition, at most one covering prefix is added to each bucket.

Theorem 9.4.3: The total number of buckets created is in the range $\left\lceil \lceil N/b \rceil, \lceil 2N/b \rceil \right\rceil$. Each bucket results in one entry in the index TCAM and one entry in the index SRAM.

Theorem 9.4.4: The index and data TCAMs populated according to the subtree-split algorithm always return the longest-matching prefix for each input address.

Finally, to split N prefixes into k buckets, subtree-split is run with parameter $b = \lceil 2N/k \rceil$. Because the maximum bucket size (including the covering prefix) is $b + 1$, we have the following theorem.

Theorem 9.4.5: Using subtree-split in a TCAM with K buckets, during any lookup at most $K + \lceil 2N/k \rceil + 1$ prefixes are searched from the index and the data TCAMs.

The post-order traversal for the 1-bit trie implies that each node in the trie is encountered at most once during the traversal. For a routing table with N prefixes, the number of nodes in the corresponding 1-bit trie is $O(N)$. Therefore, the complexity for this part of the algorithm is $O(N)$. Every time a subtree is carved out, we need to traverse the 1-bit trie all the way to the root. The number of subtrees carved out is $O(N/b)$ (from Theorem 9.4.3). If W is the maximum prefix length (hence, maximum trie depth), this gives us a complexity of $O(NW/b)$. Finally, the total work for laying out the routing table in the TCAM buckets is $O(N)$ (each routing table prefix is looked at once). Thus the total complexity of the algorithm is $O(N + NW/b)$.

Post-Order-Split. The drawback with algorithm subtree-split is that the smallest and largest bucket sizes vary by as much as a factor of 2. Zane et al. introduce another trie splitting algorithm called post-order-split that remedies this. Once again, let N be the total size of a routing table and b be the desired size of a TCAM bucket. The algorithm post-order-split partitions the routing table into buckets that each contains exactly b prefixes (except possibly the last bucket). Such an even partitioning comes at the extra cost of a larger number of entries in the index TCAM.

The main steps in the post-order-split algorithm are similar to that in the subtree-split algorithm. It first constructs a 1-bit tree from the routing table and then traverses the trie in post-order, carving out subtrees to put in TCAM buckets. However, it is possible that the entire trie does not contain $\lceil N/b \rceil$ subtrees with exactly b prefixes each. Because each resulting TCAM bucket must be of size b, a bucket here is constructed from a collection of subtrees which together contain exactly b prefixes, rather than a single subtree (as in the case of algorithm subtree-split). Consequently, the corresponding entry in the index TCAM has multiple indices that point to the same TCAM bucket in the data TCAM. Each such index is the root of one of the subtrees that constitutes the TCAM bucket.

The post-order-split algorithm is shown in Figure 9.19. The outer loop (procedure post-order-split) traverses the 1-bit trie in post-order and successively carves out subtree collections that together contain exactly b prefixes. The inner loop (procedure carve-exact) performs the actual carving—if a node v is encountered such that the count of v is b, a new TCAM bucket is created, the prefix of v is put in the index TCAM, and the covering prefix of v is put in the TCAM bucket. However, if the count of v is x such that $x < b$ and the count of v's parent is $>b$, then a recursive carving procedure is performed. Let the node next to v in post-order traversal be u. Then the subtree rooted at u is traversed in post-order, and the algorithm attempts to carve out a subtree of size $b - x$ from it. In addition, the x entries are put into the current TCAM

```
postprdr-split (b):
    i =0;
    while (there is a next node in post order)
            p = next node in post order;
        carve-exact(p, b, i)
    endwhile
carve-exact(p, b, i):
        if (count(p) == s) or (count(p) < s and (count(parent(p)) > s) then
                carve out subtree rooted at p
                put subtree in TCAM bucket bu_i
                put prefix(p) in index TCAM(index_i)
                bu_i = bu_i ∪ {cp(p)}
                foreach node u along path from root to p
                    count(u) = count(u) − count(p)
                    if (count(u) == 0) remove u endif
                endforeach
                if (count(p) < s) then
                    x = count(p)
                    q = next node in post order
                    carve-exact(q, s − x, i)
                endif
        endif
    endif
```

Figure 9.19 Algorithm post-order-split for carving the 1-bit trie into buckets of size b. Here, count(p) is the number of prefixes remaining under node p, prefix(p) is the prefix of node p, parent(p) is the parent node of p in the 1-bit trie, and cp(p) is the covering prefix of node p. Here index$_i$ is the set of entries in the index TCAM that points to the bucket bu_i in the data TCAM. (From Zane, F., Narlikar, G., and Basu, A., *Proceedings of IEEE INFOCOM*, New York: IEEE Press, San Francisco, CA, 2003. With permission.)

bucket (a new one is created if necessary), and the prefix of v is added to the index TCAM and made to point to the current TCAM bucket. The covering prefix of v is also added to the current TCAM bucket. Finally, when no more subtrees can be carved out in this fashion, the remaining prefixes, if any (they must be less than b in number), are put in a new TCAM bucket and a * entry in the index TCAM points to the last bucket. Figure 9.20 shows a sample execution of the algorithm.

Note that this algorithm may add more than one index (and covering) prefix per TCAM bucket. The number of prefixes added to the index TCAM for any given TCAM bucket is equal to the number of times that the carve-exact procedure is called recursively to create that bucket. It is possible to show that each time carve-exact is called for this bucket, we descend one level down in the 1-bit trie (except, possibly, for the last invocation of carve-exact). Therefore, the maximum number of times we can call the carve-exact procedure is $W+1$, where W is the maximum prefix length in the routing table. In other words, the algorithm post-order-split adds at most $W+1$ entries to the index TCAM and W covering prefixes to the bucket in the data TCAM.

The following properties can be proved about algorithm post-order-split when applied with parameter b to a table with N prefixes of maximum length W bits.

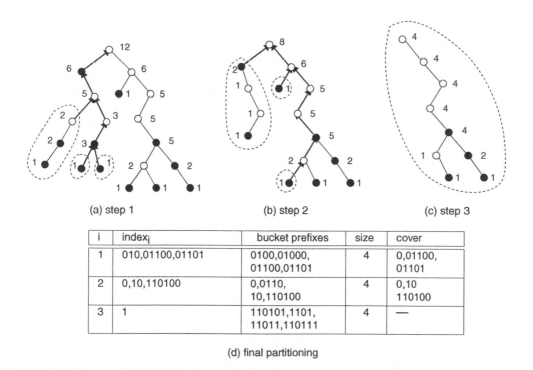

(a) step 1   (b) step 2   (c) step 3

i	index$_i$	bucket prefixes	size	cover
1	010,01100,01101	0100,01000, 01100,01101	4	0,01100, 01101
2	0,10,110100	0,0110, 10,110100	4	0,10 110100
3	1	110101,1101, 11011,110111	4	—

(d) final partitioning

Figure 9.20 **(a–c) Three iterations of the post-order-split algorithm (with parameter b set to 4) applied to the 1-bit trie from Figure 9.16. The number at each node u denotes the current value of count(u). The arrows show the path traced to the root in each iteration for decrementing count. The dashed outlines denote the set of subtrees carved out. (d) The three resulting buckets. Each bucket has $b = 4$ prefixes. The covering prefixes of each bucket that are not in the bucket are finally added to the bucket in the data TCAM. (From Zane, F., Narlikar, G., and Basu, A., *Proceedings of IEEE INFOCOM*, New York: IEEE Press, San Francisco, CA, 2003. With permission.)**

Theorem 9.4.6: The size of each bucket created by the post-order-split algorithm is b, except for the last bucket, whose size is in the range [1, b]. At most W covering prefixes are added to each bucket.

Theorem 9.4.7: The number of buckets created by post-order-split algorithm is exactly $\lceil N/b \rceil$. Each bucket contributes at most $W + 1$ entries to both the index TCAM and the index SRAM.

Theorem 9.4.8: The index and data TCAMs populated according to the post-order-split algorithm always return the longest-matching prefix for each input address.

To split N prefixes of maximum length W into K buckets, post-order-split algorithm is run with parameter $b = \lceil N/K \rceil$. Therefore, the following bound holds.

Theorem 9.4.9: Using post-order-split algorithm in a TCAM with K buckets, during any lookup at most $(W + 1)K + \lceil N/K \rceil + W$ prefixes are searched from the index and the data TCAMs.

The complexity analysis of the post-order-split algorithm is similar to that of the subtree-split algorithm, and it is possible to show that the total running time of the algorithm is $O(N + NW/b)$, where W is the maximum prefix length.

9.4.2.3 Experiments

To compare the performance of the subtree-split and the post-order-split algorithms, we show the ratio of the total number of prefixes in the routing table to the maximum number of prefixes searched during the lookup process in Figure 9.21. This maximum is computed as the sum of the total number of entries in the index TCAM (because it is always searched in full) and the number of entries in the largest bucket in the data TCAM (because only one such bucket is searched for every lookup). When the total number of buckets K in the TCAM is small (this is often a limitation on commercially available TCAMs), post-order-split algorithm performs better because it generates a more even split into buckets. However, as K grows beyond 64, the size of the index

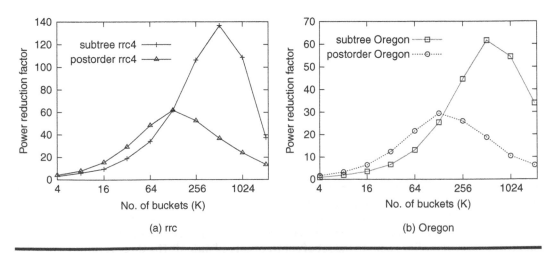

Figure 9.21 Reduction in maximum number of routing table entries searched when algorithms subtree-split and post-order-split algorithms are applied to (a) the rrc04 and (b) the Oregon routing tables. (From Zane, F., Narlikar, G., and Basu, A., *Proceedings of IEEE INFOCOM*, New York: IEEE Press, San Francisco, CA, 2003. With permission.)

TCAM begins to dominate in the case post-order-split, which may add up to $W = 32$ entries in the index TCAM for each bucket. In contrast, although subtree-split algorithm provides a less even split, it only adds one entry to the index TCAM for each bucket, and the index TCAM remains small as K increases. It is also interesting to note that for both the algorithms, this trade-off implies that there is an optimal number of buckets for which the worst-case power reduction factor is highest: $\sqrt{2N}$ buckets for subtree-split algorithm and $\sqrt{N/W}$ for post-order-split algorithm.

For the rrc04 routing table, if the data TCAM is limited to have eight buckets, bit selection using the exhaustive and greedy algorithms reduces the power consumption by factors of 7.58 and 7.55, respectively. The subtree-split and post-order-split algorithms result in power reduction factors of 6.09 and 7.95, respectively (including the power consumed by the index TCAM). Thus, in practice, a data TCAM that consumes a maximum of 15 W needs less than 2.5 W using any of these algorithms. Similarly, if the data TCAM supports 64 buckets, the power consumption can be reduced to less than 0.5 W.

However, if a worst-case power budget is required while designing the forwarding engine, the trie-based algorithms are better than the bit-selection algorithm. As an example, consider a data TCAM that can support eight buckets, and an IPv4 routing table with one million entries (here $W = 32$). Post-order-split algorithm can guarantee a split of the routing table into buckets, such that each bucket has exactly $10^6/8 = 125,000$ entries. Adding in up to 32 covering prefixes for the bucket, and an index TCAM with up to $8 \times 32 = 256$ entries, a maximum of only 125,288 prefixes need to be searched during each lookup, resulting in power reduction by a factor of $10^6/125,288 = 7.98$; this reduced power budget is independent of the distribution of prefixes in the routing table. In contrast, bit selection can guarantee a power reduction factor of only $1/.381 = 2.62$, assuming most prefixes are 16–24-bits long. Thus, with a 15-W one-million-entry data TCAM, post-order-split would result in a power budget of under 2 W, whereas bit-selection can only guarantee a power budget of around 5.7 W.

9.4.2.4 Route Updating

In the bit-selection and the trie-based architectures, adding prefixes may cause a bucket in the data TCAM to overflow, requiring a repartitioning of the prefixes into buckets and rewriting the entire table in the data TACM. Zane et al. proposed the heuristics to avoid frequent repartitions.

For the bit-selection architecture, we started by applying the brute force heuristic on the initial table, and noted the size c_{max} of the largest bucket. Using a fixed threshold t, we recomputed the hashing bits every time the largest bucket size exceeded $c_{thresh} = (1 + t) \times c_{max}$. The hashing bits were recomputed using the static (last few bits) heuristic, followed by the greedy heuristic if the threshold was not met. If the greedy heuristic also failed to bring the maximum bucket size under c_{thresh} we applied the brute force heuristic, and updated the values of c_{max} and c_{thresh}. The number of times each heuristic was applied is listed in Table 9.2; the "static" column represents the total number of recomputations required over the course of the updates.

A similar threshold-based strategy was applied to the trie-based architecture; as before, we assume buckets need to be recomputed each time any bucket overflows. The results are shown in Table 9.3. The update traces contain occasional floods of up to a few thousand route additions in a single second, where the new routes are very close to each other in the routing trie (they are often subsequently withdrawn). Although rare, these floods often cause a single bucket in the trie-based architecture to repeatedly overflow, because prefixes close together in the routing trie are placed in the same bucket in the trie-based partitioning algorithms. In contrast, the bit-selection scheme spreads out nearby prefixes across multiple buckets (because it selects a subset of prefix

Table 9.2 Number of Times Each Heuristic Is Reapplied during the Course of Route Table Updates

Routing Table		rrc04			Oregon		
#Updates Initial size Final size		3,412,540 107,195 103,873			3,614,740 119,226 113,436		
Buckets	Thresh	Static	Greedy	Brute	Static	Greedy	Brute
8	1	2	1	1	15	14	9
	5	0	0	0	3	2	1
	10	0	0	0	1	0	0
64	1	12	12	12	13	12	11
	5	6	5	2	7	5	3
	10	2	1	0	13	2	1

Note: "Buckets" are the number of buckets created: 3 (or 6) hashing bits create 8 (or 64) buckets. "Thresh" is the threshold t (in %) by which the size of the maximum bucket is allowed to grow before the bits are recomputed. "Initial" and "final" denote the number of 16–24-bit prefixes before and after the updates are applied.

Source: Zane, F., Narlikar, G., and Basu, A., *Proceedings of IEEE INFOCOM*, New York: IEEE Press, San Francisco, CA, 2003. With permission.

Table 9.3 Number of Times the Buckets Are Recomputed When the Update Traces from Table 9.2 Are Applied to the Trie-Based Architectures

Routing Table		rrc04			Oregon		
Buckets	Thresh	Subtree	Post-order	Post-Opt	Subtree	Post-order	Post-Opt
8	1	0	58	3 (0.07)	3	74	2 (0.07)
	5	0	17	2 (0.05)	1	14	1 (0.04)
	10	0	0	0 (0.01)	0	6	1 (0.03)
64	1	41	1957	236 (0.48)	14	1042	84 (0.56)
	5	24	1019	152 (0.44)	12	533	40 (0.45)
	10	20	649	109 (0.37)	11	172	11 (0.32)

Note: "Post-opt" uses the post-order-split algorithm but is optimized for updates. The numbers in parentheses denote the average additional TCAM writes per update (in addition to the minimum one data TCAM write) for transferring prefixes across neighboring buckets in the optimized scheme.

Source: Zane, F., Narlikar, G., and Basu, A., *Proceedings of IEEE INFOCOM*, New York: IEEE Press, San Francisco, CA, 2003. With permission.

bits for indexing to TCAM buckets) and therefore requires far fewer repartitions. For the subtree-split and post-order-split algorithms we used bucket sizes of $\lceil 2N/K \rceil$ and $\lceil N/K \rceil$, respectively, for N prefixes and K buckets; therefore subtree-split required significantly fewer recomputations than post-order-split.

The post-order-split algorithm partitions prefixes after arranging them in order of post-order traversal. Hence transferring prefixes between neighboring buckets is straightforward. Such a local transfer requires a small number of writes to the data and index TCAMs. Therefore, one way to mitigate the problem of frequent repartitions using the post-order-split algorithm is to simply transfer some prefixes from the overflowing bucket to one of its neighbors, and recompute the entire partitioning only when both the neighboring buckets become full. On a recomputation, we also set the size of the overflowed bucket to zero, so that it can absorb a larger number of subsequent prefix additions.

This solution, shown as "post-opt" in Table 9.3, reduces the number of recomputations. Transfers between neighboring buckets in the event of bucket overflows result in some additional TCAM writes; the numbers in parentheses in Table 9.3 denote the average TCAM writes per route update due to such transfers. For example, with eight buckets and a 1 percent recomputation threshold in the optimized post-order-split scheme, each route update results in 1.07 TCAM writes instead of 1 TCAM write required by the other schemes.

9.4.3 Compaction Techniques

The power consumption increases linearly with increasing number of entries and bits in a TCAM. Hence, the techniques proposed for compacting the routing table will result in substantial reduction in power. In this section, we will introduce the mask extension technique that can eliminate the redundancies in the routing table [7], and the prefix aggregation and expansion techniques to reduce the number of entries for comparison in a TCAM [8]. Portions are reprinted with permission (© 2003, 2005 IEEE).

9.4.3.1 Mask Extension

Each entry in forwarding table has two fields: prefix and next hop. For two entries $<P_1, N_1>$ and $<P_2, N_2>$, if P_2 is the parent of P_1 ($P_1 \subset P_2$), and they match with an IP address (D), then P_1 is the longest-matching prefix, the packets with the destination address D are forwarded by the next hop N_1. If N_1 is the same as N_2, as long as the prefix P_2 matches with the destination IP address D, the packets with D are forwarded by the correct next hop. The entry $<P_1, N_1>$ become redundant. The redundancies can be deleted, called the pruning technique [7].

To further reduce the redundancies, Liu exploits TCAM hardware's flexibility. The mask for a routing prefix stored in TCAM consists of 1's (the same number of ones as the prefix length) followed by all 0's. However, TCAM allows the use of an arbitrary mask, so that the bits of 1's or 0's need not be continuous. This technique is called mask extension because it extends the mask to be any arbitrary combination of 1's and 0's.

A simple example helps to describe the mask extension technique. Table 9.4 shows an example forwarding table on TCAM. Both P_1 and P_2 correspond to the same next hop 1, and their lengths are 6. There is only a difference at bit 4 (counting from left). According to mask extension, we can combine the two prefixes into a single entry, with the prefix set to 100011* and the mask set to 11101100, as shown in Table 9.5. The zero at bit 4 (counting from the left) in the mask prevents comparison at that bit and allows matching of P_1 and P_2 in the same entry. Table 9.5 shows the compaction of the routing table in Table 9.4. The number of entries has been reduced from 5 to 3.

Table 9.4   Forwarding Table Stored in TCAM

No.	Prefix	Mask	Next Hop
P_1	110111*	11111100	1
P_2	110011*	11111100	1
P_3	0001*	11110000	2
P_4	0011*	11110000	2
P_5	1001*	11110000	2

Table 9.5   Compacted Table Using Mask Extension

No.	Prefix	Mask	Next Hop
P_1 and P_2	110011*	11101100	1
P_3 and P_4	0001*	11010000	2
P_3 and P_5	1001*	01110000	2

The mask extension technique reduces to a logic minimization problem. Let cube refer to the combined single entry for several prefixes, and cover refer to the set of cubes that cover all prefixes. The logic minimization problem then becomes: given a set of prefixes with the same length and same next hop, find a minimal cover.

Logic minimization is an NP-complete problem, so there is little hope of finding an efficient, exact algorithm. Fortunately, there is a fast, proven, and very efficient heuristic algorithm—Espresso-II [9]—that produces a near-optimal solution with finite computing resources.

In the following, we use *minimize()* routine to represent the Espresso-II described by Brayton et al. [9]. Figure 9.22 shows the pseudo-code to compact a routing table based on the Espresso-II. For each prefix length and next hop, we find a minimal cover using the $minimize(S_{on}, S_{dc})$ routine, where S_{on} is the "on" set to be minimized, which is the set of prefixes that must be used to compute the cover, S_{dc} and the "do not care" set, which is the set of prefixes that could be used to compute a better cover.

P(l, n): The set of prefixes that have length *l* and next hop port *n*.

C(l, n): The set of cubes that cover P(l, n), which is the result of logic minimization.

Compact_routing_table ()

    For all prefix length *l*

        For all possible next hop *n*

            C(l, n) = minimize (P(l, n), *nil*);

            insertToCAM (C(l, n))

Figure 9.22   Pseudo-code to compact a routing table using mask extension. (From Liu, H., *IEEE Micro*, 22, 1, 2002. With permission.)

```
Insert_prefix (P)
    Son = { P };
    Sdc = C(l,n);
    Cp = minimize (Son,Sdc);
    For each C in C(l,n)
        /* see if C is a subset of Cp */
        If ( subset(C,Cp )
            /* remove C if duplicate */
            remove(C,C(l,n) );
            /* also remove it from TCAM array */
            removeFromCAM(C );
    C(l,n) = C(l,n) + {Cp };
    insertIntoCAM ( {Cp } );
```

Figure 9.23   Rather than the new prefix, the incremental insertion algorithm inserts a minimal cover into the TCAM array. (From Liu, H., *IEEE Micro*, 22, 1, 2002. With permission.)

For the dynamic routing table, we should find a fast update algorithm for the mask-extension technique. If a new prefix is inserted into the TCAM, this will result in the low compaction ratio. If we reconstruct the compacted routing table using mask extension technique for a new prefix, the reconstruction should be faster than the route update.

To facilitate a fast incremental update, Liu proposed a heuristic minimization algorithm. When a new prefix P is inserted, we compute a minimal cover C_p, using P as the on set of minimization and the existing compacted table as the "do not care" set. Then the minimal cover C_p is inserted into TCAM, and the existing entries that are covered by C_p are removed from TCAM. Figure 9.23 shows the pseudo-code for the insertion algorithm.

For example, a new entry E with prefix 1011* (= 1011--) and next hop 2 is inserted into the routing table (Table 9.4), S_{on} = {00-1--, -001}. After minimization, C_p = -0-1--. C_p is inserted into TCAM, 00-1-- and -001 are removed from TCAM, because they are covered by C_p. The resulting $C(4, 2)$ is {-0-1--}.

The algorithm to remove a prefix from the routing table is more complex because several cubes could cover the prefix. We must remove all cubes covering the prefix and recalculate a minimum cover from the affected prefixes. Figure 9.24 provides an example. C_1, C_2, and C_3 are cubes, and P_1, P_2, P_3, and P_4 are prefixes.

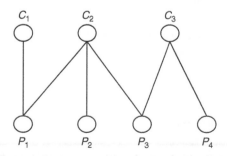

Figure 9.24   Example cube and prefix relationship. A direct line between a cube and a prefix means the cube covers the corresponding prefix. (From Liu, H., *IEEE Micro*, 22, 1, 2002. With permission.)

If P_3 needs to be removed, then C_2 and C_3 must also be removed. As a result, P_2 and P_4 no longer have any cover, so they must be included in the computation for new cover. Note that P_1 is not affected, because although C_2 is removed, C_1 still covers P_1. The incremental removal algorithm searches for prefixes no longer covered by cubes, and includes them in the computation for new cover. The pseudo-code for the incremental removal algorithm appears in Figure 9.25. As in the incremental insertion algorithm, this algorithm requires eliminating redundant cubes.

For example, if P_3 is removed from the routing table (Table 9.4), $S_{on} = \{0011\text{--}, 1001\text{--}\}$, $S_{dc} = nil$. After minimization, $set = \{0011\text{--}, 1001\text{--}\}$. Because there is no redundant cube, the resulting $C(4, 2)$ is $\{0011\text{--}, 1001\text{--}\}$.

Liu did experiments using the routing tables from IPMA project (http://www.merit.edu/ipma, March 7, 2001). Table 9.6 shows the original table size, its size after pruning, and its size after pruning and mask extension. Pruning alone consistently reduces the routing table size by roughly 25 percent. Mask extension saves roughly an additional 20 percent. The overall runtimes are shown in Table 9.6 on a Pentium III 500-MHz PC platform. The runtime for large routing tables becomes much greater because of Espresso-II's exponential behavior.

9.4.3.2 Prefix Aggregation and Expansion

The idea of prefix aggregation is based on the statistical observations that some prefixes with the same next hop have a common sub-prefix. If the prefixes are grouped based on the common

```
Withdraw_prefix (P)
        Son = nil ; /* initialize Son */
        /* search for prefix affected by removal of P */
        For each C in F (P) /* FCD is the set of cubes the cover perfix P */
                remove (C, C(l, n) ); /* remove C from the set */
                removeFromCAM (C); /* also remove from TCAM */
                For each P covered by C
                remove (C,F (P) ); /* remove C as cover for P */
                If (F (P) = nil )
                        Son = Son + {P };
        Sdc = C(l, n);
        set = minimize (Son, Sdc)
        /* remove redundant cubes */
        For each C in C(l, n)
                /* if C is a subset of set, remove it */
                If ( subset (C, set) )
                remove (C, C (l,n) );
                removeFromCAM (C);
        C(l, n) = C(l, n) + set;
        insertIntoCAM(set );
```

Figure 9.25 The incremental withdraw algorithm must find prefixes no longer covered by cubes and include them in the computation for new minimal cover. (From Liu, H., *IEEE Micro*, 22, 1, 2002. With permission.)

Table 9.6 Routing Table Compaction Result

IXP	Original Size (Prefixes)	No. of Next Hop	Size After Pruning (Ratio %)	Size After Pruning and Mask Extension (Ratio %)	Runtime (s)
Mae East	23,554	36	17,791 (24.5)	13,492 (42.7)	14.20
Mae West	32,139	40	24,741 (23.0)	18,105 (43.7)	49.30
Pacific Bell	38,791	19	28,481 (26.6)	20,166 (48.0)	77.00
Aads	29,195	37	21,857 (25.1)	16,057 (45.0)	44.10
Paix	15,906	26	11,828 (25.6)	8930 (43.9)	8.26

Source: Liu, H., *IEEE Micro*, 22, 1, 2002. With permission.

sub-prefix with few exceptions, the number of prefixes can be reduced. The largest common sub-prefix (LCS) is the sub-prefix whose length is the nearest multiple of 8 for a set of prefixes with the same next hop. For example, three prefixes 1101101101*, 1101101100*, 110110111* have the same next hop. The LCS is the prefix 11011011*. The three prefixes can be represented with the LCS and an exception prefix 1101101100*. For the real routing tables, the size of the routing table in ATT-Canada is reduced from 112,412 to 57,837, and the size of the routing table in BBN-Planet is reduced from 124,538 to 71,500 [8]. Ravikumar et al. design a two-level TCAM architecture using the prefix aggregation.

The concept of prefix expansion can be represented as follows: if P_i represents a prefix, such that $|P_i|$ is not a multiple of 8 ($|P_i|$ is the length of P_i), then the prefix expansion property expands P_i to $P_i \cdot X^m$ such that, $X =$ do not care and $m = 8 - (|P_i| \mod 8)$. The operator "$\cdot$" represents the concatenation operation.

For each $P_i \in P$, that has an LCS S, $|P_i|$ may not be a multiple of 8. However, Espresso-II algorithm will provide more compaction if all prefixes in the set are of same length. Hence, we expand that the prefixes do not confirm to these requirements.

9.4.3.3 EaseCAM: A Two-Level Paged-TCAM Architecture

Based on the prefix aggregation prefix expansion technique, Ravikumar et al. proposed *EasCAM*—a two-level TCAM architecture, as shown in Figure 9.26. Each entry in the first level contains w_1-bit sub-prefix, which is compared with w_1 most significant bits of incoming IP address. If there is a match in the first level, it enables the corresponding region in the second level to lookup the longest-matching prefix. The region is called as segment.

The compaction technique is applied as follows. At first, we use the property of the pruning technique to remove the redundant entries in the routing table. Then we use the property of prefix aggregation to form sets of these sets having LCS. Each of these sets are expanded using the prefix expansion technique and given as input to the Espresso-II minimization algorithm. The minimized set of prefixes is stored in the second-level TCAM that is of length $32 - w_1$.

The segments in the second level are implemented using a paging scheme [6]. The prefixes having the same LCS are stored in the pages that have the same page IDs, which are represented

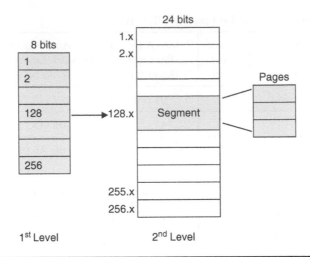

Figure 9.26 EaseCAM: two-level paged TCAM architecture. (From Ravikumar, V.C. and Rabi Mahapatra Laxmi N. Bhuyan, *IEEE Transactions*, 54, 5, 2005. With permission.)

by their LCS values. The number of prefixes in each page can range from 1 to 256. The worst-case power consumption is decided by the product of the number of pages and the number of prefixes in each page.

Because the prefix aggregation technique groups the prefixes such that they have a common prefix, which is a multiple of 8 for $w_1 = 8$, there are three such groups possible with maximum common prefix lengths of 8, 16, and 24. Thus the maximum number of pages enabled that contains the set of overlapping prefixes during a lookup process is less than 3. The upper bound on the number of entries that need to be searched during an IP lookup cannot exceed $256 * 3$.

Choice of the number of bits (w_1) in the first level affects the performance of the two-level architecture. If the page size is large, having separate pages for the smaller sized segments would result in a wastage of space. Hence, a technique is needed to compute the right page size depending on w_1 that would minimize the memory consumption. Let β be the page size; we will find the optimal β for all possible w_1, which will optimizes memory.

Let a cube represent a single entry for a set of prefixes and covering represents the set of cubes that cover all the prefixes that have the same LCS (P_i). We will find the minimal set of such cubes that combine prefixes P_i having the same length with S_i. Because $|S_i|$ can have a maximum of three values for any value of w_1 in the two-level architecture, there can be a maximum of three overlapping cubes for a given IP address if the cubes in each cover are made nonoverlapping. This ensures that the maximum number of active or enabled entries cannot exceed $256 * 3$. This will be true if all the prefixes in the cube C_i can be arranged in pages so that the total number of entries in all pages does not exceed 256. Each of these cubes will represent the page ID for the page containing the prefixes it overlaps. Further, we introduce a parameter γ, called fill factor, which represents the fraction of a page filled during reconfiguration process that covers prefixes with different LCS values. The page-filling algorithm is presented in Figure 9.27a and b. The page-filling algorithm tries to fill the entries into the pages, each of size β, efficiently by trying to find the minimal number of cubes that are nonoverlapping and cover all the prefixes using *MinimalCoverSet*. The StorePage algorithm ensures that no page has more than $\beta * \gamma$ entries.

```
// Heuristics for storing prefixes into pages

StorePage ($C_i$, $\beta$, $\gamma$, $C'$)

While   $P \neq 0$
      Create New Page with Page ID $C_i$
      Entry = 0

      While Entry <   $\beta * \gamma$

            AddToPage ($P_i$)
            $P = P - \{ P_i \}$
            Entry ++
      EndWhile
      Page ++
EndWhile
Return Page

FillPages ($P$, $w_1$, $\beta$, $\gamma$, $C'$)

$P_i \in P$, such that $|P_i| > w_1$, $\forall P_i$

$Q_{w1} = p_{i1}p_{i2}..p_{iw1}$
$Page = 0$
$C_{max} = 0$
      For all   $P_i \in P$ with same $|LCS(P_i)|$ and $Q_{w1}$

            For $|LCS(P_i)| \leq l \leq (|LCS(P_i)|/8 + 1) * 8$

                  $C_i$ = MinimalCoverSet(P, $\gamma$)
            Endfor
            For all   $P_i \in P$ covered by $C_i$
                  If ( $C_i$ covers $P_i$'s having same LCS($P_i$))
                        Page + = StorePage ($C_i$, P, $\beta$, $\gamma$)
                  Else
                        Page + = StorePage ($C_i$, P, $\beta$, $l$)
            Endfor
            If ( $C > C_{max}$) $C_{max} = C$
            $C' = C' + C$
      Endfor
Return   Page, $C_{max}$
```

Figure 9.27   Page-filling algorithm. (From Ravikumar, V.C. and Rabi Mahapatra Laxmi N. Bhuyan, *IEEE Transactions,* **54, 5, 2005. With permission.)**

To evaluate the power consumption of EaseCAM, Ravikumar et al. extended the CACTI-3.0 model to estimate the power of TCAM [8], which is a standard approach for VLSI design. For the routing tables from ATT-Canada and BBN-Planet, the power consumption is shown in Table 9.7. The EaseCAM can reduce the power as low as 0.135 mW per lookup. This is less than 1 percent of the total TCAM power requirement. For the dynamic routing table, Ravikumar et al. proposed a fast incremental update algorithm, see [8].

Table 9.7 Reduction in Power

Router	Raw Data (W)	After Compaction (W)	Effect of EaseCAM Architecture (W)
ATT-Canada	14.35	7.38	0.135
BBN-Planet	15.9	12.31	0.12

Source: Ravikumar, V.C. and Rabi Mahapatra Laxmi N. Bhuyan, *IEEE Transactions*, 54, 5, 2005. With permission.

9.4.4   Algorithms for Bursty Access Pattern

Internet traffic has received extensive analysis for a number of years, and the characteristics of traffic have been revealed. The distribution of packets per destination prefix aggregate has a heavy, Pareto-like tail, whereby some flow contain vastly more packets than others [10]. There are two fundamentally important access patterns [11]:

Independently skewed access patterns.   Certain prefixes are accessed more than other prefixes, but the access frequencies remain relatively stable over time.

Bursty access patterns.   A few prefixes get "hot" for short periods of time and are accessed more frequently. Such patterns have been observed in various applications in a number of studies, such as [12].

Wu et al. [13] design a TCAM-based forwarding engine for the bursty access pattern. The idea is based on the structure of routing table (see Section 2.5.1). The prefixes in routing table are arranged into a two-level TCAM architecture. For the independently skewed access patterns, Wu et al. proposed a static algorithm to reduce the average power consumption. For the bursty access pattern, the access frequencies of some prefixes in routing table are very high for short periods of time; Wu et al. proposed a dynamic algorithm to keep the average power consumption stable relatively.

In Section 2.5.1, the prefixes in routing table are classified into standalone prefixes, subroot prefixes, and more specific prefixes that are further partitioned by level, called Level partition. To describe the optimal scheme based on bursty access pattern, we need the following definition.

T: the number of prefixes in routing table.
N: the number of the standalone prefixes.
AP_i: a standalone prefix, $1 \le i \le N$.
Af_i: the match frequency of the prefix AP_i
M: the number of the subroot prefixes.
SP_j: is a subroot prefix, $1 \le j \le M$.
m_j: the number of the more specific prefixes of SP_j.
Sf_j: the match frequency of the prefix SP_j .
$R = T$/the number of searched prefixes: The power reduction factor.

9.4.4.1   Static Architecture

Figure 9.28 shows the static architecture. It is a two-level architecture with Index-TCAM and Sub-TCAM. The prefixes in Level 0 (standalone prefixes and subroot prefixes) are stored in Index-TCAM with next hop and index fields. Because any two prefixes in Index-TCAM have no intersection, there

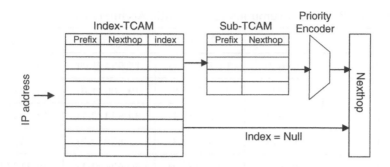

Figure 9.28 Static architecture. (From Weidong Wu, Bingxin, Shi, Jian Shi, and Ling Zuo, *IEEE Micro*, 25, 4, 2005. With permission.)

is no more than one search result for any IP address, all prefixes can be in disorder. The new prefix in Level 0 can be in any free space on Index-TCAM.

Sub-TCAM is partitioned into buckets. All more specific prefixes of a subroot prefix are stored in a bucket in chain-ancestor ordering [2].

Any destination IP address (*D*) is matched with all prefixes on Index-TCAM in parallel. If there is no prefix that matches with *D*, the packet with *D* cannot be forwarded. If there is a prefix (*P*) that matches with *D*, and the index field of *P* is null, the next hop of *P* is the next hop that can forward the packet. There is only one operation. If there is a prefix (*P*) that matches with *D*, and the index field of *P* is not null, *D* will be matched with the more specific prefixes of *P* in the bucket of Sub-TCAM, and the priority encoder gives the longest-matching prefix and its next hop. There are only two operations. The algorithm is shown in Figure 9.29. Its complexity is O(1).

Proposition 9.4.1:   In the static architecture,

a) $T = N + M + \sum_{i=1}^{M} m_i$

b) The minimum number of searched prefixes is $T_{\min} = N + M$

c) The maximum number of searched prefixes is $T_{\max} = N + M + \max_{1 \le j \le M} m_j$

d) The average number of searched prefixes is $T_{\mathrm{aver}} = N + M + \sum_{j=1}^{M} Sf_j\, m_j$

```
D;  // the destination IP-address
P[ ].prefix; // the prefix.
P[ ].nexthop // the next hop of the prefix P in
index-TCAM
P[ ].index // the pointer to Sub-TCAM

P[i] = Search( D ) in index-TCAM;
Next_hop = P[i].nexthop;
If (P[i].index == Null ) Return Next_hop;

Search( D ) in a TCAM bluck; // P[i].index pointer to.
Next_hop = Priority-Encoder( ); // to find the
longest-matching prefix
Return Next_hop;
```

Figure 9.29   Search algorithm in the static architecture. (From Weidong Wu, Bingxin, Shi, Jian Shi, and Ling Zuo, *IEEE Micro*, 25, 4, 2005. With permission.)

Table 9.8 Power Consumption for the Static Architecture

Routing Table	T	$T_{min}(R)$	$T_{max}(R)$	$T_{aver}(R)$
971108	26,669	14,714 (1.8)	15,554 (1.7)	14,716 (1.8)
981101	54,028	29,177 (1.8)	29,809 (1.8)	29,185 (1.8)
991031	71,611	32,766 (2.1)	34,524 (2.1)	32,784 (2.2)
001101	93,111	39,804 (2.3)	42,678 (2.2)	39,831 (2.3)
011101	108,520	49,402 (2.2)	50,626 (2.1)	49,410 (2.2)
021101	120,741	56,106 (2.1)	57,120 (2.1)	56,113 (2.2)

Source: Weidong Wu, Bingxin, Shi, Jian Shi, and Ling Zuo, *IEEE Micro*, 25, 4, 2005. With permission.

We analyze the real data and trace data with 1 million packets from RouteViews Project (ftp. routeviews.org/, 2003). The results are shown in Table 9.8. T_{min} is equal to the number of prefixes in Level 0. There is no much difference between T_{max} and T_{aver}. The power reduction factor (R) is about 2. From Table 2.3, the number of the *standalone* prefixes in Index-TCAM is about 90 percent of T_{min}, 85 percent of T_{max}, and 95 percent of T_{aver}. We will give a scheme to reduce the number of standalone prefixes in Index-TCAM.

9.4.4.2 Dynamic Architecture

To reduce the number of the standalone prefixes in index TCAM, the standalone bucket is added on Sub-TCAM. Some standalone prefixes will be moved from Index-TCAM into the bucket, the architecture is shown in Figure 9.30, called the dynamic architecture.

Based on the search algorithm in the static architecture, the search algorithm in dynamic architecture needs modifications: if there is no search result in Index-TCAM for a destination

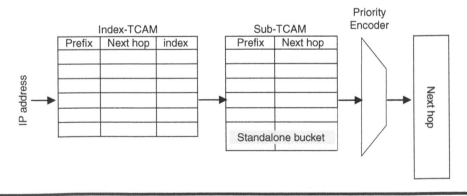

Figure 9.30 The dynamic architecture. (From Weidong Wu, Bingxin, Shi, Jian Shi, and Ling Zuo, *IEEE Micro*, 25, 4, 2005. With permission.)

IP address D, D will be matched with all prefixes in the standalone bucket. The power consumption is as follows:

Proposition 9.4.2: In the dynamic architecture, let n be the number of standalone prefixes in Index-TCAM. $N-n$ is the number of prefixes in standalone bucket.

a) The minimum number of searched prefixes is $T_{min} = n + M$
b) The maximum number of searched prefixes is $T_{max} = n + M + \max(N - n, \max_{1 \leq j \leq M} m_j)$

c) The average number of searched prefixes is $T_{aver} = n + M + \sum_{j=1}^{M} Sf_j m_j + (N - n) \sum_{i=n+1}^{N} Af_i$

The value of T_{min} increases with n. If $N - n < \max_{1 \leq j \leq M} m_j$, then $T_{max} = n + M + \max_{1 \leq j \leq M} m_j$. If $N - n < \max_{1 \leq j \leq M} m_j$, then $T_{max} = M + N$. From Table 9.8, T_{max} is about half of the number of all prefixes in the routing table. We can group some prefixes in standalone bucket to reduce the value of T_{max}. If there are prefixes in standalone bucket AP_i, AP_{i+1}, AP_{i+2}, ..., AP_{i+k}, and there is a prefix (AP') which is a subroot prefix of them, and has no intersection with all prefixes in Index-TCAM, we can store AP' in Index-TCAM, AP_i, AP_{i+1}, AP_{i+2}, ..., and AP_{i+k} are moved into a bucket of Sub-TCAM. The value of T_{min} is added 1. The value of T_{max} is reduced by $k - 1$.

T_{aver} has the relations with n and the access frequency of prefixes in a bucket. Now, we have the following problems: how much are standalone prefixes in Index-TCAM? And which *standalone* prefixes are in Index-TCAM?

For Independently skewed access patterns, there are standalone prefixes that are accessed more frequently for long periods of time. We can find the number of standalone prefixes (n_0) that are stored in Index-TCAM so as to minimize the average number of searched prefixes.

Proposition 9.4.3: Let the standalone prefixes $\{AP_i\}$ be in descending order of the access frequency $\{Af_i\}$. The average number of searched prefixes is $F(n) = n + M + \sum_{j=1}^{M} Sf_j m_j + (N - n) \sum_{i=n+1}^{N} Af_i$, then there is n_0 such that $F(n_0) = \min_{0 \leq n \leq N} F(n)$.

Proof: $F(n + 1) - F(n) = 1 - (N - n) Af_{n+1} - \sum_{i=n+1}^{N} Af_j$
$F(n + 1) - F(n)$ is a monotonically increasing function of n.

$$\because Af_N = \min_{1 \leq i \leq N} Af_n < \frac{1}{2}$$

$$\therefore F(N) - F(N - 1) = 1 - 2 Af_N > 0,$$

$$F(1) - F(0) = 1 - N Af_1 - \sum_{i=1}^{N} Af_i,$$

If $F(1) - F(0) > 0$, then $F(n + 1) > F(n)$, $F(0) = \min_{1 \leq n \leq N} F(n) = M + \sum_{j=1}^{M} Sf_j m_j + N \sum_{i=1}^{N} Af_i$

If $F(1) - F(0) < 0$, then $\exists n_0$, $F(n_0) - F(n_0 - 1) = 0$, and $F(n_0) = \min_{0 \leq n \leq N} F(n)$

From Proposition 9.4.3, the standalone prefixes with higher access frequency can be stored in Index-TCAM to minimize the value of T_{min}. The optimization algorithm is shown in Figure 9.31. Its complexity is $O(N)$, because it is N times to compute T_{aver}, where N is the number of the standalone prefixes.

Wu et al. use the real trace that contains one million TCP packets between Lawrence Berkeley Labs and the rest of the Internet. The number of power consumption is shown in Table 9.9. The power reduction factors of T_{min} and T_{aver} are more than 17 and 10, respectively.

AP[ ]; // the standalone prefixes

Af[ ]; // the match frequency.

N; // the number of the standalone prefixes

$$F0 = M + \sum_{j=1}^{M} Sf_j \, m_j \quad ; // \text{ the initial value of } T_{aver}$$

Sort AP[ ] in descending order of the match frequency;

n = 1; F1 = 0; F = F0;

While (F1<F & n<N) {

   F1 = F;

$$F = F0 + n + (N - n) \sum_{i=n+1}^{N} Af_i \quad ; // \text{ to compute } T_{aver}$$

   n++;

}

AP[1], AP[2],..., AP[n−1] are in Index-TCAM;

Figure 9.31 The optimization algorithm for independently skewed access patterns. (From Weidong Wu, Bingxin, Shi, Jian Shi, and Ling Zuo, *IEEE Micro*, 25, 4, 2005. With permission.)

For bursty access pattern, there are few standalone prefixes that are accessed more frequently for short periods of time. We can move the standalone prefixes between Index-TCAM and the standalone bucket to reduce the average number of power consumption.

A standalone prefix (AP_i) in standalone bucket is accessed more frequently than one of a standalone prefix (AP_j) in Index-TCAM ($Af_i > Af_j$). If AP_i is moved into Index-TCAM, AP_j is moved into standalone bucket, the values of T_{min} and T_{max} are not changed, because the number of prefixes in Index-TCAM is not changed. But the value of T_{aver} may be changed.

Proposition 9.4.4: Let n be the number of the standalone prefixes $\{AP_i\}$ in Index-TCAM. The average number of searched prefixes is

$$F(n) = n + M + \sum_{j=1}^{M} Sf_j \, m_j + (N - n)\sum_{i=n+1}^{N} Af_i.$$

Table 9.9 Power Consumption for the Dynamic Architecture

Routing Table	T	$T_{min}(R)$	$T_{max}(R)$	$T_{aver}(R)$
971108	26,669	1124 (23.7)	14,714 (1.8)	5637.8 (4.1)
981101	54,028	2542 (21.3)	29,177 (1.8)	4824.4 (11.2)
991031	71,611	3454 (20.7)	32,766 (2.2)	6013.3 (11.9)
001101	93,111	4832 (19.3)	39,804 (2.3)	7912.7 (11.8)
011101	108,520	6126 (17.7)	49,402 (2.2)	9870.8 (11.0)
021101	120,741	6991 (17.2)	56,106 (2.2)	11233.1 (10.7)

Source: Weidong Wu, Bingxin, Shi, Jian Shi, and Ling Zuo, *IEEE Micro*, 25, 4, 2005. With permission.

If AP_k is a prefix in Sub-TCAM, and $Af_k > Af_n$, AP_n will be moved into the *standalone* bucket and AP_k will be moved into Index-TCAM, the average number of searched prefixes is

$$F'(n) = n + M + \sum_{j=1}^{M} Sf_j m_j + (N - n)\left(\sum_{\substack{i=n+1 \\ i \neq k}}^{N} Af_i + Af_n \right)$$

Then $F(n) - F'(n) = (N - n)(Af_k - Af_n) > 0$.

From Proposition 9.4.4, for the bursty access pattern, we can exchange the prefixes between Index-TCAM and standalone bucket to keep the average power consumption stable. The optimization algorithm for the bursty access pattern is shown in Figure 9.32.

9.4.4.3 Discussions

Route Update. Because there is no intersection between the Subroot prefixes and the standalone prefixes, all prefixes in Index-TCAM and the standalone bucket can be in disorder. If the new prefix is a standalone prefix, it can be inserted into the free space in Index-TCAM or the standalone bucket in Sub-TCAM. If the new prefix is a subroot prefix, it can be inserted into the free space of Index-TCAM and its child prefix in Index-TCAM is moved into the bucket in which its more specific prefixes are stored.

If the new prefix is a more specific prefix, it should be stored in the bucket to which its Subroot prefix points. Because there may be more than one match prefix for any destination IP address, all prefixes in each bucket should be in chain-ancestor ordering [2] so as to select the longest-matching prefix. Shah and Gupta proposed a fast updating algorithm—CAO_OPT [2]. If adding prefixes cause a bucket to overflow, all prefixes in the bucket can be moved into a larger bucket.

```
AP[ ]; // the standalone prefixes in Index-TCAM

Af[ ]; // the access frequency of AP[ ].

n; // the number of the standalone prefixes in Index-TCAM

BP[ ]; // the standalone prefixes in Sub-TCAM

Bf[ ]; // the access frequency of BP[ ].

Sort AP[ ] and BP[ ] in descending order of their access frequency respectively;

k = 1;

While (Bf[k] > Af[n] ) {

Move BP[k] into Index-TCAM

Move AP[n] into Sub-TCAM

Sort( AP[ ] );

k++

}
```

Figure 9.32 Optimization algorithm for bursty access patterns. (From Weidong Wu, Bingxin, Shi, Jian Shi, and Ling Zuo, *IEEE Micro*, 25, 4, 2005. With permission.)

TCAM and Cache.   If the Index-TCAM is replaced with Cache, the power consumption will be reduced. The complex data structure should be used to store the prefixes in Cache, such as Hash table. But all prefixes are directly stored in Index-TCAM; there is no data structure to use, and it is very simple to update route.

Implementation Issues.   The level-partitioning algorithm need maintain the parent pointers between prefixes. The optimization algorithms for independently skewed and bursty access patterns need the access frequencies. We introduce an auxiliary array with following fields:

$$\begin{cases} P, & \text{// prefixes in routing table} \\ T, & \text{// pointer to its parent} \\ L, & \text{// location in TCAM} \\ f, & \text{// the access frequency} \end{cases}$$

The level-partitioning algorithm and the optimization algorithm for independently skewed access patterns are offline and there is no impact on the search algorithm. The optimization algorithm for bursty access patterns is online. If a lot of prefixes are moved between Index-TCAM and Sub-TCAM, the search algorithm maybe suspended.

9.5   A Distributed TCAM Architecture

To achieve both ultra-high lookup throughput and optimal utilization of the memory while being power-efficient, Zheng et al. devise a distributed TCAM architecture and proposed a load-balanced TCAM table construction algorithm [14]. The power efficiency is well controlled by decreasing the number of TCAM entries triggered in each lookup operation. Using four 133-MHz TCAM chips and given 25% more TCAM entries than the original route table, the lookup throughput is increased up to 533 Mpps.

In what follows, we will first present the method to divide the forwarding table into small segments and the load-balance-based table construction (LBBTC) algorithm to evenly allocate these segments into TCAMs while keeping the lookup traffic load balanced among them. Then we describe the complete architecture of the proposed ultra-high throughput and power-efficient IP lookup engine. At last we present the theoretical performance estimation based on queuing theory as well as the simulation results of the proposed scheme. This section is from [14]. Portions are reprinted with permission (© 2004 IEEE).

9.5.1   Analysis of Routing Tables

Definition 9.5.1:   An extended forwarding table is minus derived from the original forwarding table by expanding each prefix of length L ($L < 13$) into 2^{13-L} prefixes of length 13. For example, a prefix 10.0.0.0/18 is expanded into $2^{13-8} = 32$ prefixes as 10.0.0.0/13 to 10.248.0.0/13.

It is straightforward to see that the extended forwarding table is equivalent to the original one. Because the number of prefixes shorter than 13-bits is quite small (which is why we adopt 13 as the threshold here), the expanding operation introduces few additional route entries. In this section, the forwarding table is extended.

Definition 9.5.2: The extended forwarding table may contain redundant entries. Let N_0 be the number of the entries in the original forwarding table; N be the number of the entries in the extended table ($N_0 \approx N$); M be the actual number of entries in the extended forwarding table of the proposed scheme. *Prefix[i]* denotes the ith prefix in the extended table, where i is from 1 to N.

Definition 9.5.3: A concept of ID (segment) is defined to classify the prefixes in the forwarding table. We define a bit-string extracted from the first 13-bits of a prefix as its ID. For instance, if we define the 10–13th bits of a prefix its ID, then the ID of prefix 10.16.0.0/13 is "0010" (2) and the ID of prefix 10.36.5.0/24 is "0100" (4). We say *Prefix[i]* $\in j$, when the ID of *Prefix[i]* equals j. We also use the ID segment to identify the ID group, the set of the prefixes with the same ID.

Definition 9.5.4: Because the lookup traffic distribution among the IP prefixes can be derived by certain statistical method, we assume that it is known here. $D\_prefix[i]$ is defined as the ratio of the traffic load of *prefix[i]* to the total bandwidth. Therefore, $\sum_i D\_prefix[i] = 1$.

Definition 9.5.5: We introduce a concept of storing redundancy rate here. The redundancy rate of a specific prefix (or ID group) is defined as the number of duplications of this prefix (or ID group) stored in the TCAM chips. The redundancy rate of the entire scheme is M/N.

Theorem 9.5.1: In the (extended) forwarding table, two prefixes with different IDs do not need to be stored in any specific order when applying the longest prefix matching operation. As long as the prefixes in an ID group are stored in the decreasing order of prefix lengths, the TCAM(s) will return the correct matching result.

Proof: TCAM-based matching requires that *Prefix* 1 should be stored in lower address than *Prefix* 2 if *Prefix* 2 is the prefix of *Prefix* 1; but two prefixes each one of which is not a prefix of the other can be stored in arbitrary order. Because each prefix in the extended forwarding table has a length of 13 or more, the first 13-bits of the two prefixes with different IDs must not be a prefix of each other. Therefore, they need not to be stored in any specific order.

Zheng et al. have analyzed the routing table snapshot data provided by the IPMA project (http://www.merit.edu/ipma) and found out that the route prefixes can be evenly split into groups according to their IDs, so long as we make an appropriate ID bit selection. This approach will contribute to balancing the storage among the TCAM chips and result in high capacity utilizations of them (because identical-sized chips are usually adopted in most practical cases).

A brute-force approach to find the right set of ID bits would be to traverse all of the bit combination out of the first 13-bits of the prefixes, and then measure the splitting results to find out the most even one. However, this may be a quite expensive computation, because the number of prefixes may be fairly large.

There are three heuristic rules that can be adopted to significantly reduce the traversing complexity. (*i*) The width of (or, the number of bits selected as) ID need not to be large, for example, 4–5-bits are quite enough for most cases. (*ii*) The number of patterns formed by the first 6-bits of the prefixes is quite unevenly distributed according to the experiments with real-world route tables. For the sake of finding an even classification of the prefixes, we had better not adopt these (i.e., 1–6th) bits. (*iii*) If the traverse of the combination of successive bits can find reasonable even results, there is actually no need to further traverse other nonsuccessive bit combinations.

Based on the three heuristic rules, we easily found out reasonable ID bit selecting solutions for four well-known real-world prefix databases, all of which adopt the 10–13th bits and the prefixes

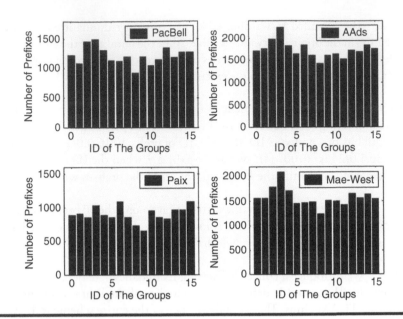

Figure 9.33   Prefix distribution among ID groups of four real-life route tables. (From Zheng, K., Hu, C.C., Lu, H.B., and Liu, B., *Proceedings of IEEE INFOCOM'04*, 3, 2004. With permission.)

are classified into $2^4 = 16$ groups, respectively. The results are depicted in Figure 9.33. All of the four tables show quite even distribution.

Notice that the ID bit selecting scheme is totally flexible. Both the number of ID bits and their positions can be freely chosen according to the practical route tables. With the three heuristic rules, we produce satisfactory even results in almost all real-world cases collected by IPMA. And according to the experimental results, it tends to be that the larger the route table is (e.g., the BGP route tables of core level routers), the better the result we can achieve.

9.5.2   Distributed Memory (TCAM) Organization

Zheng et al. have introduced a simple but efficient method to partition the forwarding table into small segments by using the ID bits of the prefixes. Because multiple TCAMs are used in our scheme, the next step is to allocate these ID segments (groups) to each TCAM. Based on the analysis of the above route table, we know that the prefixes are evenly distributed into each ID group the size of which is approximately $N/16$. Therefore, we may partition each TCAM into small blocks with a size of $N/16$ (or slightly larger) so that each block can store one ID group. Now we use the ID group as the basic unit to allocate the prefixes among the TCAMs. As Theorem 9.5.1 dictates, as long as the prefixes in an ID group are arranged in decreasing order of prefix length, the TCAM will return the LMP as the result.

9.5.3   LBBTC Algorithm

Allocating the prefixes evenly and balancing the lookup traffic among the TCAMs are the two tasks of the proposed table construction algorithm. We have addressed the first task in the previous section. For the second task, we first calculate the load distribution of the ID groups, $D\_id[j]$ ($j = 1, \ldots, 16$), by summing up the distribution of the prefixes in the same ID group:

$$D\_id[j] = \sum_{prefix[k] \in j} D\_prefix[k]$$

Two methods can then be introduced to balance the lookup traffic among the TCAMs. First, we can use $D\_id[j]$ ($j = 1, \ldots, 16$) as the weights of the ID groups and make the sum of the weights of the ID groups in each TCAM balanced. Secondly, for the ID groups with large weights, we may introduce storing redundancy into the scheme. By duplicating the prefixes in a specific ID group to multiple TCAMs, we distribute the traffic of the ID group among these TCAMs. For example, if ID group j is allocated (duplicated) to three TCAMs, then each one of these three chips bears a traffic load of $D\_id[j]/3$. The proposed algorithm is the combination of these two methods. Before we describe the table construction algorithm, a mathematical model of the problem is introduced.

9.5.3.1 Mathematical Model

Let K be the number of TCAM chips, Rd be the storing redundancy rate of the whole mechanism, T be tolerance of the discrepancy of table sizes among the TCAM chips, P be the length of the ID segment (it is assumed to be 4), and there should be 2^P ID groups. K, Rd, T, and P are given. Let S be the set of all ID groups, $S = \{1, 2, \ldots, 2^P\}$; Q_k ($k = 1, \ldots, K$), is the set of the ID groups that are allocated to TCAM chip #k, $\forall Q_k$, $Q_k \subseteq S$ and $\cup Q_k = S$($k = 1, \ldots, K$), and $|Q_k|$ denotes the number of elements (ID groups) that Q_k contains.

We define the number of TCAMs that ID group #j is allocated into as $G[j]$ ($j \in S$); $G[j]$ denotes the storing redundancy rate of the ID group #j. $W[j]$ is the incremental traffic load distributed to a TCAM chip when ID group j is allocated to this chip, so $W[j] := D\_id[j]/G[j]$ ($j \in S$), and we call $W[j]$ partition-load of ID group #j. Let $D[k]$ be the traffic load allocated to TACM chip #k, so $D[k] := \sum_{j \in Q_k} W[j]$ ($k = 1, \ldots, K$), then the optimization problem (called LBBTC problem) is given by:

Subject to:

$$Q_j \subseteq S, j = 1, 2, \ldots, K, \bigcup_j Q_j = S; \tag{9.1}$$

$$\left\| |Q_i| - |Q_j| \right\| \leq T \ (0 \leq i \leq K, 0 < j < K); \tag{9.2}$$

$$BOOL(i, j) := \begin{cases} 1 & if \ j \in Q_i \\ 0 & if \ j \in Q_i \end{cases} \tag{9.3}$$

$$G[j] := \sum_{i=1}^{k} BOOL(i, j), j \in S \tag{9.4}$$

$$\sum_{j \in S} G[j] \leq 2^P \times Rd, j \in S \tag{9.5}$$

$$1 \leq G[j] \leq K, G[j] \in Z^+, j \in S \tag{9.6}$$

$$D[k] := \sum_{j \in Q_k} \frac{D\_id[j]}{G[j]}, \quad k = 1, 2, \ldots, K;$$

(9.7)

Minimize: $\sum_{i \neq j, i, j = 1, \ldots, K} |D[i] - D[j]|$

Equation 9.1 shows the definition of Q_j, the set of ID groups allocated to a specific TCAM chip; Equation 9.2 gives the constraint of the discrepancy of the table sizes among the TCAM chips; Equations 9.3 and 9.4 represent the definition of the storing redundancy rate $G[j]$. Equation 9.5 gives the constraint of the overall storing redundancy rate; and Equation 9.6 shows that an ID group should have at least one copy (straightforward), while at most K copies (there is certainly no need to have more than one copy in a same TCAM chips).

This problem is proven to be NP-hard (please refer to [14] for detailed proofs). Therefore, we give a heuristic algorithm to solve the problem, as the following:

LBBTC algorithm

Step 1: Pre-calculation: /*calculate $G[j]$ iteratively*/
Define Rd' as the actual storing redundancy;
Let a, b be the lower and upper bound of Rd' in the iteration, respectively.
Initialize $a = 0$, $b = 2Rd$, $Rd' = (a + b)/2 = Rd$, precision = 0.001.

$$F(Rd') = \sum_{j \in S} Min\left\{ \left\lceil 2^P \times Rd' \times D\_id[j] \right\rceil, K \right\} - 2^P \times Rd$$

```
While F(Rd') ≠ 0 do
If F(Rd') < 0 then a = Rd'
else b = Rd';
        Rd' = (a + b)/2;
        if |b - a| ≤ precision then break;
end while;
G[j] = Min(⌈2^P × Rd' × D_id[j]⌉, K), j ∈ S
W[j] = D_id[j]/G[j].
```

Step 2: Sort $\{i, i = 1, 2, \ldots, 2^P\}$ in decreasing order of $W[i]$, and record the results as $\{Sid[1], Sid[2], \ldots, Sid[2^P]\}$;

```
for i from 1 to 2^P
{
    for j from 1 to G[Sid[i]]
    {
    Sort {k, k = 1, 2, ..., K} in decreasing order of D[k], and
    record the results as {Sc[1], Sc[2], ..., Sc[K]};
    for k from 1 to K do
    {
        If |Q_{Sc[k]}| < Min |Q_i| + T then
                        i=1, ... K
        Q_{Sc[k]} = Q_{Sc[k]} ∪ {Sid[i]};
```

```
      D[Sc[k]] = D[Sc[k]] + W[Sid[i]]
    break; }
    }
}
```

Step 3: Output $\{Q_k, k = 1, ..., K\}$.

The task of this algorithm is to figure out an allocation scheme ($\{Q_k\}, k = 1, ..., K$) of the ID groups, with which the lookup traffic distribution among the TCAM chips ($D[k], k = 1, ..., K$) is as even as possible.

In Step 1, the redundancy rates of the ID groups ($G[j], j \in S$) are precalculated aiming at maximizing the sum based on the traffic distribution ($D\_id[j], j \in S$), so as to maximize the TCAM space utilization (while satisfying constraint Equation 9.5). In this step, we adopt Bisection Method [15], which is simple and quickly converging, to find the solution of Rd when the total redundancy rate of all ID groups equals $2^P Rd$. Then the ID groups are allocated to TCAM chips in Step 2 following two principles: (*i*) ID groups with larger *Partition-load* ($W[j]$) are allocated more preferentially; (*ii*) TCAM chips with lower load are allocated to more preferentially. By applying the greedy algorithm, fairly good result can be achieved when all the constraints are satisfied. Step 3 outputs the result obtained after Step 2 ultimately.

9.5.3.2 Adjusting Algorithm

Because the LBBTC algorithm is a greedy algorithm, the produced result still has room for further optimization. The load-balance-based-adjusting (LBBA) algorithm presented below gives an additional simple, but efficient method to further balance the lookup traffic among the TCAM chips.

The primitive idea of this algorithm is that, further balanced traffic load distribution can be achieved by exchanging specific ID groups among the TCAM chips, based on the results presented by the LBBTC algorithm and certain principles. This action has not adverse effects on all constraints.

First of all, we would like to review some concepts. The output of the LBBTC algorithm is a sequence of sets $\{Q_k, k = 1, ..., K\}$, each representing the set of the ID groups allocated to a specific TCAM chip. Corresponding to each of the TCAM chips is the lookup traffic load ratio D_k ($k = 1, ..., K$). Now, we divide $\{Q_k, k = 1, ..., K\}$ into two sets, say Set I and Set J, $I = \{i \mid D[i] > 1/K, i = 1, 2, ..., K\}$ while $J = \{j \mid D[j] \le 1/K, j = 1, 2, ..., K\}$ ($1/K$ represents the average traffic load ratio of the K TCAM chips).

Theorem 9.5.2: Necessary conditions for exchanging ID groups are

$$i \in I, j \in J, n \in Q_i, m \in Q_j, n \notin Q_j, m \notin Q_i;$$

$$(1)\ W[n] - W[m] > 0;\ (2)\ D[i] - W[n] > D[j] - W[m]. \tag{9.8}$$

Proof: Let $D[i]$ denote the traffic load allocated to TCAM #i before the exchange and $D'[i]$ denote the traffic load allocated to TCAM #i after the exchange, then

$$|D'[i] - D'[j]| = |(D[i] - W[n] + W[m]) - (D[j] - W[m] + W[n])|$$

$$= |(D[i] - W[n]) - (D[j] - W[m]) + W[m] - W[n]|$$

$$< |(D[i] - W[n]) - (D[j] - W[m])| + |W[m] - W[n]|\ (\Delta = W[n] - W[m] > 0)$$

$$= (D[i] - W[n]) - (D[j] - W[m]) + W[n] - W[m] = D[i] - D[j] = |D[i] - D[j]|.$$

First, it is straightforward that, under the necessary conditions, the exchange does not break any constraint; then because $|D'[i] - D'[j]| < |D[i] - D[j]|$, it means that the load discrepancy between the two TCAM chips is reduced; we call $Op = |D[i] - D[j]| - |D'[i] - D'[j]|$ the optimized value of the exchanging operation. It represents the optimized value of the objective $\sum_{i \neq j,\, i, j = 1, \ldots k} |D[i] - D[j]|$ of the LBBTC problem.

LBBA algorithm:

1) Update Sets I and J;
2) Sort $\{i,\ i \in I\}$ in increasing order of $D[i]$, and record the result as $\{I[1], I[2], \ldots, I[|I|]\}$; /*$|I|$ is the number of elements in Set I*/.

Sort $\{j,\ j \in I\}$ in increasing order of $D[j]$, and record the result as $\{J[1], J[2], \ldots, J[|J|]\}$;

```
for i from 1 to |I|
 for j from 1 to |J|
        if ∃n ∈ Q_i, m ∈ Q_j, n ∉ Q_j, m ∉ Q_i satisfies W[n] -
W[m] > 0 and D[i] - W[n] > D[j] - W[m] then
            Q_i = Q_i - n + m;
            Q_j = Q_j - m + n;
            goto 1);  // Iteration.
```

Convergence of the LBBA algorithm: We define Set $\Delta = \{\delta|\delta = \sum_{n \in P_1 \cup P_2} W[n] - \sum_{m \in P_3 \cup P_4} W[m],\ P_1,$ $P_2, P_3, P_4 \subset S,\ \delta > 0\}$; it is straightforward that Set Δ is independent of the adjusting algorithm and $|\Delta|$ $< \infty$. According to the definition of the optimized value Op and Set Δ, it can be easily got that $\forall Op \in \Delta$ and $\forall Op \geq \underset{\delta \in \Delta}{Min}\, \delta > 0$. So because the objective $\sum_{i \neq j,\, i, j = 1, \ldots, K} |D[i] - D[j]| \geq 0$, and each exchanging operation reduces the objective by a value larger than $\underset{\delta \in \Delta}{Min}\, \delta$, the iteration in the algorithm is convergent.

For example, suppose that the traffic distribution among ID groups is given by Table 9.10. If $Rd = 1.25$, $T = 1$, and $K = 4$, then the allocation result given by the algorithms is shown in Table 9.11. The load ratio in each chip is about 25%.

9.5.4 Analysis of the Power Efficiency

Traditional TCAM is a fully parallel device. When the search key is presented at the input, all entries are triggered to perform the matching operations, which is the reason for its high power

Table 9.10 Traffic Distribution among ID Groups

ID	0	1	2	3	4	5	6	7
D_id%	1	2	3	5	19	7	8	8
ID	8	9	10	11	12	13	14	15
D_id%	6	3	7	2	19	2	2	6

Source: Weidong Wu, Bingxin, Shi, Jian Shi, and Ling Zuo, *IEEE Micro,* 25, 4, 2005. With permission.

Table 9.11   Prefixes Allocation (in ID Groups) Results

Group ID (Chip ID)	Partition #					Load Ratio (%)
	I	*II*	*III*	*IV*	*V*	
1	10	4	12	2	13	24.67
2	5	4	12	3	0	25.67
3	7	12	15	1	11	24.33
4	6	4	8	9	14	25.33

Source: Zheng, K., Hu, C.C., Lu, H.B., and Liu, B., *Proceedings of IEEE INFOCOM'04*, 3, 2004. With permission.

consumption. Current high-density TCAM devices consume as much as 12–15 W per chip when all the entries are enabled for search. However, the power consumption of TCAM can be reduced. We found that not all the entries need to be triggered simultaneously during a lookup operation. The entries that really need to be triggered are the ones those matching with the input key. Thus, one way of making TCAM power-efficient is to avoid the nonessential entries from being triggered during a lookup operation. To measure the efficiency of power consumption in each lookup operation, Zheng et al. introduce the concept of dynamic power efficiency (DPE), as defined below.

Definition 9.5.6:   Let V_i be the number of entries triggered during a specific lookup operation and matching the input search key i, and W be the total number of entries triggered during a specific lookup operation. We define the DPE as

$$\text{DPE}(i) := \frac{V_i}{W} \tag{9.9}$$

For instance, suppose that the input search key i is "A.B.C.D," there are five route prefixes matching this key in the route table, which are A/8, A.B/12, A.B/16, A.B.C/24, and A.B.C.D/32. The size of the total route table (all in a single chip) is 128,000, and all the entries are triggered during a lookup operation, then DPE $(i) := 5/128{,}000 = 3.90625 \times 10^{-5}$.

There are two ways to decrease the number of entries triggered during a lookup operation. One method is to store the entries in multiple small chips instead of a single large one. The other is to partition one TCAM into small blocks and trigger only one of them during a lookup operation. If the hardware (TCAM) supports the partition-disable function, the proposed scheme will benefit from both methods.

We continue using the example mentioned above. If we use four small TCAMs instead of (to replace) the large one and partition each of them into eight blocks to store the route prefixes, assuming that the storing redundancy rate is 1.25, then the number of prefixes that stored in each of the $4 \times 8 = 32$ blocks should be $128{,}000 \times 1.25/32 = 5000$. Applying the proposed TCAM partition method and supposing that the hardware supports the partition-disable feature, the DPE is now increased, meaning that the power consumption efficiency is increased by a factor of more than 25.

9.5.5 Complete Implementation Architecture

The complete implementation architecture of the ultra-high throughput and power-efficient lookup engine is shown in Figure 9.34. Given an incoming IP address to be searched, the ID bits of the IP address are extracted and delivered to the index logic for a matching operation. The index logic will return a set of partition numbers indicating which TCAMs and which block inside each TCAM may contain the prefixes matching the IP address. The Priority Selector (i.e., the adaptive load balancing logic) selects one TCAM with the shortest input queue (FIFO) from those containing the matching prefixes and sends the IP address to the input queue corresponding to the selected TCAM. To keep the sequence of the incoming IP addresses, a tag (the sequence number) is attached to the IP address being processed.

9.5.5.1 Index Logic

The function of the index logic is to find out the partitions in the TCAMs that contain the group of prefixes matching the incoming IP addresses, as depicted in Figure 9.35. It is composed of groups of parallel comparing logics. Each group of comparing logics corresponds to one TCAM, whereas each comparing logic corresponds to a partition in the TCAM. The Index field represents the ID of the prefix group and the Partition field represents the block in which this group of prefixes is stored. Each group of the comparing logics has a returning port. If the ID bits of the incoming IP address match one of the indexes in a group, the corresponding partition number is returned; otherwise, a partition number "111" (7) will be returned, which represents the

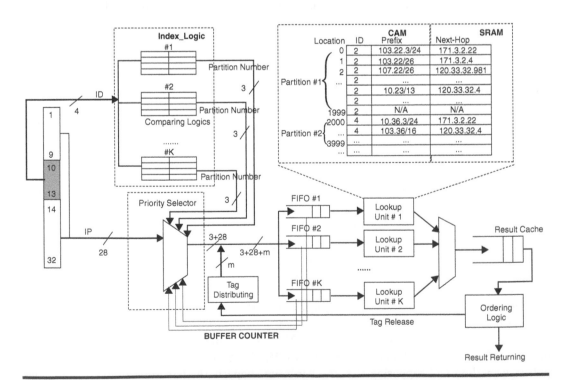

Figure 9.34 Schematics of the complete implementation architecture. (From Zheng, K., Hu, C.C., Lu, H.B., and Liu, B., *Proceedings of IEEE INFOCOM'04*, 3, 2004. With permission.)

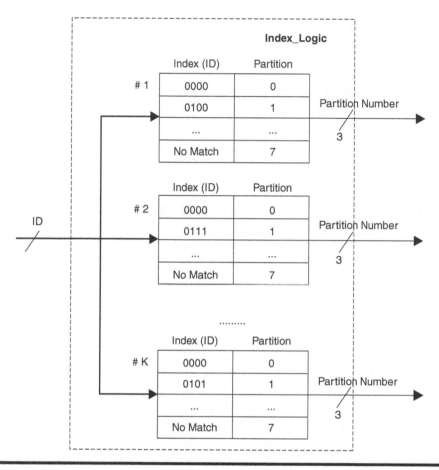

Figure 9.35 Schematics of the Index logic. (From Zheng, K., Hu, C.C., Lu, H.B., and Liu, B., *Proceedings of IEEE INFOCOM'04*, **3, 2004. With permission.)**

no-matching information. Because the data bus width of the comparing logic is narrow and fixed, and only simple "compare" operation is required, it can run at a very high speed.

9.5.5.2 Priority Selector (Adaptive Load Balancing Logic)

The function of the Priority Selector is to allocate the incoming IP address to the most "idle" TCAM that contains the prefixes matching this IP address, so that the lookup traffic is balanced among the TCAMs adaptively. As mentioned before, we introduce some storing redundancy into our scheme to guarantee the lookup throughput and one prefix may be stored in multiple TCAMs based on the traffic distribution among the ID groups. The Priority Selector uses the counter's status of the input queue for each TCAM to determine which one of the current IP address should be delivered to. The algorithm of the Priority Selector is shown as follows.

Algorithm for the priority selector:

```
Input: Counter[i], i =1,...,K; //Status of the buffer for each
TCAM.
```

```
      PN[i], i =1,...,K; //Partition numbers of the chips, which
are inputted from the Index logic.
Output: Obj // serial number of the chip that the current IP
address should be delivered to.
for i from 1 to K do //To find a feasible solution
      if (PN[i] ≠ 7) then
            Obj = i;
      Else
            break;
endfor;
for i from Obj + 1 to K do
   If (Counter[i] < Counter[obj] and PN[i] ≠ 7 ) then
      Obj = i;
endfor;
```

Because the status of the queue counters are independent of the current IP address, and high precision is not required here, this component can also be implemented in high-speed ASIC with ease.

9.5.5.3 Ordering Logic

Because multiple input queues exist in the proposed scheme, the incoming IP address will leave the result cache (as shown in Fig. 9.34) in a different sequence from the original one. The function of the ordering logic is to ensure that the results will be returned in the same order as that of the input side. An architecture based on Tag-attaching is used in the ordering logic. When an incoming IP address is distributed to the proper TCAM, a tag (i.e., sequence number) will be attached to it. Then, at the output side, the ordering logic uses the tags to reorder the returning sequence.

For example, assuming that the traffic distribution among the ID groups is given as shown in Table 9.10, we use the proposed load-balance-based algorithm to construct the tables in the TCAMs and get the result as shown in Table 9.11. When an IP address, 166.103.142.195, is presented to the lookup engine, its ID "1100" (12) is extracted and sent to the Index logic. The ID is compared with the 20 indexes in four groups simultaneously. The partition numbers are returned as "010" (2), "010" (2), "001" (1), and "111" (7), which means that TCAM #1, #2, and #3 contain the group of prefixes matching the IP address, whereas TCAM #4 does not. These results are then sent to the Priority Selector. Suppose that the counter values of the three TCAMs (#1, #2, and #3) are 6, 7, and 3, respectively. TCAM #3 is selected due to its smallest counter value. Then, the IP address "166.103.142.195," the partition number "001," and the current sequence number are pushed into the FIFO queue corresponding to chip #3. When reaching the head of the queue, the IP address is popped out and sent to partition #1 (of TCAM #3) to perform the lookup operation. The final result is sent out by the ordering logic in the original order according to its tag attached. All the above processing steps can be carried out in a pipelined fashion.

9.5.6 Performance Analysis

To measure its performance and adaptability on different types of traffic load distributions, Zheng et al. have run a series of experiments and simulations [14]. In the case of four TCAMs with a buffer depth $n = 10$ for each, the route table is the Mae-West table (http://www.merit.edu/ipma) and the arrival processes corresponding to the ID groups are all independent Poisson

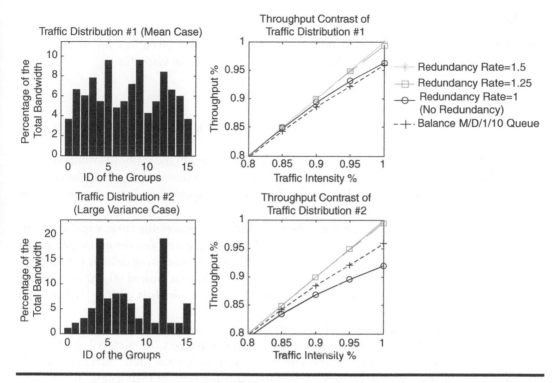

Figure 9.36 **Comparison of the throughput using different redundancy rate and different traffic distribution. (From Zheng, K., Hu, C.C., Lu, H.B., and Liu, B.,** *Proceedings of IEEE INFOCOM'04,* **3, 2004. With permission.)**

processes. The results are given in Figure 9.36 for two different traffic load distributions among the ID groups.

When the traffic load is evenly distributed (Case #1, in Fig. 9.36a), it can be found that the introduction of storing redundancy only improves the throughput slightly. The load-balance-based memory organization has already restricted the loss (block) rate within 5 percent. On the other hand, in the case of skewed distribution (Case #2, in Fig. 9.36b), the storing redundancy improves the lookup throughput distinctly when the system is heavily loaded, even though the redundancy rate is as low as 1.25. In both cases, a redundancy rate of 1.25 guarantees the throughput to be nearly 100 percent, which means that when four TCAM chips work in parallel, the LBBTC scheme improves the lookup throughput by a factor of 4 at the expense of only 25 percent more memory space.

To measure the stability and adaptability of the proposed scheme when the traffic distribution varies apart from the original one over time, Zheng et al. runs the following simulations with the redundancy rate of 1.25.

Simulation 1

The forwarding table is constructed from the traffic distribution given in Case #2 shown in Figure 9.36, the lookup traffic with the distribution is that given in Case #1.

Simulation 2

The forwarding table is constructed from the traffic distribution given in Case #1 shown in Figure 9.36, the lookup traffic with the distribution is that given in Case #2.

Simulation 3

The forwarding table is constructed from the traffic distribution given in Case #1, the lookup traffic is strictly even-distributed.

Figure 9.37 shows the results of the three simulations. Although the traffic distribution varies a lot, the lookup throughput drops less than 5 percent, meaning that the proposed scheme is not sensitive to the variation of traffic distribution. In fact, the adaptive load-balancing mechanism plays an important role in such cases.

The input queue and the ordering logic also introduce some processing latency to the incoming IP addresses. Figure 9.38 shows simulation results of the queuing and the entire processing (ordering logic included) latency of our mechanism. The entire processing delay is between $9T_s$ and $12T_s$ (service cycles). If 133 MHz TCAMs are used, the delay is around 60–90 ns, which is acceptable. The jitter of the entire processing latency is small, which is critical for hardware implementation.

TCAM table update caused by routing information modification is simply similar to that of the conventional TCAM-based lookup mechanisms. Note that prefixes are organized in ID groups in the proposed scheme. According to Theorem 9.5.1, to ensure longest prefix matching we only need to keep the prefixes in each ID group stored in decreasing order of their length. It can easily

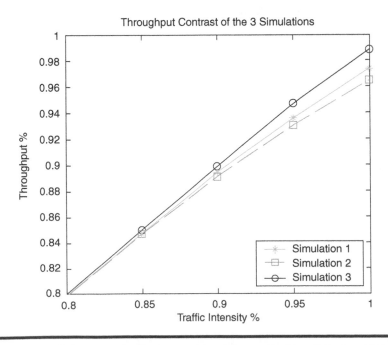

Figure 9.37 Results of the tree simulations. (From Zheng, K., Hu, C.C., Lu, H.B., and Liu, B., *Proceedings of IEEE INFOCOM'04*, 3, 2004. With permission.)

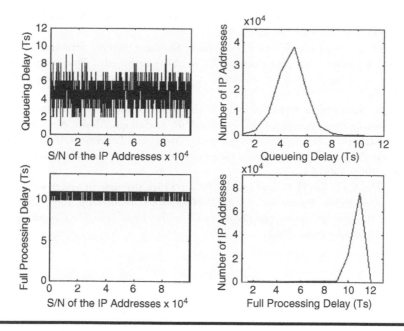

Figure 9.38 **Processing latency and latency distribution of the incoming IP addresses. (From Zheng, K., Hu, C.C., Lu, H.B., and Liu, B., *Proceedings of IEEE INFOCOM'04*, 3, 2004. With permission.)**

be implemented with incremental update using the algorithm presented in [2]. Also note that for some ID groups, there may be multiple copies among the TCAM chips, so we need to update all the copies.

References

1. Labovitz, C., Malan, G.R., and Jahanian, F., Internet routing instability, *IEEE/ACM Trans. Networking* 1999; 6(5):515–528.
2. Shah, D. and Gupta, P., Fast updating algorithms for TCAMs, *IEEE Micro* 2001; 21(1):36–47.
3. Weidong Wu, Bingxin Shi, and Feng Wang, Efficient location of free spaces in TCAM to improve router performance, *International Journal of Communication Systems* 2005; 18(4):363–371.
4. Kobayashi, M., Murase, T., and Kuriyama, A., A longest prefix match search engine for multi-gigabit IP processing. Robert Walp and Ian F. Akyildiz eds., *Proceedings of the International Conference on Communications (ICC 2000)*, New Orleans, LA, 2000, New York: IEEE Press, 1360–1364.
5. Panigrahy, R. and Sharma, S., Reducing TCAM power consumption and increasing throughput, In Raj Jain Dhabaleswar and K. Panda eds., *Proceedings of 10th Symp. High-Performance Interconnects (HOTI 02)*, Stanford, California, 2002, New York: IEEE CS Press, 107–112.
6. Zane, F., Narlikar, G., and Basu, A., CoolCAMs: Power-efficient TCAMs for forwarding engines. Eric A. Brewer and Kin K. Leung eds., *Proceedings of IEEE INFOCOM*, San Francisco, CA, 2003, New York: IEEE Press, 42–52.
7. Liu, H., Routing table compaction in ternary CAM, *IEEE Micro* 2002; 22(1):58–64.
8. Ravikumar, V.C. and Rabi Mahapatra Laxmi N., Bhuyan, EaseCAM: an energy and storage efficient TCAM-based router architecture for IP lookup, Computers, *IEEE Transactions* 2005; 54(5):521–533.

9. Brayton, R.K., et al., Logic minimization algorithms for VLSI synthesis, Kluwer Academic Publishers, Boston, 1984.

10. Eddie Kohler, Jinyang Li, and Vern Paxson, Scott Shenke, Observed Structure of Addresses in IP Traffic, In Christophe Diot and Balachander Krishnamurthy eds., *Proceedings of the 2nd ACM SIGCOMM Workshop on Internet measurement*, Marseille, France, 2002, New York: *IEEE Press*, 2002:153–166.

11. Funda Ergun, Suvo Mittra, Cenk Sahinalp, Jonathan Sharp, and Rakesh Sinha, A dynamic lookup scheme for bursty access patterns, *IEEE Infocom 2001*, Anchorage, Alaska, April 22–26, 2001.

12. Lin, S. and McKeown, N., A simulation study of IP switching. *Proceedings ACM SIGCOMM*, 1997.

13. Weidong Wu, Bingxin, Shi, Jian Shi, and Ling Zuo, Power efficient TCAMs for bursty access patterns, *IEEE Micro* 2005; 25(4):64–72.

14. Zheng, K., Hu, C.C., Lu, H.B., and Liu, B., An ultra high throughput and power efficient TCAM-based IP lookup engine. *Proceedings of IEEE INFOCOM'04*, 3, 2004, pp. 1984–1994.

15. William H. Press, et al., *Numerical Recipes in C++: The Art of Scientific Computing*, pp. 357–358, Cambridge University Press, 2002.

Chapter 10

Routing-Table Partitioning Technologies

IP-address lookup is difficult because it requires a longest-matching prefix search. That is to say, the existing exact search algorithms, such as binary search, cannot directly be used for IP-address lookup. Some researchers develop partition technologies to adapt some exact search algorithms for longest-matching prefix search. And the routing table is partitioned into small tables; we can perform IP-address lookup in parallel. In this chapter, we will introduce some partitioning technologies.

10.1 Prefix and Interval Partitioning

The primary prefix partitioning scheme was proposed by Lampson et al. [1]. The scheme is not suitable for dynamic routing tables. Lu et al. proposed a prefix partitioning scheme (multilevel partitioning) and interval partitioning to improve the performance of dynamic routing table designs [2].

10.1.1 Partitioned Binary Search Table

Binary search is one of the efficient exact search schemes. Lampson et al. [1] adapted it for the longest-matching prefix search by encoding prefixes as ranges. To improve the performance of the binary search, Lampson et al. partition the single table into multiple tables. This subsection is from [1]. Portions are reprinted with permission (© 1998 IEEE).

To illustrate the scheme from [1], we use an example in which we have 6-bit addresses and just three prefixes: 1*, 101*, and 10101*. Because the binary search does not work with variable length strings, the simplest approach is to pad each prefix to be a 6-bit string by adding zeroes. This is shown in Figure 10.1.

Now consider a search for three 6-bit addresses 101011, 101110, and 111110. Because none of these addresses are in the table, binary search will fail. Unfortunately, in a failure, all three of these addresses will end up at the end of the table because all of them are greater than 101010, which is the last element in the binary search table. Notice, however, that each of these three addresses has a different best-matching prefix.

371

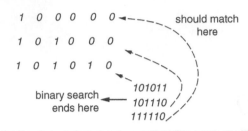

Figure 10.1   Placing the three prefixes 1*, 101*, and 10101* in a binary search table by padding each prefix to 6-bit strings. (From Lampson, B., Srinivasan, V., and Varghese, G., *Proc. IEEE INFOCOM '98*, April 1998. With permission.)

Thus, we have two problems with naive binary search: first, when we search for an address, we end up far away from the matching prefix (potentially requiring a linear search); second, multiple addresses that match to different prefixes end up in the same region in the binary table (Fig. 10.1).

10.1.1.1   Encoding Prefixes as Ranges

To solve the second problem, we recognize that a prefix like 1* is really a range of addresses from 100000 to 111111. Thus, instead of encoding 1* by just 100000 (the start of the range), we encode it using both the start and end of the range (100000 and 111111). Thus, each prefix is encoded by two full-length bit strings. These bit strings are then sorted. Figure 10.2 shows the result for the same three prefixes.

We connect the start and end of a range (corresponding to a prefix) by a line as shown in Figure 10.2. Notice how the ranges are nested. If we now try to search for the same set of addresses, they each end in a different region in the table. To be more precise, the search for address 101011 ends in an exact match. The search for address 101110 ends in a failure in the region between 101011 and 101111 (Fig. 10.2), and the search for address 111110 ends in a failure in the region between 101111 and 111111. Thus, it appears that the second problem (multiple addresses that match different prefixes ending in the same region of the table) has disappeared. Compare Figures 10.1 and 10.2.

To see that this is a general phenomenon, consider Figure 10.3. The figure shows an arbitrary binary search table after every prefix has been encoded by the low (marked L in Fig. 10.3) and its high points. The binary search for address A ends up at this point; which prefix should we map A

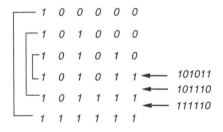

Figure 10.2   Encoding each prefix in the table as a range using two values: the start and end of the range. (From Lampson, B., Srinivasan, V., and Varghese, G., *Proc. IEEE INFOCOM '98*, April 1998. With permission.)

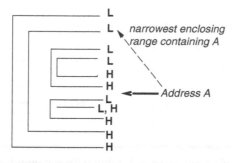

Figure 10.3 **Why longest-matching prefix correspond to narrowest encoding range, and why each range in the modified binary search table maps to a unique prefix. (From Lampson, B., Srinivasan, V., and Varghese, G.,** *Proc. IEEE INFOCOM '98,* **April 1998. With permission.)**

to? It is easy to see the answer visually from Figure 10.3. If we start from the point shown by the solid arrow and we go back up the table, the prefix corresponding to A is the first L that is not followed by a corresponding H (see dotted arrow in Fig. 10.3).

Why does this work? Because we did not encounter an H corresponding to this L, it clearly means that A is contained in the range corresponding to this prefix. Because this is the first such L, this is the smallest such range. Essentially, this works because the best-matching prefix has been translated to the problem of finding the narrowest encoding range.

10.1.1.2   Recomputation

Unfortunately, the solution depicted in Figures 10.2 and 10.3 does not solve the first problem: notice that binary search ends in a position that is far away (potentially) from the actual prefix. If we were to search for the prefix, we could have a linear time search.

However, the modified binary search table shown in Figure 10.3 has a nice property that can be exploited. Any region in the binary search between two consecutive numbers corresponds to a unique prefix. As described earlier, the prefix corresponds to the first L before this region that is not matched by a corresponding H that also occurs before this region. Similarly, every exact match corresponds to a unique prefix.

As this is the case, we can precompute the prefix corresponding to each region and the prefix corresponding to each exact match. Essentially, the idea is to add the dotted line pointer shown in Figure 10.3 to every region.

The final table corresponding to Figure 10.3 is shown in Figure 10.4. Notice that with each table entry E, there are two precomputed prefix values. If binary search for address A ends in a failure at E, it is because $A > E$. In that case, we use the ">" pointer corresponding to E. On the other hand, if binary search for address A ends in a match at E, we use the "=" pointer.

Notice that for an entry like 101011, the two entries are different. If address A ends up at this point and is greater than 101011, clearly the right prefix is $P2 = 101*$. In contrast, if address A ends up at this point with equality, the correct prefix is $P3 = 10101*$. Intuitively, if an address A ends up equal to the high point of a range R, then A falls within the range R; if A ends up greater than the high point of range R, then A falls within the smallest range that encodes range R.

With the basic binary search, the worst-case possible number of memory accesses is $\log_2 N + 1$ which for large database, could be 16 or more memory accesses. Lampson et al. use precomputed

							>	=
P1)	1	0	0	0	0	0	P1	P1
P2)	1	0	1	0	0	0	P2	P2
P3)	1	0	1	0	1	0	P3	P3
	1	0	1	0	1	1	P2	P3
	1	0	1	1	1	1	P1	P2
	1	1	1	1	1	1	–	P1

Figure 10.4   The final modified binary search table with recomputed prefixes for every region of the binary table. We need to distinguish between a search that ends in a success at a given point (= pointer) and search that ends in a failure at a given point (> pointer). (From Lampson, B., Srinivasan, V., and Varghese, G., *Proc. IEEE INFOCOM '98*, April 1998. With permission.)

table of best-matching prefixes for the first *Y* bits to improve the worst-case number of memory accesses [1]. The main idea is to efficiently partition the single binary search table into multiple binary search tables for each value of the first *Y* bits, shown in Figure 10.5. We choose *Y* = 16 for what follows, as the table size is about as large as we can afford, while providing maximum partitioning. The best-matching prefixes for the first 16-bit prefixes can be precomputed and stored in table. This table would then have *Max* = 65536 elements. For each index *X* of the array, the corresponding array element stores the best-matching prefix of *X*. Additionally, if there are prefixes of longer length with prefix *X*, the array element stores a pointer to a binary search table or tree that contains all such prefixes.

The best-matching prefixes of the elements of the array can be easily precomputed using a form of prefix expansion described in [3]. The simplest conceptual idea is to expand all prefixes, starting from the lowest length prefixes, one bit at a time until they reach 16 bits. A prefix like *P* can be expanded into *P0* and *P1*. However, if there is already a higher length prefix corresponding to either *P0* or *P1*, then we stop the expansion process as the longer length prefix captures the (expanded) lower length prefix. At the end, we have a list of 16-bit elements tagged with the appropriate prefix; this can be used to fill the table.

The above idea takes *O(Max)* time and requires storage for *Max* list elements in addition to the initial array. Another possible scheme is to build a binary search table (Fig. 10.4) on the prefixes of

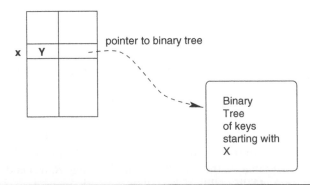

Figure 10.5   The array element with index *X* will have the best-matching prefix of *X* (say *Y*) and a pointer to a binary tree or table of all prefixes that have *X* as a prefix. (From Lampson, B., Srinivasan, V., and Varghese, G., *Proc. IEEE INFOCOM '98*, April 1998. With permission.)

lengths to no greater than 16. Then we can proceed through the array elements and use the binary search table to compute the best-matching prefix of each array element. This would require Maxlog N time, which would be too slow.

However, because the final binary search table (Fig. 10.4) is essentially a sorted table of a range of endpoints, we need not do a binary search for each element. We only need to simultaneously scan the array and the binary search table. We use two pointers: a pointer into the array (that starts at the lowest element of the array) and a pointer into the binary search table (that starts at the lowest element of the table). The current range is indicated by the current binary search entry and the following entry. The algorithm keeps incrementing the array pointer, updating each array element using the next hop corresponding to the current range; as soon as the array pointer goes beyond the current range, the algorithm increments the binary search table pointer. The algorithm terminates when all array elements have been scanned. Thus the algorithm terminates after $O(N + Max)$ time. We stress again that this recomputation needs to be done only when the prefixes are inserted and deleted, and not when a search is done.

10.1.1.3   Insertion into a Modified Binary Search Table

Because we precompute the prefix corresponding to each region and the prefix corresponding to each exact match, this can potentially slow down insertion. For a faster insertion algorithm, first we could represent the binary search table as a binary tree in the usual way. This avoids the need to shift entries to make room for a new entry. Unfortunately, the addition of a new prefix can affect the precomputed information in $O(N)$ prefixes. This is illustrated in Figure 10.6. The figure shows an outermost range corresponding to prefix P; inside this range are $N - 1$ smaller ranges (prefixes) that do not intersect. In the regions not covered by these smaller prefixes, we map to P. Unfortunately, if we now add Q (Fig. 10.6), we cause all these regions to map to Q, an $O(N)$ update process.

Thus, there does not appear to be any update technique that is faster than just building a table from scratch. Of course, many insertions can be batched; if the update process falls behind, the batching will lead to more efficient updates. We will see shortly that using an initial 16-bit table lookup can decompose the single binary search table into multiple (but smaller) binary search

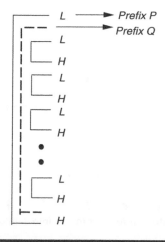

Figure 10.6   Adding a new prefix *Q* (*dotted line*) can cause all regions between an *H* and an *L* to move from prefix *P* to prefix *Q*. (From Lampson, B., Srinivasan, V., and Varghese, G., *Proc. IEEE INFOCOM '98*, April 1998. With permission.)

tables; because the decomposed binary search tables are smaller, this decomposition speeds up not only search time but also insertion.

10.1.1.4 Multiway Binary Search: Exploiting the Cache Line

Today, processors have a wide cache line. To improve the locality of access, Lampson et al. restructure data structure [1], in which keys and pointers in search tables can be laid out to allow multiway search instead of binary search. This effectively allows us to reduce the search time of binary search from $\log_2 N$ to $\log_{k+1} N$, where k is the number of keys in a search node. The main idea is to make k as large as possible so that a single search node (containing k keys and $2k + 1$ pointers) fits into a single cache line. If this can be arranged, an access to the first word in the search node will result in the entire node being perfected into cache. Thus, accesses to the remaining keys in the search node are much cheaper than a memory access.

Lampson et al. did experiments choosing $k = 5$ (i.e., doing a six-way search). If we use k keys per node, then we need $2k + 1$ pointers, each of which is a 16-bit quantity. This is because we need a pointer for the ranges between keys as well as a pointer for exact matches with each key. Thus, in 32 bytes, we can place five keys and hence can do a six-way search. For example, if there are keys $k1, ..., k8$, a three-way tree is given in Figure 10.7. The initial full array of 16 bits followed by the six-way search is depicted in Figure 10.8.

Each node in the six-way search table has five keys $k1–k5$, each of which is of 16 bits. There are equal to pointers $p1–p5$ corresponding to each of these keys. Pointers $p1–p56$ corresponding to ranges are demarcated by the keys. This is shown in Figure 10.9. Among the keys, we have the relation $k_i \le k_{i+1}$. Each pointer has a bit which says whether it is an information pointer or a next node pointer.

The search procedure is shown as follows:

Step 1: Index into the first 16-bit array using the first 16 bits of the address.
Step 2: If the pointer at the location is an information pointer, return it. Otherwise, enter the six-way search with the initial node given by the pointer, and the key being the next 16 bits of the address.

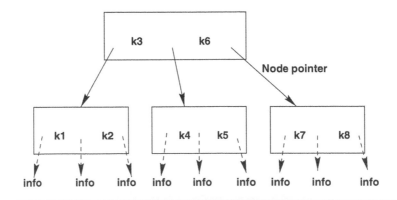

Figure 10.7 **Three-way tree for eight keys. (From Lampson, B., Srinivasan, V., and Varghese, G.,** *Proc. IEEE INFOCOM '98,* **April 1998. With permission.)**

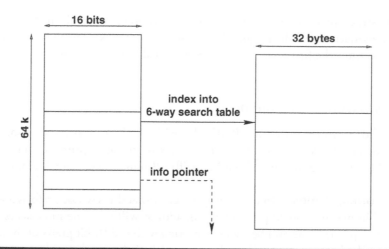

Figure 10.8 **The initial 16-bit array, with pointers to the corresponding six-way search nodes. (From Lampson, B., Srinivasan, V., and Varghese, G.,** *Proc. IEEE INFOCOM '98,* **April 1998. With permission.)**

Step 3: In the current six-way node, locate the position of the key among the keys in the six-way node. We use binary search among the keys within a node. If the key equals any of the keys key_i in the node, use the corresponding pointer ptr_i. If the key falls in any range formed by the keys, use the pointer $ptr_{i,i+1}$. If this pointer is an information pointer, return it; otherwise, repeat this step with the new six-way node given by the pointer.

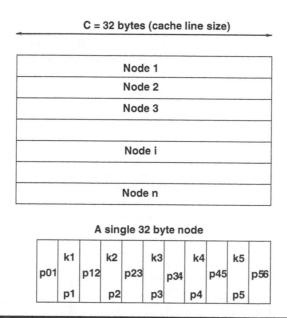

Figure 10.9 **The structure of the six-way search node. There are** *k* **keys and** *2k* + 1 **pointers. (From Lampson, B., Srinivasan, V., and Varghese, G.,** *Proc. IEEE INFOCOM '98,* **April 1998. With permission.)**

As the data structure itself is designed with a node size equal to a cache line size, good caching behavior is a consequence. All the frequently accessed nodes will stay in the cache. To reduce the worst-case access time, the first few levels in a worst-case depth tree can be cached.

10.1.1.5 Performance Measurements

Lampson et al. did experiments using a 200 MHz PC with a 8 kByte four-way set-associative primary instruction cache and a 8 kByte dual-ported two-way set-associative primary data cache. The real routing table (Mae-East) was obtained from Merit (ftp://ftp.merit.edu/statistics/ipma, 1997).

Worst-Case Lookup Times. In the six-way search, any address whose prefix is less than 16 bits will be resolved in one array lookup (depth 0); an address will take one more access to a six-way node to resolve (depth 1) if the number of keys starting with the 16-bit prefix of the address is less than 5; others will take accesses to a second six-way node if the number of keys starting with the 16-bit prefix of the address is less than 35 (depth 2); and so on. Recall that as we are doing six-way search, the depth of an address A is equal to $\lceil \log_6 n \rceil$, where $n(A)$ is the number of keys that start with the 16-bit prefix of address A. Table 10.1 describes the distribution of $n(A)$.

Because the worst-case lookup times varied with depth, Lampson et al. picked seven address lookups to display in Table 10.2. The first two rows correspond to depth 4 address, whereas the last two correspond to depth 0 address. The third, fourth, and fifth rows correspond to depths 1, 2, and 3 addresses, respectively. The search time for PATRICIA does vary a great deal (from about 1.5 to 2.5 ms). The time for basic binary search varies only a little (from around 1 to 1.3 ms). The time for binary search with the initial table clearly depends on the depth of the address. The lower rows have very small lookup times (90 ns), but the upper rows have larger times (e.g., 730 ns) corresponding to a binary search of up to 336 keys.

The time taken for the six-way search varies in the same way. The search time for depth 0 nodes are nearly identical for six-way and binary search + 16-bit table because both involve only a lookup of the initial array. However, the times at each depth greater than depth 0 are much lower than that of using binary search + the 16-bit table because of the benefits of using six-way versus two-way search. An address at depth 4 in the six-way scheme will be at a greater depth in the other binary search schemes.

Table 10.1   Distribution of $n(A)$

Number of Keys in Tree	No. of 16-Bit Prefixes Leading to Trees Containing the Number of Keys in This Range	Measured Time When Key in Tree with Total Number of Keys in This Range
0	63414	95
1..5	743	210
6..35	916	210
26..215	449	390
215..max=336	13	490

Source: Lampson, B., Srinivasan, V., and Varghese, G., *Proc. IEEE INFOCOM '98*, April 1998. With permission.

Table 10.2 Comparison of the Time Taken for a Single Address Lookup (ns)

PATRICIA	Basic Binary Search	16-bit Table + Binary Search	16-bit Table + 6-way Search
1530	1175	730	490
1525	990	620	490
1980	1440	330	210
2585	1210	400	300
1450	1140	470	390
810	1220	90	95
1170	1310	90	90

Source: Lampson, B., Srinivasan, V., and Varghese, G., *Proc. IEEE INFOCOM '98,* April 1998. With permission.

Notice also that in Table 10.2 that the lookup times for six-way search increase by around 100 ns for each memory access, yielding a lookup time of around $100*(D(A) + 1)$ for address A, where $D(A)$ is the depth of address A in the six-way scheme. This fits in well with the intuition that the primary measure of lookup speed is the number of memory accesses and that the speed of a random memory access (including memory costs, processing costs, and costs of cache misses) in the Pentium is around 100 ns.

Average Lookup Time. Table 10.1 gives the number of 16-bit prefix slots that lead to six-way trees that would take a maximum of one memory access, two memory accesses, and so on. Assuming that all IP addresses are equally likely, the data in Table 10.3 allow us to easily calculate the average time for a lookup. For the 16-bit + six-way search, the average time is 100 ns; for the 16-bit + binary search, the average time is 110 ns. This is not very different from the six-way search because a large number of initial array entries do not have any longer prefixes; thus a large fraction (in both algorithms) of addresses takes only 90 ns to resolve.

However, in practice, the distribution of addresses received by a router is unlikely to be random. To claim wire speed routing, a router must be able to repeatedly look up packets that have worst-case lookup times. Because worst-case search time is an important consideration for IP lookups, six-way search appears to be a good choice.

Memory Requirements. Table 10.3 shows the memory requirement. Six-way search has a 0.95 Mbytes memory usage that is less than the basic binary search. This is because we have added a large initial array in the six-way search algorithm. The memory requirement of six-way search is more than that of binary search with 16-bit initial table because the multiway search needs more pointers. But the multiway search can improve the search time from 730 to 490 ns. Thus both optimizations (initial array, multiway search) appear to be worthwhile.

10.1.2 Multilevel and Interval Partitioning

Multilevel partitioning [2] is the extension of the scheme in [1]. Interval partitioning applies to the alternative collection of binary search tree designs in [4]. This subsection is from [2]. Portions are reprinted with permission (© 2005 IEEE).

Table 10.3 Memory Requirement and Worst-Case Time

Algorithm	PATRICIA	Basic Binary Search	16-bit Table + Binary Search	16-bit Table + 6-Way Search
Total memory (MB)	3.2	3.6	1	1
Memory for searchable structure (MB)	3.2	1	0.75	0.95
Worst-case search time (ns)	2585	1310	730	490
Comparison with PATRICIA scheme	1	2	3.5	5

Source: Lampson, B., Srinivasan, V., and Varghese, G., *Proc. IEEE INFOCOM '98*, April 1998. With permission.

10.1.2.1 Multilevel Partitioning

In multilevel partitioning, the prefixes at each node of the partition tree are partitioned into $2^s + 1$ partitions using the next s bits of the prefixes. Prefixes that do not have s additional bits fall into partition -1; the remaining prefixes fall into the partition that corresponds to their next s bits. Prefix partitioning may be controlled using a static rule such as "partition only at the root" or by a dynamic rule such as "recursively partition until the number of prefixes in the partition is smaller than some specified threshold." In the following, we focus on two statically determined partitioning structures—one level and two level.

One-Level Dynamic Partitioning (OLDP). OLDP is a partitioning tree described as in Figure 10.10. The root node represents the partitioning of the prefixes into $2^s + 1$ partitions. Let $OLDP[i]$ refer to partition i. $OLDP[-1]$ contains all prefixes whose length is less than s; $OLDP[i]$, $i \geq 0$ contains all prefixes whose first s bits correspond to i. The prefixes in each partition may be represented using any data structures that can be used for IP-address lookup. In fact, one could use different data structures for different partitions. For example, if we knew that certain partitions would always be small, these could be represented as linear lists, whereas the remaining partitions are represented using the multibit trie in Chapter 4.

The essential difference between OLDP and the partitioning scheme in [1] is in the treatment of prefixes whose length is smaller than s. In OLDP, these shorter prefixes are stored in $OLDP[-1]$; in the scheme in [1], these shorter prefixes (along with length s prefixes) are used to determine the longest prefix for the binary representation of the first s bits. It is this difference in the way shorter

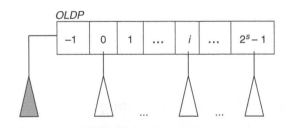

Figure 10.10 One-level dynamic partitioning. (From Haibin Lu, Kim, K., and Sahni, S., *IEEE Transactions on Computers*, 54, 5, 2005. With permission.)

Table 10.4   Statistics of One-Level Partition (s = 16)

Database	Paix	Pb	Aads	Mae-West
No. of prefixes	85988	35303	31828	28890
No. of nonempty partitions	10443	6111	6363	6026
|OLDP[−1]|	586	187	188	268
Max size of remaining partitions	229	124	112	105
Average size of nonempty partitions (excluding OLDP[−1])	8.2	5.7	5.0	4.8

Source: Haibin Lu, Kim, K., and Sahni, S., *IEEE Transactions on Computers*, 54, 5, 2005. With permission.).

length prefixes are handled that makes OLDP suitable for dynamic tables, whereas the scheme in [1] is suitable for static tables. Table 10.4 gives the results of partitioning four IPv4 router tables using OLDP with $s = 16$. These router tables were obtained from Merit (http://nic.merit.edu/ipma). As can be seen, OLDP with $s = 16$ is quite effective in reducing both the maximum and the average partition size. In all the databases, partition −1 is substantially larger than the remaining partitions.

Figure 10.11 gives the search, insert, and delete algorithms in an OLDP router table. $OLDP[i] \rightarrow x()$ refers to method x performed on the data structure for $OLDP[i]$, $first(d, s)$ returns the integer that corresponds to the first s bits of d, and length(*thePrefix*) returns the length of *thePrefix*.

```
Algorithm lookup(d){
// return the longest matching prefix (lmp) of IP-address d
if(OLDP[first(d, s)] != null && lmp = OLDP[first(d, s)]->looku
    // found matching prefix in OLDP[first(d, s)]
    return lmp;
}
return OLDP[−1]->lookup(d);
}

Algorithm insert(thePrefix){
// insert prefix thePrefix
if(length(thePrefix) >= s)
    OLDP[first(thePrefix, s)]->insert(thePrefix);
else
    OLDP[−1]->insert(thePrefix);
}

Algorithm delete(thePrefix){
// delete prefix thePrefix
if (length(thePrefix) >= s)
    OLDP[first(thePrefix, s)]->delete(thePrefix);
else
    OLDP[−1]->delete(thePrefix);
}
```

Figure 10.11   OLDP algorithms. (From Haibin Lu, Kim, K., and Sahni, S., *IEEE Transactions on Computers*, 54, 5, 2005. With permission.)

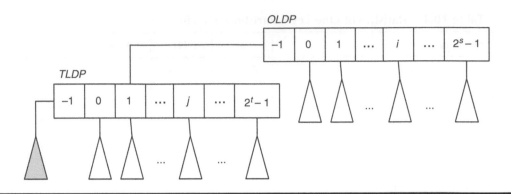

Figure 10.12 Two-level dynamic partitioning. (From Haibin Lu, Kim, K., and Sahni, S., *IEEE Transactions on Computers*, 54, 5, 2005. With permission.)

It is easy to see that when each OLDP partition is represented using the same data structure (say multibit trie), the asymptotic complexity of each operation is the same as that for the corresponding operation in the data structure for the OLDP partitions. However, a constant factor speedup is expected because each $OLDP[i]$ has only a fraction of the prefixes.

Two-Level Dynamic Partitioning (TLDP). Figure 10.12 shows the partitioning structure for a TLDP. In a TLDP, the root partitions the prefix set into partitions $OLDP[i]$, $-1 \leq i < 2^s$ using the first s bits of each prefix. This partitioning is identical to that done in an OLDP. Additionally, the set of prefixes $OLDP[-1]$ is further partitioned at node TLDP into the partitions $TLDP[i]$, $-1 \leq i < 2^t$ using the first $(< t)$, and bits of the prefixes in $OLDP[-1]$. This partitioning follows the strategy used at the root. However, t rather than s bits are used. The prefix partitions $OLDP[i]$, $0 \leq i < 2^s$, and $TLDP[i]$, $-1 \leq i < 2^t$, may be represented using any dynamic data structure. Note that the $OLDP[i]$ partitions for $i \geq 0$ are not partitioned further because their size is typically not too large (Table 10.4).

Table 10.5 gives the statistics for the number of prefixes in $TLDP[i]$ for four sample databases. Because the number of prefixes in each $TLDP[i]$ is rather small, we may represent each $TLDP[i]$ using an array linear list in which the prefixes are in decreasing order of length.

Figure 10.13 gives the TLDP search and insert algorithms. The delete algorithm is similar to the insert algorithm. It is easy to see that TLDP does not improve the asymptotic complexity of

Table 10.5 Statistics for TLDP with $s = 16$ and $t = 8$

Database	Paix	Pb	Aads	Mae-West
$\lvert OLDP[-1] \rvert$	586	187	188	268
No. of nonempty TLDP partitions	91	57	53	67
$\lvert TLDP[-1] \rvert$	0	0	0	0
Max $\{\lvert TLDP[i] \rvert\}$	33	12	15	15
Average size of nonempty TLDP partitions	6.4	3.3	3.5	4.0

Source: Haibin Lu, Kim, K., and Sahni, S., *IEEE Transactions on Computers*, 54, 5, 2005. With permission.

```
Algorithm lookup(d){
// return the longest matching prefix (lmp) of d
if (OLDP[first(d, s)] != null && lmp = OLDP[first(d, s)]->lookup(d)){
// found lmp inOLDP[first(d, s)]
    return lmp;
}
if (TLDP[first(d, t)] != null && lmp = TLDP[first(d, t)]->lookup(d)){
// found lmp inTLDP[first(d, t)]
    return lmp;
}
return TLDP[-1]->lookup(d);
}



Algorithm insert(thePrefix){
// insert prefix thePrefix
if (length(thePrefix) >= s)
    OLDP[first(thePrefix, s)]->insert(thePrefix);
else if (length(thePrefix) >=t)
    TLDP[first(thePrefix, t)]->insert(thePrefix);
else
    TLDP[-1]->insert(thePrefix);
}
```

Figure 10.13   TLDP algorithm to lookup or insert a prefix. (From Haibin Lu, Kim, K., and Sahni, S., *IEEE Transactions on Computers*, 54, 5, 2005. With permission.)

the lookup or insert or delete algorithms relative to that of these operations in an OLDP. Rather, a constant factor improvement is expected.

10.1.2.2   Interval Partitioning

Each prefix represents an IP-address range with two ends. In an IP number line, two consecutive endpoints define a basic interval. In general, n prefixes may have up to $2n$ distinct endpoints and $2n - 1$ basic intervals. Table 10.6 shows a set of five prefixes.

The interval-partitioning scheme [2] is an alternative to the OLDP and TLDP partitioning schemes that may be applied to interval-based structures such as the alternative collection of binary search tree (ACBST) [4]. In this scheme, we employ a 2^s-entry table, *partition*, to partition the basic intervals based on the first s bits of the start point of each basic interval. For each partition of the basic intervals, a separate alternative basic interval tree (ABIT) [4] is constructed; the back-end prefix trees are not affected by the partitioning. The ACBST of the example in Table 10.6 is shown in Figure 10.14. Figure 10.15 gives the partitioning table and ABITs for a five-prefix example and $s = 3$.

Notice that each entry *partition*[i] of the partition table has four fields—*abit* (pointer to ABIT for partition i), *next* (next nonempty partition), *previous* (previous nonempty partition), and *start* (smallest endpoint in partition). Figure 10.16 gives the interval-partitioning algorithm to find $lmp(d)$. The algorithm assumes that the default prefix * is always present, and the method rightmost returns the lmp for the rightmost basic interval in the ABIT.

Table 10.6 Prefixes and Their Ranges, $W = 5$

Prefix Name	Prefix	Start End	Finish End
P1	*	0	31
P2	0101*	10	11
P3	100*	16	19
P4	1001*	18	19
P5	10111	23	23

Note: Ranges: r1=[0,10], r2=[10,11], r3=[11,16], r4=[16,18], r5=[18,19], r6=[19,23], and r7=[23,31]. For each prefix and basic interval, *x*, define *next(x)* to be the smallest range prefix (i.e., the longest prefix) whose range includes the range of *x*. For example, the *next()* values for the basic intervals *r1–r7* are, respectively, P1, P2, P1, P1, P3, P4, P1, and P1.

Source: Haibin Lu, Kim, K., and Sahni, S., *IEEE Transactions on Computers*, 54, 5, 2005. With permission.

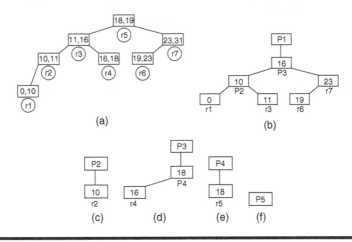

Figure 10.14 ACBST of the example in Table 10.6 (a) alternative basic integral tree, (b–e) prefix tree for P1–P5. (From Haibin Lu, Kim, K., and Sahni, S., *IEEE Transactions on Computers*, 54, 5, 2005. With permission.)

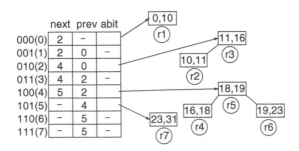

Figure 10.15 Interval-partitioned ABIT structures corresponding to Figure 10.14. (From Haibin Lu, Kim, K., and Sahni, S., *IEEE Transactions on Computers*, 54, 5, 2005. With permission.)

```
Algorithm lookup(d){
    // return Imp(d)
    p = first(d,s);
    if (partition[p].abit != null && partition[p].start <= d)
        // containing basic interval is in partition[p].abit
        return partition[p].abit->lookup(d);
    else return partition[partition[p].previous].abit->rightmost();
}
```

Figure 10.16 Interval-partitioning algorithm to find *Imp(d)*. (From Haibin Lu, Kim, K., and Sahni, S., *IEEE Transactions on Computers*, 54, 5, 2005. With permission.)

The insertion and deletion of prefixes are done by inserting and removing endpoints, when necessary, from the ABIT and adding or removing a back-end prefix tree. The use of interval partitioning affects only the components of the insert or delete algorithms that deal with the ABIT. Figure 10.17 gives the algorithms to insert and delete an endpoint. *rightPrefix* refers to the prefix, if any, associated with the right endpoint stored in a node of the ABIT.

Although the application of interval partitioning does not change the asymptotic complexity, $O(\log n)$, of the ACBST algorithm to find the longest-matching prefix, the complexity of the algorithms to insert and delete change from $O(\log n)$ to $O(\log n + 2^s)$. Because interval partitioning reduces the size of individual ABITs, a reduction in observed runtime is expected for the search algorithm. The insert and delete algorithms are expected to take less time when the clusters of empty partitions are relatively small (equivalently, when the nonempty partitions distribute uniformly across the partition table).

10.1.2.3 Experimental Results

To evaluate the performance of the multilevel and interval-partitioning schemes, Lu et al. programmed these schemes in C++ and used the data structure ACRBT to implement the schemes as a base data structure. OLDP, TLDP and interval partitioning are implemented with ACRBT. The routing table is shown in Table 10.4.

Total Memory Requirement.   Table 10.7 shows the amount of memory used by each of the structures tested. In the case of Paix, the memory required by ACRBT is more than that required by other schemes. In other cases, the memory required by ACRBT is less than that required by other schemes. Thus, the memory required by three partition schemes is related with the distribution of prefixes in the routing table.

Search Time.   To measure the average search time, Lampson et al. used two sets of destination addresses: NONTRACE and PSEUDOTRACE. NONTRACE comprised the endpoints of the prefixes corresponding to the routing table being searched. These endpoints were randomly permuted. PSEUDOTRACE was constructed from NONTRACE by selecting 1000 destination addresses. A PSEUDOTRACE sequence comprises 1,000,000 search requests. The experiments were repeated 10 times and 10 average times were obtained. The average search time is shown in Tables 10.8 and 10.9.

In Table 10.8, the use of the OLDP, TLDP, and Interval partitioning reduces the average search time in all cases. The interval-partitioning scheme outperforms the multilevel-partitioning

```
Algorithm insert(e){
    // insert the end point e
    p = first(e,s);
    if (partition[p].abit == null){
        // rn has interval that is split by e
        rn = partition[partition[p].previous].abit->rightmostNode();
        // split into 2 intervals
        partition[p].abit->new Node(e, rn->rightKey, rn->rightPrefix);
        rn->rightKey = e; rn->rightPrefix = null;
        // update next and previous fields of the partition table
        for (i=partiton[p].previous; i<p; i++) partiton[i].next = p;
        for (i=p+1; i<=partition[p].next; i++) partiton[i].previous = p;
    }
    else{
        if (partition[p].start > e){
            rn = partition[partition[p].previous].abit->rightmostNode();
            rn->rightKey = e; rn->rightPrefix = null;
        }
        partition[p].abit->insert(e);
    }
}

Algorithm delete(e){
    // delete the end point e
    p = first(e,s);
    if (partition[p].abit != null){
        if (partition[p].start == e){
            // combine leftmost interval of p with rightmost of previous
            ln = partition[p].abit->leftmostNode();
            rn = partition[partition[p].previous].abit->rightmostNode();
            rn->rightKey = ln->rightKey; rn->rightPrefix = ln->rightPrefix;
        }
        partition[p].abit->delete(e);
        if (partition[p].abit == null){
            // update next and previous fields of the partition table
            for (i=partition[p].previous; i<p; i++)
                partiton[i].next = partiton[p].next;
            for (i=p+1; i<=partition[p].next; i++)
                partiton[i].previous = partiton[p].previous;
        }
    }
}
```

Figure 10.17   Interval-partitioning algorithm to insert or delete an endpoint. (From Haibin Lu, Kim, K., and Sahni, S., *IEEE Transactions on Computers*, 54, 5, 2005. With permission.)

schemes: OLDP and TLDP. Comparing the average search time in Tables 10.8 and 10.9, there is an interesting observation: the average search time is considerably lower for the PSEUDOTRACE than for the NONTRACE. This is because of the reduction in average number of cache misses per search when the search sequence is bursty.

Table 10.7 Memory Requirement (kByte)

Schemes	Paix	Pb	Aads	Mae-West
ACRBT	12360	5120	4587	4150
ACRBT1p	12211	5479	5023	4638
ACRBT2p	12212	5482	5072	4641
ACRBTIP	12218	5344	4856	4453

Source: Haibin Lu, Kim, K., and Sahni, S., *IEEE Transactions on Computers*, 54, 5, 2005. With permission.

Table 10.8 Average Search Time (μs) for Nontrace

Schemes	Paix	Pb	Aads	Mae-West
ACRBT	1.31	0.89	0.94	0.92
OLDP	1.13	0.82	0.75	0.85
TLDP	1.13	0.89	0.87	0.86
Interval partitioning	0.87	0.67	0.62	0.61

Source: Haibin Lu, Kim, K., and Sahni, S., *IEEE Transactions on Computers*, 54, 5, 2005. With permission.

Table 10.9 Average Search Time (μs) for Psuedotrace

Schemes	Paix	Pb	Aads	Mae-West
ACRBT	1.18	0.85	0.83	0.81
OLDP	1.02	0.79	0.75	0.74
TLDP	1.01	0.78	0.75	0.74
Interval partitioning	0.71	0.53	0.49	0.48

Source: Haibin Lu, Kim, K., and Sahni, S., *IEEE Transactions on Computers*, 54, 5, 2005. With permission.

Insert Time.   To measure the average insert time for each of the data structures, we first obtained a random permutation of the prefixes in each of the databases. Next, the first 75 percent of the prefixes in this random permutation were inserted into an initially empty data structure. The time to insert the remaining 25 percent of the prefixes was measured and averaged. This timing experiment was repeated 10 times. Table 10.10 shows the average of the 10 average insert times. Three partitioning schemes can reduce the average insert time. The multilevel-partitioning schemes outperform the interval-partitioning scheme.

Delete Time.   To measure the average delete time, we started with the data structure for each database and removed 25 percent of the prefixes in the database. The prefixes to be deleted were determined using the permutation generated for the insert time test; the last 25 percent of these

Table 10.10   Average Time to Insert a Prefix (µs)

Schemes	Paix	Pb	Aads	Mae-West
ACRBT	9.86	10.73	10.37	10.20
OLDP	6.24	4.96	4.70	4.69
TLDP	6.27	4.95	4.67	4.69
Interval partitioning	9.01	7.89	7.53	7.45

Source: Haibin Lu, Kim, K., and Sahni, S., *IEEE Transactions on Computers*, 54, 5, 2005. With permission.

were deleted. Once again, the test was run 10 times and the average of the averages computed. Table 10.11 shows the average time to delete a prefix over the 10 test runs.

As can be seen, the use of OLDP, TLDP, and interval partitioning generally resulted in a reduction in the delete time. The delete time for interval-partitioning scheme was between 18 percent and 35 percent less than that for ACRBT. The delete time for multilevel-partitioning schemes was between 40 percent and 60 percent less than that for ACRBT.

10.2   Port-Based Partitioning

In forwarding table, each entry at least has two fields: (a) prefixes and (b) next hop. Prefix is unique, next hop is not null. If multiple next hops exist for an entry, we only used the highest priority next hop. Akhbarizadeh and Nourani proposed a scheme that partitions the forwarding table based on next hop [5], called IP packet forwarding based on partitioned lookup table (IFPLUT). This scheme effectively reduces the complexity of finding "the longest-prefix matching" problem to an exact match problem. In this section, we will describe the concept of partitioned forwarding table and its implementation, which is from [5]. Portions are reprinted with permission (© 2005 IEEE).

10.2.1   IFPLUT Algorithm

10.2.1.1   Primary Lookup Table Transformation

A forwarding table is an organized set of $e_i = (p_i, q_i)$, where p_i is a prefix and q_i a next hop. For two entries: $e_i = (p_i, q_i)$ and $e_j = (p_j, q_j)$ in a forwarding table, if $p_j \subset p_i$ and $q_j = q_i$, then we call p_i an identical enclosure of p_j and p_j an identical child prefix of p_i.

Table 10.11   Average Time to Delete a Prefix (µs)

Schemes	Paix	Pb	Aads	Mae-West
ACRBT	9.86	10.73	10.37	10.20
OLDP	5.64	4.33	4.03	3.97
TLDP	5.60	4.20	3.92	3.92
Interval partitioning	8.13	7.17	6.76	6.65

Source: Haibin Lu, Kim, K., and Sahni, S., *IEEE Transactions on Computers*, 54, 5, 2005. With permission.

Table 10.12   Forwarding Table

No.	Prefixes (p_i)	Next-hop (q_i)
1	11000110 100110*	1
2	11000110 10011*	2
3	11000110 10011010*	2
4	11000110 01*	3
5	10001010 11001*	3
6	01111000 0011*	1
7	11000110 011*	1
8	10001010 110011*	2
9	10001010 1011*	2
10	11000110 01110*	3
11	01111000 101	2
12	11000110 010*	1

Apparently, the identical child prefixes may be redundant. Note carefully that in general identical child prefixes cannot be eliminated from the forwarding table. For example, there is an IP address $D = 11000110\ 10011010\ 11000110\ 10011010$, there are p_1, p_2, p_3 that match with D in Table 10.12. p_3 is the longest-matching prefix, the next-hop 2 is the forwarding port. If the prefix p_3 that is an identical child of p_2 is excluded, then p_1 is the longest-matching prefix, and the next-hop 1 is the forwarding port, which is a wrong forwarding decision. Therefore, the identical child prefix cannot be excluded directly.

If the identical enclosure prefixes are expanded, we can make the correct forwarding decision. For example, in Table 10.12, the prefix p_2 is the identical enclosure of p_3. If p_2 is expanded, the new prefix is added: $p_{2.1} = 11000110\ 100111*$. p_4 is the identical enclosure of p_{10}; if it is expanded, the new prefixes are added: $p_{4.1} = 11000110\ 0110*$, $p_{4.2} = 11000110\ 01111*$, shown in Table 10.13.

Expanding Rule.   If there are two entries, (p_i, q), (p_j, q) in a forwarding table, and p_i is an identical enclosure of p_j, then (p_i, q) is expanded to d entries: $(p_{i,1}, q)$, $(p_{i,2}, q)$, ..., $(p_{i,d}, q)$ such that:

1. $p_{i,1}, p_{i,2}, ..., p_{i,d}$ and p_j are disjoint.
2. $p_{i,1} \cup p_{i,2} \cup \cdots \cup p_{i,d} \cup p_j = p_i$.

Here d is the difference in the length between p_i and p_j. The upper bound for d is the maximum prefix length W ($W = 32$ for IPv4) and, expanding an identical enclosure may generate d prefixes at most. Akhbarizadeh and Nourani [5] analyzed the real forwarding tables from Merit (http://www.merit.edu/ipma), shown in Table 10.14. The percentage of identical enclosures is below 3 percent. The range expansion is only needed for a very limited number of prefixes in a real forwarding table. The lengths of prefixes are from 8 to 24, and the differences in the length d are

Table 10.13 Expanded Forwarding Table

No.	Prefixes (p_i)	Next-hop (q_i)
1	11000110 100110*	1
2.1	11000110 100111*	2
3	11000110 10011010*	2
4.1	11000110 0110*	3
4.2	11000110 01111*	3
5	10001010 11001*	3
6	01111000 0011*	1
7	11000110 011*	1
8	10001010 110011*	2
9	10001010 1011*	2
10	11000110 01110*	3
11	01111000 101	2
12	11000110 010*	1

Table 10.14 The Number of Identical Enclosures in a Real Forwarding Table

Site	Date	No. of Prefixes	No. of Identical Enclosures	Percentage
AADS	03/15/01	22631	601	2.65
	03/14/02	21649	637	2.94
	03/15/02	21604	612	2.83
	03/19/02	21744	622	2.88
MAE-WEST	03/15/01	24012	653	2.71
	03/14/02	18619	438	2.35
	03/15/02	18681	503	2.69
	03/19/02	18768	511	2.72
PacBell	03/14/02	5222	128	2.45
	03/15/02	5239	143	2.72
	03/19/02	3875	98	2.52

Source: Akhbarizadeh, M.J. and Nourani, M., *IEEE/ACM Transactions on Networking,* 13, 4, 2005. With permission.

less than 16. Expanding all identical enclosures will generate not more than 8 percent of new prefixes. In the next section, there is no identical enclosure in all forwarding tables.

10.2.1.2 Partition Algorithm Based on Next Hops

We suppose that there is no more than one next hop for each prefix in forwarding table, the straight ideal is to partition the forwarding table into small tables based on the next hops.

Partitioning Rule. Suppose $FT = \{(p_i, q_i), p_i \in P, q_i \in Q\}$ is a forwarding table by expanding identical enclosure, P the set of prefixes, and Q the set of next hop (which is from 1 to N), we partition FT into N small tables: partitioned forwarding table (PFT) PFT^1, PFT^2, ..., PFT^N such that $PFT^k = \{(p_i, q_i)| \ \forall(p_i, q_i) \in FT$ and $q_i = k\}$.

If we apply the partitioning rule to the forwarding table shown in Table 10.12, we will get three small tables PFT^1, PFT^2, PFT^3, shown in Table 10.15. The following lemma is a direct result of the above definition.

Lemma 10.21: Having applied the partitioning rule to the forwarding table FT, every PFT^k is a disjoint set.

Table 10.15 Partial Forwarding Tables

No.	Prefixes (p_i)	Next-hop (q_i)
PFT¹		
1	11000110 100110*	1
6	01111000 0011*	1
7	11000110 011*	1
12	11000110 010*	1
PFT²		
2.1	11000110 100111*	2
3	11000110 10011010*	2
8	10001010 110011*	2
9	10001010 1011*	2
11	01111000 101	2
PFT³		
4.1	11000110 0110*	3
4.2	11000110 01111*	3
5	10001010 11001*	3
10	11000110 01110*	3

Proof: If $(p_i, k) (p_j, k) \in PFT^k$, $p_i \cup p_j \neq \emptyset$, then one of them is an enclosure of the other. Suppose p_i is an enclosure of p_j, then p_i is an identical enclosure of p_j, $\because PFT^k \subset F$, $\therefore$ there is an identical enclosure in *FT*. This is against the initial definition of *FT*.

Lemma 10.22:   Given an IP address to be looked up, searching a *PFT* will result in zero or one match.

Proof: This is an immediate conclusion of Lemma 10.21. All prefixes in each PFT are disjoint which means if an IP address matches for example p_i in PFT^k, then it cannot match another prefix, for example, p_j at the same time. If it happens, it means one of p_i or p_j is an identical enclosure of the other, an obvious contradiction.

Based on Lemma 10.22, having a partitioned forwarding table, the longest match can be performed by *N* parallel single match search operations. The IFPLUT algorithm is shown in Figure 10.18. For example, a packet has the following destination IP address: 11000110 10011010 11000001 00000001 (198.154.191.1). The first step of the algorithm looks for possible matches in PFT^1, PFT^2, and PFT^3 (Table 10.15) in parallel. As a result of this step, we will have the partial matches: $e_1 = $ (11000110 010*, 1), $e_3 = $ (11000110 10011010*, 2). This takes us to the second step (line 3) where the longest match is figured out among the partially unique matches based on their prefix length value. So, *longest_match* $= e_3$, the arrived packet should be forwarded by next-hop 2.

IFPLUT algorithm has many advantages for packet forwarding in terms of performance and cost. The most important features are as follows.

a) In each PFT, all prefixes are disjoint and there can be at most one match for every prefix search. This makes the data structure of these PFTs much simpler and the search operation fast and easy, and IFPLUT algorithm is not restricted to any specific data structure.
b) According to Lemma 10.2, having a PFT at hand, the longest-matching prefix can be found as *N* parallel first match search operations. This implies that the IFPLUT algorithm can run on parallel architecture for higher speed.
c) Simplification of data structure results in the reduction of the overall forwarding table size relatively by reducing storage complexity order.
d) The update complexity is reduced using IFPLUT. This achievement is especially helpful when content addressable memory (CAM) is employed.

IFPLUT_algorithm( )
Input: destination_ip(det_ip).
Output: destination_port(m).
01: for (i:=1 to N) Do in parallel
02: $e^k = first\_match(dest\_ip, PFT^k)$

03: longest_match$= e^m$ such that $len^m = \max_{1 \leq k \leq N}\{len^k\}$

04: Return (m)

Figure 10.18   IFPLUT algorithm. (From Akhbarizadeh, M.J. and Nourani, M., *IEEE/ACM Transactions on Networking*, 13, 4, 2005. With permission.)

10.2.2 IFPLUT Architecture

10.2.2.1 Basic Architecture

We can use a very straightforward model to implement the IFPLUT algorithm. This model is shown in Figure 10.19. The entire set of prefixes is partitioned into N subsets according to the partitioning rule. The partitions are put into memory units labeled PFT^1 through PFT^N. When an IP packet is received at input port i, its destination IP address is separated and broadcasted to all PFTs. N PFTs are searched in parallel for a possible first match for this IP address field. Each PFT outputs match length (ML) that is the length of its first (and only) prefix match, which is 5 bits (7 bits for IPv6) value of *len* field of the matched entry (if any). In second stage, a selector block reads N values (ML_1, ML_2, ..., ML_N) generated by the partitions in the first stage and outputs ML_{max} (maximal ML).

10.2.2.2 Imbalance Distribution of Prefixes

The basic architecture in Figure 10.19 works at peak efficiency for a balanced network in which the prefixes are distributed almost equally among PFTs. If a particular egress's PFT becomes much larger than the other PFTs, we say that an imbalance distribution of prefixes has occurred. This may become a serious problem as it may lead to under-utilization or even malfunctioning of IFPLUT. Examining some real data, a great imbalance distribution was observed in a sample of Mae-West routing table (http://www.merit.edu/ipma, 2001). In a roughly 19,000 entries FT that served 26 ports some partitions had as few as one entry, whereas another partition had around 12,000. For this FT, if the design of Figure 10.19 is to be employed, 26 PFTs each with a capacity of 19,000 entries (the size of the entire forwarding table) will be needed, to be prepared for the worst case. That is because in the worst case all entries of the forwarding table may point to the same egress port. This simply means only one out of 26 memory modules is really used.

To reduce the memory waste to an affordable amount and to make the design of IFPLUT generic and configurable, Akhbarizadeh and Nourani. introduced the concept of search unit (SU).

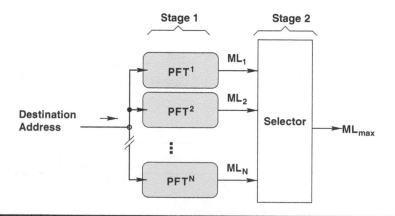

Figure 10.19 Basic IFPLUT architecture. (From Akhbarizadeh, M.J. and Nourani, M., *IEEE/ACM Transactions on Networking*, 13, 4, 2005. With permission.)

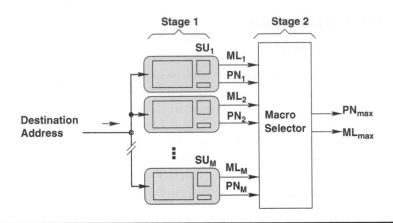

Figure 10.20 Improved IFPLUT forwarding architecture. (From Akhbarizadeh, M.J. and Nourani, M., *IEEE/ACM Transactions on Networking*, 13, 4, 2005. With permission.)

10.2.2.3 Concept of Search Unit

Due to a great diversity in the number of entries within partitions, each partition requires a different memory size. This requirement changes dynamically during the lifetime of a router, and therefore, to utilize the memory efficiently, we need a mechanism to dynamically reconfigure the distribution of memory units among the egress ports. The IFPLUT design shown in Figure 10.20 is meant to meet this important requirement while maintaining all the benefits of the original IFPLUT algorithm.

In this improved architecture, there is a pool of distinct memory units. Each partition will be assigned zero or more memory units. To keep track of which unit is assigned to which partition, a register accompanies each unit as shown in Figure 10.21. This register will store the identification (Id) of the associated partition, which can simply be the corresponding egress port index. Also, each memory unit is equipped with a search engine that, given an IP address, can search for a first

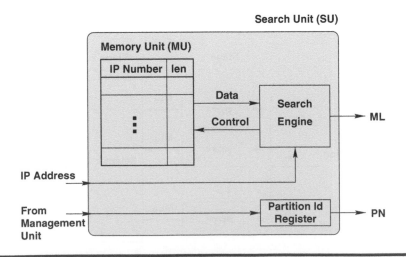

Figure 10.21 Concept of the search unit. (From Akhbarizadeh, M.J. and Nourani, M., *IEEE/ACM Transactions on Networking*, 13, 4, 2005. With permission.)

prefix match in the memory unit. Hereafter, a group of one memory unit, its corresponding search engine, and its corresponding register will be called SU.

Even though the contents of each partition may now be distributed over multiple SUs, the basic characteristics of the IFPLUT method are still valid. Even though multiple SUs may have been assigned to the same partition, no more than one of them will produce a match in each lookup operation. In other words, out of M SUs ($M > N$) that are searched simultaneously, at most N of them will generate a non-zero match. The $M - N$ other SUs will definitely generate a null match. This characteristic makes implementation of the second stage (selector block) of the IFPLUT architecture easy. All outputs of the M SUs can be taken to a selector block for the longest ML to be decided. The selector block is a multi-input maximizer.

10.2.2.4   Memory Assignment Scheme

Initially, there is a pool of M free SUs. If a partition (corresponding to an egress port) has at least one prefix, then an SU will be assigned to it. When the number of entries to a partition exceeds the SU size, it will receive a new SU from the pool of free SUs. All SUs that are assigned to the same partition carry the same partition Id value. Note that the partition Id and the egress port number (PN in Fig. 10.21) are equivalent throughout this section. When an assigned SU memory becomes empty as a result of removing entries, it will be returned to the free pool. There is a management unit that controls IFPLUT and takes care of updating issues. This unit maintains a database of the sizes of partitions and the SUs assigned to each partition. Based on this information, the management unit decides whether it needs to assign a new SU to a particular partition when a new entry is being added and whether to free an assigned SU when an entry is being removed. To assign an SU to a partition, the management unit will write the partition Id to the SU's partition Id register and then updates its own database.

10.2.2.5   Selector Block

The most straightforward way to implement the selector block is a multi-input *greater-than* comparator. Such comparator is conceptually formed a tree of cascaded 2-input *greater-than* functional blocks. Each input is a 5-bit ML bus from an SU. For K SUs, the depth of this selector block would be $\lceil \log_2(K) \rceil$. The comparator will never have more than few hundred inputs, which makes such solution viable. Because the area of such comparator is negligible compared with that the IFPLUT memory occupies on silicon.

Figure 10.22 shows a K-input selector block. The two outputs of the block are: (*i*) a 5-bit bus that provides the maximum among all *ML* inputs to that block and (*ii*) a *PN* bus that represents the corresponding unit's port number (partition Id). Figure 10.23 shows one comparator slice used in the selector block in which *greater-than* (">") function blocks are used. Each ">" block has a single bit output that is "1" only if ML_j is larger than the other input, and "0" otherwise. The port selector signal (PS_j) is the result of ANDing all those signals. In the jth comparator denoted as $comp_j$, the corresponding ML_j bus goes through a tri-state buffer that is controlled by PS_j. Therefore, the first output of each comparator block is a 5-bit tri-state bus that contains the corresponding *ML* value if this is the largest *ML* among all, otherwise it is high impedance (Hi-Z). So, among all the outputs of the comparator blocks in a selector, only one of them will have a binary value and the rest will be in Hi-Z mode. This way we can simply wire OR all those buses together and the resulting bus will be the desired output of the selector block. The same

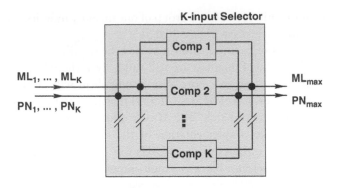

Figure 10.22 **The modified selector block. (From Akhbarizadeh, M.J. and Nourani, M., *IEEE/ACM Transactions on Networking*, 13, 4, 2005. With permission.)**

argument is valid for the *PN* outputs of each comparator block and the port number corresponding to the longest-matching unit can also be obtained by wiring all the tri-sate *PN* signals together in each selector block.

The critical path of the selector goes through only one ">" block and it is very fast but, because of a large amount of interconnect needed between the sub-blocks of the selector, this unit is not scalable beyond a limit, for example, 128 ports. When there are a large number of SUs, we need a more scalable design. To achieve this, we cascade multiple selector blocks together in a two-level tree fashion. By doing so, we can have a macro-selector unit that can handle a routing system with

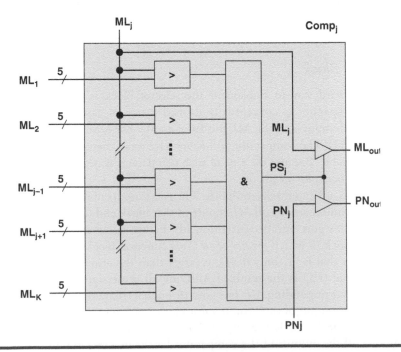

Figure 10.23 **The comparator slice. (From Akhbarizadeh, M.J. and Nourani, M., *IEEE/ACM Transactions on Networking*, 13, 4, 2005. With permission.)**

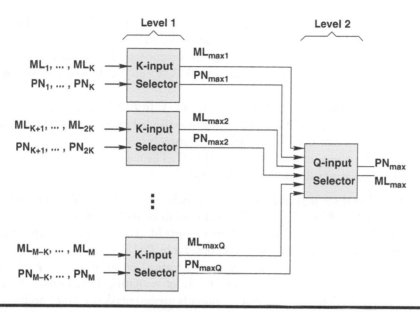

Figure 10.24 **The macro-selector architecture. (From Akhbarizadeh, M.J. and Nourani, M., *IEEE/ACM Transactions on Networking*, 13, 4, 2005. With permission.)**

large (up to 2^{14}) number of inputs by using 129 (128 in the first level and 1 in the second) selector blocks (Fig. 10.24). This will double the critical path delay, but considering the huge number of SUs that can be handled and the fact that a selector is just a combinational logic unit, this delay is still very affordable.

Figure 10.24 demonstrates how the macro-selector is formed out of cascading multiple selector blocks. In this figure, Q selector blocks, each accepting K inputs, are bound together using a Q inputs selector at the second level to construct a $M = Q \cdot K$ macro-selector block. Clearly, this structure can serve M search units.

10.2.2.6 IFPLUT Updates

Except for the identical enclosure prefixes, all other prefixes can be added to or removed from IFPLUT in a single cycle. Considering that identical enclosures are rare, one can see that IFPLUT offers a very high performance in update operations. Few identical enclosures need some special care. Because an identical enclosure is always expanded, when it is removed there should be an efficient way to detect the prefixes that represent it in the forwarding table. To keep track of the identical enclosures, we add a hashing table to the IFPLUT architecture. It can be implemented as a small CAM unit. Each entry of this table contains two fields: a *prefix* and its *Id*. The prefix part stores the identical enclosure prefix and the Id part links it to the expanded prefixes that represent it in the IFPLUT. The Id has to be unique only locally that is, only nesting identical enclosures need to have identical enclosures. Those identical enclosures, that are disjoint can have equal Id values. Because there can be a maximum of 32 (128) levels of nesting in IPv4 (IPv6), the width of the Id field has to be 5 (7) bits. In practice, the level of nesting hardly goes beyond four. Also, the situation in which multiple nesting prefixes have identical egress ports is rare. Therefore, the width of the Id field can be as small as 2 bits for all practical purposes. In spite of this observation, to maintain correctness in the worst case, we consider 5-bit Id fields. One proper choice for the Id value would be the prefix length. We refer to this table as expansion table or simply *ExpTab*. The size of this memory unit depends on the expected population of the identical enclosures. For example, this population of the identical enclosures is less than 3 percent for all practical routing tables; for a forwarding table size of 100,000 entries, this new TCAM unit can be safely of size 4000 words (1000 additional entries as a safety margin).

The *Id* field is also added to each entry in the IFPLUT. In PFT^k, prefix p_i comes with id_i. The value of this *Id* is zero for all original prefixes. For an expanded prefix the *Id* field copies the value of the longest identical enclosure in the ExpTab that parents it.

Add: Say (p_1, q_k) is going to be added to our IFLUT. The new route prefix will be written into PFT^k. Two special situations should be taken care of:

1. The system should see if p_1 has any identical enclosure in PFT^k. This is verified by looking for a first match in PFT^k. If $p_2 \in PFT^k$ is found to be such an enclosure then it should be copied together with its ExpTab and given a proper Id. It should also be expanded in PFT^k, with respect to p_1 so that it can be safely copied into PFT^k.
2. The system should check whether p_1 is an identical enclosure for any of the existing prefixes. To do so, a slightly different mode of searching is required. We call it the *Mode2Search* function. In this mode, p_1 mask is used to mask the PFT^k prefix IP numbers contrary to the normal mode where the prefix mask is used to mask the given IP address. If consequently p_1 is found to be an identical enclosure for p_3, then the former should also be copied into ExpTab with an appropriate Id. p_1 should also be expanded with respect to p_3 and all pieces will have the same Id as p_1. If p_3 has a non-zero Id, then its Id must be changed to the new Id of p_3.

Remove: To remove prefix $p_1 \in PFT^k$ from IFPLUT, the system has to take care of two special situations as well:

1. See if p_1 has caused any other prefix to expand. To do so, a longest-prefix-matching search should be conducted for p_1 in ExpTab. If p_2 is found to match p_1 and belongs to PFT^k, then rather than removing p_1, its Id in PFT^k is replaced with p_2's Id. A routine can periodically run on all PFTs to merge the expanded prefixes that have all their pieces in place.
2. See if p_1 itself was expanded. To do so, the system looks for p_1 in ExpTab. If found, it reads the associated Id (id_1). At the same time, if p_1 has a parent in ExpTab such as p_2, then its Id too must be read (id_2). Then, the system runs a Mode2 Search for p_1 on PFT^k. Any matched prefixes that have the same Id as id_1 represent p_1 and shall replace their Id with id_2 if it exists or be removed from PFT^k otherwise. The p_1 entry should also be removed from ExpTab.

The upper bound for the number of memory accesses for an *Add* or *Remove* operation is $W - 1$, which may occur only when an identical enclosure prefix is encountered. In fact, this happens in less than 3 percent of cases. In all other cases, one memory access is enough to add or remove a prefix to or from IFPLUT. Therefore, in the long run, the practical maximum number of memory accesses for updating is proportional to $0.03W + 0.97$.

10.2.2.7 Implementation Using TCAM

A TCAM-based implementation of SUs helps IFPLUT architecture use the advantages of TCAM. Such an SU will have TCAM words store the prefixes. Each word will have its Id and its length alongside the prefix. There is no need for a priority encoder at the final stage of the SU TCAM because each SU does not produce more than one result. The match signal controls a tri-state buffer that receives *len* as input. The output of the TCAM cell (*xlen*) is either *len* or $Hi - Z$, thus:

$$xlen = \begin{cases} len & if\ match = 1(IP\ address\ match\ the\ prefix) \\ Hi - Z & otherwise(match = 0) \end{cases}$$

At most one of the cells in an SU will generate a match and all others stay in Hi-Z mode. Thus, all *xlen* buses in an SU can be wire ORed to generate the *ML* output of the SU. This saves some critical time and cost on this part of IFPLUT.

The IFPLUT with such implementation functions like TCAM in search operations. It is a TCAM in which the priority encoder is replaced with a macro-selector block. Whereas the priority encoder needs to have as many inputs as the number of words in a conventional TCAM, the macro-selector in IFPLUT will never have more than a few hundred inputs and its critical path delay is relatively lower. Also, the TCAM-based IFPLUT offers more flexible table management than the traditional TCAM.

10.2.2.8 Design Optimization

Let N be the number of next hops in a router, G the size of the forwarding table, and S the size of each SU memory. Initially, there is a pool of M free SUs. After the forwarding table is partitioned according to the partition rule, the ith partition has g_i entries, $(G = \sum_i^N g_i)$, and will be assigned s_i SUs to $(s_i = \lceil g_i/S \rceil, M \geq \sum_i^N S_i)$. All SUs that are assigned to the same partition carry the same next-hop Id value. When an assigned SU memory becomes empty as a result of removing entries, the SU will be returned to the free pool. There is a problem: how to use the memory more efficiently?

For given M and G, $S \geq \lceil G/M \rceil$. The worst-case memory usage comes about when all partitions except one have only one entry each. In this case, each of those $N - 1$ partitions will get a single SU so that $N - 1$ SUs will be assigned to $N - 1$ partitions. The rest of the forwarding table entries [i.e., $G - (N - 1)$] belongs to the last partition. The $G - (N - 1)$ entries are fit in the remaining $M - (N - 1)$ SUs. Thus, we need: $M - (N - 1) > \lceil (G - (N - 1)/S \rceil$, then $S \geq (G - N + 1)/(M - N + 1)$. To be ready for the worst, we need to have the minimum value of S as $S = \max(\lceil G/M \rceil, \lceil G - N + 1/M - N + 1 \rceil)$. It is trivial to show $(G - N + 1)/(M - N + 1) \geq G/M$ as long as $G \geq M$, the condition that is always true in real forwarding table. So, S can be always determined as:

$$S = \left\lceil \frac{G - N + 1}{M - N + 1} \right\rceil \tag{10.1}$$

The goal was to have a forwarding table size of G but now the actual memory size is totally $G^* = M \cdot S = M \lceil G - N + 1/M - N + 1 \rceil$ which is always bigger or equal to G. The difference between these two values $(G^* - G)$ is the amount of memory waste that must be tolerated to be safe at the worst-case scenario. So, the memory waste (W) is given as:

$$W = G^* - G = M \cdot \left\lceil \frac{G - N + 1}{M - N + 1} \right\rceil - G \tag{10.2}$$

M always has a value between N and G, that is, $N \le M \le G$. From Equation 10.2, the memory waste is maximum when M has its minimum value and it is minimum when M is maximized. Hence:

$$W_{min} = 0 \quad when\, M = G$$

$$W_{max} = (G - N) \cdot (N - 1) \quad when\, M = N$$

So, the larger the value of M the smaller the value of S. Smaller S means lesser memory waste. However, M is restricted by other cost factors, the most important of which is the cost of selector block. Larger M means more inputs to the selector block, which increases the cost of this unit. Therefore, choosing the right value of M is a tradeoff between memory cost and other hardware costs, which depends on the technology and budget requirements. In one extreme, when $M = N$ there is not much flexibility and every partition can have only one SU. To be safe in this case we have $S = G - N + 1$ and we will face the maximum waste of memory. In the other extreme, when $M = G$ and $S = 1$, there is zero memory waste and each memory cell will have its own search engine which causes the largest macro-selector.

10.2.3   Experimental Results

For the purpose of experimentation, a system with a forwarding table capacity of $G = 20,000$ entries and a maximum of $N = 32$ ports was designed and evaluated. Table 10.16 shows the cost and delay of the IFPLUT architecture for different values of M, which is also the number of the corresponding macro-selector's inputs. The second and third columns show K (size of search units in the first stage) and Q (size of search units in the second stage). The fourth and fifth columns report the critical path delay and the cost of the selector or macro-selector block that expressed as equivalent 2-input NAND gate count. The sixth shows the memory waste (extra memory cost). The experiment was done using VHDL, in which all models were synthesized in Synopsis using

Table 10.16   IFPLUT for Different Values of M

| # of SU Units (M) | Stage 1 Inputs (K) | Stage 2 Inputs (Q) | Selector | | Extra Mem. [× 1000] |
			Delay [ns]	Cost [× 1000]	
64	64	—	6.98	197	2063
128	128	—	7.26	662	699
256	16	16	11.58	295	300
512	32	16	12.54	925	139
1024	32	32	13.50	1872	68
4096	64	64	13.96	12,854	13

Source: Akhbarizadeh, M.J. and Nourani, M., *IEEE/ACM Transactions on Networking*, 13, 4, 2005. With permission.

LSI10k which is a 0.5 micron library. For this particular case, $M = 26$ sounds like a reasonable tradeoff because its total overhead cost is the least among all other cases.

For the case of $M = 256$, the size of each SU is $S = 89$ according to Equation 10.1. So the actual total memory size will be 22,784 ($= M \cdot S$). The 256 input macro-selector block is created out of 16 blocks of 16-input selector at the first stage and a block of 16-input selector at the second stage ($K = 16$, $Q = 16$). The delay for this macro-block is 11.58 ns. The SUs are implemented as TCAM. The delay in the SU part (the comparison logic) is 3.11 ns. So, the total critical path delay is 14.69 ns.

The modeled 20K words IFPLUT was compared with a model for an equal size conventional TCAM. The priority encoder of TCAM has the cost of 524,000 gates and the critical path delay of 15.6 ns. Alternatively, the macro-selector in IFPLUT offers a critical path delay 11.58 ns with the cost of 295,000 gates. The total costs of the two entities are 3,220,000 for TCAM and 3,427,800 for IFPLUT. The overall search latency of IFPLUT is 23 percent less than TCAM, due to the speed superiority of the macro-selector over the priority encoder. Its cost is only 6 percent more and is expected to be relatively less for bigger sizes of storage.

A TCAM and an IFPLUT of equal size consume the same amount of power. We confirmed this experimentally. The same can be justified analytically by observing that the power consumption of TCAM is mainly due to the execution of parallel search in its words, and hence it is linearly proportional to the TCAM size. The same applies exactly to a TCAM-based implementation of IFPLUT. So, the only important factors to determine the relative power consumption of TCAM and IFPLUT are their memory width and depth (size).

One major advantage of IFPLUT over TCAM is its scalability with respect to the size increase. Although the delay of the priority encoder in a TCAM grows almost linearly with the size, the delay of the macro-selector grows slower than logarithmic. The depth of the macro-selector is much slower than logarithmically proportional to the number of its inputs which is equal to the number of SUs (M), which grows linearly with respect to the table size G when the acceptable memory waste is constant (see Equation 10.2).

IFPLUT is better scalable towards IPv6 than the traditional TCAM architecture. Contrary to the traditional TCAM, IFPLUT does not need to sort its content based on prefix length. Because with IPv6 the TCAM-based forwarding table would need to be sorted into 128 different priority classes instead of 32 classes in IPv4, alleviating such sorting requirement in IFPLUT is a significant scalability achievement.

10.3 ROT-Partitioning

Most of the high-performance routers available commercially these days equip each of their line cards with a forwarding engine to perform table lookups locally, for example, Cisco's 12,000 Series routers, Juniper's T-Series routers. Each line card maintains the entire set of prefixes, whose size grows steadily over time, no matter how many line cards are within the router. Tzeng proposed a partitioning scheme to fragment the routing table into N sub-tables, one for each line card, so that the number of entries in each sub-table drops as the router grows, and the table lookup is accelerated. This section is from [6]. Portions reprinted with permission (© 2006 IEEE).

10.3.1 Concept of ROT-Partitioning

Partitioning is done using appropriately chosen bits in IP prefixes, and a subset of routing table prefixes is referred to as an ROT-partitioning. For a router with N line cards, the key decision on partitioning its routing table is to choose appropriate bits for yielding N ROT-partitions in the most desirable way, namely, (*i*) each ROT-partition involving *as few prefixes as possible* and (*ii*) the size difference between the largest ROT-partition and the smallest one being minimum. Any partitioning which satisfies these two criteria is deemed optimum. Let $n = \lceil \log_2 N \rceil$, then the number of bits chosen for partitioning is n. Note that N does not have to be a power of 2 and can be any integer, say, 3, 5, 6, 7, etc. For easy explanation in the following, we consider simplified prefixes of up to 8 bits only, with the leftmost bit in a prefix denoted by b_0, the next bit by b_1, etc., despite that IP prefixes are actually sequences of up to 32 bits. The case of N being a power of 2 is explained first, followed by a generalization to any arbitrary N.

Given seven simplified prefixes in a routing table: $P_1 = 101*$, $P_2 = 1011*$, $P_3 = 01*$, $P_5 = 10010011$, $P_6 = 10011*$, $P_7 = 011001*$, if b_2 and b_4 are used for partitioning, we arrive at four ROT-partitions: $\{P_3, P_5\}$, $\{P_3, P_6\}$, $\{P_1, P_2, P_3, P_7\}$, $\{P_1, P_2, P_3, P_4\}$, where the first partition corresponds to $b_2 b_4 = 00$, the second corresponds to $b_2 b_4 = 01$, and the third (or fourth) one, to resides in the kth line card. Note that P_3 belongs to every partition, because both b_2 and b_4 of P_3 are "*." Similarly, P_1 belongs to the third and the fourth partitions, as b_4 of P_1 is "*." On the other hand, if b0 and b4 are used for partitioning, we obtain the partitions of $\{P_3, P_7\}$, $\{P_3, P_4\}$, $\{P_1, P_2, P_5\}$, $\{P_1, P_2, P_6\}$, where each partition involves two or three prefixes. It is obvious that the latter partitioning is superior to the former one, based on the two criteria listed above.

For a given set of prefixes, a chosen bit (say b_v) separates prefixes into two subsets with $(\Phi_0 + \Phi_*)$ and $(\Phi_1 + \Phi_*)$ prefixes, respectively, where Φ_0 (or Φ_1) is the number of prefixes whose b_v bits equal "0" (or "1") and Φ_* is the number of prefixes with b_v bits being "*" (because these prefixes have to appear in both subsets). According to Criterion 1 above, we need to find out the bit b_v such that $\Phi + \Phi_*$ $(= \Phi_0 + \Phi_* + \Phi_1 + \Phi_*)$ is smallest among all $0 \le v \le 31$, where Φ is the set size. This criterion is thus equivalent to locating the bit b_v which leads to a minimum Φ_*, ruling out a large v (say > 24), because the vast majority of prefixes in a routing table (e.g., more than 83 percent for the set of prefixes obtained from AS1221 routers (http://bgp.potaroo.net, Sep. 2004) have length no more than 24, and b_v in a prefix is "*" for any v larger than the prefix length. Criterion 2 requires the search of bit b_v such that $|\Phi_0 - \Phi_1|$ is minimum for all $0 \le v \le 31$. When examining bit b_v, this criterion calls for ignoring prefixes with bit $b_v = $ "*," counting only those with bit b_v being "0" or "1." Notice that for a single position bit, Criterion 1 considers only Φ_* of the position bit whereas Criterion 2 deals with Φ_0 and Φ_1 of the bit. Given N and a prefix table, the partitioning algorithm searches exhaustively for n $(= \log_2 N)$ control bits (out of 32 bits for IPv4) to get the optimum results following the aforementioned two criteria.

10.3.2 Generalization of ROT-Partition

Given an arbitrary integer N, let $N_L = \lceil N/2 \rceil$ and $N_R = N - N_L$. The above partitioning approach is still applicable under such an integer value, with a modification to Criterion (2) needed: bit b_v chosen for partitioning satisfies that Φ_0/Φ_1 (Φ_1/Φ_0) is as close to N_L/N_R as possible, if $|\Phi_0| \ge |\Phi_1|$ (or $|\Phi_0| < |\Phi_1|$), for all $0 \le v \le 31$. Each of those N nodes is labeled by $n = \lceil \log_2 N \rceil$ bits, determined according to the modified criterion shown by a binary tree in Figure 10.25. For $N = 6$, N_L is 3 and N_R is also 3, as denoted by the two nodes in the tree Level 1. Both N_L and N_R are further partitioned into two fragments each, as depicted by Level 2. A fragment with only 1 node becomes a tree leaf, whose label is determined by the path from the root to the leaf. For example, the rightmost leaf is labeled by 11*, meaning that every prefix with its 3 $(= n$ partitioning bits being 110 or 111 is homed at this line card. According to Figure 10.25, those 6 $(= N)$ line cards are labeled, respectively,

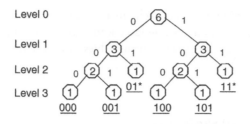

Figure 10.25   The labels of line cards for $N=6$. (From Tzeng, N.-F., *IEEE Transactions on Parallel and Distributed Systems*, 17, 5, 2006. With permission.)

by 000, 001, 01*, 100, 101, and 11*. Note that if N is a power of 2, N_L/N_R always equals 1.0 in each tree level, signifying Criterion 2 before generalization.

It should be noted that the ROT-partitioning scheme intends to minimize and equilibrate the number of prefixes held in each LC, according to the routing table of a router: The partitioning is irrespective of traffic over the router and is not meant to balance the load on line cards nor to optimize mean lookup performance. For a given routing table and N, a desirable partition is obtained according to the aforementioned method, and such a partition often gives rise to reasonably balanced load at line cards for all traces examined, regardless of the router size N. Because the load of line cards is traffic-dependent and changes dynamically, to realize a routing table partition that yields truly balanced load needs traffic profiling, which is not attempted here because it is expensive and difficult. Note also that although the routing table contents change over time as table updates take place, it is not necessary to carry out partitioning repeatedly. In practice, one may follow the ROT-partitioning scheme only once when a router gets its routing table established at the onset of its operation; the same partitioning should work almost equally well until any N-partition grows excessively (and cannot be held in the SRAM of the associated line card effectively) due to table updates. This was confirmed by examining several routing tables to select the desired control bit positions according to ROT-partitioning scheme before and after their entries being added or removed (reflecting table updates). For example, if the AS1221 routing table with 175,853 prefixes taken on 4 September 2004 (http://bgp.potaroo.net/as1221/) is partitioning into four fragments, the preferred partitioning bits determined by ROT-partitioning scheme are at positions 14 and 25, giving rise to the fragments of sizes: 51,095, 50,886, 51,029, and 49,977 (see Section 10.3.4 for details). When the routing table at AS1221 was examined on 5 September 2004, the number of table entries became 176,043 (due to additions and withdrawals of entries, plus modifications to some entries in their next-hop fields), but the most desirable partitioning bit positions obtained by ROT-partitioning scheme remain to be 14 and 25, yielding the fragments of sizes: 51,185, 50,996, 51,172, and 50,078. This partitioning result exhibits the total number of prefixes in the four fragments equal to 203,431 and the largest fragment size difference being 1107. Additionally, the routing table AS1221 was found to have 176,424 entries on 6 September 2004 (http://bgp.potaroo. net/as1221/), and its preferred partitioning bits are still of positions 14 and 25, with four fragments of sizes: 51,336, 51,122, 51,306, and 50,285. When the AS1221 routing table is partitioned into 16 fragments, the same set of four partitioning bit positions is derived using ROT-partitioning scheme for its table entries from 4 September to 6 September 2004. Consequently, the use of same partitioning bit positions even after a large number of table updates (over multiple days without recalculating a new set of bit positions) is generally seen to yield the best partitioning result.

An extra advantage resulting from the routing table partitioning is that the address space coverage of each ROT-partition increases by approximately a factor of the number of partitions, because each ROT-partition is covered by the local cache in a line card. Given an IP lookup stream toward one line card, fewer prefixes in its forwarding table lead to better lookup performance (thanks to a higher hit rate). Partitioning results of two routing tables from different ASs will be demonstrated in Section 10.3.4.

10.3.3 Complexity Analysis

Complexity analysis considers partitioning the routing table into N fragments, requiring to select n ($= \log_2 N$) control bits. Partitioning complexity involves three components:

1. Selecting n control bit positions (out of 32 under IPv4) for partitioning, and under the selected n control bit positions:
2. Deciding to which partition(s) each prefix in the routing table belongs, and
3. Identifying the smallest and the largest partitions.

Clearly, there are totally $C(32, n) = 32!/(n! \times (32 - n)!)$ choices for those n bit positions, reflecting the complexity component one above. For each of such choices, a prefix in the routing table belongs to one of the three cases, determined by the values of those n bit positions being 0, 1, or *. Let those bit positions be represented by $s_0 s_1 \cdots s_i \cdots s_{n-1}$, and the N partitions be denoted by Ω_α, where $0 \leq \alpha \leq N - 1$. A prefix belongs to Ω_α (with α expressed by the binary string of $b_0 b_1 \cdots b_i \cdots b_{n-1}$), if and if $s_i = b_i$ or $s_i = *$; totally, its $s_0 s_1 \cdots s_i \cdots s_{n-1}$ can be in any of 2^n possible forms: $b_0 b_1 \cdots b_i \cdots b_{n-1}$, $*b_0 \cdots b_i \cdots b_{n-1}$, $b_0 * \cdots b_i \cdots b_{n-1}$ etc. For $n = 3$, as an example, Ω_2 contains all prefixes with $s_0 s_1 s_2$ being 010, *10, 0*0, 01*, **0, 0**, *1*, or ***. On the other hand, a prefix with $s_0 s_1 s_2 = $ *10 belongs to Ω_6 as well. It is obvious that a prefix whose $s_0 s_1 \cdots s_i \cdots s_{n-1}$ has d *'s belongs to 2^d partitions, which can be determined immediately by $s_0 s_1 \cdots s_i \cdots s_{n-1}$ through setting each *-bit to be 0 or 1 individually for all d *'s. As a result, the complexity component (2) above is $O(2^d)$, which incurs when each prefix in the routing table is examined. Given G prefixes in the routing table, it leads to a time complexity of $O(2^d) \times G$ to go through all prefixes in the table.

The number of prefixes in each partition can be recorded in one variable, requiring N variables totally. From those N variables, one can get the total number of prefixes in all the partitions, needed for Criterion 1, and also the largest partition and the smallest one, needed for Criterion 2, by passing through those variables once. The time complexity component 3 thus is $O(N)$, bringing the time complexity for one set of n bit positions to be $O(2^d \times G) + O(N) = O(N) \times G$, as $O(2^d) \leq O(N)$, for a routing table with G prefixes. Consequently, the overall time complexity of ROT-partitioning algorithm is given by $C(32, n) \times O(N) \times G$. Because the same N variables can be used repeatedly (after their values are reset to 0) for all $C(32, n)$ choices of the n bit positions, the space complexity equals $O(N)$.

The derived time complexity expression indicates that the optimal partitioning strategy becomes more expensive as N grows. In reality, the factor $C(32, n)$ of the complexity expression can be reduced substantially because all selected control bit positions should be no larger than 24 according to Criterion 1, because the vast majority of prefixes in a routing table have length no more than 24, and should be no smaller than 8 according to Criterion 2, as those bit positions of prefixes in a routing table do not have similar numbers of 0 and 1. As a result, the time complexity in practice is reduced to $C(16, n) \times O(N) \times G$. If N equals 16 (and, thus, $n = 4$) under $G = 250K$ (which is larger than any existing routing table size), for example, the expression gives rise to 7.28×10^9. One may reduce the time complexity further by applying the two optimization criteria recursively, first to find one control bit b_v among $8 \leq v \leq 23$, and then to find a bit in each of the two subsets separately before deciding the bit for both subsets as the second control bit. Similarly, the third control bit is decided using the two criteria on the four subsets obtained using the two chosen control bits. This method leads to suboptimal partitioning with drastically reduced time complexity.

10.3.4 Results of ROT-Partitioning

Two routing tables were obtained for the evaluation use, one being the FUNET routing table with 41,709 prefixes given in [7] (called RT_1) and the other in AS1221 with 175,853 prefixes obtained on 4 September 2004 (http://bgp.potaroo.net/as1221/bgpactive.html) (called RT_2). Using ROT-partition scheme, the routing table is partitioned into sub-tables in accordance with the number of line cards (which can be of any integer, not necessary a power of 2) in the router. The results of table partitioning are exemplified by RT_1 and RT_2. Following the two criteria for table partitioning stated in Section 10.3.1, we arrived at the two desired bit positions for fragmenting RT_1 (or RT_2) into four partitions: bits 17 and 18 (or 14 and 25), giving rise to subset sizes of 10,465, 10,523, 10,493, 10,889 (or 51,095, 50,886, 51,029, 49,977). If RT_1 (or RT_2) is to be partitioned into 16 fragments, the preferred partitioning bit positions are found to be 16, 18, 19, 20 (or 17, 18, 20, 23), producing subset sizes of 2594, 2851, 2814, 2899, 2494, 2831, 2813, 2517, 2852, 2976, 2912, 2899, 2745, 2523, 2994, 2425 (or 16,258, 16,420, 17,495, 18,879, 16,676, 16,240, 17,023, 15,176, 13,431, 16,973, 19,281, 14,959, 14,667, 14,145, 13,835, 14,058).

10.3.4.1 Storage Sizes

Three distinct tries (namely, the DP trie [8], the Lulea trie [9], and the LC-trie [7]) have been implemented to assess the storage size needed for all ROT-partitions after fragmenting RT_1 and RT_2. Under the DP trie for RT_1 (or RT_2) with $N = 4$, the storage sizes of the four tries built and held in line cards after partitioning are 209, 216, 217, and 220 kBytes (or 917, 914, 917, and 898 kBytes), respectively, assuming that each node in the trie consists of one byte for the index field plus four bytes for each of the five pointers. Because the trie size before partitioning is 859 (or 3441) kBytes, the DP trie sees the amount of SRAM reduction in each line card caused by partitioning to exceed 638 (or 2524) kBytes under RT_1 (or RT_2) with $N = 4$. The DP trie after partitioning for $N = 16$ reduces its size down to the range of 50 and 62 kBytes (or of 165 and 288 kBytes) with respect to RT_1 (or RT_2). As a result, storage reduction in each line card due to partitioning is no less than 795 (or 3153) kBytes under RT_1 (or RT_2) when N equals 16. Note that the reduction amount will be much larger under IPv6. Total SRAM amounts required for the DP trie after (or without) partitioning are depicted in Figure 10.26 under DP_S (or DP_W).

For the Lulea trie [9] (whose storage requirement is often the lowest) with $N = 4$ under RT_1 (or RT_2), the partitioned tables in four line cards require 90, 91, 89, and 87 kBytes (or 504, 494, 505, and 491 kBytes), respectively, as opposed to roughly 260 (or 978) kBytes in a line card of a conventional router without partitioning. This indicates an SRAM reduction in each line card caused by partitioning for $N = 4$ under the Lulea trie to be 169 (or 473) kBytes or beyond, when RT_1 (or RT_2) is concerned. For $N = 16$, the Lulea trie sees its size to be not more than 39 (or 163) kBytes in any line card after partitioning RT_1 (or RT_2), giving rise to a savings of at least 221 (or 815) kBytes in each LC. Total SRAM amounts required for the Lulea trie after (or without) partitioning are depicted under LL_S (or LL_W) in Figure 10.26. Similarly, the LC-trie [7] under RT_1 (or RT_2) enjoys storage reduction in any line card by no less than 815 (or 1961) kBytes, for $N = 4$ with a fill factor of 0.25. The storage saving amount in an line card increases to over

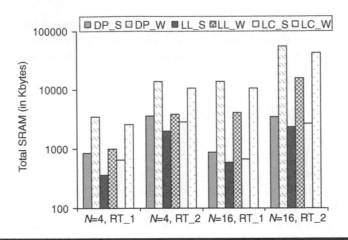

Figure 10.26 Total SRAM (in kBytes) required for different tries under IPv4. (From Tzeng, N.-F., *IEEE Transactions on Parallel and Distributed Systems*, 17, 5, 2006. With permission.)

1025 (or 2456) kBytes for RT_1 (or RT_2) under $N = 16$ with the same fill factor (of 0.25). Therefore, ROT-partitioning always leads to an SRAM savings in each line card under the three distinct tries we have implemented and examined.

10.3.4.2 Worst-Case Lookup Times

The time taken for a software lookup involves two components, one due to multiple memory accesses and the other due to program execution (involving some 100 instructions or so). The former time component dictates the worst-case lookup time for a given longest prefix matching algorithm employed in the forwarding engine of a line card, while the latter one is machine-dependent but typically is no less than 100 ns, irrespective of whether a lookup is the worst case or not. The worst-case lookup time for an existing router is thus proportional to the number of memory accesses, each of which takes, say, 12 ns when prefixes are held in off-chip SRAM (e.g., the L3 data cache, if existing; otherwise, the L2 data cache, of the forwarding engine processor). The numbers of worst-case memory accesses under different tries have been obtained via the actual implementations.

When the DP trie [8] is considered with respect to RT_1 (or RT_2) under $N = 4$, for example, the implementation results indicate that in the worst case, search through the four partitioned tries involves 38, 34, 38, 38 (or 39, 39, 35, 39) node accesses, respectively, in comparison with 42 (or 43) accesses to the DP trie without partitioning. As a result, the DP trie enjoys a reduction of no less than four node accesses under ROT-partitioning with $N = 4$ when RT_1 (or RT_2) is considered; this reduction translates to shortening the matching search time by at least 48 ns in the worst case, when a node access is assumed optimistically to involve just one SRAM access of 12 ns (because off-chip SRAM typically exhibits access time ranging from 8 to 15 ns and a node in the DP trie contains some $1 + 4 \times 5 = 21$ bytes [8]). This search time reduction well offsets the extra time incurred under SPAL. Likewise, ROT-partitioning for $N = 16$ under RT_1 (or RT_2) gives rise to no more than 34 (or 35) node accesses over any partitioned trie in the worst case according to the actual implementations, yielding a search time savings of at least 96 ns. The worst-case lookup times following the DP trie are illustrated under DP_S (or DP_W) in Figure 10.27, where the worst-case lookup times are shortened as a result of ROT-partitioning.

If the Lulea trie is adopted, a far smaller forwarding table in each forwarding engine after partitioning may avoid the dense chunks and the very dense chunks in the third level or even avoid all the third-level chunks of the trie [9], reducing two or even four memory accesses (from $4 + 4 + 4$ down to $4 + 4 + 2$ or even to $4 + 4$) for the worst case; this memory access reduction translates to a savings of 24 ns or even 48 ns (given the SRAM access time of 12 ns). Under RT_1, any Lulea trie for both $N = 4$ and $N = 16$ after partitioning sees the worst-case search to involve 10 memory accesses, in contrast to 12 memory accesses when the routing table is not partitioned, according to real implementations. The Lulea trie thus experiences search time reduction by two memory accesses in the worst case. The worst-case lookup times following the Lulea trie are shown under LL_S (or LL_W) in Figure 10.27. Similarly, if the LC-trie is employed, a smaller forwarding table under ROT-partitioning usually reduces the maximum path length in the trie constructed. Considering RT_1 (or RT_2) under a fill factor of 0.25, for example, the search path depth in the LC-trie after partitioning for $N = 4$ is bounded by 6 (or 10), whereas the maximum trie depth without partitioning equals $5 + 3 = 8$ (or $6 + 6 = 12$) according to the implementation provided in [7], yielding a savings of at least two memory accesses in the worst case; this amounts to a reduction of 24 ns. Likewise, the LC-trie with $N = 16$ under RT_1 (or RT_2) sees its maximum trie depth after partitioning to be no more than six (or nine), exhibiting the worst-case search time savings of no less than two (or three) memory accesses. This is likely to yield a smaller worst-case lookup time, as depicted in Figure 10.27 (denoted by LC_S).

10.4 Comb Extraction Scheme

Xu et al. proposed a new methodology (called comb extraction scheme (CES)) to partition the forwarding table into two smaller sub-tables [10]. The forwarding engine based on CES speeds up the best matching prefix search and can reduce storage space at the same time. This section is from [10]. Portions are reprinted with permission (© 2004 IEEE).

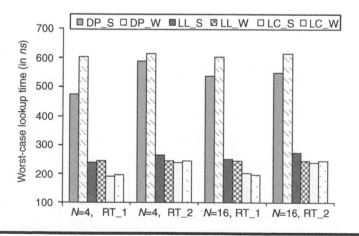

Figure 10.27 Worst-case lookup time under cycle time = 5 ns and SRAM access time = 12 ns. (From Tzeng, N.-F., *IEEE Transactions on Parallel and Distributed Systems*, 17, 5, 2006. With permission.)

10.4.1 Splitting Rule

For IPv4, an IP address A is 32-bit long. It can be decomposed into two 16-bit long sub-sequences by the following strategy: From the leftmost bit to the rightmost bit, all the bits in odd positions are extracted to form sub-sequence α, and all the bits in even positions are extracted to form subsequence β. This splitting approach is called the CES. For example, let us consider the IP address in binary bits, 10100001 00110110 11010000 11101001, in which bits in odd positions are bold. After the decomposition, α and β will be 11000101 10001110 and 00010110 11001001, respectively. Similarly, a prefix also can be decomposed two sub-sequences α and β. Both of them end by *. That is, they are two sub-prefixes. For example, let us consider the prefix 01101100 101*, the length of which is 11. After decomposition, α will be 011011*, the length of which is 6, and β will be 10100*, the length of which is 5.

Hence, a forwarding table can be converted into two extended sub-forwarding tables. For a sample forwarding table shown in Table 10.17, Tables 10.18 and 10.19 are the two sub-forwarding tables when Table 10.17 is decomposed. Each sub-entry has this kind of structure as <sub-prefix, length, port-indicator, forwarding information>.

Table 10.17   A Sample Forwarding Table

Index	Prefix	Len	Port
1	11000110100110*	14	1
2	11000110011*	11	1
3	11000110010*	11	1
4	1101011100001011*	16	1
1	1100011001*	10	2
2	011110000011*	12	2
3	11000110100*	11	2
4	10001010110011*	14	2
5	100010101011*	12	2
6	1110111101*	10	2
7	01111000101*	11	2
8	0111100011100*	13	2
1	1100011010011010*	16	3
2	1100011001110*	13	3
3	1000101011001*	13	3
4	01111000001110*	14	3

Source: Xu, Z., et al., *The 2004 International Conference on Communications (ICC 2004)*, Paris, France. With permission.

**Table 10.18   Odd Sub-Forwarding Table from Table 10.17
(Extracting Bits in Odd Positions)**

Index	Sub-Prefix	Length	Port Indicator	Forwarding Information
1	011001*	6	010	2(2)
2	0110011*	7	001	3(4)
3	011011*	6	010	2(7)
4	0110110*	7	010	2(8)
5	10010*	5	010	2(1)
6	100100*	6	100	1(3)
7	10010011*	8	100	1(4)
8	100101*	6	100	1(2)
9	1001010*	7	001	3(2)
10	100110*	6	010	2(3)
11	1001101*	7	100	1(1)
12	10011011*	8	001	3(1)
13	1011101*	7	011	2(4), 3(3)
14	101111*	6	010	2(5)
15	11110*	5	010	2(6)

Source: Xu, Z., et al., *The 2004 International Conference on Communications (ICC 2004)*, Paris, France. With permission.

In the *Forwarding Information* part of Tables 10.18 and 10.19, not only is the information about the port number noted, but also the information about the corresponding index associated with that port in Table 10.17 is involved. The forwarding information is composed of the *forwarding unit a(b)*, which implies that, in Table 10.17, the original prefix of this sub-prefix is forwarded to port a, and the associated index is b. In general, the forwarding information of each sub-entry in a sub-table consists of several forwarding units. For example, the sub-prefix of the seventh entry in Table 10.19 is 10101*, which collects the information of all original prefixes whose bits in even positions are 10101*. It is made up of three forwarding units, 2(1), 1(2), and 1(3). The forwarding unit is the union of the port and the corresponding index. Usually, a core router has no more than 128 output ports. So the length of port can satisfy that $len(\text{port}) \leq 7$ in bits. Therefore, a 20-bit long vector is enough to represent a forwarding unit, in which $len(index) = 20 - len(\text{port})$.

In each sub-table, an N–bit port indicator vector is associated with every sub-entry. A bit i is set in the bit vector if and only if the ith port occurs in its forwarding information. Usually, the width of it is no more than 128.

What is the benefit of the CES approach? We examine it from two main aspects: (*i*) Because a forwarding table is decomposed into two sub-tables, one lookup will be divided into two parallel

Table 10.19   Even Sub-Forwarding Table from Table 10.17 (Extracting Bits in Even Positions)

Index	Sub-Prefix	Length	Port indicator	Forwarding Information
1	000001*	6	010	2(5)
2	000010*	6	001	3(3)
3	0000101*	7	010	2(4)
4	10100*	5	010	2(3)
5	1010010*	7	100	1(1)
6	10100100*	8	001	3(1)
7	10101*	5	110	2(1),1(2),1(3)
8	101011*	6	010	3(2)
9	10111*	5	010	2(6)
10	11000*	5	010	2(7)
11	110001*	6	010	2(2)
12	1100010*	7	001	3(4)
13	110010*	6	010	2(8)
14	11110001*	8	100	1(4)

Source: Xu, Z., et al., *The 2004 International Conference on Communications (ICC 2004)*, Paris, France. With permission.

sub-lookups. Can CES make the two sub-lookups in balance, either in time access or memory consumption? (*ii*) After the two parallel sub-lookups, some sub-prefixes will match each sub-search key in both sub-lookups. To find the best-matching prefix, we need to combine the results, comparing the information of any reasonable pair of matching sub-prefixes from both sub-lookups. Therefore, can CES cause heavy comparison loads, which need to cost extra time?

First, CES makes the entries of two sub-tables well distributed. In comparing two bit patterns, the Hamming distance is the count of bits different in the two patterns. More generally, if two ordered lists of items are compared, the Hamming distance is the number of items that do not identically agree. Here, we give a new definition to determine the distance between two prefixes, which is similar to the Hamming distance.

Definition 10.4.1:   *a* and *b* are two prefixes in a forwarding table. |*a*| and |*b*| represent their lengths. Let $ML = \min(|a|, |b|)$. We define pseudo-Hamming distance (PHD) between two prefixes as: $PHD(a,b) = \sum_{i=0}^{ML} (L - i)|a_i - b_i|$, where a_i and b_i are the bits in the *i*th position, from the leftmost, of *a* and *b*, *L* is the maximum length of sequences (IPv4, $L = 16$ for sub-forwarding tables). MPHD is the mean of PHD of any two different prefixes in one table. PHD is affected not only by the number of bits that are not identical, but also by their positions, the leftmost bits having higher weight.

For example, assuming *a*, *b*, and *c* are 011001*, 0110010*, and 10010011* respectively. Let *L* be 16. PHD(*a*, *b*) =0, PHD(*a*, *c*) = 69, and PHD(*b*, *c*) = 79.

Lemma 10.4.1: If a and b are two prefixes, and one is a prefix of the other, then PHD(a, b) equals to zero.

The value of MPHD can stand for the distribution of entries in a table. If MPHD is a big value, it implies that, in a trie of a forwarding table, nodes spread widely, rather than just focus on several deep branches. This allows for a faster search. CES is almost the best of splitting rules to maximize the MPHD of each sub-table, and there is not much divergence between the two values, which shows that CES leads to a balanced distribution of entries in the two sub-tables.

Secondly, CES also balances the sub-prefix lengths in the two sub-tables.

Definition 10.4.2: Let mean prefix length (MPL) in any sub-table be expressed by: $\frac{1}{M} \sum_{i=0}^{SM} n_{iFO}^i$ Len_i, where M is the number of entries in the original forwarding table, SM the number of sub-entries in this sub-table, n_{iFO}^i the number of forwarding units in the ith sub-entry, and Len_i the length of the ith sub-entry.

Let a be an original prefix in a forwarding table. After having been decomposed, it will be converted into two sub-prefixes, named a_1 and a_2. By construction, the difference between the lengths of a_1 and a_2 satisfies the inequality: $0 \leq |a_1| - |a_2| \leq 1$. In fact, the difference between MPLs in CES results from the number of original prefixes, whose lengths are odd values. It is a good way to prevent big difference in searching time by making the sub-prefixes in two sub-tables almost the same.

In the example (see Tables 10.17, 10.18, and 10.19), $MPL_1 = 6.5$, $MPL_2 = 6.06$, and in short, CES is an efficient way to enable the two sub-tables keep in pace.

Thirdly, CES makes the forwarding units well distributed in each sub-table.

Definition 10.4.3: (1) Basic load of forwarding information (BLFI) of ith sub-entry in each sub-table is defined as the total number of forwarding units in the ith sub-entry.

(2) Mean load of forwarding information (MLFI) of sub-entries in each sub-table is defined by $\frac{1}{SN} \sum_{i=0}^{SN} BLFI_i$, where SN is the total number of sub-entries in this sub-table.

(3) Standard deviation of forwarding information (SDFI) of sub-entries in each sub-table is defined by $\sqrt{\frac{1}{SN} \sum_{i=0}^{SN} (BLFI_i - MLFI)^2}$.

These metrics are significant to show the performance of CES. Whether the comparing time between sub-prefixes in the second phase is reasonable or not depends on these three values. In the example, $MLFI_1 = 1.067$, $MLFI_2 = 1.143$, $DSFI_1 = 0.25$, $SDFI_2 = 0.51$.

Fourthly, CES balances the comparison cost.

Definition 10.4.4: The comparison cost factor (CCF) is used to judge whether the comparison load of those matching sub-prefixes in two sub-tables for an address lookup next is heavy or not. CCF is a statistical value from experiments, by counting the pairs really need to compare.

Actually, it is not necessary to compare every pair of matching sub-prefixes, for there are constraints among the matching sub-ones, once they are the final ones we are looking for. We know that if α_1 and α_2 are two the final matching sub-prefixes in the two-tables for an address, then they should satisfy the following:

1. $|\alpha_2|$ only can be equal to $|\alpha_1|$ or $|\alpha_1| - 1$;
2. In the two corresponding port indicator vectors (PIV), $\exists i, i < N, PIV_i^1 = PIV_i^2 = 1$.

Only if the matching sub-prefixes α_1 and α_2, which come from different sub-tables, meet the demands above, comparison is needed. CCF is a parameter to observe the number of pairs, which satisfy the conditions, and need to do real comparison. Anyway, CCF has its upper bound.

Table 10.20 Characteristics of Sub-Tables Using the CES

Forwarding Table	No. of Entries	Sub-Table	No. of Entries in Sub-Table	MPHD	MPL	Max (BLFI)	SDFI	CCF
Mae-East	47,206	1	4026	55.22	11.18	93	13.41	8
		2	5341	56.56	11.18	86	9.71	
Mae-West	77,002	1	5703	56.47	11.22	100	15.49	8
		2	6989	57.84	11.22	78	12.57	
Aads	63,980	1	5689	56.80	11.35	110	14.28	8
		2	6735	57.45	11.35	89	10.84	
Paix	22,116	1	4077	54.59	11.15	40	5.35	7
		2	4704	55.67	11.15	28	4.23	

Source: Xu, Z., et al., *The 2004 International Conference on Communications (ICC 2004)*, Paris, France. With permission.

Let $MinNum = \min(Num_1, Num_2)$, where Num_1 and Num_2 are the numbers of matching sub-prefixes from the two sub-tables. $CCF \leq 2 \times MinNum$.

For the four forwarding tables from IPMA project (http://www.merit.edu/ipma, 2004), the characteristics of sub-tables using CES are shown in Table 10.20.

10.4.2 Comparison Set

In this section, we are going to analyze the matching sub-prefixes from two sub-tables, to find the common matching prefix. This part can be implemented in an ASIC. The first step is to decide whether further comparison is necessary, which is pointed out before. The second step is to compare the forwarding units, only when the first step succeeds.

If $P1_i$ and $P2_j$ are two matching sub-prefixes coming from each sub-table, each of them contains one forwarding information set, which is a collection of forwarding units. We need to compare every unit in a set with the units in another set. Let $Info1_i$ and $Info2_j$ be the information sets of $P1_i$ and $P2_j$, which are composed of $M1$ and $M2$ information units. So, for each comparison, $M1 \times M2$ pairs of comparison units are needed.

In the comparison between $Info1_i$ and $Info2_j$, if there exists an exact match in one comparison unit, it implicates that $P1_i$ and $P2_j$ are the right decomposition parts of an original prefix in a forwarding table.

Lemma 10.4.2:   In the comparison between $Info1_i$ and $Info2_j$, there at most exists one exact match in all pairs of comparison units.

Proof: Assume that there exists two pairs of units, $(Info1_{i,k}, Info2_{j,m})$ and $(Info1_{i,l}, Info2_{j,n})$, match exactly. That is, $Info1_{i,k} = Info2_{j,m}$, and $Info1_{i,l} = Info2_{j,n}$. It means that in the original forwarding

Table 10.21   Cost for Comparison or Matching Sub-Prefix

Forwarding Table	Mae-East	Mae-West	Aads	Paix
No. of entries	47206	77002	63980	22116
Average delay (ns)	2.24	3.36	1.96	0.56
Delay (80% of comparisons) (ns)	<4.59	<7.87	<4.72	<1.21

Source: Xu, Z., et al., *The 2004 International Conference on Communications (ICC 2004)*, Paris, France. With permission.

table, there are two entries, which have the same prefix, but will be forwarded to different ports. It is impossible for unicast. So the assumption is not right. Then there at most exists one exact match in all pairs of comparison units.

Each forwarding unit is 20-bit long. Based on nowadays limitation of transforming width, it is possible to input 6 forwarding units of each matching sub-prefix at the same time. Then 36 comparison units can work in parallel. Therefore, all comparison units work in serial to the end until there is a comparison unit exact match. The delay of every 36 parallel comparison is 280ps, when using VLSI feature size of $\lambda = 0.13$ μm. Table 10.21 shows the time cost for comparing every forwarding unit of two matching sub-prefixes. We find that it varies when the forwarding table's size increases. Actually, the real cost is smaller than this, because the comparison stops when there exists an exact match.

10.4.3   Implementation Using Binary Trie

Binary trie is a basic structure for IP lookups. To evaluate the performance of CES, Xu et al. implemented the CES-based architecture using binary trie, shown in Figure 10.28. The forwarding

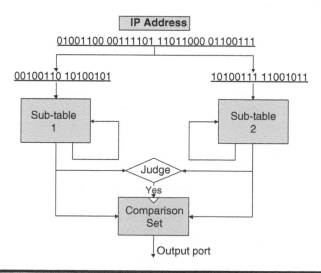

Figure 10.28   IP lookup architecture based on CES. (From Xu, Z., et al., *The 2004 International Conference on Communications (ICC 2004)*, Paris, France. With permission.)

Table 10.22   Storage Cost Comparison (kByte)

Forwarding Table	Mae-East	Mae-West	Aads	Paix
Sub-table 1	215.5	310.2	285.6	147.8
Sub-table 2	247.2	288.9	308.3	161.6
Original table	1295.3	2003.8	1657.8	718.8

Source: © 2004 IEEE.

tables are from IPMA project (http://www.merit.edu/ipma, 2004). Each forwarding table is decomposed into two half-level sub-tables based on CES. Table 10.22 gives the memory cost when we use Binary trie to construct the data structure for IP lookup, smaller than only when binary trie used. Most memory is consumed at the nodes with forwarding information. The storage cost for two 16-level tries is much smaller than one 32-level trie.

When a search starts, the first matching sub-prefixes we reach in two sub-tables are the shortest. We need to do the comparison of their forwarding information, and at the same time, we need to continue traversing the sub-tries until they are exhausted. The last exact match is the final output port of this IP packet. The total average delay in comparison is not more than 25 ns, because the CCF is less than 8 (if the total entries are not more than 80K).

References

1. Lampson, B., Srinivasan, V., and Varghese, G., IP lookups using multiway and multicolumn search. *Proc. IEEE INFOCOM '98*, Apr. 1998, pp. 1248–56.
2. Haibin Lu, Kim, K., and Sahni, S., Prefix- and interval-partitioned dynamic IP router-tables. *IEEE Transactions on Computers* 2005; 54(5):545–557.
3. Srinivasan, V. and Varghese, G., Fast address lookups using controlled prefix expansion. *ACM Transaction on Computer Systems* 1999; 17(1):1–40.
4. Sahni, S. and Kim, K., An O(log n) dynamic router-table design. *IEEE Transactions on Computers* 2004; 53(3):351–363.
5. Akhbarizadeh, M.J. and Nourani, M., Hardware-based IP routing using partitioned lookup table. *IEEE/ACM Transactions on Networking* 2005; 13(4):769–781.
6. Nian-Feng Tzeng, Routing table partitioning for speedy packet lookups in scalable routers. *IEEE Transactions on Parallel and Distributed Systems* 2006; 17(5):481–494.
7. Nilsson, S. and Karlsson, G., IP-Address lookup using LC-Tries. *IEEE J. Selected Areas in Comm.* 1999; 17(6):1083–1092.
8. Doeringer, W., Karjoth, G., and Nassehi, M., Routing on longest-matching prefixes. *IEEE/ACM Trans. Networking* 1996; 4(1):86–97.
9. Degermark, M., et al., Small forwarding tables for fast routing lookups, *Proc. ACM SIGCOMM 1997 Conf.*, Sept. 1997, pp. 3–14.
10. Zhen Xu, Ioannis Lambadaris, Yiqiang Q. Zhao, and Gerard Damm, IP packet forwarding based on comb extraction scheme. *The 2004 International Conference on Communications (ICC 2004)*, June 20–24, Paris, France, pp. 1065–1069.

Index

Printed and bound by CPI Group (UK) Ltd, Croydon, CR0 4YY

23/10/2024

01778230-0012